STRUCTURE AND FUNCTION OF GANGLIOSIDES

ADVANCES IN EXPERIMENTAL MEDICINE AND BIOLOGY

STRUCTURE AND FUNCTION OF GANGLIOSIDES

Edited by

Lars Svennerholm

Psychiatric Research Centre
Department of Neurochemistry
St. Jörgen Hospital
Göteborg, Sweden

and

Paul Mandel, Henri Dreyfus, and Paul-Francis Urban

C. N. R. S. Center for Neurochemistry
Strasbourg, France

PLENUM PRESS • NEW YORK AND LONDON

Library of Congress Cataloging in Publication Data

Symposium on Structure and Function of Gangliosides, Strasbourg, 1979.
    Structure and function of gangliosides.

    (Advances in experimental medicine and biology; v. 125)
    "Proceedings of the symposium on Structure and Function of Gangliosides, spon-
sored by C.N.R.S., held in Strasbourg (Bischenberg), France, April 23–28, 1979."
    Includes indexes.
    1. Gangliosides – Congresses. I. Svennerholm, Lars. II. France. Centre national de
la recherche scientifique. III. Title. IV. Series. [DNLM: 1. Gangliosides – Congresses.
W1 AD559 v. 125/QU85 S98906s 1979]
QP752.G3S95   591.1'88   79-25711
ISBN 978-1-4684-7846-4   ISBN 978-1-4684-7844-0 (eBook)
DOI 10.1007/978-1-4684-7844-0

Proceedings of the Symposium on Structure and Function of Gangliosides,
sponsored by C. N. R. S., held in Strasbourg (Bischenberg), France, April 23–27, 1979

Organizers
    Paul Mandel and Lars Svennerholm

General Secretaries
    Henri Dreyfus and Paul-Francis Urban

Secretary
    Henriette Urban

Organizing Committee	Scientific Advisory Committee
Louis Freysz	Shimon Gatt, *Jerusalem, Israel*
Gérard Rebel	Maurice Rapport, *New York, USA*
Louis Sarliève	Saul Roseman, *Baltimore, USA*
Guy Vincendon	Kunihiko Suzuki, *New York, USA*
Ahad Yusufi	Guido Tettamanti, *Milan, Italy*
	Herbert Wiegandt, *Marburg-Lahn, Germany*

© 1980 Plenum Press, New York
Softcover reprint of the hardcover 1st edition 1980

A Division of Plenum Publishing Corporation
227 West 17th Street, New York, N.Y. 10011

PREFACE

     This volume records the proceedings of an International
Symposium on The Structure and Function of Gangliosides, held
at Le Bischenberg, Alsace, France, in April 23-27, 1979. The
meeting was convened to get a comprehensive view of the immense
activity that had occurred in the field since the previous
conference on gangliosides held at Mont Sainte-Odile, Alsace,
France, in April 1973. At a conference on Enzymes of Lipid
Metabolism held at the same place in April, 1977, several of
the participants from the first ganglioside conference in 1973
met again. All previous participants agreed that the first
meeting with its many frank and stimulating lectures, round
tables and informal discussions had been of tremendous importance
for the activity in the field and led to many personal contacts
and a warm friendship among the ganglioside researchers. The
success of the first meeting must be ascribed largely to one
single man, PAUL MANDEL. Therefore, we decided to dedicate the
next ganglioside conference to him and I was given the privilege
to arrange the meeting together with him and staff members at
Centre de Neurochimie, Strasbourg.

     The Symposium on The Structure and Function of Gangliosides
was arranged to honour PAUL MANDEL for his unique and never-
failing efforts to promote and strengthen international collaboration in all fields of neurochemistry. He has worked harder
than most scientists to raise money for international meetings,
and we are many researchers who have had the privilege to
participate in more than one conference arranged by Centre de
Neurochimie. PAUL MANDEL has been able to give all the con-
ferences a large fund of his own spirit and rich humanity, and
nobody who has had the opportunity to participate in a meeting
organized by PAUL MANDEL will ever forget his great generosity
and appreciation of others' research work. PAUL MANDEL is a
great scientist and humanist with the most precious quality,
goodness.

The elucidation of the chemical structure of the ganglioside stored in Tay-Sachs disease and the development of an assay procedure for the enzymic lesion had a tremendous clinical implication. It initiated prenatal diagnosis of inherited disorders and a world screening programme for carriers of lethal traits. The discovery that one single ganglioside is the specific receptor for cholera toxin stimulated an intense search for the receptor function of other gangliosides for bacterial toxins, viruses, hormones and transmittor substances. All these aspects are reviewed. A recent practical development which can be of fundamental importance for the preparation of effective vaccines is also described - the use of ganglioside GM1 for the large scale affinity chromatographic isolation of cholera toxin. The book also gives the reader a vision about the role of gangliosides in synaptic transmission. The receptor role of gangliosides has created urgent need for a better isolation and separation methods and more refined techniques for determination of their chemical structure. The comprehensive reviews on the metabolism of gangliosides and on their localization in various tissues, cells and subcellular organelles are a must for the understanding of the physiological role of the gangliosides. Many of the findings are novel and have not yet appeared in the scientific journals. It is therefore hoped that the reviews will be up to date when the book becomes available to the profession although the area is under such active investigation.

The success of a meeting depends not only on eminent speakers and topics of current interest but also on planning and, unfortunately, sufficient funds. Henry Dreyfus and Paul-Francis Urban were the two persons who planned the symposium together with Paul Mandel and me and who solved all the practical problems in the most excellent manner. Our task has been greatly facilitated by the devoted and expert assistance with the organizational, secretarial and editorial work of Mrs. Henriette Urban.

The symposium was sponsored by a very generous support from CNRS, which is gratefully acknowledged. Without such generosity the meeting could not have been held. We would also like to acknowledge the generous support from Institut Mérieux, Lyon, whose contribution made it possible to invite a large number of speakers from countries in Europe, U.S.A. and Japan.

Finally I am glad and proud of having had the privilege to organize this symposium as a hommage to my friend PAUL MANDEL in one of the most charming places in Europe, the lovely Alsace of France.

Lars Svennerholm

CONTENTS

CHEMICAL STRUCTURE OF GANGLIOSIDES

IMMUNOLOGICAL METHODS
FOR THE IDENTIFICATION OF GANGLIOSIDES

CONTENTS

CHEMICAL STRUCTURE OF GANGLIOSIDES

INTRODUCTORY REMARKS ON CHEMICAL STRUCTURES OF GANGLIOSIDES

Herbert Wiegandt

Department of Biochemistry, University Marburg-GFR

Ever since their discovery by Ernst KLENK, the gangliosides among all glycosphingolipids have received particular attention. Interest in gangliosides was early stimulated by speculations on their possible involvement in specific functions of the brain, where they have been found most highly concentrated. In addition it was learned that gangliosides were stored in certain hereditary disorders affecting the central nervous system. Another finding that seemed to indicate a role of gangliosides in nerve conduction was their specific binding to tetanus toxin. It was in this way that W.E. VAN HEYNINGEN (1959) could explain the fixation of this toxin by brain tissue that much earlier, in 1898, had been described in WASSERMANN's classical paper "Über eine neue Art Künstlicher Immunität". Futhermore, many biological properties of sialic acid were recognized, for which gangliosides having this sugar acid as characteristic constituent of their carbohydrate residue, had to be considered as glycolipid carrier. But gangliosides also appeared to be the glycosphingolipids with the most complex chemical structures. Therefore they provided a challenge to the chemist in the determination of their molecular constitution.

Glycosphingolipids are distinguished by their ceramide-linked carbohydrate residue. Significant differences in their structure are seen according to animal species, cell types and cell maturation.

Glycosphingolipids can be classified into several series that reflect their biogenic relations (Fig. 1).

Except for the gangliosides NeuAcα→6Glc→Cer and NeuAcα→3Gal→Cer, all gangliosides that were found so far belong to the ganglio- or lacto-series.

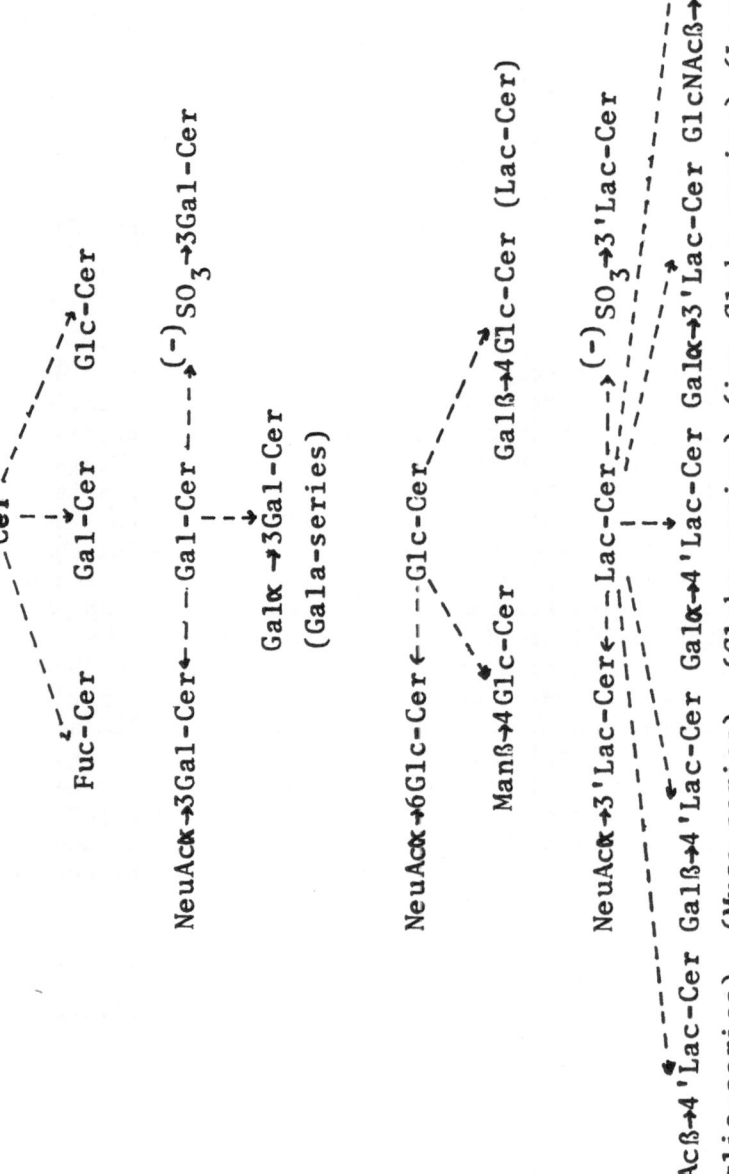

Fig. 1   Glycosphingolipids classified according to carbohydrate series

Following the historical development of ganglioside research a few landmarks that show characteristic features of their chemical structures could be cited. Studies in this field began with the elucidation of hematoside, II^3NeuAc-Lac-Cer isolated from red blood cells, the ganglioside II^3NeuAc-GgOse$_3$-Cer that was stored in Tay-Sachs disease and the four major ganglioside components of normal mammalian brain:

	Abbrev.	Ref.
Gal[3←αNeuAc]β→4Glc→Cer	G_{Lac}1, GM3	a
GalNAcβ→4Gal[3←αNeuAc]β→4Glc→Cer	G_{Gtri}1, GM2	b
Galβ→3GalNAcβ→4Gal[3←αNeuAc]β→4Glc→Cer	G_{Gtet}1, GM1	c
Gal[3←αNeuAc]β→3GalNAcβ→4Gal[3←αNeuAc]β→4Glc→Cer	G_{Gtet}2a GD1	c
Galβ→3GalNAcβ→4Gal[3←αNeuAc8←αNeuAc]β→4Glc→Cer	G_{Gtet}2b, GD1b	c
Gal[3←αNeuAc]β→3GalNAcβ→4Gal[3←αNeuAc8←αNeuAc]β→4Glc→Cer	G_{Gtet}3b, GT1b	c

Ref. (a) KLENK and GIELEN, 1963; SVENNERHOLM, 1963; KUHN and WIEGANDT, 1964b; (b) SVENNERHOLM, 1962; KLENK et al., 1963; KUHN and WIEGANDT, 1963a; (c) KUHN and WIEGANDT, 1963b.

It might have appeared at first that these gangliosides presented a closed group with no further modifications. In 1964 however a new ganglioside was discovered from bovine erythrocytes and human spleen that contained N-acetylglucosamine. This ganglioside IV^3NeuAc-nLcOse$_4$-Cer, belonged to a different series, the lacto-series of type II, with N-acetyllactosamine besides lactose as core carbohydrate unit:

	Abbrev.
Gal[3←αNeuAc]β→4GlcNAcβ→3Galβ→4Glc→Cer	G_{Lntet}1a, GlcNAc-GM1

Ref. KUHN and WIEGANDT, 1964a; WIEGANDT and SCHULZE, 1969; LI et al., 1973; WIEGANDT, 1973.

Even though this ganglioside G_{Lntet}1a is also found as a trace component in the brain, it is considered a major ganglioside of mammalian peripheral nerve and extraneural tissues. In 1973 one more ganglioside of the lacto-series, type II, was characterized from human spleen, IV^6NeuAc-nLcOse$_4$-Cer:

	Abbrev.
Gal[6←αNeuAc]β→4GlcNAβ→3Galβ→4Glc→Cer	G_{Lntet}1b

Ref. WIEGANDT, 1973.

Also in the year 1973 yet another structural component of a ganglioside was recognized, the possible presence of fucose as in ganglioside IV^2Fuc- , $II^3NeuAc-GgOse_4-Cer$:

	Abbrev.
Gal[2←αFuc]β→3GalNAcβ→4Gal[3←αNeuAc]β→4Glc→Cer	G_{Gfpt}1

Ref. WIEGANDT, 1973; SUZUKI, ISHIZUKA and YAMAKAWA, 1975; GHIDONI *et al.*, 1976.

The latter fuco-ganglioside discovered in bovine liver (WIEGANDT, 1973), also occurs in small amounts in bovine brain (GHIDONI *et al.*, 1976). It is a characteristic component of pig testes (SUZUKI *et al.*, 1975).

During the years that followed other gangliosides were isolated all classified to belong to the ganglio- or lacto-series.

The gangliosides of the ganglio-series, in particular those isolated from fish (chondroichties, teleost), reptile and mammalian brain were found generally more highly sialylated as compared to ganglioside of the lacto-series. This was due to the frequent occurrence of multiple sialic acid residues linked to one another by α2→8 ketoside. In all "ganglio"-sialo-glycolipids the sialic acid residues always were only in 3-position of galactose. A penta-sialo-ganglioside IV^3NeuAc_2-, $II^3NeuAc_3-GgOse_4-Cer$ from fish brain so far is the most highly sialylated component that was structurally identified:

```
Galβ→3GalNAcβ→4Galβ→4Glc→Cer                      Abbrev.
3                   3
↑                   ↑                             G     5c, G
αNeuAc8←αNeuAc  αNeuAc8←αNeuAc8←αNeuAc                Gtet      Plc
```

Ref. ISHIZUKA and WIEGANDT, 1972.

Taking the structure of this penta-sialo-ganglio-tetraosyl-ceramide, all such gangliosides have by now been identified from brain tissues by various research teams, whose structure may be derived at by a stepwise subtraction of one sialic acid or one mono-hexosyl unit at a time from ganglioside $G_{Gtet}5c$.

It was shown by SVENNERHOLM *et al.* (1973), that ganglioside can be isolated from brain having ganglio-pentaose as neutral carbo-hydrate backbone:

GalNAcβ→4Gal→3GalNAcβ→4Galβ→4Glc→Cer   Abbrev.
     3   3
     ↑   ↑ $G_{Gpt}2a$, GalNAc-GD1a
  αNeuAc   αNeuAc

Ref. SVENNERHOLM, MANSSON and LI, 1973.

In ganglioside IV[3]NeuAc-, II[3]NeuAc-GgOse$_5$-Cer, both internal sialic acid residues resist *Vibro cholearae* neuraminidase.

A ganglioside with a unique hematoside structure that never-theless also might be classified in this series, was reported from the starfish *Asterina pectinifora* by SUGITA and HORI (1976):

  Galβ→4Glcβ→Cer
   3
   ↑
  αNeuAc8←βGal
     4
     ↑
   βGal6←βAra

In addition more ganglioside components were also found in recent years that belong to the lacto-series of type II. Such gangliosides show lacto-neotetraosyl-residues extended by the typical N-acetyl-lactosaminyl (Galβ→4GlcNAcβ) core unit either in straight or branched chain position:

```
Galβ→4GlcNAcβ→3Galβ→4GlcNAcβ→3Galβ→4Glc→Cer        Abbrev.     Ref.
  3
  ↑                                                   G    1a      a
 αNeuAc                                                 Lnhex
          Fucα
            ↓                                                       b
            3
    Galβ→4GalcNAcβ
               ↓
               6
 Galβ→4GlcNAcβ→3Galβ→4Glc→Cer
   3
   ↑
  αNeuAc
```

Ref. (a) WIEGANDT, 1974; (b) WATANABE *et al.*, 1974.

Both types of extension are met in a monosialo-, monofuco-octa-
hexaosyl-ceramide described by WATANABE *et al.* (1978):

```
    Fucα
      ↓
      2
      Galβ→4GlcNAcβ
                 ↓
                 6
  Galβ→4GlcNAcβ→3Galβ→4GlcNAcβ→3Galβ→4Glc→Cer
    3
    ↑
   αNeuAc
```


It is to be expected that in future many more gangliosides will
still be discovered and their chemical structures identified.

In view of the high multiplicity of these seemingly characteristic
and specific lipid-sialo-glycoconjugates, it will definitely also have
to be one of the primary goals in ganglioside research to correlate
such structural diversities with biological meaning and significance.

REFERENCES

GHIDONI R., SONNINO S., TETTAMANTI G., WIEGANDT H. and ZAMBOTTI V.
  (1976): On the structure of two new gangliosides from beef brain.
  J. Neurochem. 27, 511-515.

ISHIZUKA I. and WIEGANDT H. (1972): An isomer of trisialoganglioside and the structure of tetra- and pentasialogangliosides from fish brain. Biochim. Biophys. Acta 260, 279-289.

KLENK E. and GIELEN W. (1963): Über ein zweites hexosaminhaltiges Gangliosid aus Menschengehirn. Z. Physiol. Chem. 330, 218-226.

KLENK E., LIEDTKE U. and GIELEN W. (1963): Das Gangliosid des Gehirns bei der infantilen amaurotischen Idiotie vom Typ Tay-Sachs. Z. Physiol. Chem. 334, 186-192.

KUHN R. and WIEGANDT H. (1963a): Die Konstitution der Gangliotetraose und des Gangliosids G_I. Chem. Ber. 96, 866-880.

KUHN R. and WIEGANDT H. (1963b): Die Konstitution der Ganglioside G_{II}, G_{III} und G_{IV}. Z. Naturforsch. 18b, 541-543.

KUHN R. and WIEGANDT H. (1964a): Über ein glucosaminhaltiges Gangliosid. Z. Naturforsch. 19b, 80-81.

KUHN R. and WIEGANDT H. (1964b): Weitere Ganglioside aus Menschenhirn. Z. Naturforsch. 19b, 256-257.

LI Y.T., MANSSON J.E., VANIER M.T. and SVENNERHOLM L. (1973): Structure of the major glucosamine-containing ganglioside of human tissues. J. biol. Chem. 248, 2634-2636.

SUGITA M. and HORI T. (1976): New types of gangliosides in starfish with sialic acid residues in the inner part of their carbohydrate chains. J. Biochem. 80, 637-640.

SUZUKI A., ISHIZUKA I. and YAMAKAWA T. (1975): Isolation and characterisation of a ganglioside containing fucose from boar testis. J. Biochem. 78, 949-954.

SVENNERHOLM L. (1962): The chemical structure of normal human brain and Tay-Sachs gangliosides. Biochem. Biophys. Res. Commun. 9, 436-441.

SVENNERHOLM L. (1963): Chromatographic separation of human brain gangliosides. J. Neurochem. 10, 613-623.

SVENNERHOLM L., MANSSON J.E. and LI Y.T. (1973): Isolation and structural determination of a novel ganglioside, a disialopenta-hexosylceramide from human brain. J. biol. Chem. 248, 740-742.

VAN HEYNINGEN W.E. (1959): The fixation of tetanus toxin by nervous tissue. J. Gen. Microbiol. 20, 291-300.
Chemical assay of the tetanus toxin receptor in nervous tissue. J. Gen. Microbiol. 20, 301-309.

Tentative identification of the tetanus toxin receptor in nervous tissue. J. Gen. Microbiol. 20, 310-320.

WASSERMANN J. (1898): Über eine neue Art künstlicher Immunität. Berl. klin. Wschr. 35, 4-5.

WATANABE K., STELLNER K., YOGEESWARAN G. and HAKOMORI S. (1974): A branched, long chain neutral glycolipid and gangliosides of human erythrocytes membranes. Fed. Proc. 33, 1225, Abstr. 3.

WATANABE K., POWELL M. and HAKOMORI S.I. (1978): Isolation and characterization of a novel fucoganglioside of human erythrocyte membranes. J. biol. Chem. 253, 8962-8967.

WIEGANDT H. (1973): Gangliosides of extraneural organs. Z. Physiol. Chem. 354, 1049-1056.

WIEGANDT H. (1974): Monosialo-lactoisohexaosyl-ceramide: a ganglioside from human spleen. Eur. J. Biochem. 45, 367-369.

WIEGANDT H. and SCHULZE B. (1969): Spleen gangliosides: the structure of ganglioside $G_{Lntet}1$. Z. Naturforsch. 24b, 945-946.

GANGLIOSIDE DESIGNATION

Lars Svennerholm

Psychiatric Research Centre, Department of Neurochemistry,
University of Göteborg, St. Jörgen Hospital,
S-422 03 HISINGS BACKA, Sweden

LIPID DOCUMENT (1977)	SVENNERHOLM (1963)
II^3NeuAc-LacCer	GM3
II^3NeuGc-LacCer	
II^3(NeuAc)$_2$-LacCer	
II^3NeuAc,NeuGc-LacCer	GD3
II^3(NeuGc)$_2$-LacCer	
II^3NeuAc-GgOse$_3$Cer	GM2
II^3(NeuAc)$_2$-GgOse$_3$Cer	GD2
II^3NeuAc-GgOse$_4$Cer	GM1
IV^3NeuAc-nLcOse$_4$Cer	LM1
IV^3NeuAcII^{3}NeuAc-GgOse$_4$Cer	GD1a
II^3(NeuAc)$_2$-GgOse$_4$Cer	GD1b
IV^3(NeuAc)$_2$-nLcOse$_4$Cer	LD1
IV^3(NeuAc)$_2$II^{3}NeuAc-GgOse$_4$Cer	GT1a
IV^3NeuAc,II^3(NeuAc)$_2$-GgOse$_4$Cer	GT1b
II^3(NeuAc)$_3$-GgOse$_4$Cer	GT1c
IV^3(NeuAc)$_2$,II^3(NeuAc)$_2$-GgOse$_4$Cer	GQ1b
IV^3NeuAcII^3(NeuAc)$_3$-GgOse$_4$Cer	GQ1c
IV^3(NeuAc)$_3$II^3(NeuAc)$_2$-GgOse$_4$Cer	GP1b
IV^3(NeuAc)$_2$II^3(NeuAc)$_3$-GgOse$_4$Cer	GP1c

A NEW APPROACH TO THE ANALYSIS OF GANGLIOSIDE MOLECULAR SPECIES

Yoshitaka Nagai and Masao Iwamori

Department of Biochemistry
Tokyo Metropolitan Institute of Gerontology
35-2 Sahaecho, Itabashi-ku, Tokyo 173   Japan

Recent advancement in structural analysis of gangliosides has revealed that the world of ganglioside molecular species is unexpectedly full of variety (1). Gangliosides having a novel structure are increasing in number and most of them are present as minor components. This situation suggests that gangliosides are uniquely qualified for the specific and diversified functions of the cell surface membranes. As regards gangliosides of the nervous tissues, they occur in the highest concentration in these tissues, particularly in the synaptosome, and are thought to be participating in a synaptic transmission. A recent observation that gangliosides play a role in a calcium-mediated release of serotonin from synaptosome (2) supports the presumable function in the nerve ending. The other suggestive observations that gangliosides are closely related to the synaptic transmission have been obtained from the experiments using bacterial neurotoxins. The toxins from Clostridium tetani and Clostridium botulinum, that are known to inhibit releasing of transmitter, have been shown to bind the specific gangliosides in vitro (3,4). But the exact function of gangliosides in the nervous tissues is still not clear. On the other hand, arising from the in vitro experiments for binding of bacterial toxins to gangliosides the functional significance of gangliosides as receptors or antigen molecules has been emphasized by a number of investigators. Such experiments that are aiming to identify the specificity of membrane function by the definite carbohydrate entity require the use of absolutely pure gangliosides as a prerequisite, and very frequently necessitate to make a careful survey of molecular species of the membrane carrying that function. For such purpose, an improved method for the purification and characterization of gangliosides has been developed in this laboratory.

13

Scheme 1.   Isolation and purification of gangliosides

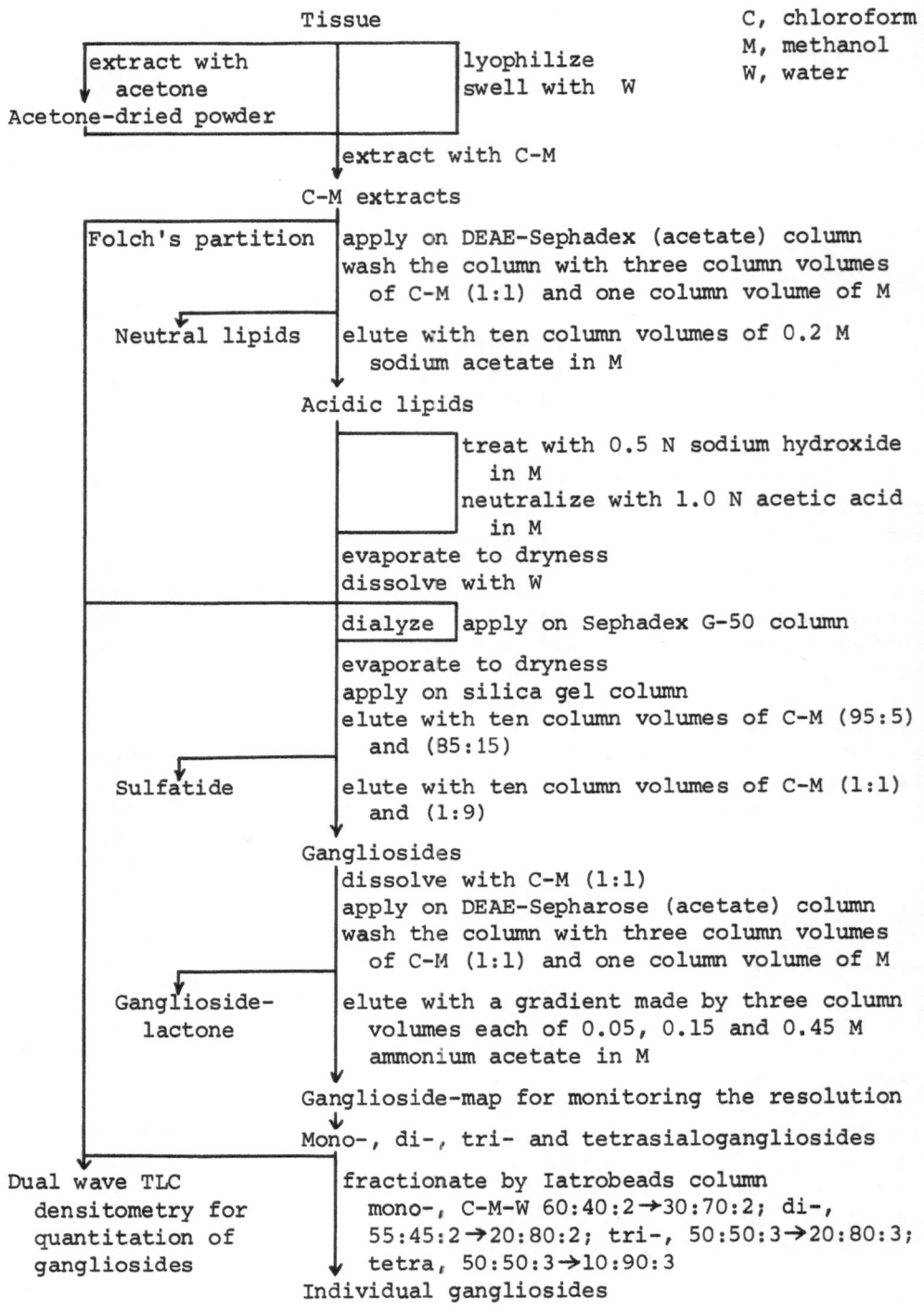

PURIFICATION OF GANGLIOSIDE INDIVIDUALS AND GANGLIOSIDE-MAPPING

        The process for the isolation and purification of ganglio-
sides are grossly shown in Scheme 1. For a preparation of gangliosides from large amounts of tissue, they should be first homogenized
in acetone in order to remove neutral lipids and tissue water. The
preparation of acetone-dried powder prior to chloroform-methanol
extraction facilitate the extraction not only in reducing the solvents used,but in increasing the recovery of gangliosides. However,
for an analytical purpose, tissues should be extracted with chloroform-methanol mixture directly or after lyophilization. Then, the
chloroform-methanol extracts are directly applied on DEAE-Sephadex
(acetate form) column (5) and the elution of gangliosides from the
column is achieved with ten column volumes of 0.2 M sodium acetate
in methanol after equilibrating the column with one column volume of
methanol (6). The colored compounds originated from red blood cells
would remain in the column under the condition using 0.2 M sodium
acetate in methanol and can be removed from gangliosides. Also, in
so far as the complete recovery can not be assured with five column
volumes of 0.2 M sodium acetate in methanol, the elution with ten

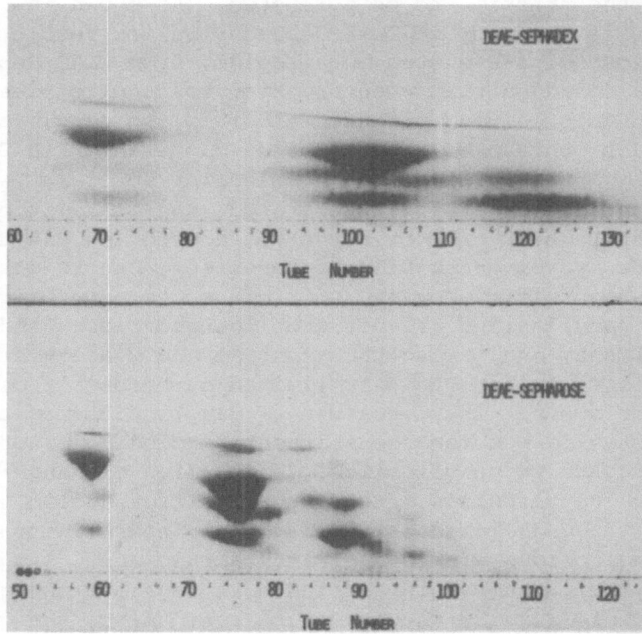

Fig. 1.   Comparison by the mapping procedure of the separation of
          bovine brain gangliosides between DEAE-Sephadex and DEAE-
          Sepharose column chromatography. The plates were developed
          with chloroform-methanol-2.5 N ammonia (60:40:9, v/v/v).

column volumes of the solvent is necessary. Under the condition,
tetrasialogangliosides can be quantitatively recovered, but for the
tissues containing polysialogangliosides, like fish brain, 0.5 M
sodium acetate in methanol must be used. Except that the esterified
gangliosides are the objective compounds, the acidic lipids thus
obtained should be treated with alkali in order to cleave the lipids
containing ester linkage such as phosphatidylserine and ganglioside·
lactone and the resulting fatty acid methyl esters are removed by
partitioning with n-hexane. For a preparative purpose in a large
scale, Folch's partitioning is still useful to simplify the step.
Then, to remove inorganic salts and small molecule contaminants,
dialysis or gel permeation chromatography on Sephadex G-50 (8) can
be applied. For the dialysis of gangliosides, the materials must be
dissolved in water that is free from methanol. The presence of
methanol during dialysis causes considerable loss of gangliosides
(9). The next step for removal of sulfatide can be easily and
quantitatively performed by silica gel column chromatography using
Unisil, Silica gel 60 or Iatrobeads (6,7). The crude gangliosides
dissolved in chloroform-methanol (1:1, v/v) are then applied on
DEAE-Sepharose (acetate form) column, eluted with strictly controlled gradient system, and then ganglioside-map (10) should be made
to identify the individual components by their location on the maps
and to decide the fraction to be collected. As shown in Fig.1,
DEAE-Sepharose is superior to DEAE-Sephadex in its resolution (10)
and the gangliosides are eluted more rapidly from DEAE-Sepharose.
It should be also added that when erythrocyte gangliosides are chromatographed on this column, pigment materials and gangliosides do
not overlap each other. With ganglioside-map using DEAE-Sepharose
that enables high resolution of each molecular species, two spots
around GT1a ganglioside are well separated, and in addition many
minor components, which can not be detected with the one dimensional
TLC or with the system using DEAE-Sephadex (6), can be detected and
resolved with high efficiency and sufficient reproducibility. The
maps of brain gangliosides are not much changed before and after
alkaline treatment, indicating that lactone and O-acetylated types
of gangliosides contribute to only limited numbers of those newly
appearing minor spots. Thus, the use of ganglioside-map allow us to
compare the detailed membrane constituents including minor molecular
species with regard to gangliosides. With anion exchange column
chromatography, gangliosides with a subtle difference in charge can
be separated: the gangliosides with N-glycolylneuraminic acid elute
behind the gangliosides with N-acetylneuraminic acid, suggesting
that N-glycolylneuraminic acid carries more negative charge than
N-acetylneuraminic acid by an unknown reason. Also, the gangliosides differing in molecular size can be separated by the column:
for example, GM4, GM3, GM2 and GD3 are separated in that order from
GM1 and GD1a. It is true that the fractionation on anion exchange
column chromatography is essential for the purification of ganglioside individuals. After the fraction depending on the differences
in charge and molecular size of gangliosides, they are further

fractionated on a unique porous silica sphere, Iatrobeads (6,7).
The resolution of Iatrobeads (6RS-8060, 60 μm of average particle
diameter) is the best among the other silica gels tested, that is,
Unisil and Silica gel 60, and the recovery of gangliosides is
quantitative, although the latter two have their own merits in some
cases. By combined use of those DEAE-Sepharose and Iatrobeads
columns with strictly controlled gradient systems, the gangliosides
in very high purity can be obtained in a sufficient amount and in a
short time. In addition, after a suitable washing, DEAE-Sephadex,

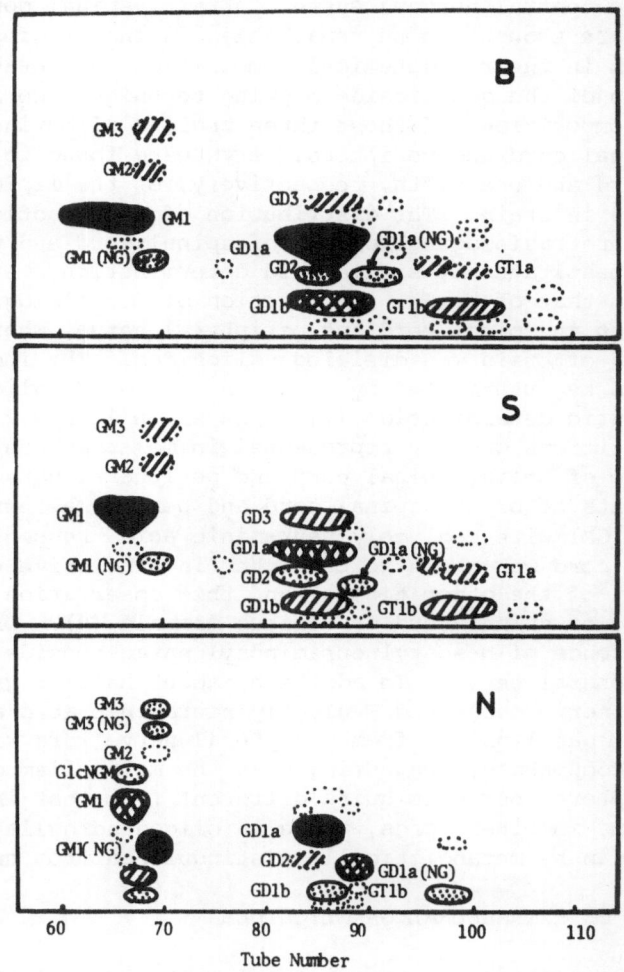

Fig. 2. Ganglioside compositions of bovine brain (B), spinal cord
         (S) and peripheral nerve (N). Molar percentages of indi-
         vidual gangliosides are represented by the following sym-
         bols: ●, 21-50 %; ⊗, 11-20 %; ⊘, 5-10 %; ☺, 1-4 %; ⊘,
         0.1-0.9 %; ⊙, <0.1 %. GlcNGM, NeuNAc-paragloboside.

DEAE-Sepharose and Iatrobeads can be used repeatedly.

GANGLIOSIDE COMPOSITIONS OF BOVINE BRAIN, SPINAL CORD AND PERIPHERAL
NERVE

Since gangliosides seem to be involved in the essential function of nervous system, it is of interest to compare the ganglioside
compositions of the different regions of nervous system. As is well
known, myelination of central nervous system and that of peripheral
nerve are conducted by a different cell, and the composition and
metabolism of spinal cord are also different from those of other
regions of the central nervous system. Brain, spinal cord and peripheral nerve are thought to be remarkably distinct regions of the
nervous system in their biochemical composition and metabolism.
By application of the ganglioside-mapping technique, we analyzed the
ganglioside compositions of those three regions of bovine nervous
tissues. Spinal cord and peripheral nerve were found to contain
about one-third and one-fifth, respectively, of the lipid-bound
sialic acid as in brain. The distribution of the fractions of mono-
di-, tri- and tetrasialogangliosides of spinal cord and peripheral
nerve, when quantitated by sialic acid determination, were also
different from that of brain: the fraction of monosialogangliosides
was predominant in spinal cord and peripheral nerve, whereas the
major fraction of brain was disialogangliosides. The individual
components can be quantitated by a combination of ganglioside-map
and densitometric determination (11). As shown in Fig.2, the ganglioside compositions grossly represented in molar percentage were
characteristic of brain, spinal cord and peripheral nerve. The
major components of brain, spinal cord and peripheral nerve were
GDla, GMl and GMl with N-glycolylneuraminic acid, respectively.
GD3 of spinal cord was shown to be higher in a relative concentration than that of the other tissues and this observation seems to be
characteristic of spinal cord of various mammals (8). On the other
hand, the presence of N-acetylneuraminosyl paragloboside was remarkable in peripheral nerve. In addition, about half of gangliosides
in peripheral nerve possessed N-glycolylneuraminic acid and the
fatty acids of gangliosides from peripheral nerve were rich in
longer chain components, suggesting that the metabolism of gangliosides in peripheral nerve is quite different from that in central
nervous system, in other words, those in oligodendroglia cell and
Schwann cell can be metabolitically distinguished from each other.

A FEW REMARKS ON EXTRANEURAL GANGLIOSIDES

A novel ganglioside, monosialosyl pentahexaosyl ceramide, was
isolated from human brain (12). The structure is:
GalNAc(β,1-4)Gal(β,1-3)GalNAc(β,1-4)Gal(β,1-4)Glc(β,1-1)ceramide.
                                        3
                                        ‾
                                        2
                                      αNeuNAc

And up to present almost 28 molecular species of gangliosides have
been isolated by our technique from various organs including brain
tissues. By comparing these compounds, it was found that gangliosides with N-glycolylneuraminosyl residue give in general a retarded
migration from those with N-acetylneuraminosyl residue on TLC developed with alkaline solvent (Fig.3). However, on TLC with neutral
solvents, both types of gangliosides migrated closely, suggesting a
special behavior of N-glycolylresidue in secondary structure of
ganglioside molecules.

A GD3 ganglioside having a novel disialosyl residue was isolated from rabbit thymus (13), whose structure was identified to be:

NeuNGlyc(α,2-8)NeuNAc(α,2-3)Gal(β,1-4)Glc(β,1-1)ceramide.

Besides such a molecular species, thymic organ was found to be
unique in ganglioside composition. For example, in thymus of rabbit
almost 90 % of ganglioside sialic acids was comprised of N-glycolyl
type, whereas in other organs N-acetyl type was the principal sialic
acid of gangliosides (13,14). Furthermore our studies currently in
progress reveals the presence of gangliosides full of molecular

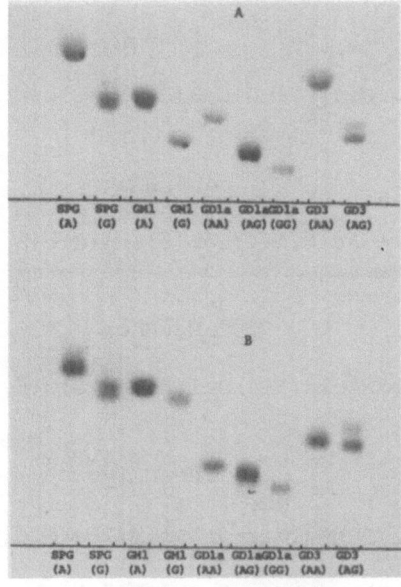

Fig.3. Comparison of the mobilities of gangliosides with N-acetylneuraminic acid and N-glycolylneuraminic acid. Abbreviations used in the Figure are presented at the end of the
text. A, chloroform-methanol-2.5 N ammonia (60:40:9)
         B, chloroform-methanol-0.5 % calcium chloride (55:45:10)

variety, suggesting their special roles in this unique immunological
organ as well as the related cells (15).

ABBREVIATIONS

SPG(A): NeuNAc(α,2-3)Gal(β,1-4)GlcNAc(β,1-3)Gal(β,1-4)Glc(β,1-1)
        ceramide

SPG(G): NeuNGlyc(α,2-3)Gal(β,1-4)GlcNAc(β,1-3)Gal(β,1-4)Glc(β,1-1)
        ceramide

GM1(A): Gal(β,1-3)GalNAc(β,1-4)Gal(β,1-4)Glc(β,1-1)ceramide
                                3
                                2
                            αNeuNAc

GM1(G): Gal(β,1-3)GalNAc(β,1-4)Gal(β,1-4)Glc(β,1-1)ceramide
                                3
                                2
                            α NeuNGlyc

GD1a(AA): Gal(β,1-3)GalNAc(β,1-4)Gal(β,1-4)Glc(β,1-1)ceramide
          3   3
          2   2
        αNeuNAc   αNeuNAc

GD1a(AG): Gal(β,1-3)GalNAc(β,1-4)Gal(β,1-4)Glc(β,1-1)ceramide   or
          3   3
          2   2
        αNeuNAc   αNeuNGlyc

          Gal(β,1-3)GalNAc(β,1-4)Gal(β,1-4)Glc(β,1-1)ceramide
          3   3
          2   2
        αNeuNGlyc   αNeuNAc

GD1a(GG): Gal(β,1-3)GalNAc(β,1-4)Gal(β,1-4)Glc(β,1-1)ceramide
          3   3
          2   2
        αNeuNGlyc   αNeuNGlyc

GD3(AA): NeuNAc(α,2-8)NeuNAc(α,2-3)Gal(β,1-4)Glc(β,1-1)ceramide

GD3(AG): NeuNGlyc(α,2-8)NeuNAc(α,2-3)Gal(β,1-4)Glc(β,1-1)ceramide
        Two spots of GD3(AG) in Fig.3 are due to the difference of
        fatty acid composition.

REFERENCES

1. T. Yamakawa and Y. Nagai, Glycolipids at the cell surface and their biological functions, Trends in Biochem. Sci., 3:128 (1978).

2. G. A. Dette and W. Weseman, On the significance of sialic acid in high affinity 5-hydroxytryptamine uptake by synaptosomes, Hoppe-Zeyler's Z. Physiol. Chem., 359:399 (1978).

3. T. B. Helting, O. Zwisler and H. Wiegandt, Structure of tetanus toxin, J. Biol. Chem., 252:194 (1977).

4. M. Kitamura, M. Iwamori and Y. Nagai, Interaction between Cl. botulinum neurotoxin and gangliosides, in Abstracts for the XIth International Congress of Biochemistry, (1979).

5. R. W. Ledeen, R. K. Yu and L. F. Eng, Gangliosides of human myelin : sialosylgalactosylceramide (G_7) as a major component, J. Neurochem., 21:829 (1973).

6. T. Momoi, S. Ando and Y. Nagai, High resolution preparative column chromatographic system for gangliosides using DEAE-Sephadex and a new porous silica, Iatrobeads, Biochim. Biophys. Acta, 441:488 (1976).

7. S. Ando, M. Isobe and Y. Nagai, High performance preparative column chromatography of lipids using a new porous silica, Iatrobeads : I. Separation of molecular species of sphingoglycolipids, Biochim. Biophys. Acta, 424:98 (1976).

8. K. Ueno, S. Ando and R. K. Yu, Gangliosides of human, cat, and rabbit spinal cords and cord myelin, J. Lipid Res., 19:863 (1978).

9. R. Ghidoni, S. Sonnino and G. Tettamanti, Behavior of gangliosides on dialysis, Lipids, 13:820 (1978).

10. M. Iwamori and Y. Nagai, A new chromatographic approach to the resolution of individual gangliosides : ganglioside mapping, Biochim. Biophys. Acta, 528:257 (1978).

11. M. Iwamori and Y. Nagai, Ganglioside-composition of brain in Tay-Sachs disease : increased amounts of GD_2 and N-acetyl-β-D-galactosaminyl GD_1a ganglioside, J. Neurochem., 32:767 (1979).

12. M. Iwamori and Y. Nagai, Isolation and characterization of a novel ganglioside, monosialosyl pentahexaosyl ceramide from human brain, J. Biochem., 84:1601 (1978).

13. M. Iwamori and Y. Nagai, Isolation and characterization of GD_3 ganglioside having a novel disialosyl residue from rabbit thymus, J. Biol. Chem., 253:8328 (1978).

14. M. Iwamori and Y. Nagai, GM_3 ganglioside in various tissues of rabbit : tissue-specific distribution of N-glycolylneuraminic acid-containing GM_3, J. Biochem., 84:1609 (1978).

15. M. Iwamori and Y. Nagai, Ganglioside compositions of various tissues of rabbit : uniqueness of thymus, Proc. Japanese Conference on Biochem. Lipids, 20:323 (1978).

ISOLATION AND SEPARATION OF GANGLIOSIDES ON A NEW FORM OF GLASS

BEAD ION EXCHANGER *

Pam Fredman

Psychiatric Research Centre, Department of Neurochemistry,
University of Göteborg, St. Jörgen Hospital,
S-422 03 HISINGS BACKA, Sweden

ABSTRACT

A new anion exchange resin, Spherosil-DEAE-Dextran, built on
porous glass beads covered with crosslinked DEAE-Dextran, has been
compared with other resins, DEAE-Sephacel, DEAE-Sephadex and DEAE-
Sepharose, regarding its value in the isolation and separation of
gangliosides from a crude ganglioside extract. In comparison with
the best commercial resin, DEAE-Sepharose, Spherosil-DEAE-Dextran
had higher binding capacity, higher binding specificity and gave a
better separation of gangliosides with four and five sialic acids.
The constant bed volume facilitated regeneration.

Gangliosides were separated according to the number of sialic
acids by stepwise elution with potassium acetate in methanol. The
elution scheme was used for both large and small scale separation.

Separation of gangliosides on anion exchange resins according
to their number of sialic acids results in a better discrimination
and identification of the many closely allied gangliosides. This
property is particularly important in the separation of gangliosides
with a wide variation in the substitution of the sialic acid,
N-acetyl, N-glycolyl or N,O-diacetyl groups. An N,O-diacetyl sialic
acid will exhibit chemical properties very similar to those of a
ganglioside with the N-acetyl sialic acid in lacton form, but they
could be separated on the anion exchanger.

*This study was performed in collaboration with O. Nilsson and
L. Svennerholm, Psychiatric Research Centre, and J.L. Tayot,
Institute Mérieux, Lyon. It was supported by grants from the
Swedish Medical Research Council (3X-627) and the Medical Faculty,
University of Göteborg.

INTRODUCTION

A large number of gangliosides have been isolated and
characterized and new ganglioside structures are regularly being
reported. Very little is known about their biological function, but
gangliosides have been found to be receptors for certain bacterial
toxins, cholera, botulinus and tetanus (HOLMGREN et al., 1973;
SIMPSON and RAPPORT, 1971; VAN HEYNINGEN, 1963). Any investigation
of the role of gangliosides as receptors requires the use of gangliosides with a uniform carbohydrate composition.

Separation of gangliosides on silica gel is difficult owing to
the complexity of the gangliosides. Anion exchange chromatography,
separating the gangliosides according to the number of sialic acids,
combined with silica gel chromatography increases the possibility
of isolating gangliosides with a uniform carbohydrate moiety.

LEDEEN et al. (1973) described the use of DEAE-Sephadex for
small scale isolation of gangliosides from a total lipid extract.
In those experiments large amounts of resins were necessary to
achieve complete attachment of the gangliosides. Several different
resins, DEAE-Sephadex, QAE-Sephadex, DEAE-Sepharose and the new
Spherosil-DEAE-Dextran were tested for large scale isolation of
gangliosides from total lipid extracts, but the capacity of all the
resins proved unsatisfactorily low.

In the present study the new resin, Spherosil-DEAE-Dextran was
compared with DEAE-Sephacel, DEAE-Sephadex and DEAE-Sepharose regarding its value in the isolation and separation of gangliosides.
A general scheme for stepwise elution of gangliosides according to
the number of sialic acids with KAc in methanol is described.

The method has also been applied to separation of N,O-di-
acetylneuraminic acid containing ganglioside from the same ganglioside in lacton form.

MATERIALS

Chemicals

Sephadex G-25, DEAE-Sepharose CL-6B, DEAE-Sephadex A-25,
DEAE-Sephacel and DEAE-Dextran were purchased from Pharmacia Fine
Chemicals, Uppsala, Sweden. The gels were swollen and rinsed with
distilled water until no carbohydrates were found in the eluate.
Spherosil, porous glass beads, with a mean pore diameter 125 nm,
particle size 100-200 μm, surface area 25 m^2/g and pore volume
1 ml/g was obtained from Rhone Poulence Industries, Division Chimie
Fine, Paris. To prepare Spherosil-DEAE-Dextran (TAYOT, 1978), 1 kg
of Spherosil was poured into 2.3 l of a 7.5% aqueous solution of

DEAE-Dextran at pH 11.5 and then dried 15 h at 80°C. The DEAE-Dextran was crosslinked with 1.4 butanedioldiglycidyl ether (Aldrich Chemicals Company, Milwakuee, WI). Precoated thin-layer plates, Kiselgel 60, 20x20 cm were obtained from Merck AG, Darmstadt, West Germany.

Organic solvents and other chemicals were of analytical quality. Gangliosides and N-acetylneuraminic acid were prepared in this laboratory.

Tissue

Human brains were obtained through Dr. Kerstin Boström, Head, Department of Forensic Medicine. Written permission to examine these brains was given to Dr. Lars Svennerholm by the Royal Swedish Medical Board.

METHOD

Isolation of a crude ganglioside extract from human forebrain

Brain material was homogenized with 3 volumes (1 g = 1 ml) of water in a Waring blendor at maximum speed for 2 minutes at +4°C. The homogenate was poured into 16 volumes of chloroform-methanol 1:2. After centrifugation at 2000xg for 30 min., the residue was re-extracted with 10 volumes of chloroform-methanol-water 4:8:3. The supernatants were pooled and then partitioned by adding water to a final ratio chloroform-methanol-water 4:8:5.6. The lower phase was repartitioned by adding methanol and 0.01 mol KCl/l in water to a final ratio chloroform-methanol-water 2:1:0.7 and the two upper phases were evaporated. This extract was then dialysed against running tap water for 48 h and against two changes of distilled water for 24 h. After evaporation the extract was dissolved in chloroform-methanol-water 60:30:4.5.

Separation of crude gangliosides on Spherosil-DEAE-Dextran

Spherosil-DEAE-Dextran was suspended in distilled water and the fines were decantated off. The resin was then slurried into a glass column and converted into acetate form by elution with 2 mol NaAc/l in water, until no chloride was detectable in the eluate; 5 bed volumes were sufficient. The resin was rinsed with 5 bed volumes of distilled water, 1 volume of methanol and 1 volume of chloroform-methanol-water 60:30:4.5.

The other resins were converted to acetate form by elution with 2 mol NaAc/l in water as described above. They were then dried, by

elution with ethanol, to constant volume and equilibrated with
chloroform-methanol-water 30:60:8 before loading.

Spherosil-DEAE-Dextran in acetate form was packed in glass
columns. The crude gangliosides, dissolved in one bed volume of
chloroform-methanol-water 60:30:4.5 were passed through the column
with a flow rate 0.5 ml/min/100 ml resin. Rinsing was then performed
with 1 volume of the same solvent and 1 volume of methanol. Gangliosides were then eluated with KAc in methanol according to the
following scheme, flow rate ca 2 ml/min/100 ml resin.

```
8 bed volumes of 0.02 mol KAc/l in methanol
10   "      "      "  0.10 "   "    "  "
10   "      "      "  0.20 "   "    "  "
 5   "      "      "  0.50 "   "    "  "
 3   "      "      "  1.0  "   "    "  "
```

The fractions collected were evaporated and dialysed as
described above.

Assay methods

The yield of gangliosides from each isolation was determined
by the assay of lipid bound NeuAc with the resorcinol method
(SVENNERHOLM, 1957). Quantification of the ganglioside pattern was
performed by scanning the TLC plates, sprayed with resorcinol and
heated for 5 minutes at 140°C, on a Zeiss KM3 chromatogram scanner
at 620 nm.

Following solvent systems were used for TLC: chloroform-
methanol-0.25% KCl in water 60:35:8 for 2 h, 1-propanol-0.25% KCl
in water 3:1 or 7:3 for 4 h and chloroform-methanol-2.5 mol NH$_3$/l
in water 60:40:9 for 2 h.

RESULTS AND DISCUSSION

Separation of gangliosides according to the number of sialic acids

Isolation of gangliosides from other brain lipids using DEAE-
cellulose was described by ROUSER et al. (1964) and WINTERBOURN
(1971) used DEAE-cellulose for separation of gangliosides according
to the number of sialic acids. The gangliosides were separated by
stepwise elution with ammonium acetate in methanol. Our intention
was to use anion exchange chromatography for block separation of
gangliosides according to the number of sialic acids. The following
resins were tested, DEAE-Sephacel, DEAE-Sephadex, DEAE-Sepharose
and Spherosil-DEAE-Dextran.

All the resins retained 100% of the gangliosides. With DEAE-Sephacel it was not possible on elution to achieve any separation according to the number of sialic acids. Chromatography on DEAE-Sephadex separated the gangliosides, but under considerable tailing compared with DEAE-Sepharose and Spherosil-DEAE-Dextran. In the following studies only DEAE-Sepharose and Spherosil-DEAE-Dextran were used.

NAGAI and co-workers reported a method for separating gangliosides on DEAE-Sephadex (MOMOI et al., 1976) and DEAE-Sepharose (IWAMORI and NAGAI, 1978) with the use of linear gradient elution with ammonium acetate in methanol. They also found that DEAE-Sepharose gave sharper elution peaks than DEAE-Sephadex.

The binding capacity, defined as the amount of gangliosides that could be retained to 100% on the column, was found to be 25 μmoles NeuAc/10 ml resin for DEAE-Sepharose, but as much as 50 μmoles/10 ml resin for Spherosil-DEAE-Dextran.

For evalution of the separation of especially higher gangliosides, crude fish brain ganglioside extract, which contains large proportions of tetra- and pentasialogangliosides (ISHIZUKA, 1970) was added to human brain ganglioside extract. The best resolution was achieved with the use of ammonium acetate on DEAE-Sepharose and potassium acetate on Spherosil-DEAE-Dextran. The mono- and disialogangliosides fractions on the two resins were similar, while the trisialoganglioside fraction from DEAE-Sepharose was contaminated with GQ 1b (Fig. 1). On DEAE-Sepharose all the remaining gangliosides were eluted with 0.5 mol acetate/1, but on Spherosil-DEAE-Dextran also tetra- and pentasialogangliosides were separated. On both resins trisialogangliosides remained in the 0.5 mol/1 fraction. By increasing the volume of 0.2 mol acetate/1, all trisialogangliosides on Spherosil-DEAE-Dextran were eluted, but on DEAE-Sepharose the increase in volume resulted also in elution of tetrasialogangliosides. The elution volumes were modified according to the volume of the column, large column required smaller volumes, and to the composition of the gangliosides extract. Summerized, Spherosil-DEAE-Dextran had higher binding capacity, higher binding specificity and gave a better separation of higher gangliosides. Constant bed volume facilitated regeneration and the properties of the resin were the same also after five running cycles.

An advantage with stepwise elution was the block separation of the gangliosides according to number of sialic acids, which simplified the further separation on silica gel. The procedure also facilitated quantification and identification of the individual gangliosides in each group, monosialogangliosides, disialogangliosides etc. When continuous gradient elution (IWAMORI and NAGAI, 1978) is used, all fractions have to be tested, which

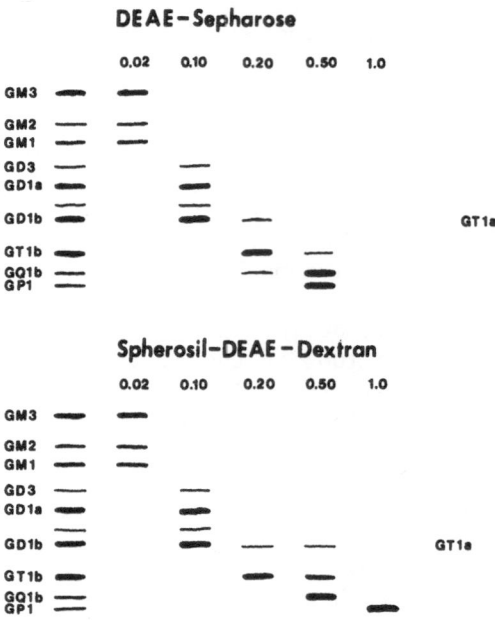

Fig. 1.   Separation of gangliosides on anion exchange resin.

causes losses of material which is a considerable drawback, especially when only small amounts are available. All fractions eluted with a higher concentration of acetate have to be desalted before TLC. Gradient elution does not completely separate the gangliosides; it only gives a concentration of individual gangliosides in the various fractions collected. Really quantitative collection of a single ganglioside is also difficult, since the beginning and end of the elution curve will very often be missed.

Application of the method to a special problem

Large scale separation of gangliosides using only silica gel chromatography is complicated because some gangliosides differing in the carbohydrate moiety are eluted with a broad overlap. The separation could be improved by using anion exchange chromatography before silica gel chromatography. With this combination we have succeeded in isolating large amounts of GD3. On silica gel the major part of GD3 is eluted together with GM1, but on anion exchange chromatography they were completely separated into the

TABLE I. R_{GM1}-VALUES

Ganglioside	Chloroform Methanol 0.25% KCl 60:35:8	Chloroform Methanol 2.5% mol NH_3/l 60:40:9	n-Propanol- 0.25% KCl 3:1
GD3 N-acetyl	0.78	1.21	0.86
GD3 Lacton	1.62	-	1.28
GD1b N-acetyl	0.33	0.32	0.54
GD1b N,O-diacetyl	0.44	0.69	0.62
GD1b Lacton	0.88	-	0.94
GT1b N-acetyl	0.16	0.29	0.37
GT1b N,O-diacetyl	0.23	0.56	0.40
GT1b Lacton	0.38	-	0.69

P.FREDMAN, O.NILSSON and L.SVENNERHOLM (unpublished results)

di- and monosialoganglioside fractions respectively. Ganglioside GD3 was then easily separated from the other disialogangliosides on silica gel.

Gangliosides containing fucose have R_f-values similar to those of the main brain gangliosides on silica gel. We encountered this problem when fucose-containing gangliosides were to be isolated from the nervous system of mini-pig. Using anion exchange chromatography for the first separation of the crude ganglioside extract, three fucose-containing gangliosides were isolated. Two of them were monosialogangliosides, one with N-acetyl- and one with N-glycolylneuraminic acid, and the third was a disialoganglioside with two N-acetylneuraminic acids (in the brains). Final separation of those was performed by silica gel chromatography of the mono- and disialoganglioside fractions with ammonia solvent. Gangliosides containing fucose were eluted after the other gangliosides.

It has always been difficult to distinguish between a ganglioside with a sialic acid in lacton form and the same ganglioside containing N,O-diacetylneuraminic acid. Both are alkali labile, give a positive ester reaction and have higher R_{GM1}-values than the ganglioside with just N-acetylneuraminic acid (Table I). O-acetyl groups are easily lost during the hydrolysis for isolation and identification of the sialic acids by GLC or TLC. These two gangliosides were separated on Spherosil-DEAE-Dextran, and, to avoid conversion of the lacton or loss of the O-acetyl group, the elution was performed with potassium acetate in methanol buffered to pH 6.8, at which the lacton was stable. The ganglioside containing

N,O-diacetylneuraminic acid was eluted according to the total
number of sialic acids, while the ganglioside with one sialic
acid in lacton form was eluted together with the gangliosides
containing one sialic acid less.

REFERENCES

HOLMGREN J., LÖNNROTH I. and SVENNERHOLM L. (1973): Tissue
    receptor for cholera exotoxin: Postulated structure from studies
    with GM1 ganglioside and related glycolipids. Infect. Immunity 8,
    208-214.

ISHIZUKA I., KLOPPENBURG M. and WIEGANDT H. (1970): Characterization
    of gangliosides from fish brain. Biochim. Biophys. Acta 210,
    299-305.

IWAMORI M. and NAGAI Y. (1978): A new chromatographic approach to
    the resolution of individual gangliosides. Biochim. Biophys.
    Acta 528, 257-267.

LEDEEN R.W., YU R.K. and ENG L.F. (1973): Gangliosides of human
    myelin: Sialosylgalactosylceramide (G_7) as a major component.
    J. Neurochem. 21, 829-839.

MOMOI J., ANDO S. and NAGAI Y. (1976): High resolution preparative
    column chromatographic system for gangliosides using DEAE-
    Sephadex and a new porous silica, Iatrobeads. Biochim. Biophys.
    Acta 441, 488-497.

ROUSER G., GALLI C., LIEBER E., BLANK M.L. and PRIVETT O.S. (1964):
    Analytical fractionation of complex lipid mixtures: DEAE-cellulose
    column chromatography combined with quantitative thin layer
    chromatography. J. Am. Oil Chem. Soc. 41, 836-840.

SIMPSON L.L. and RAPPORT M.M. (1971): Ganglioside inactivation of
    botulinum toxin. J. Neurochem. 18, 1341-1343.

SVENNERHOLM L. (1957): Quantitative estimation of sialic acids
    II: A colorometric resorcinol-hydrochloric acid method. Biochim.
    Biophys. Acta 24, 604-611.

TAYOT J.-L., TARDY M., GATTEL P., PLAN R. and ROUMIANTZEFF M.
    (1978): Industrial ion exchange chromatography of proteins on
    DEAE-Dextran derivatives of porous silica beads, in "Chromato-
    graphy of Synthetic and Biological Polymers", EPTON R., Ed.,
    Vol. 2, Ellis Horwood Ltd. (Chichester, West Sussex), pp. 96-110.

VAN HEYNINGEN W.E. (1963): The fixation of tetanus toxin, strychnine, serotonin and other substances by ganglioside. J. Gen. Microbiol. 31, 375-387.

WINTERBOURN C.C. (1971): Separation of brain gangliosides by column chromatography on DEAE-cellulose. J. Neurochem. 18, 1153-1155.

STRUCTURES OF SOME NEW COMPLEX GANGLIOSIDES OF FISH BRAIN

Robert K. Yu and Susumu Ando*

Yale University School of Medicine, New Haven, CT 06510
U.S.A.
(*Present address-Tokyo Metropolitan Institute of
Gerontology, Tokyo-173, Japan)

SUMMARY

Three novel trisialogangliosides of fish brain, G_{T3}, G_{T2} and G_{T1c}, have been isolated in their intact forms and their structures characterized. The discovery of these ganglioside species provides essential links for a new possible biosynthetic pathway leading to the major tetrasialoganglioside, G_{Q1c}, of the fish brain.

The structure of the oligosaccharide, tetrasialoganglio-N-tetraose (desphingosino-G_{Gtet4}), isolated from the ozonolysate of a fish brain ganglioside mixture was established in 1972 by Ishizuka and Wiegandt (1972). They suggested that fish brain tetrasialoganglioside possessed a G_{Q1c} structure (1). It was further presumed that mammalian brain tetrasialoganglioside might have the same arrangement of sugar residues (Ishizuka and Wiegandt, 1972; Wiegandt, 1973). We have recently isolated the major tetrasialoganglioside from human, bovine, and chicken brains as an intact glycolipid, and proposed a G_{Q1b} structure for this ganglioside (Ando and Yu, 1977a). We subsequently isolated the major cod

---

(1) The nomenclature is based on the system of Svennerholm (1963, 1970). Other gangliosides which have not been officially designated by Svennerholm include the following: $G_{M4} = I^3NeuAc$-GalCer; $G_{T3} = II^3(NeuAc)_3$-LacCer; $G_{T2} = II^3(NeuAc)_3$-GgOse$_3$Cer; $G_{T1c} = II^3(NeuAc)_3$-GgOse$_4$Cer; and $G_{Q1c} = IV^3NeuAc, II^3(NeuAc)_3$-GgOse$_4$Cer.

fish brain tetrasialoganglioside in its intact form. Structural
analyses on the intact glycolipid confirmed the G_{Q1c} structure
first proposed by Ishizuka and Wiegandt (Yu and Ando, 1978; Ando
and Yu, 1979). Interestingly, G_{Q1c} does not appear to be present
in the brains of higher vertebrates, and no G_{Q1b} has been found
in cod fish brains. In addition to G_{Q1c}, we have also isolated
several trisialogangliosides, termed G_{T3}, G_{T2} and G_{T1c}, from cod
fish brain and studied their structures. In this paper, we describe the isolation and characterization of these gangliosides from
cod fish brain and discuss their possible relationship to G_{Q1c}
synthesis.

Ganglioside isolation and fractionation

The total lipids of cod fish brain (126g) were extracted by
homogenizing the tissue with 10 volumes of chloroform-methanol
(1:2). Total gangliosides were then prepared from the lipid extract
by a scaled-up version of the method of Ledeen et al. (1973).
The sample was then fractionated into six major peaks on a DEAE-
Sephadex column (A-25, 1.4 cm x 110 cm) by employing continuous
gradient elutions with increasing concentrations of ammonium
acetate in methanol (methanol, 1200 ml; 0.35 M ammonium acetate
in methanol, 1200 ml; and 0.60 M ammonium acetate in methanol,
1200 ml) (Momoi et al. 1976; Ando and Yu, 1979). The
elution profile is shown in Fig. 1. The first and second peaks

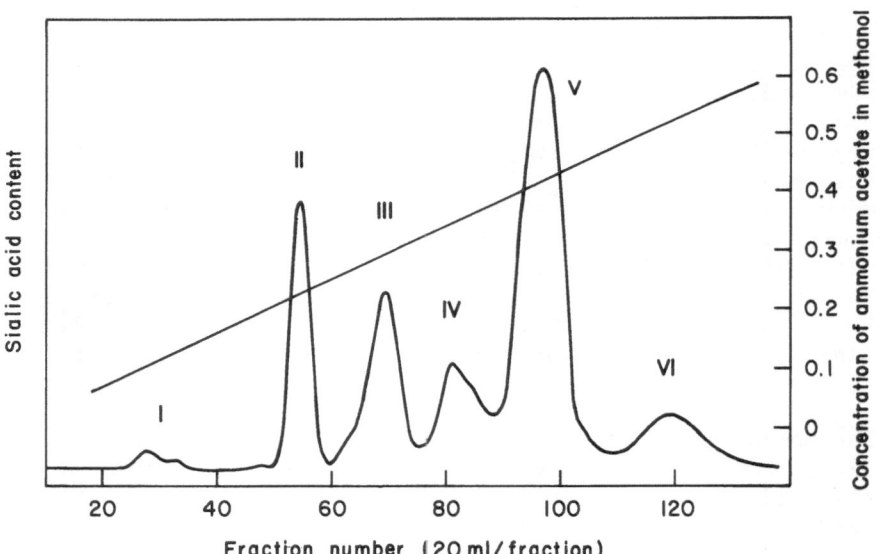

Fig. 1. The elution profile of fish brain gangliosides

consisted of mono- and di-sialoganglioside species, respectively. The third peak contained G_{T2}, G_{T1c} and G_{D3}; and then the fourth one contained G_{T3}, G_{T1a} and G_{T1b}. The major components of the fifth and sixth peaks were G_{Q1c} and G_{P1}, respectively. The third and fourth fractions were each dialyzed against water, 0.2M sodium acetate, and finally water to replace ammonium with sodium. Each fraction was then lyophilized. G_{T3}, G_{T2} and G_{T1c} were isolated and purified from the corresponding ganglioside fractions by chromatography on an Iatrobead column using gradient elutions (Ando et al. 1976; Ando and Yu, 1977b). Fig. 2 shows the chromatographic behavior of these trisialogangliosides and G_{Q1c} on a silica gel plate.

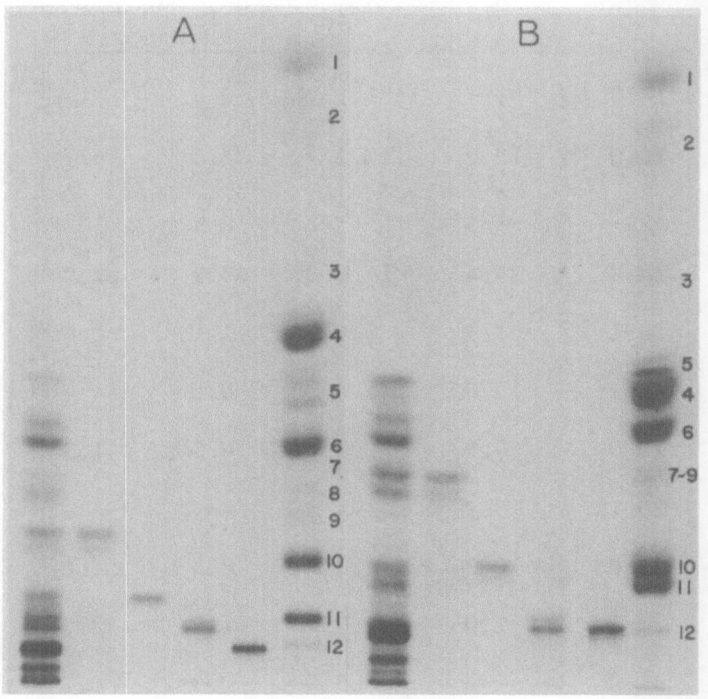

F G_{T3} G_{T2} G_{T1c} G_{Q1c} H   F G_{T3} G_{T2} G_{T1c} G_{Q1c} H

Fig. 2. Thin-layer chromatograms of G_{T3}, G_{T2}, G_{T1c}, and G_{Q1c}. Silica gel 60 HPTLC plates (Merck) were used. Plate A was developed with chloroform-methanol-water (55:45:10) containing 0.02% $CaCl_2$; and plate B with chloroform-methanol-2.5N ammonia (60:40:9). Gangliosides were visualized with resorcinal-HCl reagent. F, fish whole brain gangliosides; H, human brain white matter gangliosides. 1, GM4; 2, GM3; 3, GM2; 4, GM1; 5, GD3; 6, G_{D1a}; 7, G_{D1a}-GalNAc; 8, G_{T1a}; 9, G_{D2}; 10, G_{D1b}; 11, G_{T1b}; and 12, G_{Q1b}.

Each ganglioside appeared as doublets due to its particular fatty acid compositions (Table 1). The gangliosides contained primarily 4-sphingenine and no 4-eicosasphingenine was detected. Table 2 shows the unusual pattern of the fish brain gangliosides as compared with human brain gangliosides.

Table 1. Fatty Acid Composition of Fish Brain Gangliosides*

	G_{T3}	G_{T2}	G_{T1c}	G_{Q1c}
14:0	1.7	1.1	0.4	---
16:0	12.0	12.2	5.3	5.7
16:1	1.1	1.0	0.6	---
18:0	25.4	16.6	63.9	22.0
18:1	7.9	2.8	1.1	0.6
18:2	14.0	1.5	0.5	---
20:0	0.7	0.3	0.3	1.3
20:1	0.7	---	---	---
22:0	0.3	0.7	0.2	2.0
22:1	2.0	2.1	1.7	3.2
24:0	---	0.5	0.3	0.7
24:1	34.2	61.2	25.7	60.7
26:1	---	---	---	3.8

*Fatty acids were analyzed as their methyl esters by gas-liquid chromatography using 10% SP222PS (Supelco Co.) column. Values expressed as percent distributions.

Table 2. Ganglioside Composition of Fish and Human Brains

	Cod fish[*] whole brain	Human brain[**] white matter	Human brain[**] gray matter
Total gangl. S.A. μg/g w.w.	193	275	875
		% distribution of sialic acid	
G_{M4}	---	8.6	1.5
G_{M3}	0.4	4.8	2.7
G_{M2}	0.4	2.5	4.1
G_{M1}	0.1	21.6	14.9
G_{D3}	3.5	8.8	5.4
G_{D2}	1.5	3.1	8.0
G_{D1a}	10.1	16.6	21.7
G_{D1b}	0.9	16.9	18.2
$G_{D1a-GalNAc}$	---	1.1	0.4
G_{T3}	5.7	---	---
G_{T2}	4.6	---	---
G_{T1a}	3.5	2.2	1.8
G_{T1b}	2.5	11.1	16.3
G_{T1c}	7.4	---	---
G_{Q1b}	---	2.7	5.0
G_{Q1c}	44.0	---	---
G_{P1}	10.6	---	---
Other (G_{H1}?)	3.8	---	---

[*]Total gangliosides were separated into 6 fractions by a DEAE-Sephadex column, and the individual gangliosides in each fraction were analyzed by HP-TLC followed by densitometric scanning (Ando et al. 1978).
[**]Total gangliosides were directly applied on HP-TLC plates and quantitated by densitometry (Ando et al. 1978).

Structural analysis of G_{T3}, G_{T2} and G_{T1c}

The carbohydrate compositions of the purified gangliosides, determined by gas-liquid chromatography (Ando and Yamakawa, 1971), were as follows: G_{T3} contained glucose, galactose, and N-acetyl-neuraminic acid in the molar ratios of 1:1:3; G_{T2}, glucose, galactose, N-acetylgalactosamine and N-acetylneuraminic acid in the ratio of 1:1:1:3; and G_{T1c}, glucose, galactose, N-acetylgalactosamine and N-acetylneuraminic acid in the ratio of 1:2:1:3 (Table 3). The data indicated that these were indeed trisialogangliosides.

Table 3. Molar ratios of the chemical constituents of gangliosides before (B) and after (A) periodate oxidation and boro-hydride reduction

		Glc	Gal	GalNAc	Sialic Acid		LCB
					Intact	7-C	
G_{T3}	B	1.0	1.0	---	2.8	---	1.2
	A	1.0	1.2	---	1.5	1.0	1.2
G_{T2}	B	1.0	0.9	0.9	2.7	---	1.1
	A	1.0	0.8	---	1.6	1.0	1.1
G_{T1c}	B	1.0	1.8	1.1	2.7	---	1.1
	A	1.0	1.1	0.9	1.6	1.0	0.9

Mild neuraminidase (Clostridium perfringens, Type IV, Sigma Chemicals, St. Louis, Mo.) treatment of a polysialoganglioside has been previously shown to yield a series of partially desialylated glycolipid products (Ando and Yu, 1977b). By the similar treatment G_{T3} gave a mixture of G_{D3}, G_{M3} and lactosyl ceramide, which were identified with the corresponding authentic samples on thin-layer chromatograms in different solvent systems. G_{T2} produced G_{D2} and G_{M2}. G_{T1c} yielded G_{D1b} and G_{M1}, but no detectable G_{D1a} (Fig. 3). Exhaustive digestion of G_{T3}, G_{T2} and G_{T1c} with the enzyme gave lactosyl ceramide, G_{M2} and G_{M1} as their only glycolipid products, respectively.

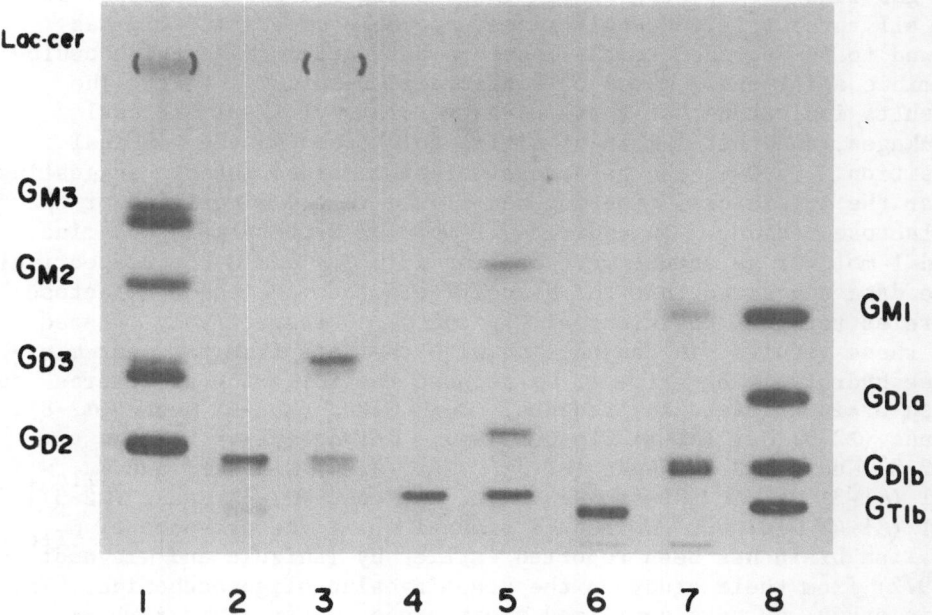

Fig. 3. Thin-layer chromatogram of hydrolysis products by mild
neuraminidase treatment.
        The conditions for the mild enzyme treatment were
described previously (Ando and Yu, 1977b). After
removing non-lipid salts and the liberated free sialic
acid by Sephadex G-50 column (Ueno et al. 1978), the
glycolipid products were chromatographed on thin-layer
plates. The solvent system employed was chloroform-
methanol-water (55:45:10) containing 0.02% $CaCl_2$. 1 and
8, authentic glycolipid standards; 2, G_{T3} (with small
amount of contaminant below) before treatment; 3, G_{T3}
after treatment; 4, G_{T2} before treatment; 5, G_{T2} after
treatment;   6, G_{T1c} before treatment; 7, G_{T1c} after
treatment. All spots were purple to resorcinal-HCl
reagent except lactosyl ceramide which was brownish-
yellow.

In order to delineate the sugar linkages in these trisialogangliosides, each ganglioside was subjected to sodium metaperiodate oxidation followed by sodium borohydride reduction under
conditions as described previously (Ando and Yu, 1977b). The gangliosides were then subjected to methanolysis and the resulting
carbohydrate components analyzed as their trifluoroacetyl derivatives
by gas-liquid chromatography. The results are shown in Table 3.
In all three trisialogangliosides, one mole of sialic acid was
found to be degraded to the 7-carbon derivative (N-acetyl-heptulo-
saminic acid) and 2 moles of sialic acid remained intact. The
results indicated that there were two sialosyl (2-8) sialosyl
linkages, and that 1 mole of sialic acid occupied the terminal
position. In G_{T3}, the galactose moiety remained intact, suggesting
that the trisialosyl grouping was linked to the 3 position of the
galactose residue. In addition, 1 mole of N-acetylgalactosamine
and 1 mole of galactose were destroyed in G_{T2} and G_{T1c}, respectively.
The data suggested that the N-acetylgalactosamine and a galactose
were at terminal positions of G_{T2} and G_{T1c}, respectively. Based
on these results, in conjunction with the data from the neuraminidase hydrolysis experiment, we propose the following structures for
fish brain trisialogangliosides: G_{T3} = NeuAc (α 2-8) NeuAc (α2-8)
NeuAc (α2-3) Gal (β1-4) Glc-Cer; G_{T2} = GalNAc (β1-4) [NeuAc
(α2-8) NeuAc (α2-8) NeuAc (α2-3)] Gal (β1-4) Glc-Cer; and G_{T1c}=
Gal (β1-3) GalNAc (β1-4) [NeuAc (α2-8) NeuAc (α2-8) NeuAc (α2-3)]
Gal (β1-4) Glc-Cer. We should emphasize that the presence of G_{T1c}
in fish brain has been reported earlier by Ishizuka and Wiegandt
(1972) from their study on the desphingosino-oligosaccharide. In
this study, we have confirmed their proposed structure for G_{T1c}
and characterized additional trisialogangliosides, G_{T3} and G_{T2}.

Possible biosynthetic pathway for G_{Q1c}

The present study and studies by others (Ishizuka et al.
1970; Avrova, 1971; McCluer and Argranoff, 1972; Ishizuka and
Wiegandt, 1972) have clearly demonstrated that fish brains contain
a preponderance of polysialogangliosides and relatively small
amounts of monosialo species. In addition, the major ganglioside
in fish brain, G_{Q1c}, is also structurally different from the tetrasialoganglioside, G_{Q1b}, of the brains of higher vertebrates.
Rahmann (1978) recently suggested that the differences in brain
ganglioside distribution observed between adult homeothermic (e.g.,
mammals and avian species) and poikilothermic (e.g., fish) vertebrates could be attributed to the thermal adaption of fish species
to lower environmental temperatures. He postulated that lower body
temperature would favor polysialisation of brain gangliosides.
However, our own study on gold fish acclimatized at 30° and 10° did
not reveal any significant differences in brain ganglioside contents
and patterns (Yu, R.K. and Agranoff, B.W., unpublished results).
Furthermore, it is difficult to explain the apparent difference
in ganglioside structures (G_{Q1b} and G_{Q1c}) among different vertebrate

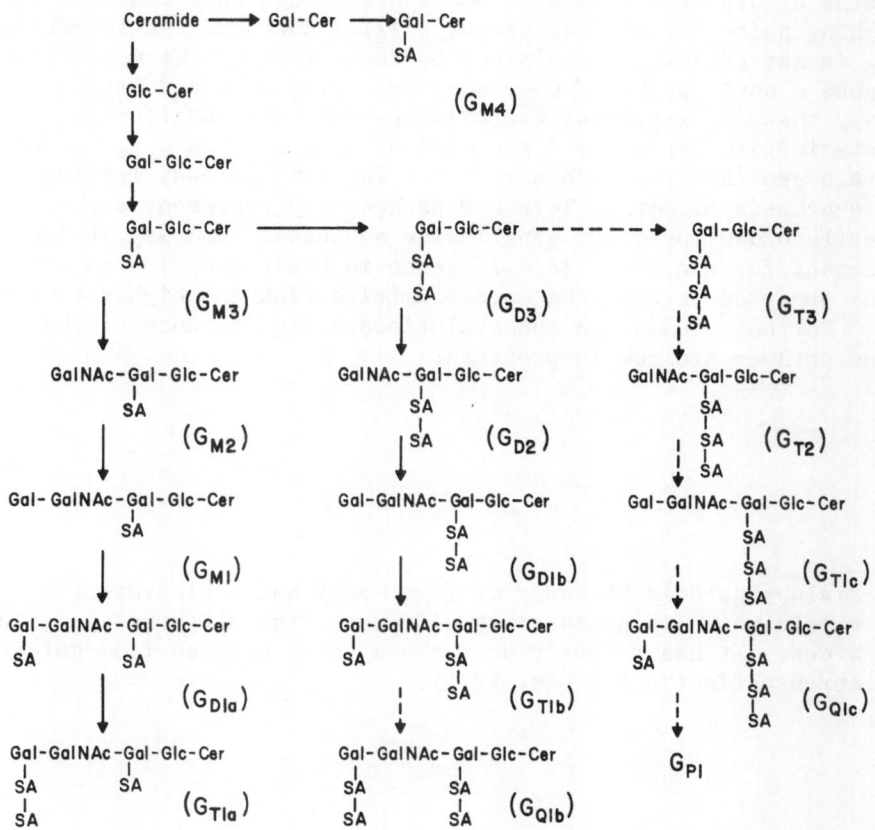

Fig. 4. The proposed pathways for ganglioside synthesis.
Abbreviations are: Cer, ceramide; Glc, glucose;
Gal, galactose; GalNAc, N-acetylgalactosamine; and
SA, sialic acid. Solid arrows denote established path-
ways. Broken arrows represent reactions yet to be
established.

species by the effect of polysialisation. Alternative explanations
seem necessary.

Two major pathways for the biosynthesis of gangliosides
starting from lactosyl ceramide have been presented (Roseman, 1970;
Fishman and Brady, 1976). One is the pathway through G_{M3}, G_{M2},
G_{M1} and G_{D1a} (Kaufman et al. 1968; Dicesare and Dain, 1971; Yip
and Dain, 1970) and finally G_{T1a} (Yohe and Yu, 1979). The other is
through G_{M3}, G_{D3}, G_{D2}, and G_{D1b} (Cumar et al. 1971; Arce et al. 1971)
(Fig. 4). It has further been demonstrated that the latter pathway
may lead to the synthesis of G_{T1b}, but not G_{T1c} (Mestrallet et al.
1977). Presumably G_{Q1b} may also be derived from this pathway
(Svennerholm, 1970). In the above scheme, all gangliosides present in

the brains of higher vertebrates emenate from G_{M3} that constitutes a branching point for the two pathways (2). However, the formation of G_{Q1c} is not adequately explained by these routes. We would like to propose a novel pathway, G_{M3}, G_{D3}, G_{T3}, G_{T2}, G_{T1c} and G_{Q1c} (Fig. 4), that may represent the major pathway for ganglioside synthesis in fish brain. The presence of G_{T3}, G_{T2} and G_{T1c} in cod fish brain provides the necessary links for this pathway leading to the synthesis of G_{Q1c}. This new pathway may represent a phylogenetically older route for ganglioside synthesis, and may, therefore account for the dramatic difference in brain ganglioside patterns observed between the more primitive fish and higher vertebrates. Further studies on the evolutionary significance of the proposed pathway are now in progress.

---

(2). Sialosylgalactosyl ceramide (G_{M4} = G_7) has a biosynthetic origin which is necessarily different from all other gangliosides. It has recently been shown to be derived from galactocerebroside (Yu and Lee, 1976).

ACKNOWLEDGEMENT

We thank Dr. T. Itoh for performing the carbohydrate analyses. This work is supported by a USPHS grant NS-11853 and a grant from the Kroc Foundation.

BIBLIOGRAPHY

ANDO S. and YAMAKAWA T. (1971): Application of trifluoroacetyl derivatives to sugar and lipid chemistry. I. Gas chromatographic analysis of common constituents of glycolipids. J. Biochem. 70, 335-340.

ANDO S., and YU R.K. (1977a): Isolation and characterization of human and chicken brain tetrasialoganglioside. Proc. Int. Soc. Neurochem. 6, 535.

ANDO S. and YU R.K. (1977b): Isolation and characterization of a novel trisialoganglioside, G_{T1a}, from human brain. J. Biol. Chem. 252, 6247-6250.

ANDO S. and YU R.K. (1979): Isolation and characterization of two isomers of brain tetrasialogangliosides. J. Biol. Chem. (submitted)

ANDO S., CHANG N.-C. and YU R.K. (1978): High performance thin-layer chromatography and densitometric determination of brain ganglioside composition of several species. Anal. Biochem. 89, 437-450.

ANDO S., ISOBE M. and NAGAI Y. (1976): High-performance preparative column chromatography of lipids using a new porous silica, Iatrobeads. I. Separation of molecular species of sphingolipids. Biochim. Biophys. Acta 424, 98-105.

ARCE A., MACCIONI H.J. and CAPUTTO R. (1971): The Biosynthesis of gangliosides. Biochem. J. 121, 483-493.

AVROVA N.F. (1971): Brain ganglioside patterns of vertebrates. J. Neurochem. 18, 667-674.

CUMAR F.A., FISHMAN P.H. and BRADY R.O. (1971): Analogous reactions for the biosynthesis of monosialo-and disialo-gangliosides in brain. J. Biol. Chem. 246, 5075-5084.

DICESARE J.L. and DAIN J.A. (1970): The enzymic synthesis of ganglioside. IV. UDP-N-acetylgalactosamine: (N-acetylneuraminyl)-galactosylglucosyl ceramide N-acetylgalactosaminyltransferase in rat brain. Biochim. Biophys. Acta 231, 385-393.

FISHMAN P.H. and BRADY R.O. (1976): Biosynthesis and function of gangliosides. Science, 194, 906-915.

ISHIZUKA I. and WIEGANDT H. (1972): An isomer of trisialoganglioside and the structure of tetra-and pentasialogangliosides from fish brain. Biochim. Biophys. Acta 260, 279-289.

ISHIZUKA I., KLOPPENBURG M. and WIEGANDT H. (1970): Characterization of gangliosides from fish brain. Biochim. Biophys. Acta
210, 299-305.

KAUFMAN B., BASU S. and ROSEMAN S (1968): Enzymatic synthesis of
disialogangliosides from monosialogangliosides by sialyltransferases from embryonic chicken brain. J. Biol. Chem. 243, 5804-
5807.

LEDEEN R.W., YU R.K. and ENG L.F. (1973): Gangliosides of human
myelin: Sialosylgalactosylceramide (G7) as a major component.
J. Neurochem. 21, 829-839.

MCCLUER R.H. and AGRANOFF B.W. (1972): Studies on gangliosides of
goldfish brain. J. Neurochem. 19, 2307-2315.

MESTRALLET M.G., CUMAR F.A. and CAPUTTO R. (1977): Trisialoganglioside synthesis by a chicken brain sialyltransferase. Comparative study with a similar reaction for the synthesis of
disialoganglioside. Molec. Cell. Biochem. 16, 63-70.

MOMOI T., ANDO S. and NAGAI Y. (1976): High resolution preparative
column chromatographic system for gangliosides using DEAE-Sephadex
and a new porous silica, Iatrobeads. Biochim. Biophys. Acta
441, 488-497.

RAHMANN H. (1978): Gangliosides and thermal adaptation in vertebrates. Japan. J. Exp. Med. 48, 85-96.

ROSEMAN S. (1970): The synthesis of complex carbohydrates by
multiglycosyltransferase systems and their potential function
in intercellular adhesion. Chem. Phys. Lipids 5, 270-297.

SVENNERHOLM L. (1963): Chromatographic separation of human brain
gangliosides. J. Neurochem. 10, 613-623.

SVENNERHOLM L. (1970): Gangliosides, in "Handbook of Neurochemistry", LAJTHA A, Eds., Vol. 3, Plenum Press (New York), pp.
425-452.

UENO K., ANDO S. and YU R.K. (1978): Gangliosides of human, cat
and rabbit spinal cords and cord myelin. J. Lipid Res. 19, 863-
871.

WIEGANDT H. (1972): Recent advances on the chemistry and localization of brain gangliosides and related glycosphingolipids.
Adv. Exp. Med. Biol. 25, 127-140.

YIP M.C.M. and DAIN J.A. (1970): Frog brain uridine diphosphate galactose-N-acetylgalactosaminyl-N-acetylneuraminylgalactosyl-glucosylceramide galactosyltransferase. Biochem. J. 118, 247-252.

YOHE H.C. and YU R.K. (1979): Biosynthesis of the novel trisialo-ganglioside, G_{T1a}. Trans. Amer. Soc. Neurochem. 10, 93.

YU R.K. and ANDO S. (1978): Novel gangliosides of fish brain. Trans. Amer. Soc. Neurochem. 9, 135.

YU R.K. and LEE S.H. (1976): In vitro biosynthesis of sialosyl-galactosylceramide (G7) by mouse brain microsomes. J. Biol. Chem. 251, 198-203.

STRUCTURAL FINGERPRINTING OF GANGLIOSIDES AND OTHER

GLYCOCONJUGATES BY MASS SPECTROMETRY

Karl-Anders Karlsson

Department of Medical Biochemistry
University of Göteborg
Göteborg, Sweden

INTRODUCTION

In view of the already long use of permethylation for structural studies of carbohydrate it is surprising that the first mass spectra of non-degraded permethylated derivatives of glycolipids appeared rather late (KARLSSON, 1973). Since then several laboratories have adopted this technique as an important supplement to conventional degradation methods. However, only one other research group (see WATANABE et al., 1978) has added the use of permethylated-reduced derivatives (see below), which we consider necessary for a safe conclusion. One purpose of the present communication is to illustrate the importance of a combined use of the two derivatives. Secondly, the potent application on mixtures of glycolipid antigens will be shown. In this case mass spectrometry is the only microchemical method available today able of a specific detection of separate sequences in a mixture.

DERIVATIVES AND EQUIPMENT

Glycolipids were permethylated in one step according to HAKOMORI (1964). Reduction of permethylated derivatives with $LiAlH_4$ followed by silylation was done as described (KARLSSON, 1974). Substitutions produced are shown in Figs. 4 and 5. The mass spectrometer used was MS 902 (AEI Ltd., England) with a heatable direct inlet probe and a data system (Datamass One, Instem Ltd., England). The handling of sample has been discussed (KARLSSON, 1976).

47

IMPORTANT SPECTRAL CHANGES UPON REDUCTION

Blood Group A Fucolipid

In the low mass region there are important ions indicative of sugar sequence (one, two, three, etc., sugars; S ions of Figs. 4 and 5). Fig. 1 shows the specific S ions produced from an A-active

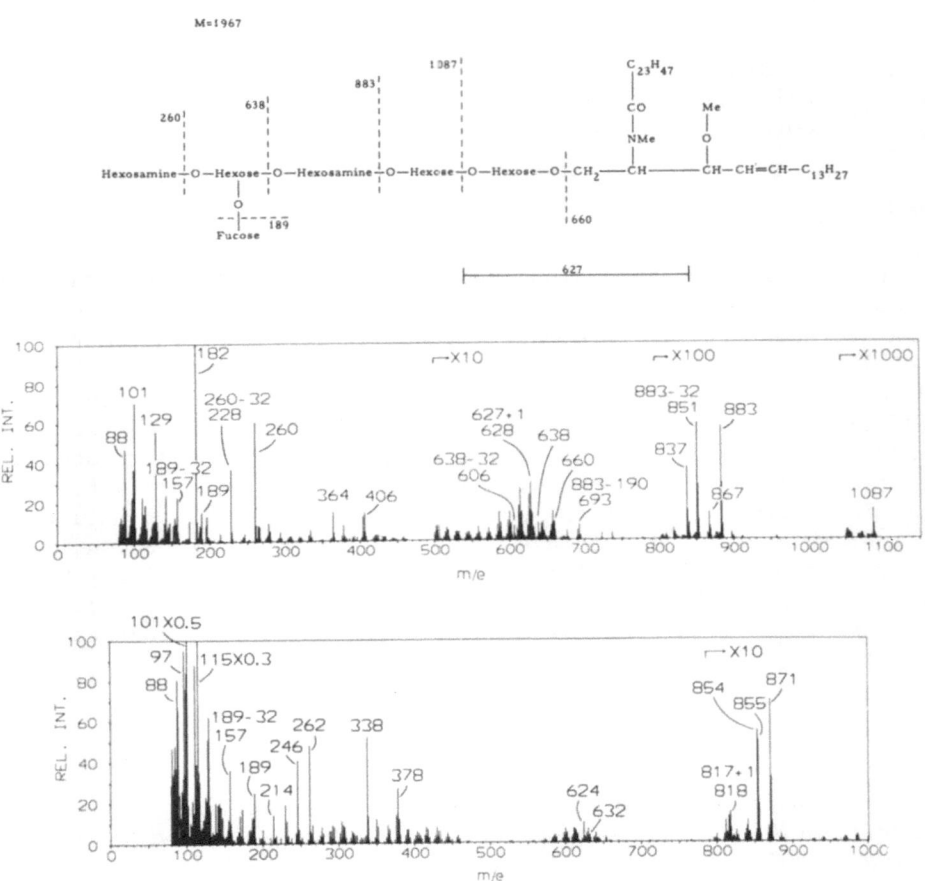

Fig. 1. Comparison of sequence ions of permethylated (top) and permethylated-reduced (bottom) derivatives of a blood group A hexaglycosylceramide of human erythrocyte. The simplified formula corresponds to the permethylated derivative. The analogous formula of the reduced derivative is shown in Fig. 2.

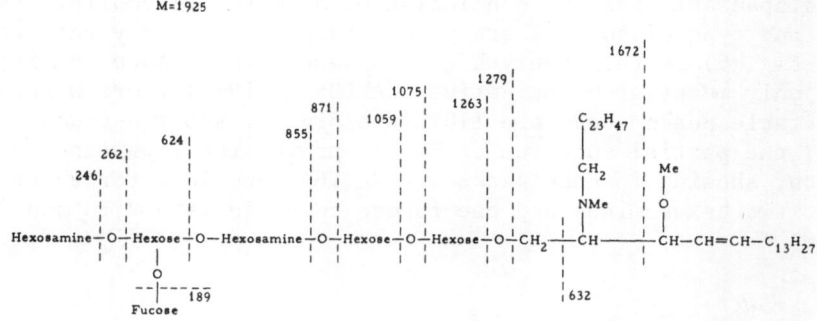

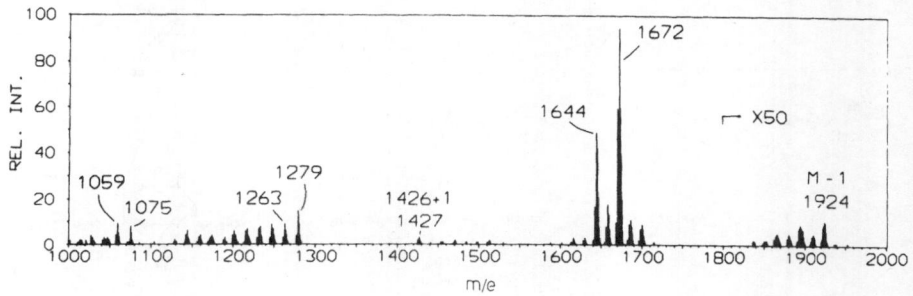

Fig. 2. Partial spectrum demonstrating abundant F ions (compare Fig. 5) at m/e 1644 (22:0 fatty acid) and 1672 (24:0 fatty acid) of a permethylated-reduced derivative of a blood group A hexaglycosylceramide of human erythrocyte. The intesity scale was multiplied 10 times. The corresponding interval (1100-1900) of the only permethylated derivative lacked readable fragments.

hexaglycosylceramide of human A_1 erythrocyte membrane (KARLSSON, 1976). The top formula represents schematically the permethylated derivative. Terminal sugars of hexosamine (GalNAcα1→3) and of fucose (Fucα1→2) appear distinctly at m/e 228 (260 minus methanol) and 260, and at 157 and 189, respectively (top spectrum). Ions indicative of the sequence are found at m/e 606, 638, 851, 883 and 1087. Upon reduction (compare Fig. 5) each reduction point (amides) loses 14 mass units, which is documented in the spectrum of Fig. 1, bottom. Terminal hexosamine is now 246 (and 262 due to cleavage on the other side of the glycosidic oxygen) but fucose is unchanged at 157 and 189. Other sequence ions are consequently found at 624 and 855 (and 871). The mass changes found upon reduction (2x14 for tetrasaccharide, from 883 to 855) are therefore further evidence for the presence of hexosamines and their location in the sequence detected. One may note the relatively high intensity of peaks produced by cleavage at the glycosidic oxygen of hexosamine (851 and 883 of top, and 854, 855 and 871 of bottom spectrum), which is of importance for selected ion monitoring (see below).

Indispensable for the conclusion of a certain glycolipid species (number and type of sugars) are F ions (sugar plus fatty acid ions, compare Fig. 5) of relatively high abundance from reduced derivatives. For the only methylated derivative (KARLSSON, 1976) there were no interpretable peaks above m/e 1100 (see Fig. 1, top spectrum). However, the partial spectrum of Fig. 2 proves the importance of the reduction, showing intense peaks at m/e 1644 and 1672 for three hexoses, two hexosamines and one fucose combined with 22:0 and 24:0

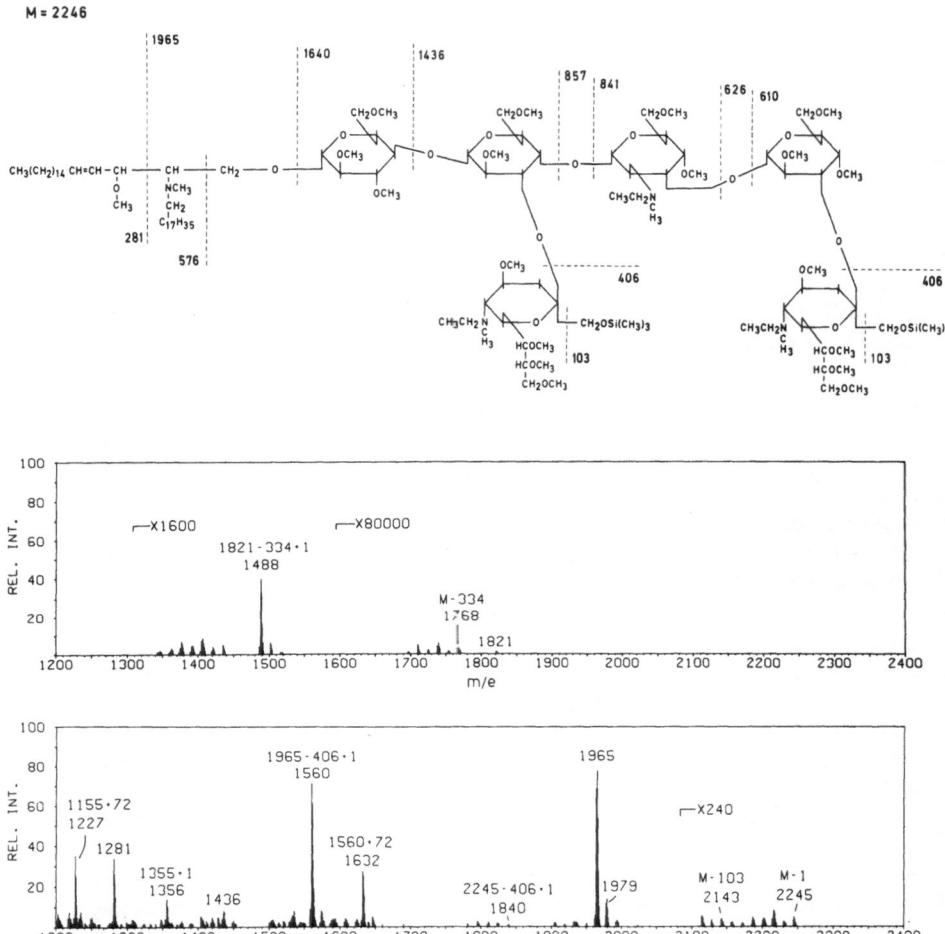

Fig. 3. High mass region of spectra documenting the importance of silylation of reduced ganglioside derivatives. The top spectrum is from permethylated-reduced and the bottom spectrum from permethylated-reduced-silylated derivative (see top formula) of disialoganglioside of bovine brain. For the lower spectrum the intensity scale was multiplied 60 times. Before silylation there are weak peaks and poor information.

fatty acid, respectively.

Disialoganglioside

The important choice of suitable derivatives is further illustrated in Fig. 3 for disialoganglioside of bovine brain (KARLSSON,
1974). The top spectrum shows the high mass region of the reduced
derivative, with peaks of very low intensity and poor information.
After silylation, however (top formula and bottom spectrum), of the
sialic alcohol (two substitution points) the F ions are surprisingly
stable. Although the only reduced derivative (top spectrum) usually
adds little information in the mass interval 1200-3000, less heavy
ions are of diagnostic importance (not shown). Therefore, gangliosides are often analyzed in three forms, as permethylated, permethylated-reduced and permethylated-reduced-silylated derivatives.

GENERAL FRAGMENTATION CHARACTERISTICS OF SEPARATE DERIVATIVES

A summary of fragmentation patterns is presented for methylated
(Fig. 4) and for methylated-reduced (Fig. 5) derivatives (and in case
of gangliosides also silylated derivatives, Fig. 5). The primary
characteristic of a glycolipid species (number and type of sugars
and fatty acid) is obtained by the series of abundant F fragments
from the reduced (and silylated) derivative (Fig. 5). When properly
recorded the F peaks give a semiquantitation of fatty acids (compare
Fig. 7). Sequence information of S fragments is selected from both
derivatives although the methylated one is preferred (compare Figs.
1 and 8). As discussed above, loss of 14 mass units upon reduction
is further evidence for a hexosamine in the chain (S_4 and S_4' in the
two spectra, respectively). Other peaks shown or not shown in the
two spectra may support the above conclusions (A+1 of Fig. 4; M-1
and D+1 of Fig. 5) or add information on fatty acid and long-chain
base (C of Fig. 4; B, D+1 and M-1 of Fig. 5).

PRESENT MASS LIMIT FOR ANALYSIS

By a successive modification of our earlier introduction and
evaporation techniques (KARLSSON, 1976) we are now able to analyze
glycolipids with up to twelve sugars (in preparation). One example
is shown in Fig. 6 where F ions are shown in the region m/e 2835-2977
for a reduced blood group A-active dodecaglycosylceramide (in
preparation). This is perhaps the most promising methodological
result in recent time for future microchemical studies of mixtures
of saccharide receptors or antigens.

SUBSTITUTIONS

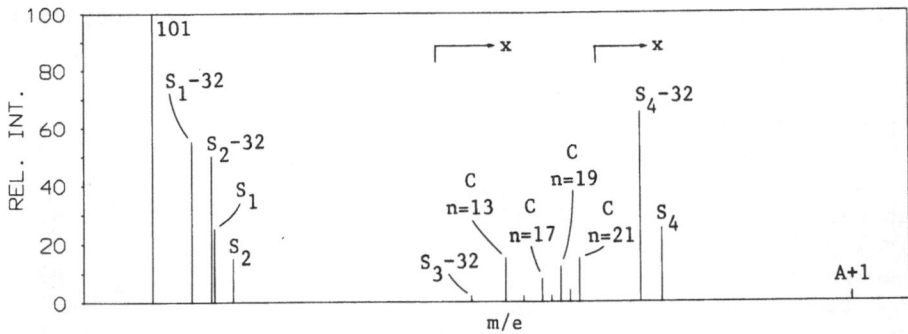

Sugar Type	Mass of Terminal Residue	R_1	R_2	R_3	R_4	Mass of Internal Nonbranched Residue
Hexose	219	$-CH_2OMe$	$-OMe$	$-OMe$	$-H$	204
Hexosamine	260	$-CH_2OMe$	$-OMe$	$-NCOCH_3$ ... Me	$-H$	245
Deoxyhexose	189	$-CH_3$	$-OMe$	$-OMe$	$-H$	174
Pentose	175	$-H$	$-OMe$	$-OMe$	$-H$	160
Neuraminic Acid N-Acetyl	376	$-CHOMe$ $CHOMe$ CH_2OMe	$-NCOCH_3$ ... Me	$-H$	$-COOMe$	361
N-Glycoloyl	406	$-CHOMe$ $CHOMe$ CH_2OMe	$-NCOCH_2OMe$ ... Me	$-H$	$-COOMe$	391

FRAGMENTATION

Fig. 4. Simplified mass spectrum to illustrate fragment patterns of permethylated glycolipids. Substitution steps and masses of separate terminal and internal sugars are shown above.

SUBSTITUTIONS

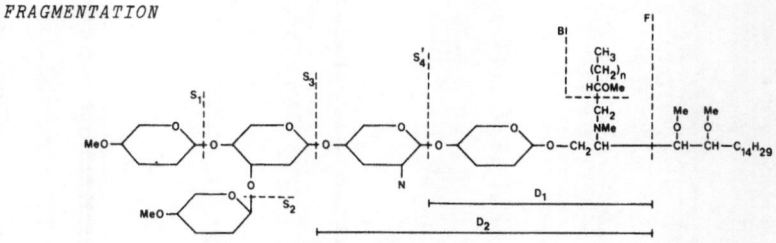

Sugar Type	Mass of Terminal Residue	R_1	R_2	R_3	R_4	Mass of Internal Nonbranched Residue
Hexose	219	$-CH_2OMe$	$-OMe$	$-OMe$	$-H$	204
Hexosamine	246	$-CH_2OMe$	$-OMe$	$-NCH_2CH_3$ $\quad$ Me	$-H$	231
Deoxyhexose	189	$-CH_3$	$-OMe$	$-OMe$	$-H$	174
Pentose	175	$-H$	$-OMe$	$-OMe$	$-H$	160
Neuraminic Acid N-Acetyl	406	$-CHOMe$ $CHOMe$ CH_2OMe	$-NCH_2CH_3$ $\quad$ Me	$-H$	$-CH_2OSiMe_3$	391
N-Glycoloyl	436	$-CHOMe$ $CHOMe$ CH_2OMe	$-NCH_2CH_2OMe$ $\quad$ Me	$-H$	$-CH_2OSiMe_3$	421

FRAGMENTATION

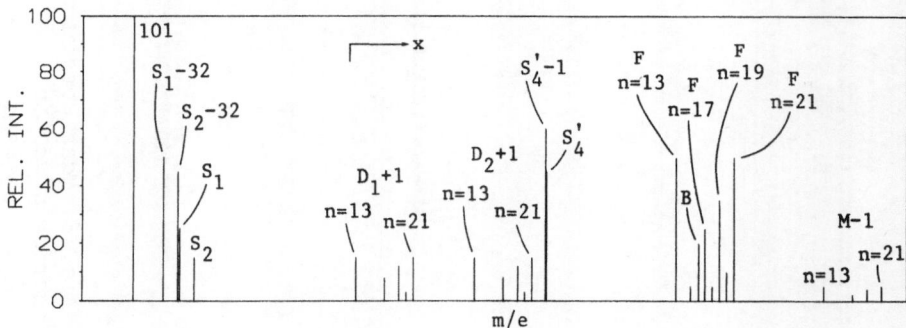

Fig. 5. *Simplified mass spectrum to illustrate fragment patterns of permethylated-reduced-silylated glycolipids. Substitution steps and masses of separate terminal and internal sugars are shown above.*

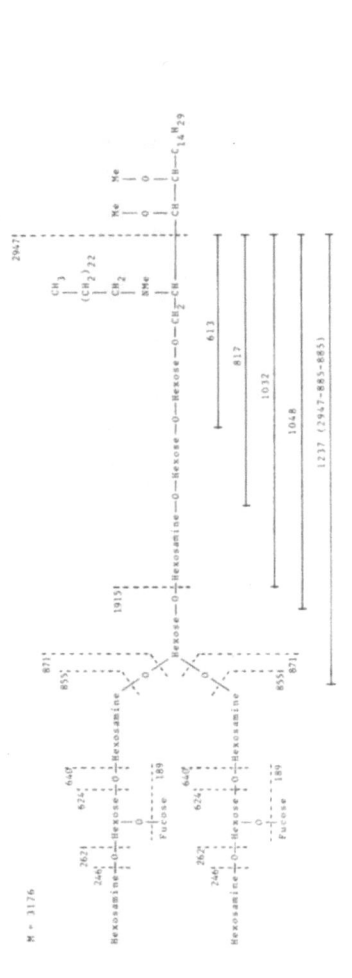

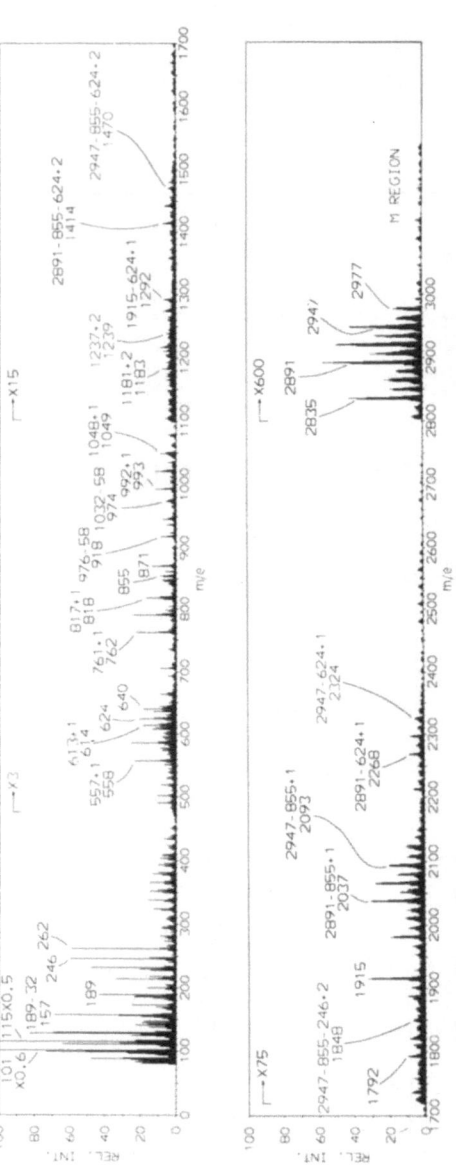

Fig. 6 Mass spectrum of permethylated-reduced derivative of a blood group A-active glycolipid of epithelial cells of rat small intestine. The amount of substance used was 90 µg, the electron energy 29 eV and probe temperature 305°C.

SELECTED ION MONITORING OF MIXTURES OF GLYCOLIPID ANTIGENS

The specificity of information from separate sequences as is demonstrated for the two kinds of derivatives has permitted the analysis of complex mixtures of glycolipids (BREIMER et al., 1978; 1979). I will briefly illustrate this with a non-acid glycolipid fraction of rabbit small intestine and limit the presentation to the high-temperature fucolipid interval. In Fig. 7 the result from recording from the reduced mixture of F and B ions (compare Fig. 5) of four different fucolipids is shown as a function of evaporation temperature. During these conditions species with 5 to 7 sugars evaporate at about the same temperature but are present in different ratios. Of course the formulas as written here with settled sequences and ceramide composition cannot be deduced from these curves but must also be taken from other ions and the only methylated derivative.

Due to rather intense peaks from cleavage at the anomeric carbon of hexosamines (compare discussion above) ions of specific

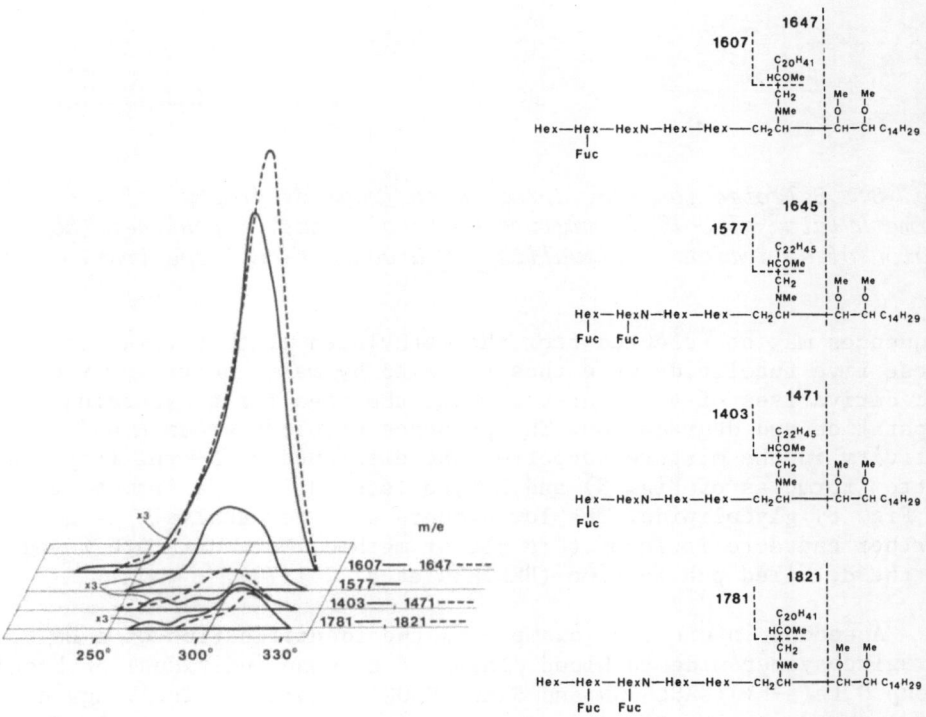

Fig. 7. Selected ion monitoring (mass fragmentography) of a permethylated-reduced glycolipid mixture of rabbit small intestine. F and B ions (compare Fig. 5) have been fragmentographed for 4 separate fucolipid species in the high temperature evaporation region.

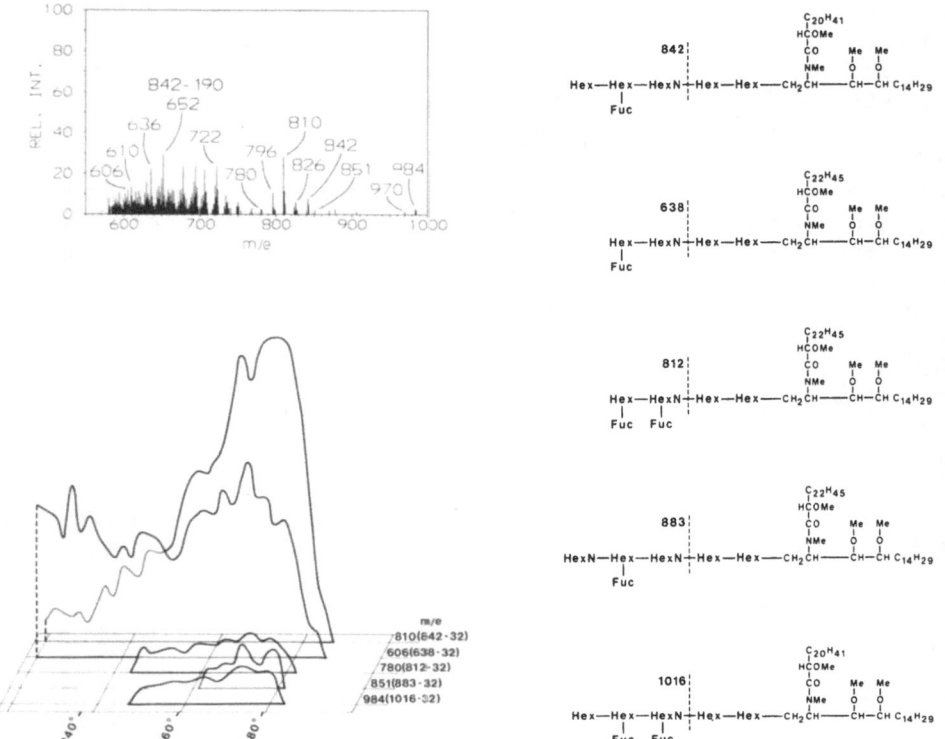

Fig. 8. Selected ion monitoring (mass fragmentography) of a permethylated glycolipid mixture of rabbit small intestine. The region of selection is visualized by a conventional spectrum.

sequences may be selected from the methylated mixture (Fig. 8). These five fucolipids were thus revealed by mass spectrometry of two derivatives of a mixture, without the need for a preceding separation and degradation. The presence of blood group A and B activity in the mixture supported the existence of B-type (top and bottom formulas of Fig. 8) and A-type (second formula from bottom of Fig. 8) glycolipids. The low amounts used for analysis is a further characteristic of this potent method. More data are found in the detailed publication (BREIMER et al., 1979).

A second interesting example is the identification of a Le[b] hexaglycosylceramide in blood plasma of a human individual of blood group O,Le(a-b+)(KARLSSON and SAMUELSSON, in preparation). Again total non-acid glycolipid derivatives (corresponding to 4 ml of plasma) were distilled off in the ion source and separate glycolipids successfully recorded. An almost pure species of Le[b] glycolipid was detected after evaporation of the dominating mono- to tetraglycosylceramides (Fig. 9).

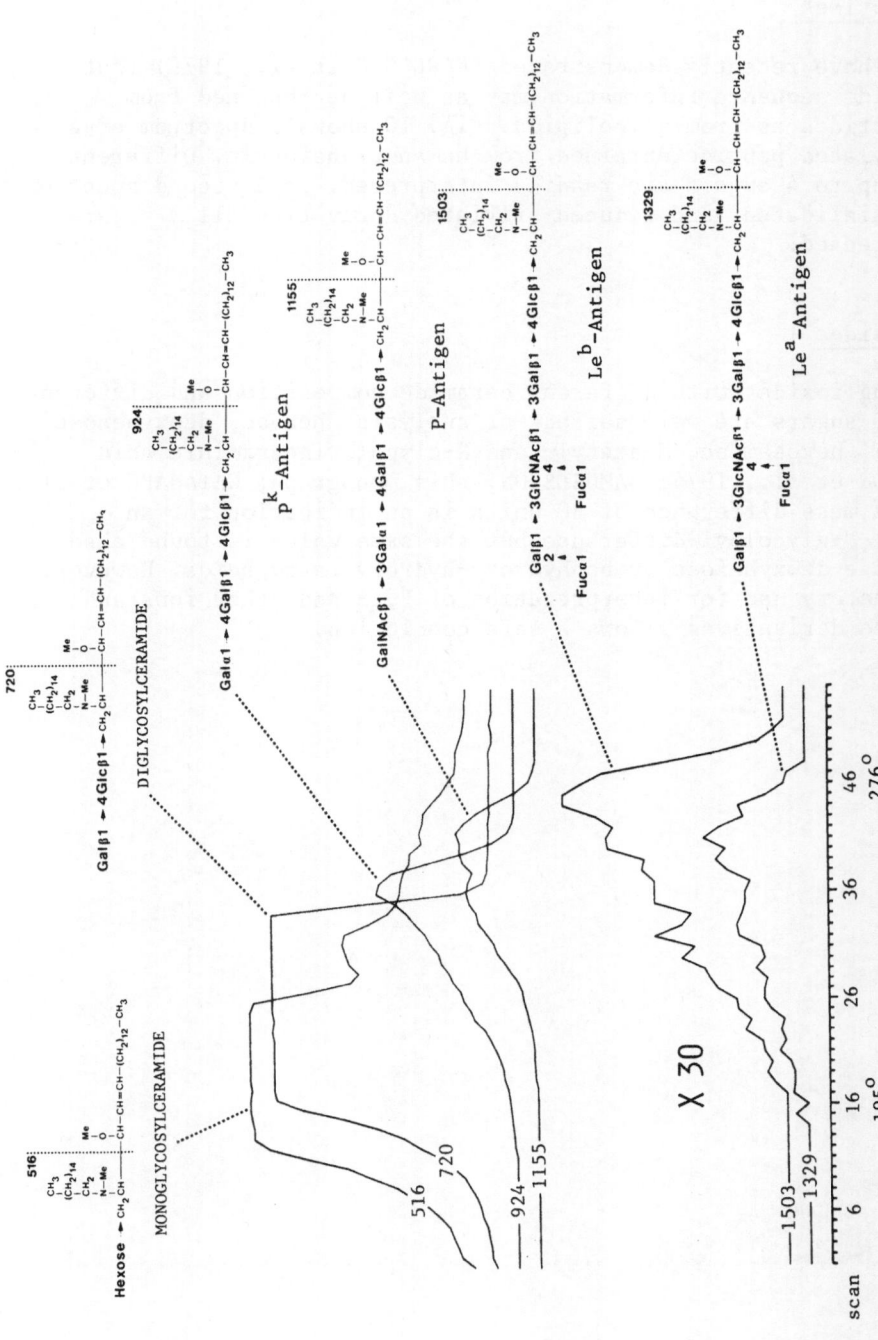

Fig. 9. Selected ion monitoring (mass fragmentography) of F ions (see Fig. 5) of a permethylated mixture of non-acid glycolipids of human plasma. The structures reproduced were made probable after analysis also of the non-reduced derivative, and NMR spectroscopy.

ASPECTS ON THE ANALYSIS OF SIALOSACCHARIDES

Glycopeptides

We have recently demonstrated (KARLSSON et al., 1978) that
saccharide sequence information may as well be obtained from
glycopeptides as from glycolipids. Fig. 10 shows a spectrum of a
permethylated peptide obtained from human transferrin. Different
S ions up to 4 sugars are readily interpreted. In a second spectrum
from desialidated, and reduced-silylated derivative all 9 sugars
were detected.

Gangliosides

Gangliosides with different ceramide composition and different
kinds of sugars are well suited for analysis: hexose, deoxyhexose
(fucose), hexosamine, N-acetyl- and N-glycoloylneuraminic acid
(KARLSSON et al., 1974; SAMUELSSON, this monograph; WATANABE et al.,
1978). A mass difference of 30 units is an indication for an
N-acetyl-N-glycoloyl difference but the same value is found also
for hexose-deoxyhexose or nonhydroxy-hydroxy fatty acids. However,
a combinatory use for interpretation of F, S and other ions and
all three derivatives allows a safe conclusion.

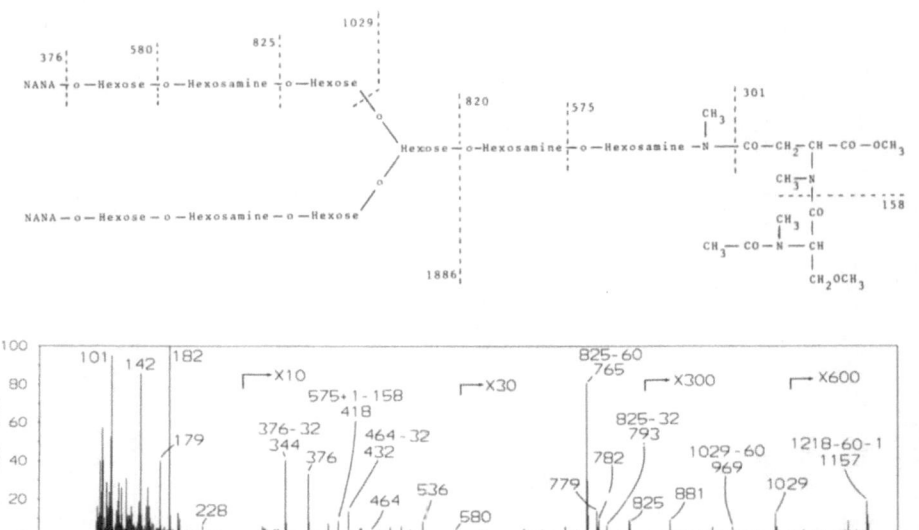

*Fig. 10. Mass spectrum of the permethylated derivative of a
glycopeptide derived from human transferrin.*

The present limit of analysis is a tetrasialoganglioside of human brain (work together with SVENNERHOLM et al.) with F ions at m/e 2747, containing 4 sialic acids, three hexoses and one hexosamine plus stearic acid. Other ions of all three derivatives allow a conclusion of the correct sequence of the basic saccharide hexose-hexosamine-hexose-hexose with two sialic acids in sequence bound to the second and fourth hexose from ceramide.

Important help may be gained from mass spectrometry in the case of modified ganglioside prepared for receptor studies. Cholera toxin binding to G_{M1} ganglioside was studied after chemical change of the ceramide part and glucose (WIEGANDT et al., 1976). The expected result is almost impossible to document on a microscale with conventional methods. However, mass spectra easily answered the question and in some cases also indicated the presence of byproducts (KARLSSON, 1977). Result from one successful modification is shown in Fig. 11, which proves an intact receptor saccharide as well as the nature of the modified glucose and lipophilic part.

Concerning the analysis of mixtures of gangliosides we will later present results from an application on rat small intestine. In this case the interpretation may be somewhat less clear than for non-acid glycolipids due to an apparent temperature-dependence of fragmentation pathways. This means that the same ganglioside species may present somewhat different spectra at different points of the evaporation scale (compare Figs. 7 and 8). However, in spite of this possible drawback this kind of analysis will certainly be of great help in the future.

CONCLUSIONS

Mass spectrometry is at present the only chemical micromethod (compare immunology) that is able of directly and specifically recognizing a saccharide sequence, even in a complex mixture. The present range of analysis (dodecasaccharide and use of only a few hundred micrograms) is a challenge for glycolipid studies in cell biology and medicine. Improvements of the technique of selected ion monitoring will certainly open up new fields of research or make available tissues from single individuals not possible to study before. An example mentioned here is the analysis of human plasma for glycolipid antigens (Fig. 9). Although most of the work done so far is in the fucolipid field and the analysis of ganglio-sides may create special difficulties, there is no doubt that the technique will also be of importance in studies on the structure and function of gangliosides.

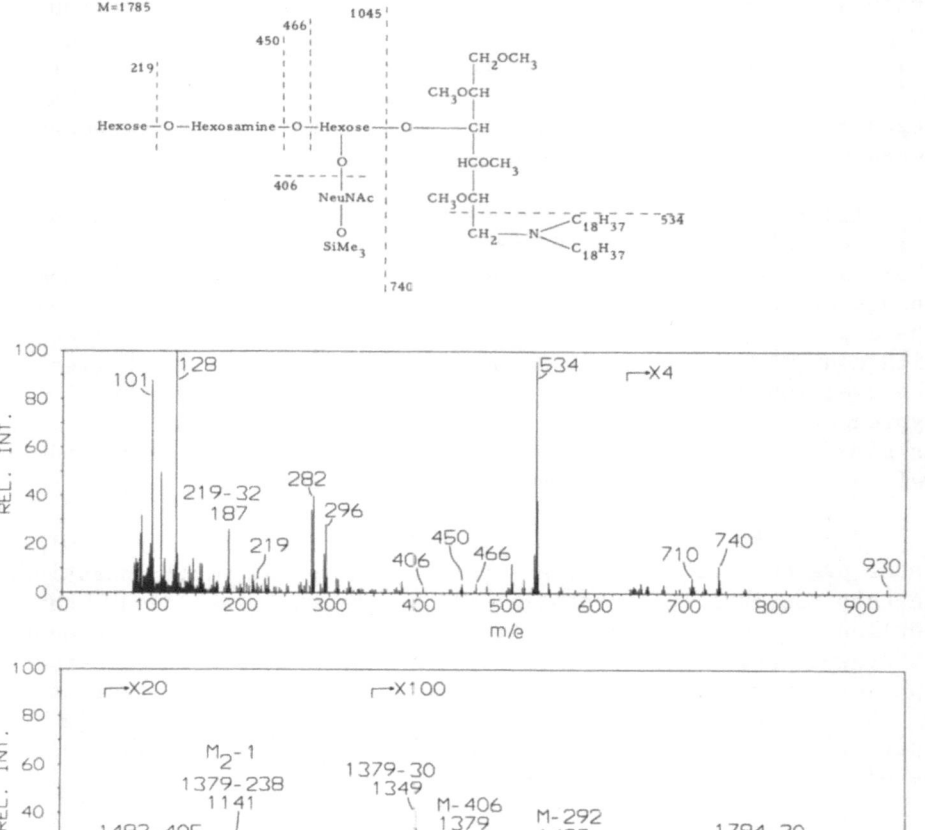

Fig. 11. Mass spectrum of the permethylated-reduced-silylated derivative of a chemically modified receptor of cholera toxin.

ACKNOWLEDGEMENTS

This work was done in collaboration with M.E. BREIMER, G.C. HANSSON, H. LEFFLER, I. PASCHER, W. PIMLOTT and B.E. SAMUELSSON. A. PAKVIS contributed to the layout and typing of manuscript. Financial support was obtained from the Swedish Medical Research Council (grant No. 3967).

REFERENCES

BREIMER, M.E., HANSSON, G.C., KARLSSON, K.-A., LEFFLER, H., PIMLOTT, W. and SAMUELSSON, B.E. (1978): Structure determination of blood group type glycolipids of cat small intestine by mass fragmentography. FEBS Lett. 89, 42-46.

BREIMER, M.E., HANSSON, G.C., KARLSSON, K.-A., LEFFLER, H., PIMLOTT, W. and SAMUELSSON, B.E. (1979): Selected ion monitoring of glycosphingolipid mixtures. Identification of several blood group type glycolipids in the small intestine of an individual rabbit. Biomed. Mass Spectrom. 6, in press.

HAKOMORI, S.-I. (1964): A rapid permethylation of glycolipid and polysaccharide, catalyzed by methylsulfinyl carbanion in dimethyl sulfoxide. J. Biochem. 53, 205-208.

KARLSSON, K.-A. (1973): Carbohydrate composition and sequence analysis of cell surface components by mass spectrometry. Characterization of the major monosialoganglioside of brain. FEBS Lett. 32, 317-320.

KARLSSON, K.-A. (1974): Carbohydrate composition and sequence analysis of a derivative of brain disialoganglioside by mass spectrometry, with molecular weight ions at m/e 2245. Potential use in the specific microanalysis of cell surface components. Biochemistry 13, 3643-3647.

KARLSSON, K.-A. (1976): Microscale fingerprinting of blood-group fucolipids by mass spectrometry, in "Glycolipid Methodology", Witting, L.A. ed., American Oil Chemists' Society (Champaign, Illinois), pp. 97-122.

KARLSSON, K.-A. (1977): Mass-spectrometric sequence studies of lipid-linked oligosaccharides. Blood-group fucolipids, gangliosides and related cell-surface receptors. Progr. Chem. Fats Other Lipids 16, 207-230.

KARLSSON, K.-A., PASCHER, I. and SAMUELSSON, B.E. (1974): Analysis of intact gangliosides by mass spectrometry. Comparison of different derivatives of a hematoside of a tumor and the major monosialoganglioside of brain. Chem. Phys. Lipids 12, 271-286.

KARLSSON, K.-A., PASCHER, I., SAMUELSSON, B.E., FINNE, J., KRUSIUS, T. and RAUVALA, H. (1978): Mass spectrometric sequence study of the oligosaccharide of human transferrin. FEBS Lett. 94, 413-417.

WATANABE, K., POWELL, M. and HAKOMORI, S.-I. (1978): Isolation and characterization of a novel fucoganglioside of human erythrocyte membranes. J. Biol. Chem. 253, 8962-8967.

WIEGANDT, H., ZIEGLER, W., STAERK, J., KRANZ, T., RONNEBERGER, H.J., ZILG, H., KARLSSON, K.-A. and SAMUELSSON, B.E. (1976): Studies of the ligand binding to cholera toxin. I. The lipophilic moiety of sialoglycolipids. Hoppe-Seyler's Z. Physiol. Chem. 357, 1637-1646.

MASS SPECTRA AND NMR SPECTRA OF GANGLIOSIDES CONTAINING FUCOSE

Bo E. Samuelsson

Department of Medical Biochemistry
University of Göteborg
Fack, S-400 33 Göteborg, Sweden

SUMMARY

Two gangliosides containing fucose, prepared from minipig posterior root ganglion, were analyzed as methylated, methylated-reduced and methylated-reduced-trimethylsilylated derivatives by mass spectrometry and NMR spectroscopy and shown to have the following structures:

Fucα1→Hexβ1→3HexNAc(←NeuAc)β1→Hexβ1→Hexβ1→1Ceramide

Fuc-Hex-HexNAc(NeuGc)-Hex-Hex-Ceramide

INTRODUCTION

The present status of structural fingerprinting of glycoconjugates by mass spectrometry has recently been reviewed (KARLSSON, 1977) and is also the subject of a paper by Karlsson in this monograph.

Mass spectra of pure methylated derivatives of gangliosides containing fucose were first presented in 1977 (KARLSSON et al., 1977; OHASHI and YAMAKAWA, 1977). Recently, mass spectrometry of the methylated-reduced-trimethylsilylated derivative was used in the structural characterization of an earlier unknown fucoganglioside containing 10 sugars in a branched carbohydrate chain (WATANABE et al., 1978). The present report will show the use of mass spectrometry in the structural characterization of two fucogangliosides prepared from minipig posterior root ganglion.

An improved micromethod of NMR spectroscopy of glycolipids using the same type of derivatives has recently been reported (FALK et al., 1979). With this method it has been possible to assign all anomeric protons in glycolipids with up to 8 sugars. This improved method is now applied for the first time on gangliosides.

MASS SPECTRA

Mass spectra of the methylated-reduced-trimethylsilylated derivatives of the two gangliosides are shown in Figs. 1 and 2.

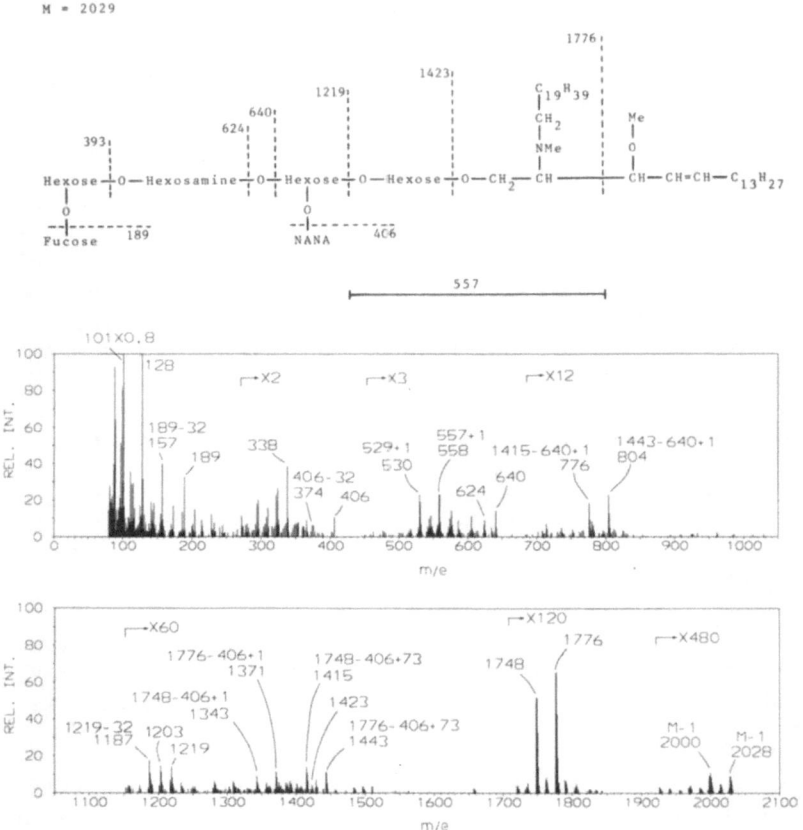

Fig.1. Mass spectrum of the methylated-reduced-trimethylsilylated fucoganglioside containing N-acetylneuraminic acid. The simplified formula corresponds to the major molecular species found. NANA = N-acetylneuraminic acid. The conditions of analysis were: electron energy 75 eV, trap current 500 microA, acceleration voltage 4 kV, ion source temperature 300°C, probe temperature 280°C.

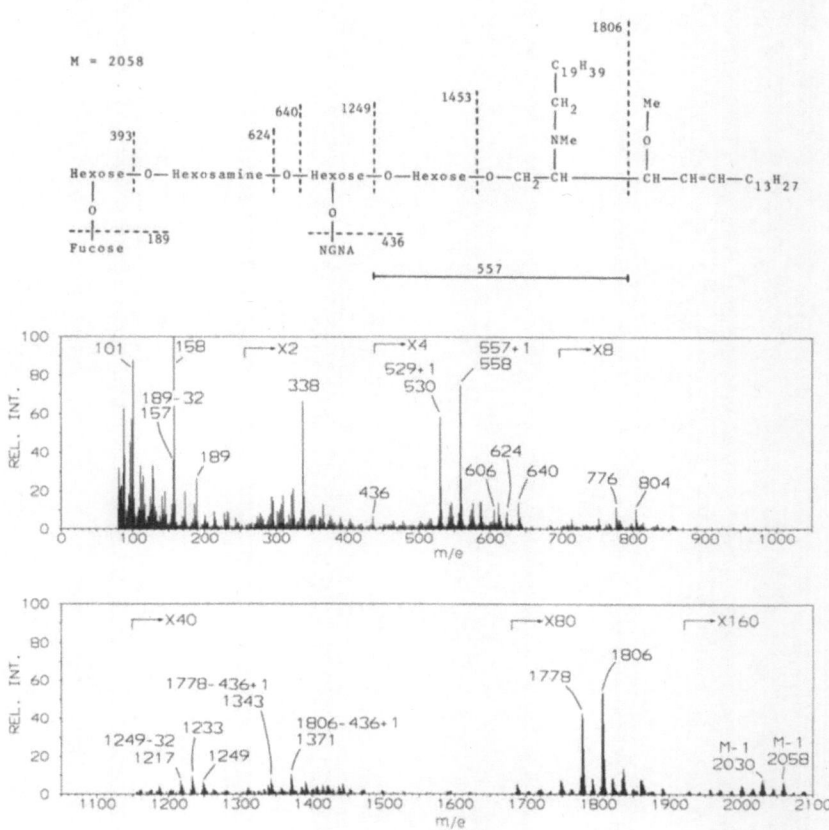

Fig.2. Mass spectrum of the methylated-reduced-trimethylsilylated
fucoganglioside containing N-glycoloylneuraminic acid. The
simplified formula corresponds to the major molecular species
found. NGNA = N-glycoloylneuraminic acid. The conditions of
analysis were: electron energy 46 eV, trap current 500 microA,
acceleration voltage 4 kV, ion source temperature 290°C and
probe temperature 270°C.

The two samples produce very similar spectra except for the specific
mass shifts due to a N-acetyl substituted neuraminic acid in one
sample and a N-glycoloyl substituted neuraminic acid in the other.
This specific mass shift is the same for all three derivatives (Figs.
1-4; spectra of methylated-reduced derivatives are not reproduced).
The type and ratio of sugars are clearly shown by the series of very
intense peaks in the spectra of methylated-reduced-silylated derivatives at m/e 1748 and 1776 for the N-acetyl compound (Fig.1) and
m/e 1778 and 1806 for the N-glycoloyl compound (Fig.2). These correspond to the hexasaccharides shown and a fatty acid composition

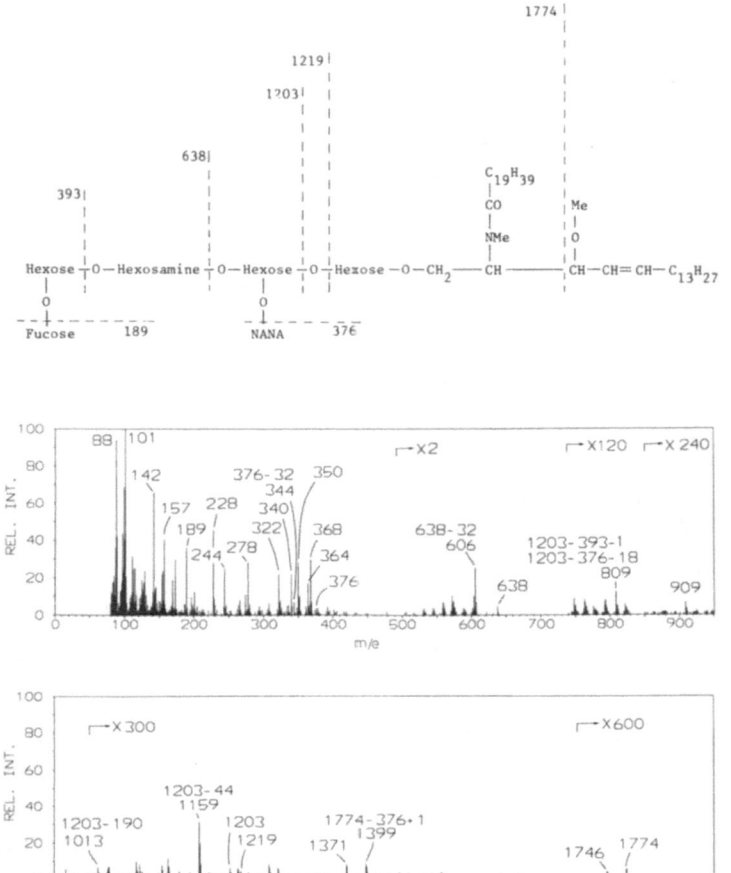

Fig.3. Mass spectrum of the methylated fucoganglioside containing
N-acetylneuraminic acid. Concerning formula, compare Fig.1.
The conditions of analysis were: electron energy 38 eV,
trap current 500 microA, acceleration voltage 6 kV, ion
source temperature 300°C, and probe temperature 265°C.

with primarily C_{18} and C_{20} fatty acids. In both glycolipids,
sphingosine is the major long-chain base as shown by peaks at m/e
364 for the methylated derivatives (Figs. 3 and 4) and at m/e 338
for the reduced derivatives (Figs. 1 and 2). Thus the two glycolipids
are hexaglycosylceramides with sphingosine and C_{18} to C_{20} fatty
acids as major ceramide components. The sugar ratios are hexose:
hexosamine:fucose:N-acetylneuraminic acid, 3:1:1:1 and hexose:
hexosamine:fucose:N-glycoloylneuraminic acid, 3:1:1:1, respectively.
Furthermore, relatively intense sequence ions in spectra of all

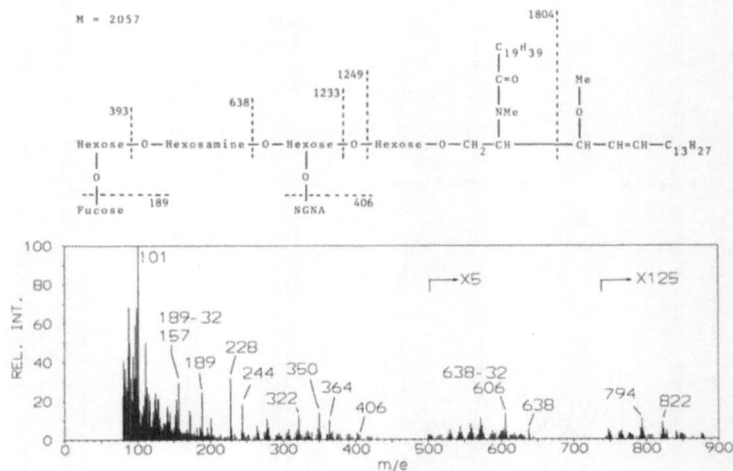

Fig.4. Mass spectrum of the methylated fucoganglioside containing
N-glycoloylneuraminic acid. Concerning formula, compare
Fig.2. The conditions of analysis were: electron energy
62 eV, trap current 500 microA, acceleration voltage 6 kV,
ion source temperature 290°C, and probe temperature 280°C.

three types of derivatives confirm conclusively the sequences
proposed.

The composition and sequence of the terminal trisaccharide is
determined by the presence of definite peaks at m/e 638 and 606
(638-32) for methylated derivatives (Figs. 3 and 4) and at m/e 640
and 624 for reduced derivatives (Figs. 1 and 2). The presence of an
ion at m/e 228 for both compounds as methylated derivatives proves
that the internal hexosamine is monosubstituted (no branch), and
the absence of a peak at m/e 219 (terminal hexose) is further
evidence for the sequence given.

The base peak of Fig.2 is at m/e 158 and is diagnostic for the
N-glycoloyl group. Gangliosides with N-acetylneuraminic acid have
m/e 128 as a major peak (compare Fig.1).

Molecular ions (M-1) were only recorded for the methylated-
reduced-silylated derivatives (m/e 2000, 2028 and m/e 2030, 2058
respectively).

NMR SPECTRA

High resolution NMR spectra (270 MHz Bruker WH-270 spectrometer operating in the Fourier-Transform mode) were recorded for the

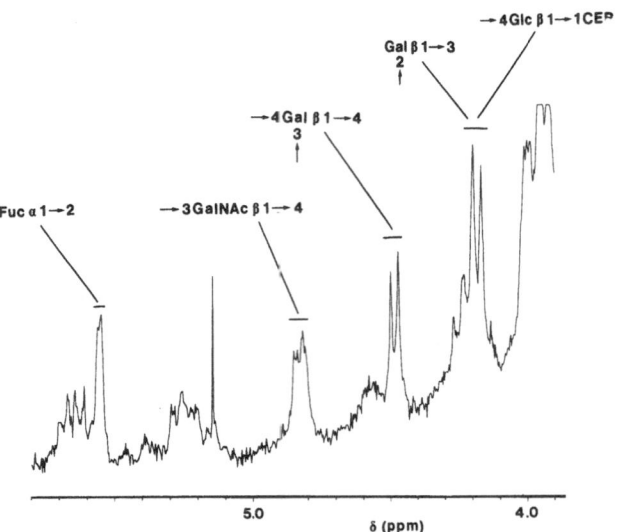

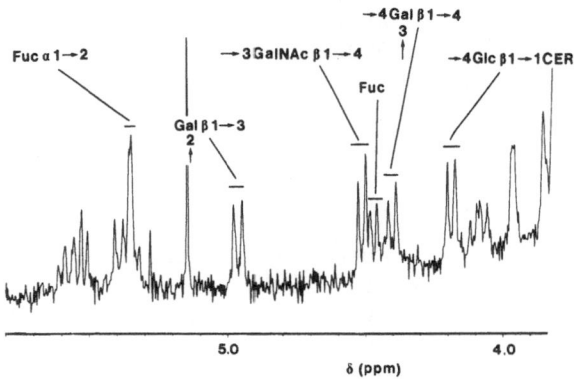

Fig.5. <u>Top</u>: NMR spectrum of permethylated fucoganglioside containing
N-acetylneuraminic acid; 1 mg in 0.5 ml C^2HCl_3, 1600 pulses
at 40°C. (The peak at 5.15 ppm is due to contaminant in
chloroform).
<u>Bottom</u>: NMR spectrum of methylated-reduced fucoganglioside
containing N-acetylneuraminic acid; 1 mg in 0.5 ml C^2HCl_3,
1800 pulses at 40°C. (The peak at 5.15 ppm is due to
contaminant in chloroform).

fucoganglioside containing N-acetylneuraminic acid as methylated
and methylated-reduced derivatives. The spectrum of the methylated-
reduced derivative (Fig.5, bottom) shows 5 groups of anomeric signals
with an approximate relative area of 1:1:1:1:1. From earlier published results (FALK et al., 1979) one may conclude that the doublet at
4.19 ppm ($J_{1,2}$=8.1 Hz) is due to H-1 of β-hexose linked to ceramide

and the doublet at 4.40 ppm ($J_{1,2}$=8.1 Hz) comes from H-1 of the next
β-hexose. The signal at 4.52 ppm ($J_{1,2}$=8.4 Hz) is produced by H-1
of β-hexosamine and the doublet at 5.34 ppm, with a small coupling
constant, ($J_{1,2}$=2.4 Hz) comes from H-1 of the terminal α-fucose.
The doublet at 4.96 ppm with a coupling constant characteristic of
β-protons ($J_{1,2}$=8.9 Hz) is due to the β-hexose bound to hexosamine.
The low-field location of this signal in the reduced derivative
has been discussed before (FALK et al., 1979) and is expected from
a deshielding effect of the amine nitrogen. This effect is only
present when the hexose is bound in position 3 of hexosamine. The
partly overlapped quartet seen at 3.48 ppm is probably due to H-5
of fucose. The spectrum of the same glycolipid before reduction
with $LiAlH_4$ is more complex, Fig.5, top. However, in comparison with
reference spectra (FALK et al., 1979 and unpublished) the different
anomeric signals could be assigned.

DISCUSSION

By mass spectrometry alone the two gangliosides prepared from
minipig posterior root ganglion were shown to have the following
structures:

I   Fucose-Hexose-Hexosamine-Hexose-Hexose-Ceramide
                         |
                       NeuAc

II   Fucose-Hexose-Hexosamine-Hexose-Hexose-Ceramide
                         |
                       NeuGc

Furthermore, the NMR spectra established all binding configurations
in compound I except for the sialosyl residue. Signals defining one
α-fucose, three β-hexoses and one β-hexosamine were found. In
addition, a β1→3 linkage to hexosamine was established.

The study reported here thus presents evidence for structural
identity with two gangliosides earlier identified in boar testis
(SUZUKI et al., 1975) and bovine liver (WIEGANDT, 1973) but different from fucogangliosides of human kidney (RAUVALA, 1976).

The presented results illustrate the potential of the combined
use of mass spectrometry and NMR spectroscopy on the same type of
chemical derivatives. Most of the overall structural data will
thus be available without the need for substance-consuming degradations (in the present case about 1 mg of sample was used for
NMR analysis; sample recoverable).

The work presented here was done in collaboration with
K.-E. Falk, P. Fredman, K.-A. Karlsson, G. Klinghardt, I. Pascher,

W. Pimlott, L. Svennerholm, J. Ångström, and will be published in
detail elsewhere.

ACKNOWLEDGEMENTS

    The work was supported by a grant from The Swedish Medical
Research Council (03X-3967).

REFERENCES

FALK K.-E., KARLSSON K.-A. and SAMUELSSON B.E. (1979): Proton
    NMR analysis of anomeric structure of glycosphingolipids. Lewis-
    active and Lewis-like substances. Archs Biochem. Biophys. 192,
    191-202.
KARLSSON K.-A. (1977): Mass-spectrometric sequence studies of
    lipid-linked oligosaccharides. Blood-group fucolipids, ganglio-
    sides and related cell-surface receptors. Prog. Chem. Fats
    Lipids 16, 207-230.
KARLSSON K.-A., PIMLOTT W. and SAMUELSSON B.E. (1977): Mass-
    spectrometric sequence studies of oligosaccharides. Blood-group
    fucolipids, gangliosides and related cell-surface receptors.
    11th FEBS Meeting, Copenhagen, Denmark, Abstr. No. B4-501.
OHASHI M. and YAMAKAWA T. (1977): Isolation and characterization
    of glycosphingolipids in pig adipose tissue. J. Biochem.(Tokyo)
    81, 1675-1690.
RAUVALA H. (1976): Gangliosides of human kidney. J. biol. Chem.
    251, 7517-7520.
SUZUKI A., ISHIZUKA I. and YAMAKAWA T. (1975): Isolation and
    characterization of a ganglioside containing fucose from boar
    testis. J. Biochem. (Tokyo) 78, 947-954.
WATANABE K., POWELL M. and HAKOMORI S.-i. (1978): Isolation and
    characterization of a novel fucoganglioside of human erythrocyte
    membranes. J. biol. Chem. 253, 8962-8967.
WIEGANDT H. (1973): Gangliosides of extraneural organs. Hoppe-
    Seyler's Z. physiol. Chem. 354, 1049-1056.

TRIFLUOROACETOLYSIS, A NEW METHOD FOR STRUCTURAL STUDIES OF GLYCOLIPIDS

Sigfrid Svensson

Department of Clinical Chemistry
University Hospital
S-221 85 Lund, Sweden

INTRODUCTION

Trifluoroacetolysis is a reaction which is carried
out in mixtures of trifluoroacetic acid (TFA) and trifluoroacetic anhydride (TFAA) in various proportions and
at different temperatures. 2-Acetamido-2-deoxy functions
in sugar units are converted into 2-deoxy-2-trifluoro-
acetamido groups, by trifluoroacetolysis, employing
TFA/TFAA in proportions varying from 1:1 to 1:50 at 100°C
for 48 h (1). Under these conditions of trifluoroacetolysis most reducing sugars (2) and glycosides (3) are
stable due to the stabilizing effects excerted by the
O-trifluoroacetyl groups rapidly formed by the action of
TFA/TFAA on the hydroxyl groups in the sugar residues.
Peptide bonds are cleaved by transamidation and thus
proteins and the protein part of glycoproteins will be
degraded, by trifluoroacetolysis, whereas the carbohydrate portion of glycoproteins remains virtually intact,
apart from some degradation of 2-acetamido-2-deoxy sugars
situated at the reducing end (2,4,5). N-Glycosidically
linked carbohydrate chains are cleaved off from glycoproteins by transamidation (4,5) and O-glycosidically
linked carbohydrate chains are cleaved off from serine
or threonine residues by an acid catalyzed elimination
reaction (5,6).

STABILITY OF SUGARS TOWARDS TRIFLUOROACETOLYSIS

When reducing sugars and glycosides are subjected to
trifluoroacetolysis they are rapidly converted into their

pertrifluoroacetylated derivatives. Due to the strong
inductive effects excerted by the O-trifluoroacetyl groups
adjacent to acetalic oxygens, acid catalyzed degradation
of reducing sugars and solvolysis of glycosides are prevented. The relative importance of O-trifluoroacetyl
groups situated in different positions of the sugar ring
has been studied in detail using partially methylated
methyl α-D-glucopyranosides (7), methyl α-D-xylopyrano-
sides and D-xyloses (8) as models. It was found that
2-O-, 4-O- and 6-O-trifluoroacetyl groups were the most
effective in preventing acid catalyzed degradation of
reducing sugars as well as preventing solvolysis of glycosides. This is consistent with the idea that both reactions are initiated by ionization of the ring oxygen
(Scheme 1).

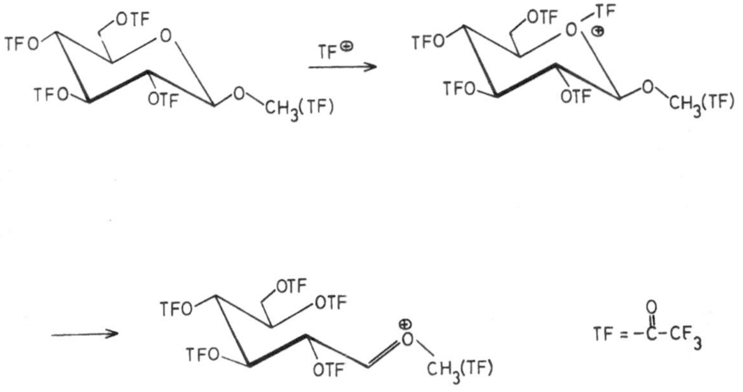

Scheme 1

The solvolysis of glycosides may also be initiated
by ionization of the exocyclic aglycon oxygen (Scheme 2).

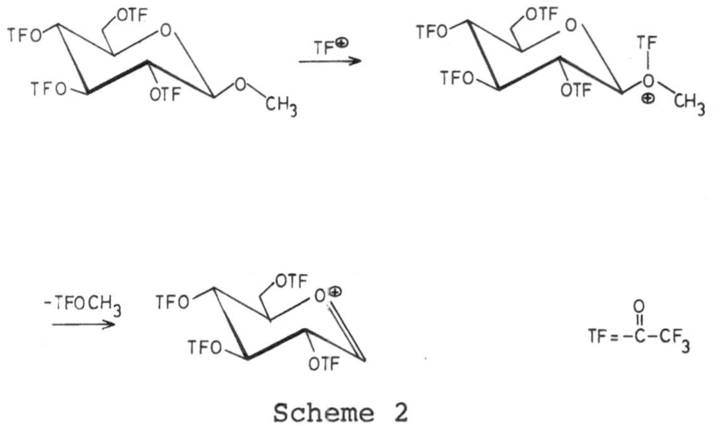

Scheme 2

Thus reducing sugars lacking O-trifluoroacetyl groups in the 2-, 4- or 6-position are most sensitive to acid catalyzed degradation reactions (cf degradation of reducing 2-acetamido-2-deoxy sugars (2)). In oligosaccharides further protection of the glycosidic bond toward solvolysis should be furnished by any adjacent O-trifluoroacetyl group in the aglycon moiety. Model compound studies (8) have shown that the protective action of such O-trifluoroacetyl groups is surprisingly small.

TRIFLUOROACETOLYSIS OF GLYCOLIPIDS

Trifluoroacetolysis of glycolipids under conditions that will effect transamidation (TFA/TFAA, 1:1 or 1:50, 100°C, 48 h) results in near quantitative cleavage of the glycosidic bond linking the sugar moiety to the ceramide portion (9). This cleavage could possibly be initiated by elimination of an O-trifluoroacetyl group from the sphingosine unit yielding an allyl cation (Scheme 3).

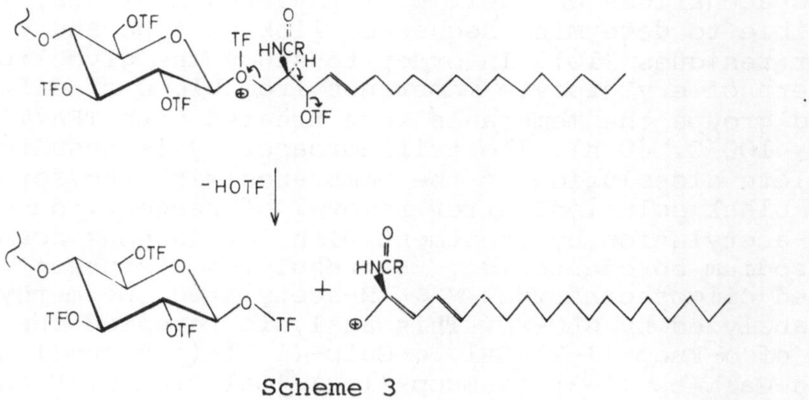

Scheme 3

Trifluoroacetolysis of D-glucosyl-, D-galactosyl-ceramide and trihexoside thus results in the formation of pertrifluoroacetylated D-glucose, D-galactose and α-Galp-(1-4)-β-Galp-(1-4)-Glc. Globoside will besides eliminating the ceramide portion also undergo transamidation by trifluoroacetolysis and thus the pertrifluoroacetylated derivative of β-GalNTF-(1-3)-α-Galp-(1-4)-β-Galp-(1-4)-Glc is formed.

N-Acetyl-neuraminic acid is rapidly destroyed during trifluoroacetolysis but N-acetyl-neuraminic acid linked to the 3-position of a pyranosidic sugar residue is not cleaved off but undergoes modification. An N-acetyl-neuraminic acid residue linked to the 6-position of a sugar

unit is under transamidation conditions cleaved off and
destroyed.

DE-O- AND DE-N-TRIFLUOROACETYLATION

The O-trifluoroacetyl groups introduced in sugar
residues during trifluoroacetolysis can be cleaved off
by treatment of the per-O-trifluoroacetate with methanol
followed by 50% aqueous acetic acid at room temperature
(1,2). N-Trifluoroacetyl groups are quite stable under
acidic conditions but they can be removed by treatment
with M ammonia in methanol/water at room temperature or
by reduction conditions using sodium borohydride in
ethanol/water (1).

STUDIES OF GLYCOLIPID PATTERNS

The finding that trifluoroacetolysis releases the
carbohydrate moiety from glycolipids has made it feasible
to analyze the carbohydrate portion by gas chromatography-
mass spectrometry (GLC-MS) methodologies. By GLC-MS of
oligosaccharides as their permethylated alditols, it is
possible to determine sequence, linkage type and class of
sugar residues (10). In order to study the glycolipid
pattern of erythrocyte membranes from blood of different
blood groups the membranes were treated with TFA/TFAA
(1:1, 100°C, 48 h). The trifluoroacetolysis resulted in
complete dissolution of the membranes with the formation
of a black solution. After removal of reagent, de-O- and
de-N-acetylation by treatment with 50% aqueous acetic acid
and sodium borodeuteride, the resulting mixture of released oligosaccharides was N-acetylated, permethylated
and analyzed by GLC-MS. This analysis revealed the presence of α-Fucp-(1-2)-Gal, α-Galp-(1-3)-(α-Fucp-(1-2)-)Gal
and α-GalNAcp-(1-3)-(α-Fucp-(1-2)-)Gal from O, B and A
erythrocytes, respectively. These oligosaccharides represent the immunodominant oligosaccharides of the O, B
and A antigens and are released from glycolipids and/or
glycoproteins presumably as a result of solvolysis and
degradation of a branched GNAc residue to which they had
been linked. Similar results have been obtained by trifluoroacetolysis of soluble blood group substances (11)
(Fig. 1).

Erythrocytes of different blood group P phenotypes
have also been subjected to trifluoroacetolytic degradation and released oligosaccharides were analyzed by
GLC-MS. It was found that blood from individuals with
the p phenotype releases a large amount of lactose
deriving from the known excess of lactosyl ceramide in
these membranes. Cells belonging to the P_1^k phenotype

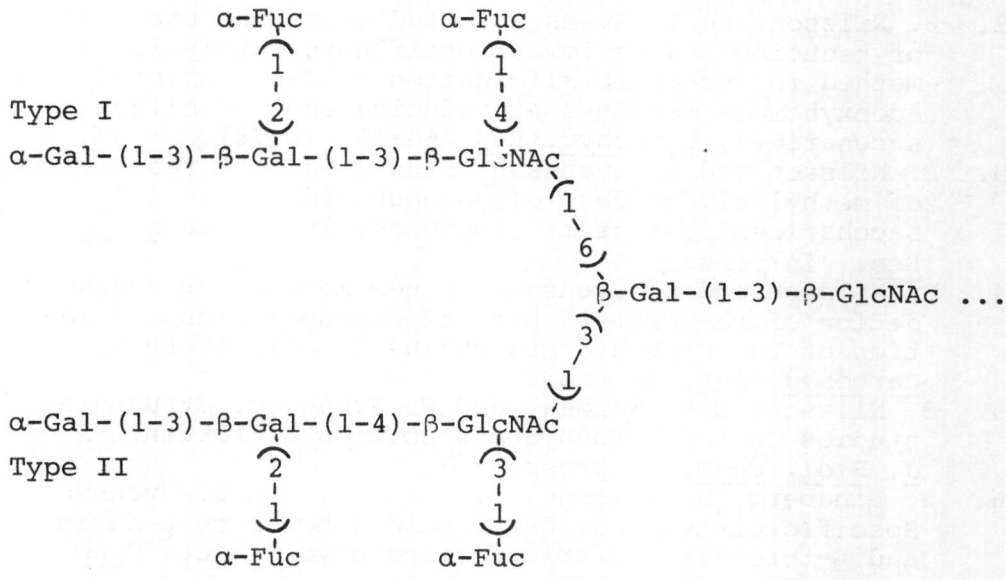

Fig. 1. Suggested composite structure for the outer portion of a B-specific megalosaccharide in soluble blood
group substances (12).

gave small amounts of lactose but the major component
was α-Galp-(1-4)-β-Galp-(1-4)-Glc. This trisaccharide is
most likely derived from the P^k antigen (trihexosyl
ceramide). Membranes from blood of the phenotype P_1
yielded small amounts of both lactose and the P^k trisaccharide together with a large quantity of the tetrasaccharide β-GalNAcp-(1-3)-α-Galp-(1-4)-β-Galp-(1-4)-Glc.
The tetrasaccharide represents the carbohydrate portion
of the P antigen (globoside) (13).
        The above technique can also be used as a screening
methodology for quantitation and identification of accumulated glycolipids in sphingolipidoses.

ACKNOWLEDGEMENTS

        This work was supported by grants from the Swedish
Medical Research Council (03X-4956), the Medical Faculty,
University of Lund, and Magnus Bergvall's Foundation.

REFERENCES

1. B. Nilsson and S. Svensson, A new method for N-deace-
    tylation of 2-acetamido-2-deoxy sugars, Carbohyd.
    Res. 62:377 (1978).

2. B. Nilsson and S. Svensson, Studies of the stability
    of reducing sugars towards trifluoroacetolysis: a
    method for specific elimination of 2-acetamido-2-
    -deoxyhexose residues at reducing ends of oligo-
    saccharides, Carbohyd. Res. 65:169 (1978).
3. B. Nilsson and S. Svensson, Studies of the reactivity
    of methyl glycosides, oligosaccharides, and poly-
    saccharides towards trifluoroacetolysis, Carbohyd.
    Res. in press.
4. B. Nilsson and S. Svensson, A new method for degra-
    dation of the protein part of glycoproteins. Isola-
    tion of the carbohydrate chains of asialofetuin,
    Carbohyd. Res. in press.
5. B. Nilsson, N.E. Nordén, and S. Svensson, Structural
    studies on the carbohydrate portion of fetuin,
    J. Biol. Chem. in press.
6. B. Lindberg, B. Nilsson, T. Norberg, and S. Svensson,
    Specific cleavage of O-glycosidic bonds to L-serine
    and L-threonine by trifluoroacetolysis, Acta Chem.
    Scand. Ser B in press.
7. L.-E. Franzén and S. Svensson, The stability of
    partially methylated methyl α-D-glucopyranosides
    towards trifluoroacetolysis, Carbohyd. Res. in press.
8. L.-E. Franzén and S. Svensson, The stability of
    partially methylated methyl α-D-xylopyranosides and
    partially methylated D-xyloses towards trifluoro-
    acetolysis, Carbohyd. Res. submitted.
9. J. Lundsten, S. Svensson, and L. Svennerholm, A method
    for specific cleavage of the glycosidic bond to the
    ceramide portion of glycolipids, FEBS Lett. in press.
10. J. Lundsten, S. Svensson, and P.A. Öckerman, GC/MS
    screening of urinary oligosaccharides, in: 'Proc.
    Mass Spectrometry and Combined Techniques in Medi-
    cine, Clinical Chemistry and Clinical Biochemistry,
    Tubingen (1977), p. 193.
11. M.A. Chester and S. Svensson, unpublished results.
12. K.O. Lloyd, E.A. Kabat, and E. Licerio, Immuno-
    chemical studies on blood groups. XXXVIII. Structures
    and activities of oligosaccharides produced by alka-
    line degradation of blood-group Lewis[a] substance.
    Proposed structure of the carbohydrate chains of
    human blood-group A, B, H, Le[a], and Le[b] substances,
    Biochemistry 7:2976 (1968).
13. A. Lundblad, S. Svensson, B. Löw, L. Messeter,
    B. Cedergren, and H.R. Nevanlinna, Release of oligo-
    saccharides from human erythrocyte membranes of
    different blood group P phenotypes by trifluoro-
    acetolysis, J. Biol. Chem. submitted.

HIGH RESOLUTION ^{1}H-NMR SPECTROSCOPY OF CARBOHYDRATE STRUCTURES

Johannes F.G. Vliegenthart

Department of Bio-Organic Chemistry, State University

Croesestraat 79, 3522 AD Utrecht, The Netherlands

High resolution ^{1}H-NMR spectroscopy has obtained a firm position among the techniques available for structural studies of bio-molecules. The instrumentation has enormously been improved during the last decade. New methods have been developed to optimize the information which can be deduced from the spectra[1-5]. ^{1}H-NMR spectroscopy has provided valuable data on the structure, conformation and inter- and intra-molecular interactions of various types of bio-molecules.

Only recently, the application of high resolution ^{1}H-NMR spectroscopy to complex carbohydrates started. In the beginning mainly ^{1}H-NMR data were described for derivatives of mono- and oligosaccharides. We investigated by 220-MHz and/or 300-MHz ^{1}H-NMR pertrimethylsilyl[6-10] and permethyl derivatives[11-14]. The application of derivatives has the drawback that chemical modifications have to be carried out, followed by sometimes elaborate purifications or fractionations. On the other hand, it has the advantage that for reducing compounds, the different anomeric forms can be isolated and separately investigated. For derivatives of relatively simple compounds it turned out to be possible to give a complete interpretation of the spectra. In general the definite assignment and refinement of the NMR-parameters could be obtained by spectral simulation in an iterative procedure. The resulting NMR data can be used to derive detailed information on many structural aspects like configuration of the glycosidic bonds, substitution pattern, type and conformation of the constituting monosaccharides. Going to larger oligomers, the spectra become more complex, thereby making a complete

interpretation virtually impossible. However, already a partial
interpretation of the spectra can furnish relevant information
which is difficult to obtain along other routes.

   Similar interpretation problems as mentioned for derivatives
are encountered in the study of free saccharides dissolved in
D_2O. In the case of reducing saccharides, there is the additional
complication that anomeric mixtures have to be investigated
unless a chemical modification of the reducing end is carried out.
In the following a few examples will be presented of the scope
and limitations of 360-MHz ^{1}H-NMR spectroscopy of carbohydrate
structures. In the spectra the signals can roughly be subdivided
into resonances of:

a. the anomeric protons
b. the bulk of the non-anomeric protons, which usually gives rise
   to a broad, poorly resolved signal
c. special non-anomeric protons, which resonate outside of
   "the big hump". These important signals can be conceived as
   "structural reporter groups".
d. protons of substituents like N-acetyl, N-glycolyl, O-acetyl,
   O-lactyl groups etc.

Let us first consider the simple disaccharide β-Gal$p$(1→3)-Gal,
which gives rise to a highly complex spectrum as shown in Fig. 1.

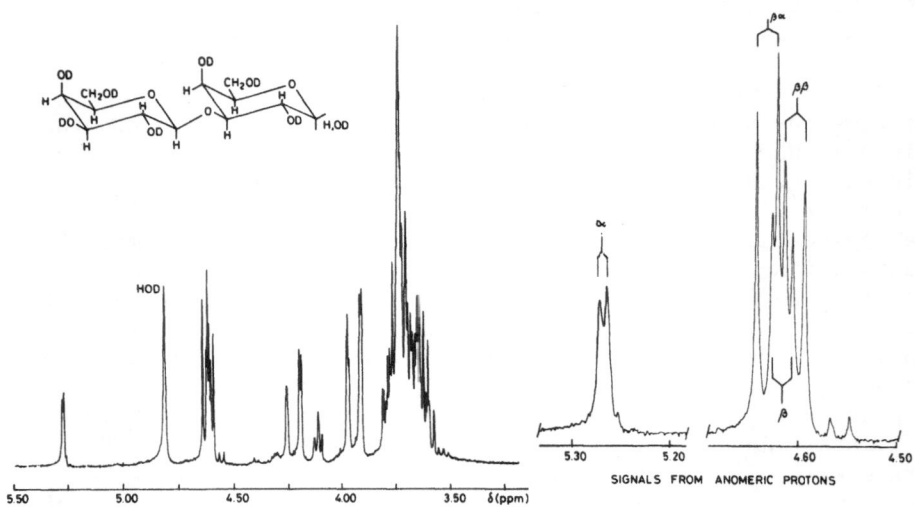

Fig. 1.   360-MHz ^{1}H-NMR spectrum of β-Gal$p$(1→3)-Gal in D_2O

This spectrum is mainly a superposition of the spectra of
β–Galp(1→3)–α–Galp and β–Galp(1→3)–β–Galp. As indicated in Fig. 1,
the anomeric signals can easily be recognized. It has to be noted
that differences in anomeric configuration at the reducing end
are also reflected in the chemical shifts in all of the skeleton
protons of the reducing unit and in some of the ring protons of
the non-reducing unit.

The spectrum of the alditol obtained after reduction with
NaBH₄ of β–Galp(1→3)–Gal is much simpler and can be interpreted
in more detail as demonstrated in Fig. 2.

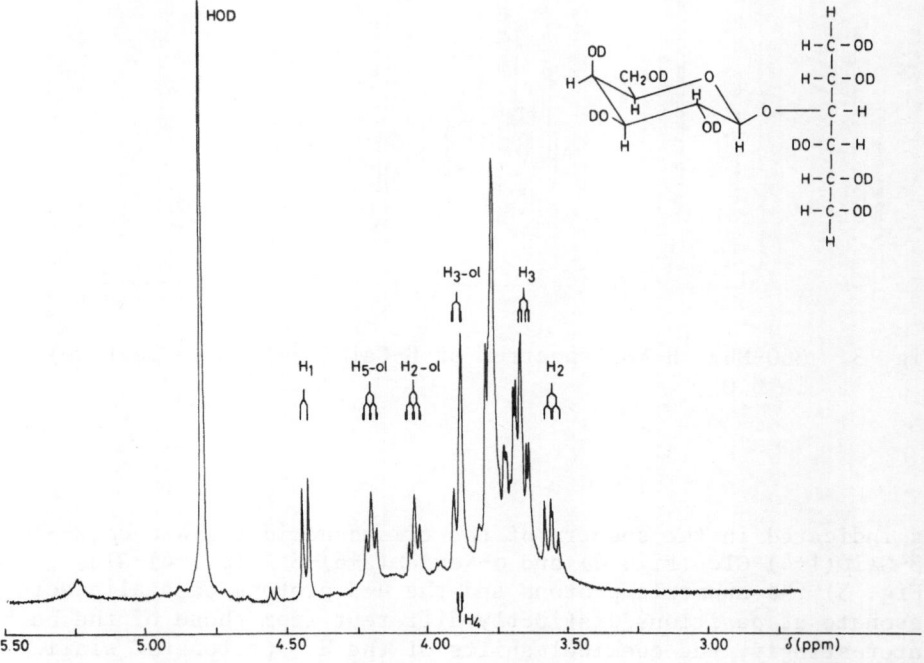

Fig. 2.   360–MHz ¹H-NMR spectrum of β–Galp(1→3)–Gal–ol in D₂O

The substitution of galactitol and the configuration of the glycosidic linkage can be derived from the NMR data. Spectra of
reducing disaccharides are not always as complicated as given in
Fig. 1. Anomerization does not necessarily affect the chemical
shifts of so many protons. For example the spectrum of lactose as
shown in Fig. 3 is somewhat less complex. The resonance position
of the H-1 of the non-reducing galactose residue is identical
for both anomers.

As mentioned before, for larger compounds only a partial
interpretation of the spectrum can be achieved.

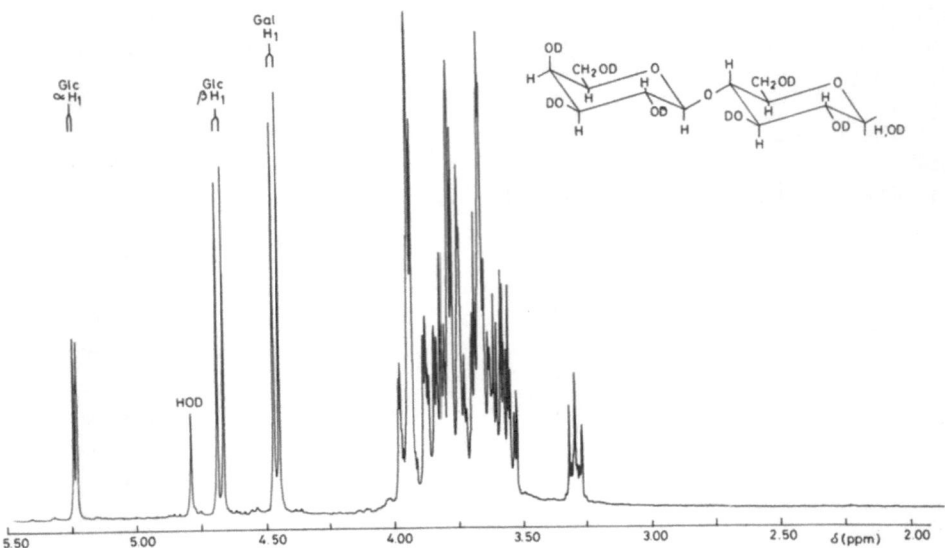

Fig. 3.   360-MHz ^{1}H-NMR spectrum of β-Gal$p$(1→4)-Glc (= lactose)
          in D$_2$O

As indicated in the spectra of the trisaccharides α-NeuAcp(2→3)-
-β-Gal$p$(1→4)-Glc (Fig. 4) and α-NeuAcp(2→6)-β-Gal$p$(1→4)-Glc
(Fig. 5) the anomeric protons and the H-3 protons of sialic acid
resonate at positions distinctly different from those of the bulk.
Interestingly, the chemical shifts of the H-3 protons of sialic
acid are characteristic for the position and configuration of
the linkage of NeuAc to Gal. To some extent the chemical shifts
of the H-3 protons are also influenced by the type of sugar
residue to which Gal is glycosidically attached. E.g. replacement
of Glc by GlcNAc introduces shift increments as shown in Table 1.
On the other hand, the chemical shifts of H-1 of Gal are significantly affected by the presence and the position of attachment
of sialic acid as demonstrated in Table 2.

    The chemical shifts of the H-3 protons of sialic acid are
extremely useful in structural studies. For example in horse
pancreatic ribonuclease we have found that (2→3) as well as (2→6)
linked sialic acid residues can occur in one glyco chain (Fig. 6).

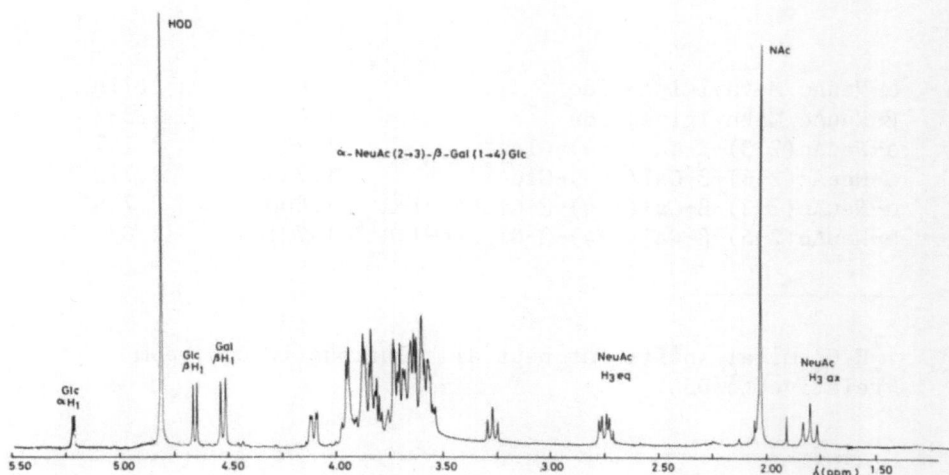

Fig. 4.   360-MHz ¹H-NMR spectrum of α-NeuAc*p*(2→3)-β-Gal*p*(1→4)-Glc
          (= sialyl(2→3)lactose) in D$_2$O

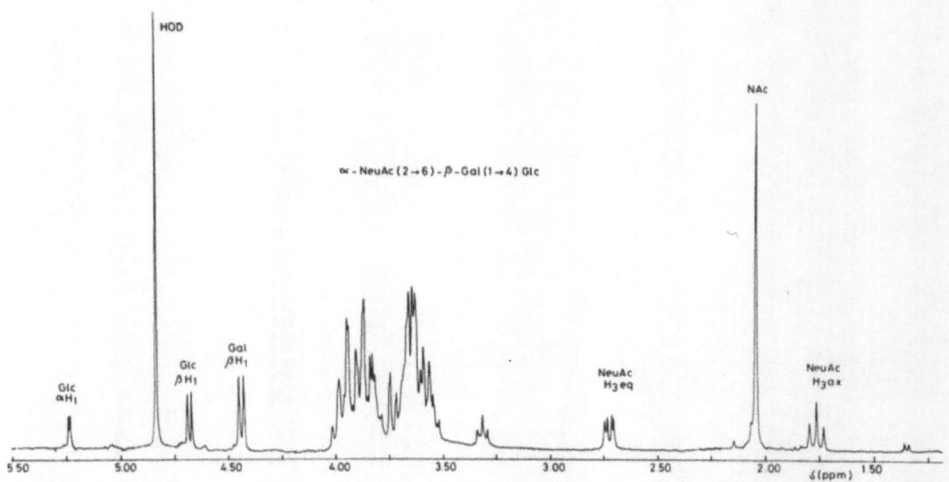

Fig. 5.   360-MHz ¹H-NMR spectrum of α-NeuAc*p*(2→6)-β-Gal*p*(1→4)-Glc
          (= sialyl(2→6)lactose) in D$_2$O

Table 1.   Chemical shifts of sialic acid H-3 protons for
various sialic acid glycosides*

Compound	δH-3ax	δH-3eq
α-NeuAc Methylglycoside[15]	1.626	2.718
β-NeuAc Methylglycoside	1.645	2.337
α-NeuAc(2→3)-β-Gal(1→4)-Glc[16]	1.799	2.757
α-NeuAc(2→6)-β-Gal(1→4)-Glc[16]	1.739	2.712
α-NeuAc(2→3)-β-Gal(1→4)-β-GlcNAc-1→R[16]	1.800	2.758
α-NeuAc(2→6)-β-Gal(1→4)-β-GlcNAc-1→R[16]	1.721	2.670

*[1]H Chemical shifts for neutral solutions in D_2O; ppm
relative to DSS.

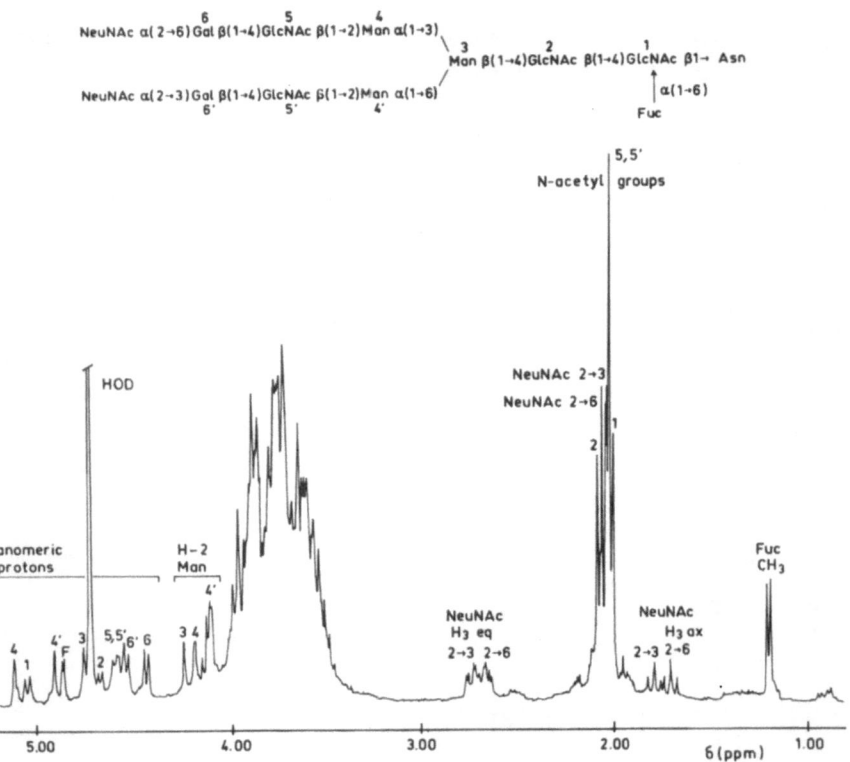

Fig. 6.   360-MHz ^{1}H-NMR spectrum of glycopeptide Eq. GPI isolated
from horse pancreatic ribonuclease.

Table 2. Chemical shifts of the galactose H-1
          proton for lactose, sialyl(2→3)lactose
          and sialyl(2→6)lactose

Compound	δH-1 Gal
β-Gal(1→4)-Glc	4.443
α-NeuAc(2→3)-β-Gal(1→4)-Glc[16]	4.531
α-NeuAc(2→6)-β-Gal(1→4)-Glc[16]	4.427

On the basis of the peak areas and the chemical shifts of the H-3
protons, the molar ratios of the (2→3) and (2→6) linked sialic
acids could be derived[17]. The location of the differently linked
sialic acids could be inferred from the chemical shifts of the
H-1 protons of the mannose residues (vide infra).

For compounds containing the mannotriosido branching core as
indicated in Fig. 7 the H-2 protons of mannose can act as structural
reporter groups.

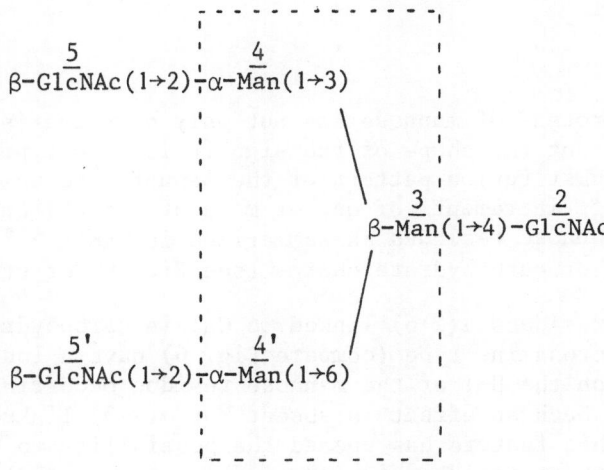

Fig. 7. Structure of a hexasaccharide (A), possessing the manno-
         triosido branching core (within dotted lines).

This oligosaccharide is a structural unit of many N-glycosidically
linked carbohydrate chains of glycoproteins. In Fig. 8 the most
relevant part of the spectrum is given[18].

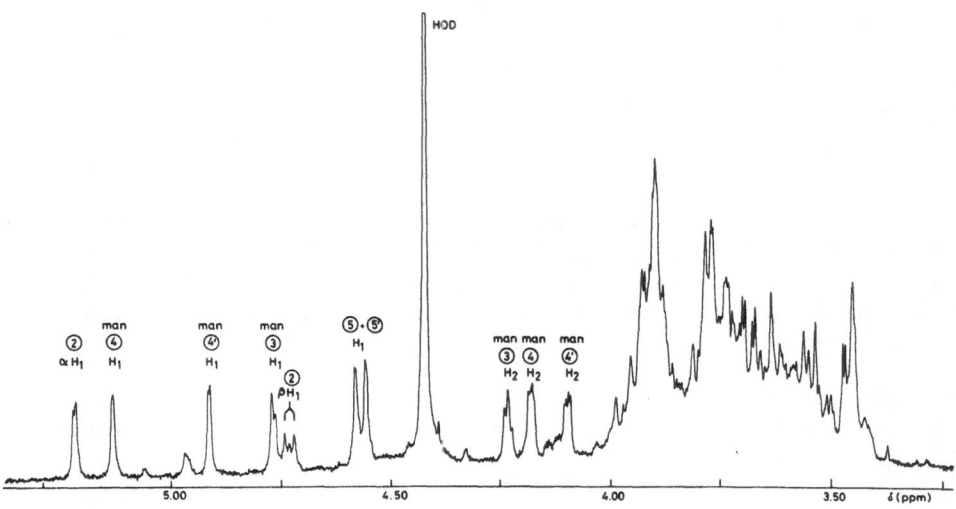

Fig. 8.   360-MHz ^{1}H-NMR spectrum of hexasaccharide A (see Fig. 7)
in D_2O

The set of H-2 protons of mannose has not only characteristic
chemical shifts, but the shape of the signals is also typical.
Changes in the substitution pattern of the mannotriose unit are
reflected in shift increments of one or more of the H-1 and H-2
protons of the mannose residues as summarized in Table 3[19] for the
lactosamine type of carbohydrate chains (see Fig. 9 for structures).

Sialic acid residues α(2→6) linked to Gal in carbohydrate
chains of the lactosamine type (compare Fig. 6) have a long
distance effect on the H-1 of the mannose residue occurring in
the same branch. Such an effect is absent for α(2→3) linked sialic
acid residues. This feature has opened the possibility to determine
the location of these residues in compounds wherein both types of
sialic acid linkages occur. This approach is effective because the
chemical shifts of the H-1 protons of the various mannose residues
in the asialo compounds are different (Table 4) [17,19]. Evidently
the shift increments of the H-1 of Gal and GlcNAc from the lactosamine units cannot be used because the chemical shifts of the H-1
protons are almost identical for the different branches.

```
      6            5            4
   βGal(1→4)βGlcNAc(1→2)αMan(1→3)
                                      \  3            2            1
   bi-antenna                         βMan(1→4)βGlcNAc(1→4)βGlcNAc1→Asn
      6'           5'           4'   /
   βGal(1→4)βGlcNAc(1→2)αMan(1→6)


      8            7
   βGal(1→4)βGlcNAc(1→4)
      6            5        \  4
   βGal(1→4)βGlcNAc(1→2)αMan(1→3)
                                      \  3            2            1
   tri-antenna                        βMan(1→4)βGlcNAc(1→4)βGlcNAc1→Asn
      6'           5'           4'   /
   βGal(1→4)βGlcNAc(1→2)αMan(1→6)


      8            7
   βGal(1→4)βGlcNAc(1→4)
      6            5        \  4
   βGal(1→4)βGlcNAc(1→2)αMan(1→3)
                                      \  3            2            1
   tetra-antenna                      βMan(1→4)βGlcNAc(1→4)βGlcNAc1→Asn
      6'           5'           4'   /
   βGal(1→4)βGlcNAc(1→2)αMan(1→6)
      8'           7'          /
   βGal(1→4)βGlcNAc(1→6)


               5            4
            βGlcNAc(1→2)αMan(1→3)
                                 \  3            2
   hexasaccharide A              βMan(1→4)GlcNAc
               5'           4'  /
            βGlcNAc(1→2)αMan(1→6)


               5            4
            βGlcNAc(1→2)αMan(1→3)
                     9          \  3            2
   heptasaccharide B   βGlcNAc(1→4)βMan(1→4)GlcNAc
               5'           4'  /
            βGlcNAc(1→2)αMan(1→6)
```

Fig. 9. Structures of the asparagine-linked glycans of the
         N-acetyllactosamine type, hexasaccharide A and
         heptasaccharide B.

Table 3. Chemical shifts of mannose H-1 and H-2 protons for bi-,
tri- and tetra-antennary asparagine-bound glycan chains
of the N-acetyllactosamine type, and for oligosaccharides
A and B (see Fig. 9).

Structure	δH-1 of residue			δH-2 of residue		
	3	4	4'	3	4	4'
bi-antenna	4.764	5.121	4.928	4.247	4.189	4.110
tri-antenna	4.757	5.119	4.924	*4.215*	*4.215*	4.109
tetra-antenna	4.754	5.127	*4.866*	*4.215*	*4.215*	*4.092*
hexasaccharide A	~4.77	5.119	4.922	~4.25	4.192	4.112
heptasaccharide B	*~4.70*	*5.062*	*5.004*	*~4.18*	*4.250*	*4.151*

Table 4. Chemical shifts of mannose H-1 protons for an asialo-
biantennary asparagine-bound glycan chain of the N-acetyllactosamine type (see Fig. 9), and of glycopeptide
Eq. GPI from horse pancreatic ribonuclease (see Fig. 6).

Structure	δH-1 of residue		
	3	4	4'
asialo-biantenna	4.764	5.121	4.928
Eq. GPI	4.766	*5.135*	4.923

Table 5. Chemical shifts of H-1, H-5 and H-6 protons of fucose in
different linkages to N-acetylglucosamine.

type of linkage	δH-1	δH-5	δH-6
α-Fuc(1→3)-β-GlcNAc(1→	5.11	4.83	1.17
α-Fuc(1→4)-β-GlcNAc(1→	5.02	4.87	1.18
α-Fuc(1→6)-β-GlcNAc(1→	4.90	4.12	1.21

Another example of a reporter group is the H-5 proton of fucose. In the carbohydrate chains of glycoproteins various types of fucose linkages can occur. As shown in Table 5, the $\alpha(1\rightarrow3)$-, $\alpha(1\rightarrow4)$- and $\alpha(1\rightarrow6)$-linkage of fucose to N-acetylglucosamine can be distinguished on the basis of the chemical shifts of the H-5 and H-1 protons of fucose[20,21].

Apart from this structural information, which can be derived from the spectra, it is important to stress that the spectra are useful for determination of the homogeneity of the sample and of the quantitative carbohydrate composition of pure compounds. These apparently trivial data are not always easy to obtain along classical routes. Structural studies on heterogeneous samples may lead to erroneous results. In view of the non-destructiveness of NMR investigations, these types of experiments have to be carried out before chemical and/or enzymic degradations are performed.

With regard to the ^{1}H-NMR studies of intact, underivatized gangliosides we are still in a very preliminary stage. An extremely important aspect is the choice of the solvent. As shown in Fig. 10, it is evident that D_2O is unsuitable, due to the formation of aggregates.

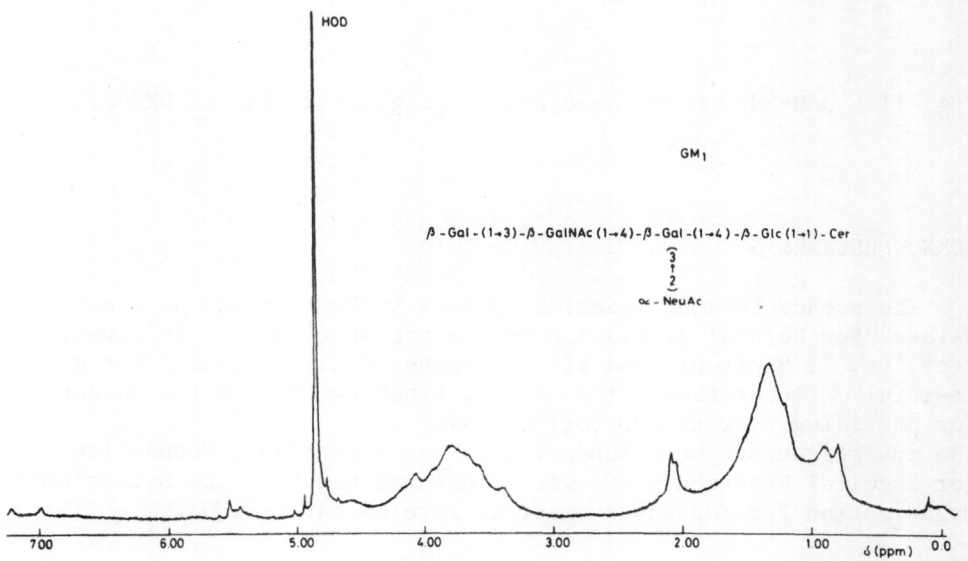

Fig. 10.   360-MHz ^{1}H-NMR spectrum of ganglioside GM_1 in D_2O

The spectrum provides hardly any structural information. However, it seems that spectra recorded in DMF-d$_7$ (Fig. 11) can yield valuable data. The resolution of the significant signals is at least comparable to that of the permethyl compounds[22-24]. It is likely, that also for the analysis of gangliosides high resolution ^{1}H-NMR spectroscopy can become a very powerful tool.

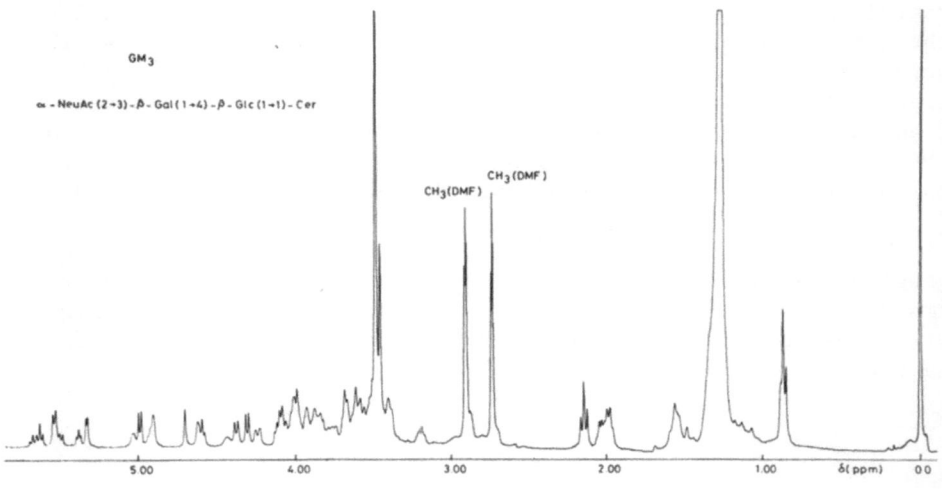

Fig. 11.   360-MHz ^{1}H-NMR spectrum of ganglioside GM$_3$ in DMF-d$_7$

ACKNOWLEDGEMENTS

The author is much indebted to Drs. L. Dorland and H. van Halbeek for helpful discussion in the preparation of this paper, Prof. Dr. J. Montreuil and his colleagues (Lille, France) for a fruitful collaboration, Prof. Dr. B. Lindberg (Stockholm, Sweden) for providing us with β-Galp(1→3)-Gal. The investigations were supported by the Netherlands Foundation for Chemical Research (SON) with financial aid from the Netherlands Organization for the Advancement of Pure Research (ZWO).

REFERENCES

1. B.Coxon, Proton Magnetic Resonance Spectroscopy: Part I,
   Adv. Carbohyd . Chem. Biochem. 27:7 (1972).
2. R.A. Dwek, "Nuclear Magnetic Resonance (N.M.R.) in Biochemistry;
   Applications to Enzyme Systems", Clarendon Press,
   Oxford (1973).
3. L.D. Hall, Solutions to the Hidden-Resonance Problem in Proton
   Nuclear Magnetic Resonance Spectroscopy, *Adv. Carbohyd. Chem.
   Biochem.* 29:11 (1974).
4. P.F. Knowles, D. Marsh and H.W.E. Rattle, "Magnetic Resonance
   of Biomolecules", John Wiley and Sons, New York (1976).
5. K. Wüthrich, "NMR in Biological Research: Peptides and Proteins",
   North-Holland Publishing Company, Amsterdam (1976).
6. D.G. Streefkerk, M.J.A. de Bie and J.F.G. Vliegenthart,
   Conformational studies on pertrimethylsilyl derivatives of
   some mono- and disaccharides by 220 MHz PMR spectroscopy,
   Tetrahedron 29:833 (1973).
7. D.G. Streefkerk, M.J. A. de Bie and J.F.G. Vliegenthart,
   Conformational studies on some pertrimethylsilyl-(TMS)-6-
   -deoxy-aldohexo-pyranoses and four less common TMS-aldohexo-
   pyranoses by 220 MHz spectroscopy. Substitutional and
   configurational effects on the chemical shifts of the ring
   protons, *Carbohyd. Res.* 38:47 (1974).
8. D.G. Streefkerk, M.J.A. de Bie and J.F.G. Vliegenthart,
   Conformational studies on pertrimethylsilyl derivatives of
   2-acetamido-2-deoxy-aldohexopyranoses by 220 and 300 MHz
   PMR spectroscopy, *Carbohyd. Res.* 33:339 (1974).
9. D.G. Streefkerk, M.J.A. de Bie and J.F.G. Vliegenthart,
   Conformational studies on pertrimethylsilyl derivatives
   of D-fructoses and oligosaccharides containing β-D-fructo-
   furanose units by 220 and 300 MHz PMR spectroscopy,
   Carbohyd.Res. 33:249 (1974).
10. D.G. Streefkerk, M.J. A. de Bie and J.F.G. Vliegenthart,
    The conformation of pentakis-OTMS-β-D-galactofuranose
    in solution as determined by 220 MHz PMR spectroscopy,
    Carbohyd. Res. 33:350 (1974).
11. J. Haverkamp, J.P.C.M. van Dongen and J.F.G. Vliegenthart,
    PMR and CMR spectroscopy of methyl-2,3,4,6-tetra-O-methyl-
    -α and β-D-glucopyranoside. An application to the identi-
    fication of partially methylated glucoses,
    Tetrahedron 29:3431 (1973).
12. J. Haverkamp, J.P.C.M. van Dongen and J.F.G. Vliegenthart,
    ¹³C and ¹H nuclear magnetic resonance spectroscopy of
    permethylated α- and β-galactopyranoses,
    Carbohyd. Res. 33:319 (1974).
13. J. Haverkamp, M.J.A. de Bie and J.F.G. Vliegenthart,
    ¹³C and ¹H NMR spectroscopy of permethylated gluco-, galacto-
    and mannopyranoses and 6-deoxy-analogues,
    Carbohyd. Res. 39:201 (1974).

14. J. Haverkamp, M.J.A. de Bie and J.F.G. Vliegenthart,
    ^{13}C and ^1H nuclear magnetic resonance spectroscopy of
    permethylated disaccharides, *Carbohyd. Res.* 37:111 (1974).

15. J.Haverkamp, L. Dorland, J.F.G. Vliegenthart, J. Montreuil
    and R. Schauer, Abstr. IXth Int. Symp. Carbohyd. Chem.,
    London, 281 (1978).

16. L. Dorland, J. Haverkamp, J.F.G. Vliegenthart, G. Strecker,
    J.-C. Michalski, B. Fournet, G. Spik and J. Montreuil,
    360-MHz ^1H nuclear magnetic resonance spectroscopy of
    sialyl-oligosaccharides from patients with sialidosis
    (mucolipidosis I and II), *Eur. J. Biochem.* 87:323 (1978).

17. B.L. Schut, L. Dorland, J. Haverkamp, J.F.G. Vliegenthart and
    B. Fournet, Structure of a complex asparagine-bound
    glycopeptide from horse pancreatic ribonuclease containing
    (2→3) and (2→6) linked sialic acid residues,
    Biochem. Biophys. Res. Comm. 82:1223 (1978).

18. G. Strecker, M.-C. Herlant-Peers, B. Fournet, J. Montreuil,
    L. Dorland, J. Haverkamp, J.F.G. Vliegenthart and
    J.-P. Farriaux, Structure of seven oligosaccharides
    excreted in the urine of a patient with Sandhoff's disease
    (GM_2 gangliosidosis-variant O), *Eur. J. Biochem.* 81:165
    (1977).

19. L. Dorland, J. Haverkamp, J.F.G. Vliegenthart, B. Fournet,
    G. Strecker, G. Spik, J. Montreuil, K. Schmid and
    J.-P. Binette, Determination by 360 MHz ^1H-NMR spectroscopy
    of the type of branching in complex asparagine-linked
    glycan chains of glycoproteins, *FEBS Lett.* 89:149 (1978).

20. B. Fournet, J. Montreuil, G. Strecker, L. Dorland, J. Haver-
    kamp, J.F.G. Vliegenthart, J.-P. Binette and K. Schmid,
    Determination of the primary structures of 16 asialo-
    -carbohydrate units derived from human plasma α_1-acid
    glycoprotein by 360-MHz ^1H-NMR spectroscopy and permethylation
    analysis, *Biochemistry* 17:5206 (1978).

21. L. Dorland, B.L. Schut, J.F.G. Vliegenthart, G. Strecker,
    B. Fournet, G. Spik and J. Montreuil, Structural studies on
    2-acetamido-1-N-(4-L-aspartyl)-2-deoxy-β-D-glucopyranosyl-
    amine and 2-acetamido-6-O-(α-L-fucopyranosyl)-1-N-(4-L-
    -aspartyl)-2-deoxy-β-D-glucopyranosylamine by 360 MHz
    proton magnetic resonance spectroscopy, *Eur. J. Biochem.*
    73:93 (1977).

22. K.-E. Falk, K.-A. Karlsson and B.E. Samuelsson, Proton nuclear
    magnetic resonance analysis of anomeric structure of glyco-
    sphingolipids; The globo-series (one to five sugars),
    Arch. Biochem. Biophys. 192:164 (1979).

23 K.-E. Falk, K.-A. Karlsson and B.E.Samuelsson, Proton nuclear
    magnetic resonance analysis of anomeric structure of glyco-
    sphingolipids; Blood group ABH-active substances,
    Arch. Biochem. Biophys. 192:177 (1979).

24. K.-E. Falk, K.-A. Karlsson and B.E. Samuelsson, Proton nuclear magnetic resonance analysis of anomeric structure of glyco-sphingolipids; Lewis-active and Lewis-like substances, *Arch. Biochem. Biophys.* 192:191 (1979).

ENZYMIC DEGRADATION OF GANGLIOSIDES

Yu-Teh Li, May-Jean King and Su-Chen Li

Department of Biochemistry
Tulane University School of Medicine and Delta
Primate Center, New Orleans, La. 70112 U.S.A.

As in the case for the enzymic degradation of sugar chains in other glycoconjugates, the following considerations should be made in investigating the enzymic degradation of saccharide chains in gangliosides. One should be aware of the fact that the same glycosidase isolated from different sources may vary considerably in its substrate specificity. In some instances, the reactions are greatly affected by the amount and the nature of the detergent or the activator to be included in the reaction mixture. In addition, one should consider the substrate concentration and the ionic strength of the buffer, as they may severely affect enzyme activity. It is also very important to realize that O-acetylated or O-methylated glycons are often resistant to glycosidases. Finally, one should bear in mind that the glycosidases used may be contaminated with one or several unexpected enzyme activities.

Gangliosides can be roughly classified into three categories. They are: a) hexosamine-free gangliosides such as sialylgalactosylceramide and sialyllactosylceramide, b) galactosamine-containing gangliosides such as GM1-ganglioside and GM2-ganglioside, and c) glucosamine-containing gangliosides such as sialylparagloboside. The purpose of this paper is to use specific examples to illustrate the enzymic degradation of these three categories of gangliosides.

Enzymic Hydrolysis of Hexosamine-Free Gangliosides

Generally sialic acids in hexosamine-free gangliosides are easily cleaved by bacterial as well as mammalian neuraminidases (Drzeniek, 1973). However, the substitution in the sialyl residue may render it resistant to neuraminidase. *Fig. 1* shows the

hydrolysis of two hematosides isolated from horse erythrocytes by
clostridial neuraminidase. Of these two hematosides, the fast-
moving one migrates slightly faster than sialylgalactosylceramide,
and the slow-moving one has been identified to be N-glycolylneuraminyllactosylceramide. The N-glycolylneuraminyllactosylceramide
can be easily converted into lactosylceramide by clostridial neuraminidase; however, the fast-moving ganglioside is completely resistant to neuraminidase because the sialyl residue is 4-O-acetylated.
Resistance of 4-O-acetylated sialic acid to bacterial neuraminidase

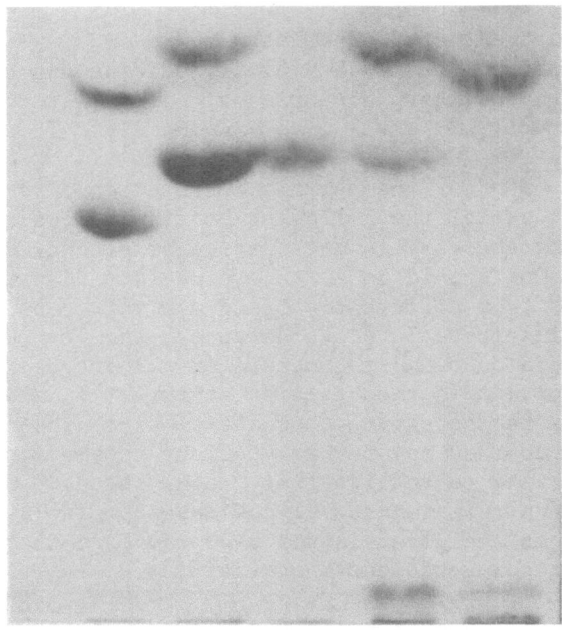

Fig. 1. Enzymic hydrolysis of equine hematosides. Incubation
          mixture (100 µl) contained 20 nmoles hematosides, 25 mM
          sodium acetate buffer (pH 5.5) and clostridial neura-
          minidase (6 munits). Amount of sodium taurodeoxycho-
          late (TDC) used was 100 µg. After overnight incubation
          at 37°C, the reaction mixture was dialyzed and lyophi-
          lized. The plate was sprayed with aniline/diphenyla-
          mine reagent. Solvent=CHCl$_3$/MeOH/H$_2$O (65:25:4).
          1, Standard GM4 and GM2; 2, two GM3 from equine
          erythrocytes; 3, Standard N-glycolyl GM3; 4, incubation
          of 2 with neuraminidase; 5, incubation of 2 with neura-
          minidase in the presence of TDC.

has been described by Schauer and Faillard (1968), and Hakomori and Saito (1969). Thus the O-acetylation of neuraminyl residue may play a role in controlling the removal of sialic acids from gangliosides. It should be noted that TDC stimulated the enzymic conversion of GM3 into lactosylceramide by clostridial neuraminidase (*Fig. 1*, lane 5).

Hydrolysis of Sialic Acids from Galactosamine-Containing Gangliosides

The galactosamine-containing gangliosides with polysialyl residues such as GD2, GD1a, GT etc., can be converted into GM2 or GM1 by bacterial as well as mammalian neuraminidases. However, the sialic acid in GM1 and GM2 is rather resistant to the neuraminidases from a wide variety of sources (Drzeniek, 1973). Wenger and Wardell (1973), and Li and Li (1973) showed that the sialic acid in GM1 or GM2 ganglioside can be hydrolyzed by clostridial neuraminidase in the presence of bile salts. We would like to show further that the hydrolysis of sialic acid from GM1 ganglioside in the presence of a bile salt is greatly affected by the ionic strength of the buffer. *Fig. 2* shows that in the presence of sodium taurodeoxycholate (TDC), GM1 can be easily converted into asialo GM1 by clostridial neuraminidase at buffer concentrations lower than 2.5 mM. When the buffer concentration is above 20 mM, this conversion is greatly reduced. The addition of TDC to the reaction mixtures increased their respective pH by 0.1 to 0.2 unit. Under our incubation conditions the enzyme activity detected is directly related to the ionic strength of the buffer and not the pH. In the absence of TDC very little GM1 could be converted into asialo GM1 by clostridial neuraminidase in the same range of ionic strength.

The effect of buffer concentration on the conversion of GM2 into asialo GM2 in the presence of taurodeoxycholate is shown in *Fig. 3*. Complete conversion of GM2 into asialo GM2 occurred when the buffer concentration was lower than 4 mM. At buffer concentrations higher than 8 mM the rate of conversion was greatly reduced. The thin-layer chromatogram shown in *Fig. 3* was sprayed with H_2SO_4. With this spray, the glycolipids GM2 and asialo GM2 appeared as violet bands while taurodeoxycholate with greenish-brown color moved immediately behind asialo GM2. After incubation, some of the taurodeoxycholate was converted into a compound having a faster mobility. During the course of using neuraminidases for the structural analysis of glycoconjugates, we found that clostridial neuraminidase preparations obtained from commercial sources often contain a number of enzymes such as protease and endo-β-N-acetyl-glucosaminidase (Chien et al., 1975). This study showed that the clostridial neuraminidase preparation obtained from Sigma Chemical Company was contaminated with an enzyme activity which converted

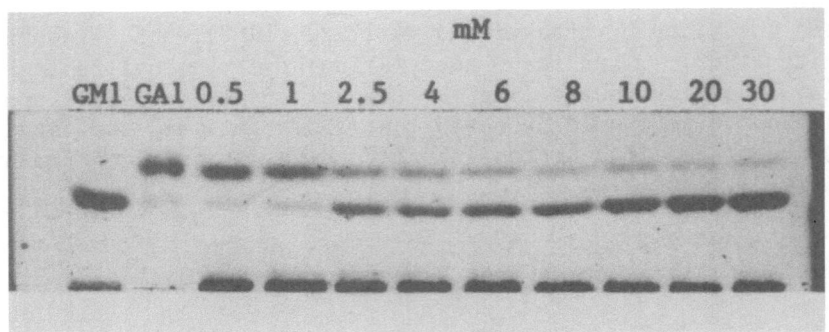

Fig. 2.   Effect of buffer concentration on the enzymic hydrolysis
of GM1 in the presence of sodium taurodeoxycholate (TDC).
Incubation mixture (200 µl) contained 4.2 nmoles GM1,
0.5 mM to 30 mM sodium acetate buffer (pH 5.5), clostridial neuraminidase (12 munits), and 200 µg TDC. Upon the
addition of TDC, the pH of the reaction mixture increased
by 0.1 to 0.2 unit. After overnight incubation at 37°C,
the reaction mixtures were processed according to the
condition described in *Fig. 1*. The plate was sprayed
with aniline/diphenylamine reagent. Solvent=$CHCl_3$/MeOH/
H_2O (60:35:8).

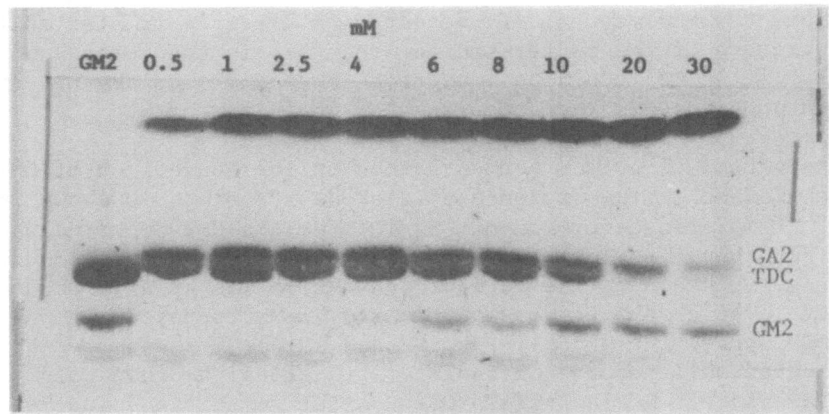

Fig. 3.   Effect of buffer concentration on the enzymic hydrolysis
of GM2 in the presence of sodium taurodeoxycholate (TDC).
For incubation conditions, refer to the legend for *Fig. 2*.
The concentration of GM2 was 3.3 nmoles. After overnight
incubation at 37°C, the reaction mixtures were processed
according to the condition described in *Fig. 1*. The plate
was sprayed with 50% H_2SO_4 in MeOH. Solvent=$CHCl_3$/MeOH/H_2O
(60:35:8).

taurodeoxycholate into deoxycholate (*Fig. 4*). From this example,
it is very clear that one should be alerted to the possible contamination by an unexpected enzyme activity in the enzyme to be used.

In the absence of taurodeoxycholate, there was considerable
conversion of GM2 into asialo GM2 at buffer concentrations between
1 mM and 10 mM (*Fig. 5*). This result may suggest that the sialic
acid in GM2 is more accessible for cleavage by clostridial neuraminidase than that in GM1.

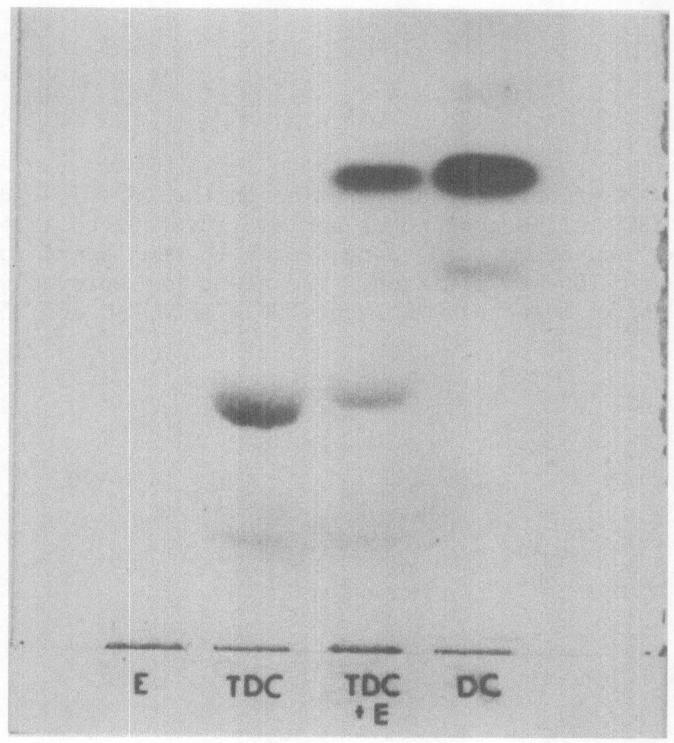

Fig. 4.   Enzymic conversion of sodium taurodeoxycholage (TDC) to
            sodium deoxycholate (DC). Incubation mixture (100 μl)
            contained 100 μg TDC, clostridial neuraminidase (6 munits),
            and 10 mM sodium acetate buffer (pH 5.5). Controls were
            set up with 100 μg TDC or 100 μg DC and no enzyme was
            added. After overnight incubation at 37°C, the reaction
            mixtures were processed according to the condition
            described in *Fig. 1*. The plate was sprayed with 50%
            H_2SO_4 in methanol. Solvent=$CHCl_3$/MeOH/H_2O (60:35:8).

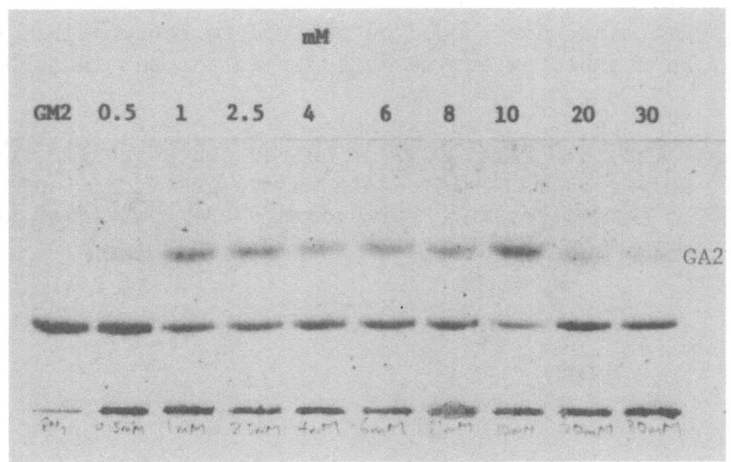

Fig. 5. Effect of buffer concentration on the enzymic hydrolysis
of GM2 in the absence of sodium taurodeoxycholate. For
the incubation conditions, refer to the legend for *Fig. 3*
except TDC was ommitted. The plate was sprayed with 50%
H_2SO_4 in methanol. Solvent=$CHCl_3$/MeOH/H_2O (60:35:8).

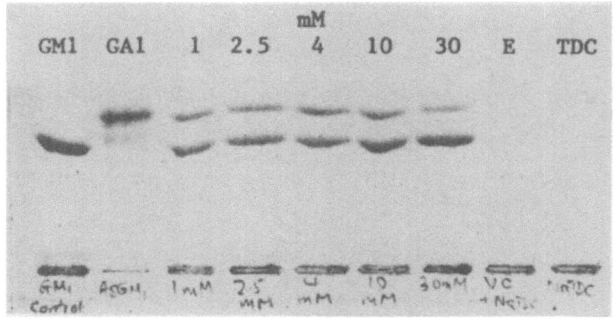

Fig. 6. Hydrolysis of GM1 by neuraminidase from *Vibrio cholerae*
at varying buffer concentrations in the presence of
sodium taurodeoxycholate. Incubation mixture (200 µl)
contained 4.2 nmoles of GM1, 1 mM to 30 mM sodium acetate
buffer (pH 5.5), 5 units of neuraminidase (*Vibrio cholerae*,
Grade B, 500 units/ml, Calbiochem), 1 mM $CaCl_2$ and 200 µg
of sodium taurodeoxycholate. The plate was sprayed with
aniline/diphenylamine reagent. Solvent=$CHCl_3$/MeOH/H_2O
(60:35:8).

Hydrolysis of Sialic Acid from GM1-Ganglioside by the Neuraminidase Isolated from *Vibrio cholerae*

In contrast to the report by Wenger and Wardell (1973) that the sialic acid in GM1-ganglioside is resistant to the neuraminidase isolated from *Vibrio cholerae*, we found that this neuraminidase could convert GM1 into asialo GM1 in the presence of sodium tauro-deoxycholate. As shown in *Fig. 6*, the hydrolysis of GM1 ganglioside by cholera neuraminidase is relatively insensitive to the ionic strength.

Effect of Ionic Strength on the Hydrolysis of GM2 Catalyzed by Human Hepatic β-N-Acetylhexosaminidase A

In order to investigate the chemical pathology of Tay-Sachs disease, the hydrolysis of GM2 ganglioside by β-N-acetylhexosamini-dases has been extensively studied. For some unknown reason, the rate of this hydrolysis differs considerably among the reported values. Recently we found that this inconsistency is in fact due to the differences in the conditions used for the *in vitro* hydrolysis. *Fig. 7* shows the effect of ionic strength on the enzymic conversion of GM2 into GM3 catalyzed by human hepatic β-N-acetylhexosaminidase A. The ionic strengths of the buffers used by different investigators are also included in the figure. It is evident that by lowering the ionic strength of the buffer from 0.2 to 0.01, the rate of GM2 hydrolysis increased about five times. This result would explain partly, if not completely, why the data obtained in various laboratories are so different from one another. It is of interest to note that the hydrolysis of the synthetic substrate, p-nitrophenyl-β-N-acetylglucosaminide by β-N-acetylhexosaminidase and the hydrolysis of GM1 by β-galacto-sidase are not affected by ionic strength.

The Activator Proteins for Enzymic Degradation of GM1 and GM2 Gangliosides

Some glycoconjugates are susceptible to only those glycosidases isolated from mammalian tissues. It has been reported that the hydrolysis of sphingoglycolipids by mammalian exoglycosidases requires the presence of a protein activator. The activator for glucocerebrosidase was reported by Ho and O'Brien (1971). Mraz, Fischer, and Jatzkewitz (1976) reported an activator which stimu-lates the enzymic hydrolysis of cerebroside sulfatide. We iso-lated an activator which stimulates the enzymic hydrolysis of GM1, GM2, and ceramide trihexoside (Li and Li, 1976). Recently, Hechtman and LeBlanc (1977) reported the presence of an activator which stimulates the enzymic hydrolysis of GM2 ganglioside. The following points concerning the role of the activator remain to be clarified: 1) Does one activator serve many glycosidases or does each

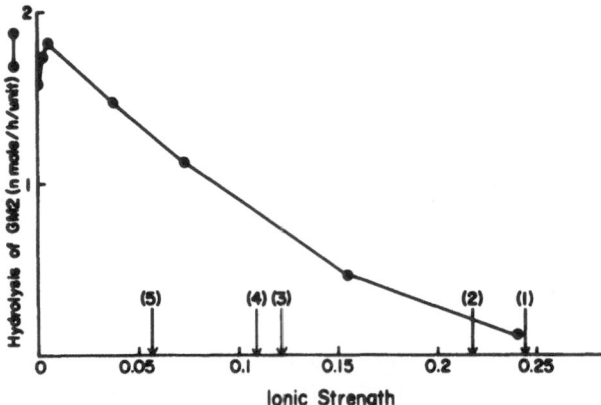

(1) Tallman, Brady, Quirk, Villalba and Gal (1974) JBC 249, 3489.

(2) Sandhoff (1970) FEBS Lett, 11, 342-344.

(3) Wenger, Okada and O'Brien (1972) ABB, 153, 116-129.

(4) Bach and Suzuki (1975) JBC, 250, 1328-1332.

(5) Kolodny, Brady and Volk (1969) BBRC, 37, 526-531.

Fig. 7.   Effect of ionic strength on the hydrolysis of GM2 catalyzed
by human hepatic β-N-acetylhexosaminidase A. Tritium
labeled GM2 (12 nmoles) was incubated with 0.5 unit of
human hepatic β-hexosaminidase A in 0.5 ml of 0.01 M acetate
buffer pH 4.6 containing 100 µg of sodium taurodeoxycholate
per 100 µl. Incubation was carried out at 37°C for 16 hrs.

glycosidase require a specific activator? 2) Does the activator
act upon enzyme or substrate? and 3) What is its physiological
function? In this section we present some of our most recent
results which may answer the first two questions.

    By careful fractionation of the human liver extract on DEAE-
Sephadex A-50 column, we were able to show that the hydrolyses of
GM1 and GM2 gangliosides by β-galactosidase and β-N-acetylhexosaminidase respectively were stimulated by two different activators.
Fig. 8(A) shows the elution profile of a crude activator fraction
on a DEAE-Sephadex A-50 column after it has been initially processed
through ammonium sulfate fractionation and Sephadex G-200 filtration. *Fig. 8(B)* shows that the activator which stimulated the GM2
hydrolysis was eluted by 0.05 M sodium phosphate buffer, pH 7.0,
while the activator for GM1 hydrolysis was eluted by 0.05 M sodium
citrate buffer, pH 6.0. We further found that the antiserum against
the activator for GM1 hydrolysis did not cross-react with the
activator for GM2 hydrolysis. This result suggests that each glycosidase may require a specific activator. However, it does not
exclude the possibility of one activator being able to serve more
than one enzyme.

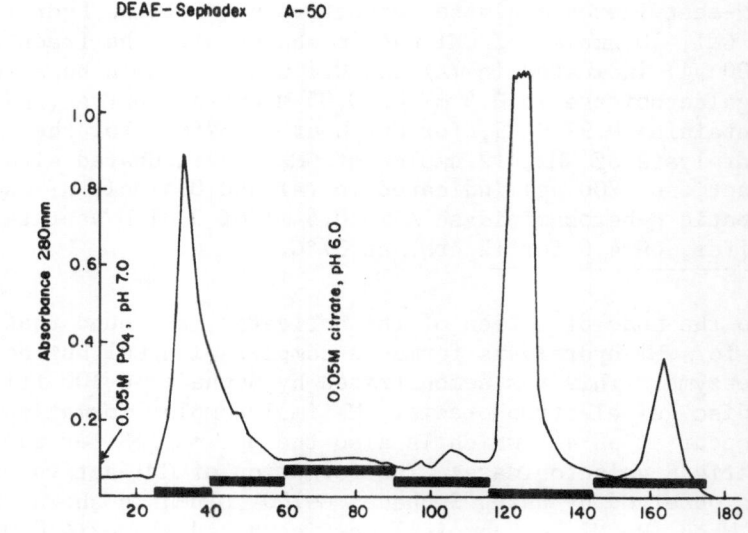

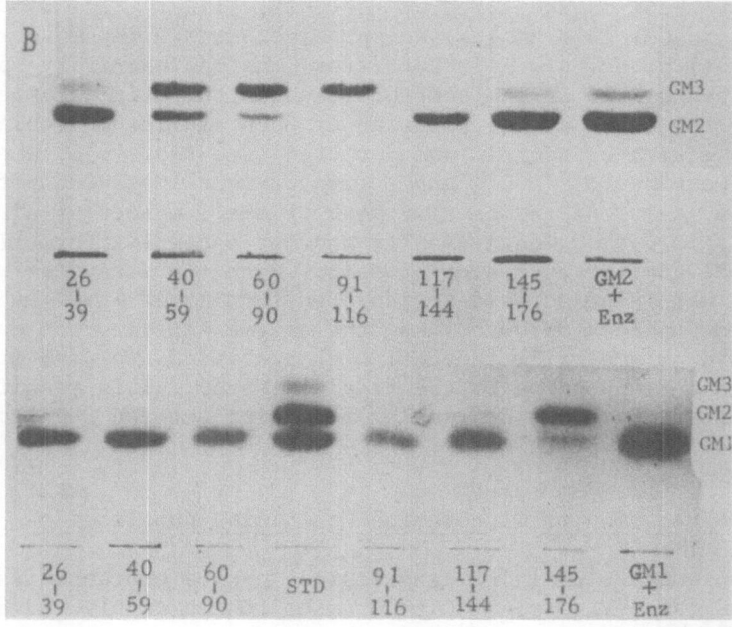

Fig. 8(A) — Fractionation of a crude activator fraction on
    DEAE–Sephadex A-50 column.

   (B) — Thin-layer chromatogram showing the effect of various
           fractions as shown in *(A)* on the hydrolysis of GM1 and
           GM2 catalyzed by human hepatic ß–galactosidase and

β–N–acetylhexosaminidase respectively. For the hydrolysis
of GM1, 10 nmoles of GM1 was incubated with the fractions
(200 μg) indicated in *(A)* and 0.1 unit of human hepatic
β–galactosidase in 0.5 ml of 0.05 M citrate buffer, pH 4.6
containing 0.9% NaCl, for one hour at 37°C. For the
hydrolysis of GM2, 12 nmoles of GM2 was incubated with the
fractions (200 μg) indicated in *(A)* and 0.5 unit of human
hepatic β–hexosaminidase A in 0.5 ml of 0.01 M acetate
buffer, pH 4.6 for 12 hrs. at 37°C.

As to the mode of action of the activator, we found that the
activator for GM1 hydrolysis formed a complex with GM1 but not
with the enzyme. This was demonstrated by Sephadex G–100 filtration and disc gel electrophoresis. Maximal complex formation was
found to occur at pH 4.5 which is also the optimal pH for the
human hepatic β–galactosidase. The formation of GM1–activator
complex as observed by using Sephadex G–100 column is shown in
Fig. 9a. When ^{3}H–GM1 (8.8 nmoles) was incubated alone in 0.05 M
sodium citrate buffer, pH 4.6, it was eluted as a sharp peak
immediately after the void volume of the column, indicating the
micellar form of this ganglioside. In this figure, the position
of the activator is indicated by the dotted line. *Fig. 9b* shows
that when ^{3}H–GM1 (8.8 nmoles) was pre-incubated with 40 μg
(1.8 nmoles) of GM1-specific activator, the radioactivity split
into two peaks: one at the same position as the original micellar
form, and the other at the position of much smaller molecular size
where the activator protein was detected (see *Fig. 9a*). When equal
molar proportions of ^{3}H–GM1 and activator were incubated before
filtration, only one radioactive peak of small molecular size was
detected (*Fig. 9c*). These results suggest a complex formation
between GM1 and the activator protein. When the activator was
incubated with β–galactosidase isolated from human liver and then
applied to Sephadex G–100 column, the enzyme activity was detected
at the void volume while the activator protein was eluted at the
same place corresponding to the free activator. This result
suggests that there was no complex formation between β–galactosidase
and the activator protein.

Enzymic Degradation of Glucosamine–Containing Gangliosides

Glucosamine-containing gangliosides can be expressed as sialyl→
[3Galβ1→4GlcNAc]₁₋₃3Galβ1→4Glc→Ceramide (Chien et al., 1978). The
sugar chain contains one to three keratan sulfate type repeating
units, →3Galβ1→4GlcNAcβ→. This repeating unit is in turn attached
to lactosylceramide. Sialic acid is usually attached to the C–3
position of the galactose at the non-reducing terminal. It is well
known that sialic acids in glucosamine-containing gangliosides are
easily hydrolyzed by bacterial neuraminidases. The sialic acid
free neutral sugar chain can in turn be cleaved by different

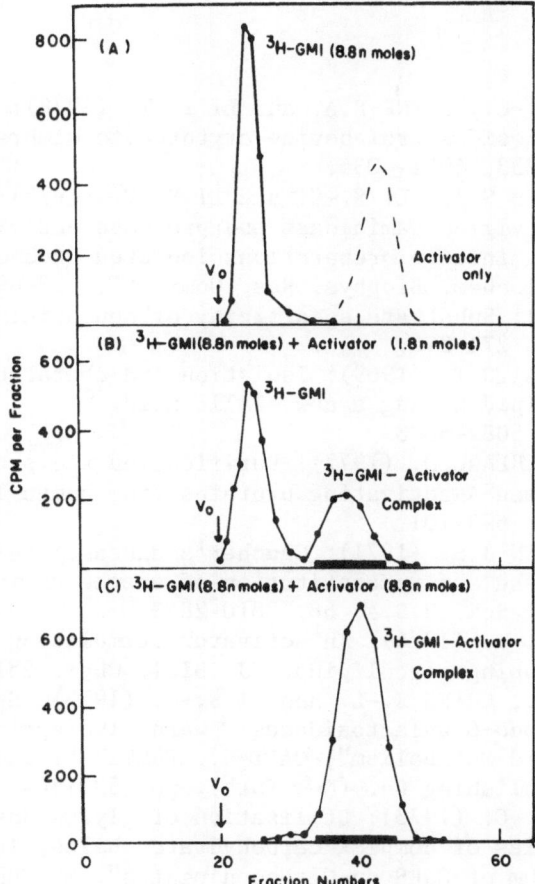

Fig. 9. Demonstration of complex formation between the activator and GM1 by Sephadex G-100 filtration.

exo-glycosidases such as β-galactosidase and β-N-acetylhexosamini-dase. Besides the actions of neuraminidase and exo-glycosidases on non-reducing terminals, the endo-β-galactosyl linkages in glucosamine-containing gangliosides can be cleaved by an endo-β-galactosidase isolated from *Escherichia freundii* (Nakagawa et al., 1979). For example, the sialylparagloboside, sialyl→3Galβ1→4Glc-NAcβ1→3Galβ1→4Glc→Cer can be converted into GlcCer and an oligo-saccharide sialyl→3Galβ1→4GlcNAcβ1→3Gal in the presence of tauro-deoxycholate. I have discussed the specificity of the endo-β-galactosidase at this symposium two years ago (Li et al., 1978). Recently we have extended our study to the structural analysis of polyglycosylceramide. We found that the polyglycosylceramide iso-lated from human erythrocytes was converted into glucosylceramide by endo-β-galactosidase in the presence of taurodeoxycholate. In the absence of a detergent the polyglycosylceramide can only be converted into several shorter chain glycosylceramides (Nakagawa et al., 1979).

BIBLIOGRAPHY

CHIEN S.-F., LI S.-C., LAINE R.A. and LI Y.-T. (1978): Characterization of gangliosides from bovine erythrocyte membranes.
J. Biol. Chem. 253, 4031-4035.

CHIEN S.-F., YEVICH S.J., LI S.-C. and LI Y.-T. (1975): Presence
of endo-β-N-acetylglucosaminidase and protease activities in the
commercial neuraminidase preparations isolated from *Clostridium
perfringens*. Biochem. Biophys. Res. Comm. 65, 683-691.

DRZENIEK R. (1973): Substrate specificity of neuraminidases.
Histochem. J. 5, 271-290.

HAKOMORI S. and SAITO T. (1969): Isolation and characterization of
a glycosphingolipid having a new sialic acid.
Biochemistry 8, 5082-5088.

HECHTMAN P. and LEBLANC D. (1977): Purification and properties of
the hexosaminidase A-activating proteins from human liver.
Biochem. J. 167, 693-701.

HO M.W. and O'BRIEN J.S. (1971): Gaucher's disease: Deficiency of
acid β-glucosidase and reconstitution of enzyme activity *in vitro*.
Proc. Nat. Acad. Sci. U.S.A. 68, 2810-2843.

LI S.-C. and LI Y.-T. (1976): An activator stimulating the enzymic
hydrolysis of sphingoglycolipids. J. Biol. Chem. 251, 1159-1163.

LI Y.-T., WAN C.C., CHIEN J.-L. and LI S.-C. (1978): Specificity of
some exo- and endo-β-galactosidases toward glycosphingolipids, in
"Enzymes of Lipid Metabolism", GATT S., FREYSZ L. and MANDEL P.,
Eds., Plenum Publishing Co. (New York), pp. 537-544.

LI Y.-T. and LI S.-C. (1973): Utilization of glycosidases for the
structural studies of complex carbohydrate chains, in "International Colloquium of CNRS on Glycoconjugates", MONTREIL J., Ed.,
Vol. 1, CNRS International Symposia (Lille, France), pp. 339-350.

MRAZ W., FISCHER G. and JATZKEWITZ H. (1976): The activator of human
cerebroside sulfatase. Hoppe-Seyler's Z. Physiol. Chem. 357,
201-206.

NAKAGAWA H., YAMADA T., KITAMIKADO M., LI S.-C. and LI Y.-T. (1979):
Isolation and characterization of an endo-β-galactosidase from a
new strain of *Escherichia freundii*. Federation Proc. 38, 631.

SCHAUER R. and FAILLARD H. (1968): Zur Wirkungsspezifität der
Neuraminidase. Das Verhalten isomerer N.O-Diacetyl-neuramin-
säureglykoside im Submaxillarismucin von Pferd und Rind bei
Einwirkung bakterieller Neuraminidase. Hoppe-Seyler's Z.
Physiol. Chem. 349, 961-968.

WENGER D.A. and WARDELL S. (1973): Action of neuraminidase from
Clostridium perfringens on brain gangliosides in the presence
of bile salts. J. Neurochem. 20, 607-612.

This work was supported by NSF Grant PCM 76-16881 and NIH Grant
NS 09626.

CHARACTERIZATION OF MICELLAR AND LIPOSOMAL DISPERSIONS

OF GANGLIOSIDES AND PHOSPHOLIPIDS

Y. Barenholz[1,3], B. Ceastaro[4], D. Lichtenberg[2], E. Freire[3], T. E. Thompson[3] and S. Gatt[1]
Dept. of Biochemistry and pharmacology[2], The Hebrew University-Hadassah Medical School, Jerusalem, Israel; Dept. of Biochemistry, University of Virginia School of Medicine, Charlottesville, Virginia, USA[3], Institute of Biological Chemistry, University of Milano School of Medicine, Milano, Italy[4].

Gangliosides are minor surface components of most mammalian cells, where they are located mainly in the outer leaflet of the lipid bilayer of the plasma membrane (1,2,3). They are also present in membranes of some enveloped viruses (4). In membranes they serve as receptors for various toxins, viruses, hormones and their pattern is often drastically changed in neoplasia (1,3,5). The gangliosides, which are part of the lipid bilayer of the membranes differ in their lyotropic behaviour from the phospholipids and cholesterol which constitute the main lipid components of the membrane (6,38). The membrane phospholipids have two long hydrophobic chains, an interface region and a relatively small inogenic head group. They are classified as "non-soluble swelling amphipaths (6), implying that they do not form micelles but disperse spontaneously in

ABBREVIATIONS: CMC - Critical micellar concentration, PCS-Photon correlation spectroscopy, DPH - 1.6 diphenyl, 1.3.5. hexatriene, DSC - Differential scanning calorimetry, Egg PC - Egg phosphatidyl choline, Egg PE - Egg phosphatidyl ethanolamine, DMPC - Dimyristoyl phosphatidylcholine, DPPC - Dipalmitoyl phosphatidylcholine, DSPC - Distearoyl phosphatidylcholine, NANA - N-acetyl neuraminic acid, NMR-Nuclear magnetic resonance, MLV - Multimaller large vesicles, SUV - Small, sonicated unilamellar vesicles, TNBS - Trinitrobenzene sulfonic acid. The ganglioside numeclature suggested by Svennerholm was used throughout this paper.

water, forming bilayered multilamellar large liposomes (MLV) or, upon
ultrasonic irradiation small unilamellar vesicles (SUV). In contrast,
the gangliosides are "soluble amphipaths" (6) which form micelles in
water (7). The fact that, in spite of their two long hydrophobic chains
the gangliosides are classified as "soluble" amphipaths is explained by
their large and highly negatively charged polar head group. Since
gangliosides and phospholipids differ in their state of aggregation, their
coexistance in membranes may affect or even disturb the bilayered
structure. This study aimed to investigate the mutual relations between
gangliosides and the main lipid components of the membranes. For this
purpose the structural and dynamic properties of dispersions composed
of well-characterized membrane lipids (either synthetic or of natural
sources) and well defined, pure gangliosides were studied with the aid of
physical and enzymatic methods.

EXPERIMENTAL PROCEDURE

MATERIALS

   GM1 was a gift of Prof. Svennerholm; GD1a was a gift of Prof.
Tettamanti; Egg Phosphatidylcholine and egg phosphatidylethanolamine
were purchased from Makor Chemicals, P.O. Box 6570 Jerusalem, Israel;
Cholesterol and trinitrobenzene sulfonic acid were purchased from Sigma
Chemical Co. The cholesterol was further purified either by recrystaliz-
ation from ethanol or using the procedure described by Cohen & Barenholz
(8). Dimyristoyl, dipalmitoyl and distearoyl phosphatidylcholine were
prepared, analyzed and purified as described by Lentz et al (9). All lipids
were analyzed by thin layer chromatography on silica gel plates heavily
loaded with the lipid sample to check for as low as 1% impurities (9). All
lipids were more than 99% pure. Neuraminidase of V. cholerae was
purchased from Serva, all other reagents were of analytical grade.
Organic solvents (analytical grade) were glass-distilled.

METHODS

   Micelles of GM1 in aqueous solutions were prepared as follows: A
solution in organic solvent (methanol, or chloroform: methanol: H_2O
60:35:5 by volume) was evaporated under reduced pressure, and the
solvents were further removed in a vacuum oven at room temperature.
The residue was dispersed in H_2O, D_2O or in the desired buffer. GD1a
micelles and mixed dispersions of GD1a with other lipids were prepared by
ultrasonic irradiation as described by Barenholz et al (10) except that the
lipids were mixed in chloroform: methanol 2: 1(v/v). The concentration

of phospholipids was determined according to Bartlet   (11), the concen -
tration of cholesterol was determined as described by Barenholz et al (12).
The concentration of ganglioside was derived from the N-acetyl neuraminic
acid content (13).   The total PE as well as that portion which interacts
with the non penetrating reagent TNBS was determined as described
elsewhere (10).

Hydrolysis of GD1a by Vibrio cholerea neuraminidase was followed
using dispersions containing 0.4mM GD1a and the desired concentration
of phospholipids in 25 mM Tris HCl pH 6.7 containing 0.5mM $CaCl_2$.
0.25ml. of this dispersions was incubated at $37^{\circ}C$ with 0.2 mUnits of
enzyme for 10 min for initial rates or 2mUnits and the specified time for
complete hydrolysis.   Reaction rates or total GD1a content available to
the enzyme were estimated from the released NANA (14, Ceastaro et al,
submitted).   The CMC of GM1 was determined as described by Yedgar
et al (15).   Using perylene as the flurophore, the CMC was independent
of the perylene concentration.   Fluorescence anisotropy of perylene
and 1.6 diphenylhexatriene (DPH) was measured as described elsewhere
(9,18,19).   All hydrodynamic measurements were performed as described
in the legend to table I.

RESULTS & DISCUSSION

Aqueous dispersion of pure gangliosides

The phase behaviour of hydrated bovine brain gangliosides was recently
studied by Curmatolo et al (38).   Over the hydration range of 18-50 wt %
water, mixed brain gangliosides exhibit a hexagonal mesophase structure
with the sugar group on the surface of the cylinder in contact with the
water.   At higher water contents an isotropic micellar solution is formed.
Our studies of pure ganglioside are limited to the latter range of the
isotropic micellar solution.

GM1 was used as a model for the behaviour of a pure ganglioside in
aqueous medium.   Previous data on the critical micellear concentration
of gangliosides were obtained with mixtures of brain gangliosides which also
contained traces of other lipids (21,22).   Yohe et al, determined the CMC
of pure GT1, GD1a, GM2 and GM1 and obtained the following similar
values of 9.9×10^{-5}M, 9.4×10^{-5}M, 7.4×10^{-5}M, 8.4×10^{-5}M respectively (7,23).   Since CMC of gangliosides are very low, their exact values
depend greatly on the sensitivity of the method used.   Using insensitive
methods may result in obtaining values of CMC which are too high.

In this study we used the flurophore perylene to determine the CMC of pure gangliosides in aqueous solution.   The main advantage of this method is its high sensitivity which permits using small quantities of the probe.   The measured CMC value was $1-2 \times 10^{-6}$ M and was almost identical for GM1 in pure water, or in 0.1M sodium or potassium acetate buffer, pH 5.6.   It also was only very little affected by 1 or 100 mM $CaCl_2$.   The fluorescence anisotropy of the perylene was dependent on GM1 concentration.   It was reduced from an anisotropy value of 0.22 at 1×10^{-6} M reaching a plateau of anisotropy value of 0.13×10^{-6} M at about 2×10^{-5} M GM1.   This suggests a change in the structure and size of the GM1 micelles in the above concentration range. The sensitivity of the procedures is not sufficient to rule out the formation of very small micelles below the above measured CMC which can not be monitored by the perylene due to the small aggregation number; therefore the true CMC may be even smaller and more than one step of aggregation may occur.   The formation of ganglioside micelles having a small aggregation number at lower concentration than 10^{-7} M was recently suggested by Kohn (personal communication).

The physical characteristics of GM1 micelles in pure water was calculated from the hydrodynamic parameters described in table I. All the values described in the table were obtained for the 10^{-3} M concentration range of GM1.

The population of the GM1 micelles in water is monodisperse and homogeneous as determined by the following parameters.

a)   The Schlieren pattern during the sedimentation in the analytical ultra-
       centrifuge.
b)   The shape and good fitting to a single exponent of the curve describing
       autocorrelation of light scattering as function of correlation time.
c)   The lack of dependency of the diffusion coefficient (D) obtained by the
       latter method on the scattering angle (24).
d)   The excellent agreement between the D values obtained by analytical
       ultracentrifuge and by PCS.

The above results depend on the procedure by which the GM1 was purified as well as the way the micelles were prepared (Barenholz & Gatt submitted).   The variation in the calculated values of the molecular weight may vary up to 2.5 times, variation occurs also in the degree of homogeneity of the dispersion.

The presence of Na^+, K^+ or Ca^{+2} ions had but a minor effect on the sedimentation coefficients, suggesting a relatively small effect of these ions on the state of aggregation of the ganglioside.

TABLE I: PHYSICAL CHARACTERIZATION OF GM$_1$ MICELLES
IN WATER

SEDIMENTATION COEFFICIENT	$(S^o_{20}, w)^a$	8.2 ± 0.3
DIFFUSION COEFFICIENT	$(D^o_{20}, w)^b$	$3.7 \pm 0.3 \times 10^{-7}$
" "	$(D^o_{20}, w)^c$	$3.7 \pm 0.5 \times 10^{-7}$
PARTIAL SPECIFIC VOLUME	$(\overline{V}, ml\ g^{-1})^d$	0.78 ± 0.002
MOLECULAR WEIGHT	$(daltons)^e$	$244,000$
VOLUME OF MICELLE	$(\overset{o}{A}^3)^f$	3.16×10^5
AGGREGATION NUMBER [g]		158
STOKES RADIUS	$(Rs, \overset{o}{A})^h$	$57.6\ \overset{o}{A}$
EQUIVALENT RADIUS[i]	$(Req, \overset{o}{A})^i$	$42.3\overset{o}{A}$
Rs/Req		1.36

a) Measured and calculated from the intercept of the concentration dependent S in 1-5mM GM1 concentration range in water as described by Yedgar et al (36).

b) Measured and calculated as described by Cooper et al from auto-correlation spectroscopy of Rayleigh scattered light (24). This will be referred to as photon correlation spectroscopy (PCS)

c) Determined using analytical ultra-centrifuge (28).

d) Determined by using a Paar density meter using GM1 in 1-10mM concentration range (10).

e) Calculated as described elsewhere (36)

f) Calculated from the following equation: $V = W \times \overline{V}$ where V is Volume of a micelle; W is the weight of one micelle in grams; W is obtained by dividing the molecular weight (in daltons) by Avogadro number (36). V is the partial specific volume in ml/gr.

g)h)i) Obtained as described by Yedgar et al 1974 (36).

The dynamics of the above GM1 micelles in D_2O was studied using proton NMR as described by Lichtenberg, Barenholz, Gat & Gatt (submitted). GM1, dissolved in DMSO-d_6 was used as a reference for a monomolecular (or very small aggregate) dispersion of GM1.

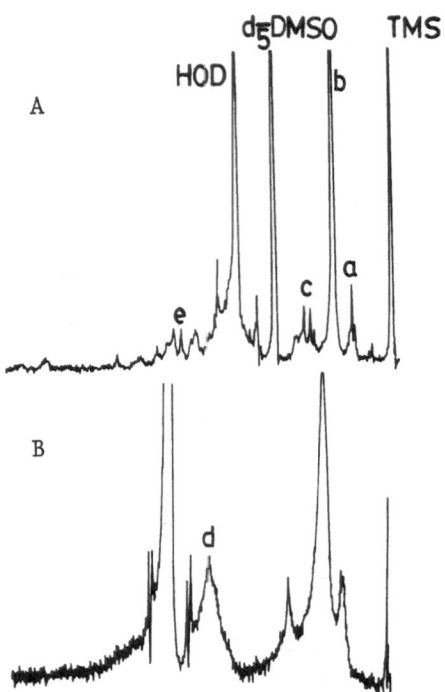

Fig. 1. High resolution proton NMR spectra of monomolecular solution of GM1 in DMSO-d6 (Fig. 2A) and of GM1 micelles in D_2O (Fig. 2B). The measurements were done on a Jeol MH-100 using a probe temperature at 31° C. For assignment of the signals see text. For more details see ref. 25.

Fig. 1 shows a comparison between the high resolution proton NMR spectra of GM1 in DMSO-d_6 (1A) and D_2O (1B). In DMSO, the observed peaks described from right to left were due to the terminal methyl protons (at 0.82 ppm, peak a), the protons of the methylene of the hydrocarbon chains (at 1.22 ppm, peak b) those of the carbohydrate N-acetyl protons (at 1.73 and 1.85 ppm, peaks c) and of the protons of the other sugar moieties (part of whom were obscured by overlapping with the HOD signal in DMSO and another part appeared at a field of 4.17 to 5.85 ppm, peaks e).

All proton NMR signals were broader in the aqueous dispersions than in those of the DMSO solutions (Fig. 1B). Thus, the linewidth of peak b increased from 5 to 20 Hz, peak a was broadened from 1 to 5 Hz and peaks c were unresolved (unless a higher field of 270 M Hz was applied (Lichtenberg et al submitted). One distinct broad signal (at 3.80 ppm, peak d), was derived from the sugar protons in D_2O, which were not observed in DMSO, probably because of overlapping with the HOD peak. Comparison of figures 1A and 1B suggests that micellization was accompanied by changes in the chemical shifts of at least some of the sugar protons, and a downfield shift of peaks a, b and c of up to 0.20 ppm. Although the reasons for the differences between the spectra obtained for micelles and monomers cannot be defined precisely, the observation that the signals of protons located within the hydrophobic as well as the polar regions of the GM_1 molecule similarly shifted and broadened indicates that micellization results in changes of the environment of all the segments of the GM1 molecules. Similar results were obtained for GD1a, GD1b & GT1 (not shown in the figure.) This suggests that, dissimilar to the polyolthoxyethylene chains in Triton-x-100 micelles (25) the motion of the polar groups in micelles of pure GM1 is somewhat restricted probably due to the tendency of ganglioside oligosacharide chains to interact with each other. The possibility of strong interaction, in the polar region of the ganglioside micelles, between the carbohydrage mpeities of adjacent molecules is also supported by ESR data (26) and by ^{13}C NMR and proton NMR data obtained for various pure gangliosides (27). No distinct thermotropic first order phase transition was observed for GM1 micelles in the range of 10-70°C using high sensitivity differential scanning calorimetry (20) or fluorescence depolarization of DPH (19).

Interaction of gangliosides with membrane lipids

Studies were done on the effect of lipid composition in dispersions composed of well defined phospholipids and pure gangliosides (with and without cholesterol) on the organization, dynamics and state of agreggation of the dispersion. Previous studies (39) using hand-shaken aqueous dispersions of mixtures of ox brain gangliosides and egg PC suggested that the state of aggregation of the dispersion depended on the mole ratio of its components. In this study, two different procedures for preparing the lipid dispersion were used. The first is the same one used for preparing MLV of phospholipids (39) the second was previously used for preparing SUV (10).

A crude "phase diagram" was obtained for systems where egg PC or egg PC : egg PE, 8:1 (M/M) were codispersed with GD1a by ultrasonic irradiation (second procedure). The presence of egg PE permits measuring bilayer integrity by exposure to the non penetrating reagent TNBS. This method can be used to determine the percentage of intact vesicles in the dispersion (Ceastaro, Barenholz & Gatt submitted). This was also confirmed from the size and molecular weight obtained by hydrodynamic measurements (see experimental procedure and legend to Table 1). It should be pointed out that it is not known if the above systems are in equilibrium.

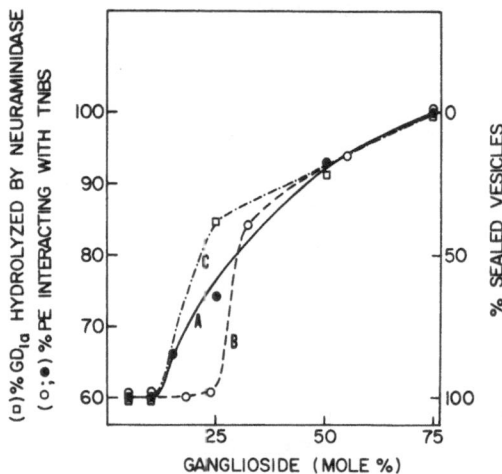

Fig. 2 "Phase diagram" like of phospholipids and gangliosides dispersions. Comparison of the effects of increasing content of GD1a on the exposure of egg PE to TNBS in dispersions made of egg PC:egg PE 8:1 (mole ratio) and GD1a (Curve A) and its effect on the availability of GD1a to Vibrio cholerea neuraminidase (Curve C). The latter was obtained from the plateau range of the time dependent enzymatic activity curves (Ceastaro et al submitted). Curve B describes the effect of increasing GM1 content on the exposure of egg PE to TNBS in lipid dispersions made of egg PC: egg PE: GD1a 8:1:1 (mole ratio) and GM1. All dispersions were made by ultrasonic irradiation (see experimental procedure).

The results are shown in Fig. 2 curves A and C. For dispersions composed of egg PC: egg PE, 8:1 (M/M) and increasing GD1a proportion, upto 10 mole %, both the exposure of egg PE to TNBS and the hydrolysis of GD1a by bacterial neuraminidase was 60% of the total. The molecular weight was $2 \pm 0.2 \times 10^{6}$ daltons. These values suggested that all lipid molecules are present in small, unilamellar vesicles (SUV) which are similar(inmolecular weight, size and lipid distribution between the two faces of the vesicle bilayer) to egg PC SUV (28) or egg PC: egg PE 9:1 SUV (Litman, unpublished results). This shows that, up to 10 mole % GDla has but a minor effect of the integrity and size of the SUV. Also, the GDla molecules are distributed between the two faces of the vesicle bilayer similar to the lipid molecules distribution. Increasing the mole fraction of GDla increased the following parameters a) the hetrogeneity of particles size b) the exposure of egg PE to TNBS c) the availability of GDla to hydrolysis by neuraminidase. When GDla reached a value of 25 mole % the population of particles became very hetrogeneous (based on PCS) with an average molecular weight of $9 \pm 2.5 \times 10^5$ daltons which indicated that a large portion of the lipid is present in the form of mixed micelles. The discrepancy between the GDla availability to neuraminidase (86%) and PE exposure to TNBS (71%) suggests that in the mixed micelles the relative content of GDla is higher than in the residual SUV. Studies by Harris & Thornton (27) using gel filtration chromatography and ^{13}C NMR are in good agreement with the above results.

As was shown by Fig. 2, curve B, GM1 is less potent in its capability to solubize bilayers than GDla (curve A). These dispersions were prepared by ultrasonic irradiation of increasing mole fractions of GM1, in the presence of a mixture of egg PC: egg PE: GDla in 8:1:1 mole ratio. The curves show that the integrity of the vesicles was disrupted only when the total gangliosides (GM1 plus GDla) were above 25 mole % while with GDla as the sole ganglioside the vesicles were already partly disrupted at 15 mole %. This conclusion was further supported, in a similar system composed of egg PC and GM1, prepared for studies of proton NMR (Lichtenberg, et al submitted). The integrity of the SUV was calculated from the degree of exposure of protons of N methyl egg PC choline to the paramagnetic ion Pr (29). Lichtenberg et al (submitted) have shown that only above 25 mole % GM1, the ratio between N-methyl protons exposed to Pr^{+3} (Iout) to those unexposed to these ions (Iin) exceeded the value of 2.1 ± 0.2 (which is characteristic for SUV of egg PC (28, 29) or mixed vesicles of egg PC and egg PE 9:1). At 30 mole % GM1 the I out/I in ratio increased to 4, suggesting that 80% of the PC molecules are exposed to Pr^{+3} ions as a consequence of formation of mixed micelles. It should be noted that this is similar to the degree of exposure of egg PE to TNBS (Fig. 2 curve B).

When cholesterol was included in the bilayered structures, more GDla was required to increase the exposure of the PE to TNBS than without cholesterol. Fig. 3 shows that without any cholesterol in the bilayer, the vesicle became leaky to TNBS above 10% mole of GDla. In comparison, when the lipid bilayer contained 37-50 mole % cholesterol all the vesicles

were intact up to at least 37.5 mole percent GD1a.   These results must
be attributed to the interaction between GD1a and cholesterol possibly by
hydrogen bonding between the 3-β-hydroxyl group of the cholesterol and
the amide bond carbonyl of the N-acyl sphingosine moiety of the GD1a.
However a possible effect of the carbohydrate moiety cannot be ruled out.

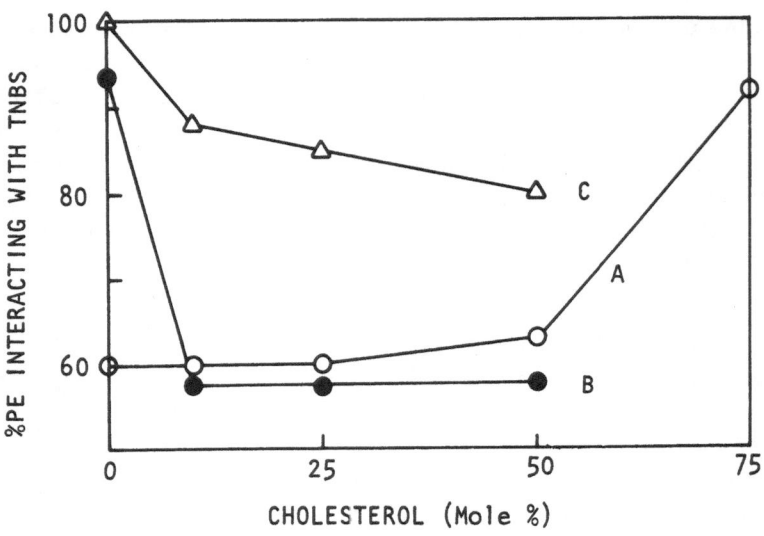

Fig. 3   The effect of increasing cholesterol content on the exposure of
egg PE to TNBS. In all cases the lipid dispersion was prepared by ultrasonic
irradiation as described in the experimental procedure.   The basic
composition of the dispersion composed of phospholipid mixture (composed of
egg PC: egg PE 8:1 mole ratio) and GD1a in the following phospholipids to
GD1a mole ratio curve A-9:1; Curve B-1:1; Curve C-1:9 (for more details
see experimental procedure).

    The above data might be used to explain the findings that gangliosides
are present mainly in the plasma membranes which have the highest
cholesterol content of all cell membranes.   Since it is very likely that
transmembrane dislocation of gangliosides is very slow (probably due to
their carbohydrates moieties) their presence in internal membranes of
cells which are poorer in cholesterol is rather small. This might also
explain the mutual accumulation of cholesterol together with the gangliosides
in the gangliosidoses (40).

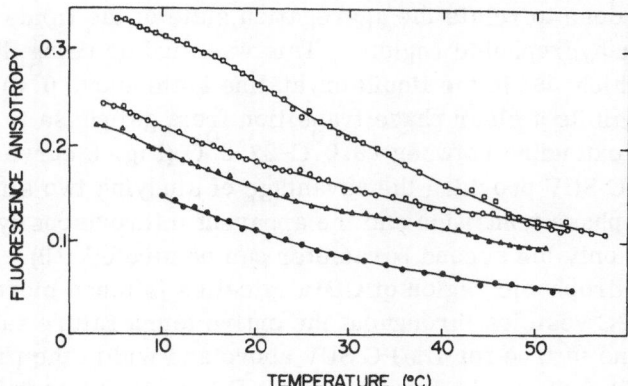

<u>Fig. 4</u>   The Temperature dependent fluorescence anisotropy of DPH
in lipid dispersions of egg PC:GD1a prepared by ultrasonic irradiation
(for more details see text).   The following lipid dispersions were used
egg PC:GD1a ● 100:0; △ 80:20; ○   50:50; ▢   0:100.

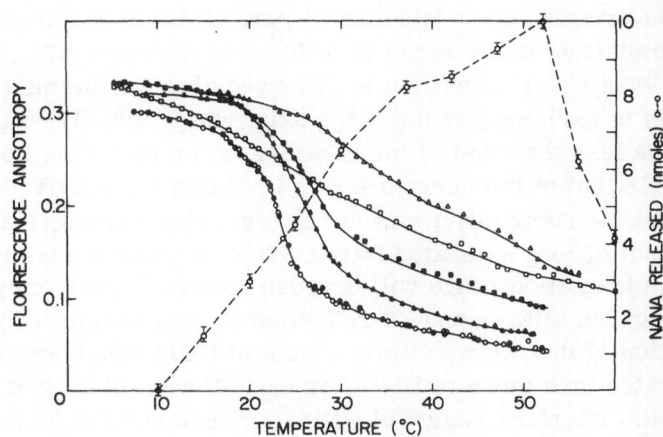

<u>Fig. 5</u>   Temperature dependent fluorescence anisotropy of DPH in
lipid dispersions of DMPC:GD1a and its relation to initial rate of GD1a
hydrolysis by Vibrio cholerea neuraminidase.   The lipid dispersions were
prepared as described in "Methods" in the medium used for neuraminidase
treatment.   The following lipid dispersions were prepared: DMPC: GD1a
(M/M) Curve A ○——○ ; – 100:0; Curve A'●——●100:0 without Ca^{+2} ; curve
B ▲—▲ ; – 92:8; curve C ■——■   82:18; curve D ▢——▢ 58:42;
curve E △—△ 0:100   Curve B'○---○ describes   the effect of temperature
on the initial reaction rates of GD1a hydrolysis by V. cholerea neuraminidase
using the lipid dispersion described in curve B. For more details see
"Methods" and "Results."

Studies were done to relate the aggregation state of the lipids and the dynamics of their hydrophobic region.   This was studied using SUV of either egg PC (which is in the liquid crystaline state above 0° C) or DMPC, which exhibits a clear phase transition from gel phase to liquid crystaline phase extending between 19.0° C-27.3° C (Fig. 5 curve A & A'). Analysis of DMPC SUV provided the advantage of studying two simultaneous parameters, the phase transition and the apparent microviscosity, while with egg PC SUV only the second parameter can be studied (19).   Fig. 4 shows that the hydrophobic region of GD1a micelles is much more rigid than that of egg PC vesicles throughout the entire temperature range of $0-60^\circ$ C;   the same is true for DMPC SUV above and within the phase transition (Fig. 5).   When the bilayer of egg PC or DMPC contained 10-20 mole % of GD1a the only effect was an increase in the apparent microviscosity without much effect on the shape of the temperature-dependent fluorescene anisotropy curves (Fig. 4 & 5).   For mixed dispersions of egg PC and GD1a, even at 50 mole % GD1a the only effect of this compound was an increase in the anisotropy and therefore in the apparent microviscosity (19).   For mixed dispersions of DMPC and GD1a at 50 mole percent GD1a the phase transition was broadened considerably and became less pronounced.   Micelles of pure GD1a do not show any distinct phase transition in the range of $5-60^\circ$ (see also ref. 27), but when mixed with DMPC a phase transition is observed also for the mixed micelles.   This is explained by the large axial ratios (30,31), their curvature will be less than that of the micelles of the pure gangliosides and will resemble that of the curvature of a lipid bilayer in SUV.   This will enable to get the cooperativity required for a phase transition.   Due to the high content of long saturated fatty acids of the GD1a (mainly stearic acid) the transition range will broaden toward higher temperature. A rigidification of the bilayer and/or reduction in free volume (32) is due to the contribution of the two paraffinic chains of GD1a (which are longer and more staurated than those of DMPC or egg PC) as well as to a possible contribution of the interface region of GD1a.   These results indicate that GD1a and DMPC mix   without the occurrence of a distinct phase separation between the two.

These effects are a result of mixing the ganglioside with the phospholipid and not due to the radius of curvature.   This is deduced from the fact that similar results were obtained for MLV composed of DPPC and GD1a.   As measured from the thermotropic behaviour using high sensitivity differential scanning calorimetry (20,33), micelles of GD1a do not show any thermotropic phase transition in the range of $10-70^\circ$ C.   Fig. 6 shows that the endotherm of DPPC is characterized by a ΔH of **8.5** Kcal/mole, a T_m (the temperature of maximum change) at 41.4° C and a $T_{\frac{1}{2}}$ (the width of transition at half the endotherm height) of 0.5° C (Fig. 6 endotherm 1).   By having 6 mole % GD1a this is shifted to a T_m of 42.1° C and broadened to a $T_{\frac{1}{2}}$ of 1.4° C without much effect of the ΔH (Fig. 6 endotherm 2A) Increasing GD1a to 13 mole% results in a further shift in Tm to 43.5°C. $T_{\frac{1}{2}}$ broadened considerably to 5°C and the ΔH was not much affected although the endothern became asymmetric.

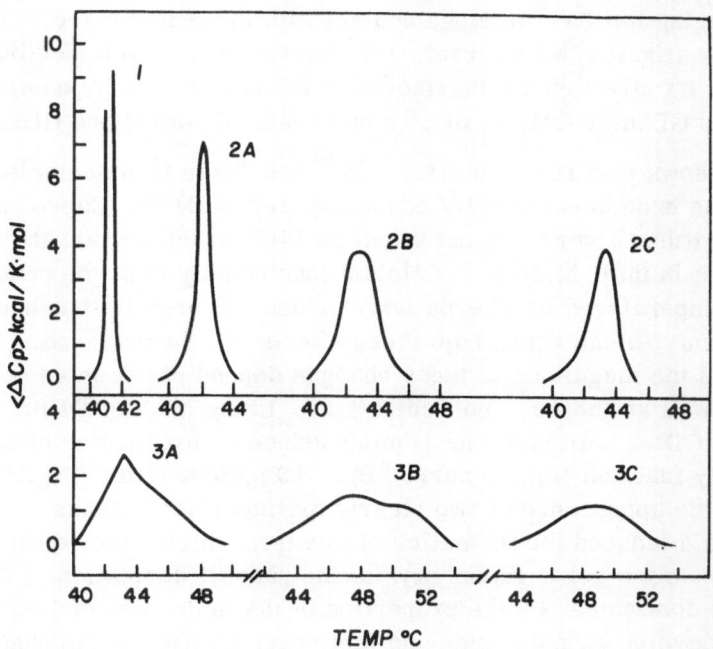

Fig. 6 High sensitivity DSC of MLV of the following composition
   1) Pure DPPC (10 mM) with or without 10 or 20 mM $CaCl_2$ 2A-DPPC: GD1a 94:6, 2B-As 2A + 10mM $CaCl_2$,2C As 2A+20mM $CaCl_2$ 3A-DPPC: GD1a 87:13; 3B - As 3A + 10mM $CaCl_2$, 3C as 3A + 20 mM $CaCl_2$.
Total lipid concentration was 10 mM throughout.

The results can again be explained by assuming that GD1a induced a reduction in the cooperativity of the transition by introducing acyl chain hetrogeneity, through the addition of longer acyl chain to the bilayer. The endotherm asymmetry at 13 mole % GD1a suggests a phase transition which is more complex than a simple first order transition.   (Fig. 6 1B, 1C, also see ref. 34).   It also obviates the possibility that a new species, such as mixed micelles,is formed in the dispersion (as was suggested for egg PC: GD1a dispersions having more than 10 mole % GD1a (See Fig 2 Curve A).   The possibility of a phase separation cannot be ruled out, though the degree of asymmetry of the endotherm is much too high to account for a simple separation of domains of GD1a.   The fact that Δ H was unaffected by the presence of GD1a suggests that either component (namely DPPC and GD1a) participates in the phase transition.

Effect of polyvalent ions on ganglioside_ phospholipid mixtures.

Ca^{+2} is considered as a "second messenger" and its permeability through membranes is a key step in many biological activities.   Gangliosides have a   prominent negative charge and therefore may be a vehicle bringing the Ca^{+2} in close contact with the cell surface.   In this study we investigated the interaction of calcium ions with a ganglioside by following its effect on the thermotropic behaviour of MLV composed of DPPC and GD1a or GM_1, or of SUV composed of DMPC and GD1a.

Fig. 6 shows that 10 or 20 mM of Ca^{+2} ions have no detectable effects on the endotherm of MLV containing 10mM DPPC (Curve 1).   In comparison it has a very distinct effect on MLV which contain DPPC as well as GD1a in their bilayer.   Calcium ions usually shift the endotherm to higher temperatures or alternatively induce a change in the shape of the endotherm;   some times both these effects are encountered.   In the present case the magnitude of these changes depend on the mole percent of GD1a as well as the Ca^{+2} concentration.   Using MLV of DPPC containing 6 mole % of GD1a, calcium ions (10mM) induced a broadening of the excess heat capacity function (Fig. 6 curve 2B).   13 mole % GD1a (Fig. 6 curve 3B) caused the appearance of two clearly distinguishable peaks.   Raising Ca^{+2} to 20mM induced the formation of clusters which affected the transition from the gel to liquid-crystalline phase;   formation of the latter seems to be dominated by the properties of the molecules of DPPC. The presence of GD1a molecules and their participation in the phase transition reduces the overall enthalpy change to the value expected for a system containing only DPPC molecules.   This effect of Ca^{+2} is carried out isothermally.   (Fig. 6 curves 2C, 3C; Freire & Barenholz submitted).   Morphological changes in the lipid structure could not be detected by freeze fracture electron microscopy thereby ruling out the formation of large domains due to the presence of GD1a.

The above results may be interpreted in several possible ways, as follows: a) The calcium-induced change in the bilayer dynamics observed in the presence of GD1a is a consequence of the formation of mini aggregates which are in a dynamic equilibrium with individual GD1a molecules lying in the bilayer plane.   The calcium ions interact with the two sialic acid residues of a GD1a molecule thereby neutralizing its negative charge and permitting interaction, by hydrogen bonds, in both the polar headgroup and the interface region of adjacent GD1a molecules.   Such intermolecular interactions are impossible in the absence of Ca^{+2} because of the strong repulsion between adjacent GD1a molecules in the bilayer plane.   b) Calcium ions induce a change in the structure of the GD1a molecules, which remain in the all-trans-configuration without participating directly in the phase transition, though they might still have an indirect effect on the DPPC molecules which is expressed by a shift of the phase transition to higher temperatures(Freire & Barenholz submitted).   The effect of Ca^{+2}, observed when GD1a was present in the lipid bilayer is not due to changes in the radius of curvature.   This is borne out from the findings that similar effects on the membrane dynamics were observed for DMPC SUV of minimal size (200-250A diameter), as measured by the fluorescence depolarization of DPH (Barenholz et al submitted).   The effect of Ca^{+2} on the fluorescence anisotropy occurred only when GD1a was present in the bilayer.   Thus, at 25°C for 0.4mM DMPC SUV containing 8 mole % GD1a the fluorescence anisotropy increased from 0.196 without Ca^{+2} to 0.216 in the presence of 0.5mM Ca^{+2} and the Tm was elevated by 1.5° suggesting a reduction of the free volume of the hydrophobic region.   Our preliminary results with DPPC MLV containing the ganglioside GM1 indicate that the Ca ions can interact only intermolecularly by bridging two adjacent GM1 molecules similar to the case of phosphatidyl serine (41).

It is very likely that the interaction of Ca^{+2} with the bilayer is primarily due to its strong interaction with the sialic acid residues of the ganglioside.   This increases the concentration of Ca^{+2} ions near the plane of the PC head groups forming a fixed Ca^{+2} layer which further enables a stronger interaction of these ions with the phosphorus of the phosphorylcholine moiety whose interaction is usually weak.   This results in increased packing density of the PC molecules in the bilayer thereby reducing the free volume in the bilayer and shifting the phase transition to higher temperatures.   Direct effects of ganglioside on the choline N-methyl protons (Fig. 7) and phosphorus group of the phosphorylcholine moiety was suggested by the finding of stronger interaction with Pr^{+3} ions, when ganglioside was present in egg PC SUV (Lichtenberg et al, submitted).

The organization of GD1a in lipid bilayers can also be studied with the aid of enzymes.   It was shown that the rate of hydrolysis by bacterial neuraminidases was enhanced 1-2 orders of magnitude when the GD1a molecules were part of an intact lipid bilayer as compared to similar activity on GD1a micelles.   This effect was explained by the even and random distribution of GD1a molecules in the bilayer plane, which reduces the interactions between adjacent GD1a molecules. Such interactions and therefore also neuraminidase activity are a function of the relative content of GD1a in the lipid dispersion (Ceastaro et al. submitted.) See also paper by Gatt et al., this book .

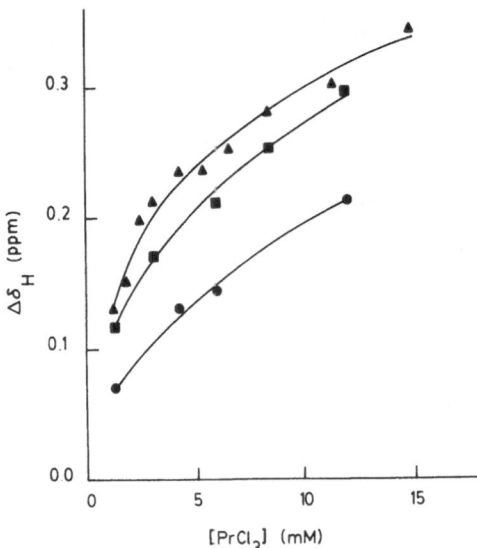

Fig. 7. The effect of GM1 on the praseodymium induced shift ($\Delta\delta_H$) of
the $-N^+(CH_3)_3$ protons of the egg phosphatidyl choline molecules in the
outer surface of the small unilamellar vesicles.   Egg PC concentration
was 20 mM throughout.   The symbols ●——● represent pure egg PC SUV
(without GM1)■——■ and ▲——▲ represent the shifts obtained when
GM1 (7.7 and 14.3 mol. %) respectively was added to the premade SUV
prior to the PrCl₃ addition.   All spectra were measured on a Jeol
MH-100 using a prob temperature at 31°C. For more details see ref. 25.

    It was shown that the activity of some lipolytic enzymes is enhanced
at the phase transition of the lipid bilayer (8, 35).   A dependence of
phospholipase A₂ activity on the phase transition of the lipid bilayer
could be induced by including sphingomyelin molecules in PC bilayers
(Cohen et al., submitted).   The experiments shown in Fig. 5 curve B'
describe the activity of neuraminidase on GD1a codispersed with DMPC.
At 8 mole % GD1a the enzymatic activity on these SUV's was unrelated
to the phase transition of the phospholipid.   Thus optimal rates were
obtained at 48°C while the phase transition was in the range of 19.0°C-
27.2°C.   This seeming discrepancy might have two separate explanations
a s follows:-

a) There is a phase separation, GD1a forming a separate phase; the
temperature - dependent enzymatic activity mirrors the properties of
the separate GD1a phase.
b) GD1a and PC molecules from a mixed phase without a phase separation
but the N-acetyl neuraminyl linkage which is hydrolysed by the enzyme,

being deep in the aqueous medium, is unaffected by the phase transition. The experimental curves mirror the properties of the enzymatic protein. Though we have no direct information on the way GD1a is distributed in the plane of the lipid bilayer numerous indications support the second interpretation.   Thus, the temperature-dependent neuraminidase activity on GD1a, codispersed with DPPC or DSPC as SUV, is almost identical to that obtained when the GD1a was a part of a DMPC SUV (50° C). This obviates a temperature-dependent enzyme activity.   Bridging by calcium ions of two sialic acid residues of the same GD1a molecule permits a random distribution of the ganglioside in the plane of the bilayer.   In the absence of Ca^{+2}, the electrostatic repulsion the GD1a molecules will cause a distribution in which the distance between adjacent GD1a molecules is maximal.   In the presence of Ca^{+2} and at low GD1a concentration in the bilayer, the latter will distribute evenly in the bilayer plane.   This is further supported by studies on spin-labelled gangliosides as a function of mole % ganglioside codispersed with PC (26).   The possibility should not be overlooked that interaction with proteins in a biomembrane might induce aggregation of ganglioside molecules.

Conclusions

1) When present in a lipid bilayer gangliosides may reduce bilayer integrity. The threshold-value of this effect depends on the molecular species of the ganglioside.
2)   Inclusion of cholesterol stabilizes the membrane and a higher ganglioside content is required to attain the above threshold.
3)   When present in a lipid bilayer, even a small mole fraction of ganglioside may affect the dynamics of the membrane.   It may also modulate the properties of the membrane when external effectors, such as calcium ions are present.   The latter effect depends on the molecular species of the ganglioside.

Acknowledgements

This work was supported in part by United States- Israel Binational Science foundation 1688 and by U.S. Public Health Service Grants NS02967 and GM 23573.

The authors thank Dr. C. Huang for stimulating discussions.

REFERENCES

1) Fishman, P.H., & Brady, R.O., (1976) Science 194, 906.

2) Hansson, H.A., Holmgren, J. & Svennerholm, L., (1977) Proc. Natl. Acad. Sci. (US) 74, 3782.

3) Wallach, D.F., (1975) in Membrane Molecular Biol. of Neoplastic Cells, Elsevier Pub. N.Y.

4) Klenk, H.D. (1973) in Biological Membranes, Chapman, D. and S. Wallach, D.F.H., eds. Academic Press N.Y. P. 145.

5) Yamakawa, T., & Nagai, Y., (1978) Trends in Biochemical Sciences 3, 128.

6) Small, D.M., (1970) Fed. Proc. 29, 1320.

7) Yohe, H.C., Roarck, D.E., Rosenberg, A., (1976) J. Biol. Chem. 251, 7083.

8) Cohen, R., Barenholz, Y., (1978) Biochem. Biophys. Acta 509, 181.

9) Lentz, B.R., Barenholz, Y., Thompson, T.E., (1976) Biochemistry 15, 4521.

10) Barenholz, Y., Gibbs, D., Litman, B.J., Goll, J., Thompson T.E., & Carolson, F.D., (1977) Biochemistry 16, 2806.

11) Bartlett, G.R., (1959) J. Biol. Chem. 134, 466.

12) Barenholz, Y., Patzer, E.J., Moore, N.F. & Wagner, R.R. (1978) in the Enzymes of Lipid Metabolism, Gatt. S., Freysz, L., & Mandel, P., ed. Plenum N.Y. P. 45.

13) Svennerholm E. & Svennerholm, L., (1963) Biochim. Biophys. Acta 70 688.

14) Warren, L., (1959) J. Biol. Chem. 234, 1971.

15) Yedgar, S., Hertz, R. & Gatt, S., (1974) Chem. Phys. Lipids 13, 404.

16) Shinitzky, M., Dianoux, A.C., Gitler, C. & Webber, G., (1971) Biochemistry 10, 2106.

17) Shinitzky, M. & Barenholz, Y., (1974) J. Biol. Chem, 249, 2652.

18) Hertz, R., & Barenholz, Y., (1977) J. Coll & Iner. Sci. (1977) 60, 188.

19) Shinitzky, M., & Barenholz, Y., (1978) Biochim. Biophys. Acta 515, 367.

20) Suurkuusk J., Lentz, B.R., Barenholz, Y., Biltonen, R.L., & Thompson T.E., (1976) Biochemistry 15, 1393.

21) Gammack, D.B., (1963) Biochem. J. 88 373.

22) Howard, R.E., Burton, S., (1964) Biophys. Biochem. Acta 84, 435.

23) Yohe, H., Rosenberg, A., (1972) Chem Phys. Lipids 9 279.

24) Cooper, V.G., Yedgar, S., & Barenholz, Y., (1974) Biochim. Biophys. Acta 363, 86.

25) Lichtenberg, D., Yedgar, S., Cooper, V.G., & Gatt, S., (1979) Biochemistry In Press

26) Sharom, F.J. & Grant, C.W.H., (1978) Biochim. Biophys. Acta 507, 280.

27) Harris, P.L., & Thronton, E.R., (1978) J. Amer. Chem. Soc. 100, 6737.

28) Mason, J.T. & Huang, C., (1978) Ann. N.Y. Acad. Sci. 308, 29.

29) Bergelson, L.K., & Bursukov, L.I. (1977), Science, 197, 224.

30) Israelachvili, J.N., Mitchell, D.J., & Ninham B.W., (1976) Chem. Soc. Faraday Trans. II 72, 1525.

31) Winsor, A., (1968) Chemical Rev. 68, 1.

32) Lakowitz, J.R., Prendergast, F.G., & Hogen, D., (1979) Biochemistry 18, 508.

33) Barenholz, Y., Suurkuusk, J., Monncastle, D., Thompson, T.E., & Biltonen, R.L., (1976) Biochemistry 15, 2441.

34) Jacobson, M.B., & Sturtevant, J.M., J. Biol. Chem. 252, 4749.

35) Op den Kamp, J.A.R., Kauerz, M. Th., & Van Deenen L.L.M., (1975) Biochim. Biophys. Acta 406, 169.

36) Yedgar, S., Barenholz, Y., & Cooper V.G., (1974) Biochim. Biophys. Acta 363, 111.

37) Quinn, P.J., & Barenholz, Y., (1975) Biochem., J., 149, 199.

38) Curmatolo, W., Small, D.M., & Shipley, G.G., (1977) Biochim. Biophys. Acta 468, 11.

39) Hill, M.W., & Lester, R., (1972) Biochim. Biophys. Acta 282, 18.

40) O'brian, J.S., in the metabolic basis of inherited disease, Stanbury, J.B., Wyngaarden, J.B., & Fredrickson, D.S., McGraw Hill Book Com. N.Y. P.841

41) MacDonald, R.C., Simon, S.A., & Bare, E., (1976) Biochemistry 15, 885.

STRUCTURAL MODIFICATIONS OF GANGLIOSIDES IN SYNAPTIC MEMBRANES

K.C. Leskawa and A. Rosenberg

Department of Biological Chemistry
The Milton S. Hershey Medical Center
The Pennsylvania State University
Hershey, PA 17033

INTRODUCTION

It has been hypothesized that neuronal gangliosides are involved in receptors for biogenic amines (1-3) and certain neurotoxins (4,5), and it is generally, although not universally (6), believed that, in the central nervous system, gangliosides are to be found concentrated in synaptic membranes. For further information, we have prepared purified intact and hypoosmotically ruptured synaptosomes from beef brain in order to probe the structural disposition of gangliosides in these preparations by surface labelling techniques for sialosyl and galactosyl components, and we have made preliminary observations on the modifying influence of calcium ions.

RESULTS AND DISCUSSION

The Nature of Structural Sialic Acid in the Synaptosomal Membrane

A zonal rotor procedure for the preparation of synaptosomes from bovine brain (7,8), based directly upon a commonly-used sucrose density gradient arrangement (9), has revealed detailed information concerning the synaptosomal (P_2B) fraction. This technique has enabled us to separately analyze particles along the P_2B profile from the leading, more bouyant, edge to the trailing, less bouyant, edge. Much of the contamination within this fraction was found in the two "tails" of the P_2B peak. Exclusion of these areas and collection of only the densest fractions from the zonal rotor yields a purified synaptosome preparation. Myelin contamination was restricted to the leading, more bouyant, and free mitochondria mainly to the trailing, less bouyant, tail. The entire P_2B peak entered into the next dense step in the gradient, that is, the center of the P_2B profile did not correspond to the 0.8/1.2 M sucrose interface. When separate samples along the gradient were analyzed for total bound sialic acid (10), the maximum was found in the area of the sucrose density interface. Analysis (11) demonstrated sialic acid bound to glycoprotein. Sialic acid bound to glycolipid, however, closely followed the bell-shaped curve of the synaptosomal profile. The finding that the concentration of lipid-bound sialic acid parallels the synaptosomal profile encouraged us to more closely examine the individual gangliosides in this fraction. The results are presented in Figure 1. Ganglioside G_{M1} content reaches a maximum before the P_2B peak, i.e. the interfacial area. This result may reflect the fact that this area also contains the myelin contamination in the P_2B fraction (7,8). Ganglioside G_{D3} appears to decrease as one examines toward the less bouyant tail of the peak. A slight decrease in G_{D1a} content is seen in the area highest in G_{M1} content, after which this ganglioside becomes more constant along the profile. The concentration of gangliosides G_{D1b}, G_{T1}, and G_{Q1} all roughly correspond to the bell-shaped P_2B curve. These results support previous observations that the higher sialosylated gangliosides are more specifically associated in cortical gray matter (12) with synaptosomal membranes (13).

Surface Labelling of Gangliosides and the Effects of Ca^{++}

The disposition of the gangliosides in the synaptosomal membrane was investigated by several surface labelling techniques. Synaptosomes were prepared by zonal rotor centrifugation as described above. A portion of the preparation was lysed by quickly freeze-thawing three times. Membranes were collected by centrifugation and washed in 10 mM HEPES buffer, pH 7.4. Intact synaptosomes were pelleted and resuspended in 175 mM NaCl, 5 mM KCl, and 10 mM HEPES buffer, pH 7.4. Assaying supernatants for lactate dehydrogenase activity revealed that the former preparation was 95% lysed; the latter, intact, sample was at least 90% osmotically intact. These samples (10 mg protein) were treated with galactose oxidase, (30 U, 1 hour at 37°C) followed by sodium borotritide (2 mCi). Samples were extracted (11) and separated by TLC (14). The greatest difference in labelling between intact and lysed synaptosomes was seen in gangliosides G_{M1} and G_{D1b}, suggesting a partial cytoplasm-facing orientation for these lipids in this membrane.

In order to more closely examine sialic acid labelling, we employed a surface labelling technique reportedly specific for N-acetylneuraminic acid (17). Again, lysed and intact preparations were used, and 5 mM Ca^{++} was added to some experimental groups. Samples were oxidized with periodate for 5 min. at 0°C, pelleted, and resuspended in $NaB[^3H]_4$. After 30 min at room temperature

NaBH$_4$ was added and the membranes were collected and washed by centrifugation. The labelling pattern of the individual gangliosides is given in Table 1.

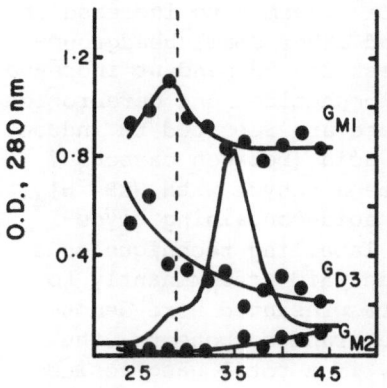

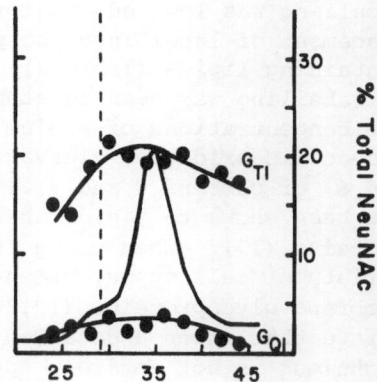

FRACTION NUMBER

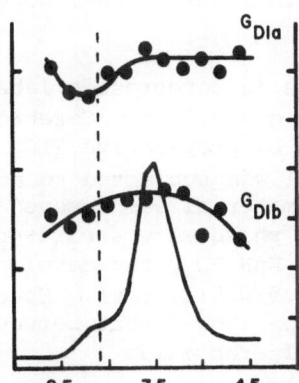

Figure 1. Ganglioside extracts (11) were subjected to thin-layer chromatography (19), visualized with resorcinol-HCl (15), and subjected to monochromatic dual beam transmittance spectrodensitometry using a Schoeffel SD3000 densitometer (16). Solid line, O.D. at 280 nm of particles along the P$_2$B profile after diluting and vortexing; circles, % total sialic acid within each lane of the chromatogram; dashed line, the sucrose interface (0.8 M, left; 1.2 M, right).

TABLE 1

LABELING OF GANGLIOSIDES BY PERIODATE-NaB[^{3}H]$_4$ (CPM/mg. PROTEIN)

Ganglioside*	Intact	Intact +Ca^{++}	Lysed	Lysed +Ca^{++}
G$_{M3}$	170	130	220	280
G$_{M2}$	970	280	1480	600
G$_{M1}$	3380	1230	5570	1660
G$_{D3}$	450	250	570	300
G$_{D1a}$	640	260	950	410
G$_{D1b}$	1270	430	2240	650
G$_{T1}$	290	170	230	190
G$_{Q1}$	200	190	120	130

*notation according to Svennerholm (24)

The greatest difference between lysed and intact preparations, in the absence of Ca^{++}, was seen in gangliosides G_{M1} and G_{D1b}. The presence of 5 mM Ca^{++} reduced labelling in all gangliosides; in some cases up to a 75% decrease was observed. Although ganglioside labelling was lowered, the presence of Ca^{++} seemed to increase the placement of label into phospholipids and other small headgroup-containing lipids (Table 2). The greatest Ca^{++}-dependent increase in labelling was seen in ethanolamine phosphatides and cerebroside. Low concentrations of sodium metaperiodate are reported to induce a specific oxidative cleavage of sialic acid (between carbons 7 and 8, or carbons 8 and 9). Reducing the aldehyde with $NaB[^3H]_4$ has been shown to yield label in sialic acid-containing glyco-proteins (17). When using this surface labelling technique on a variety of cell types, attention has been paid predominantly to membrane glycoproteins (19,20). The data presented here demon-strate that, upon a detailed study of membrane components, the technique is not entirely specific, and, if a total ganglioside preparation from membranes treated in this manner is subjected to acid hydrolysis, up to 20% of the tritium label can be found in sugar residues other than sialic acid. This percentage, however, does not differ in the presence of Ca^{++}.

The apparent Ca^{++}-dependent increase in cerebroside labelling prompted us to more closely examine this glycolipid. Cerebroside from synaptosomal membranes was isolated by preparative TLC. After elution from the silica gel, the material was subjected to acid hydrolysis, after which the hydrophobic material was removed with chloroform. The sugars from the aqueous phase were separated on silica TLC plates impregnated with 0.3 M NaH_2PO_4, and developed in acetone:n-butanol:acetic acid:water (8:0.5:0.5:1, v/v). Spectro-densitometric analysis after visualization showed that between 50-75% of this lipid fraction is glucosylcerebroside.

TABLE 2

LABELLING OF LOWER PHASE LIPIDS

BY PERIODATE-$NaB[^3H]_4$ (CPM/mg. PROTEIN)

Lipid*	Intact	Intact +Ca^{++}	Lysed	Lysed +Ca^{++}
Lyso PC	80	100	110	250
PS	100	140	120	280
PC	80	100	130	180
Cerebroside	300	630	290	840
PI	90	110	120	220
Sulfatide	110	130	140	250
PE	220	630	250	660

*notation - Lyso PC, lysophosphatidylcholine; PS, phosphatidyl-serine; PC, phosphatidylcholine; PI, phosphatidylino-sitol; PE, phosphatidylethanolamine, or the cor-responding plasmalogen forms.

Ca^{++} had no effect on the labelling introduced by tritide reduction following galactose oxidase treatment of the membranes.

These experiments were repeated with lysed membranes using identical incubation conditions, but omitting periodate or galactose oxidase. Samples were extracted and partitioned according to Suzuki (11). The resulting aqueous (ganglioside) and organic (phospholipid) phases were brought up to the same volume and aliquots were taken from each and counted. As shown in Table 3, the addition of 5 mM calcium ion caused an increase in label found in the lower phase. Again, the greatest increase in labelling was found in the ethanolamine phosphatides. This nonspecific labelling produced by NaB[^{3}H]$_4$ may be due to the reduction of plasmalogens (i.e. phosphatidalethanolamine, phosphatidalcholine, and phosphatidalserine). It has been suggested that hydrophobic regions of the bimolecular membrane leaflet are partially exposed at phase boundaries (21). A great percentage of the ethanolamine phosphatides exist in the plasmalogen form in brain tissue, with lesser amounts of phosphatidalcholine and phosphatidalserine (22).

Intrinsic Synaptosomal Sialidae Activity and Alterations in Ganglioside Structure by Ca^{++}.

Subcellular fractionation of mammalian brain has revealed that sialidase (N-acetylneuraminosyl glycohydrolase, EC 3.2.1.18) is a membrane-bound enzyme associated with the synaptic region (23,24). Developmental studies support this (25,26), and the enzyme may come to be recognized as an intrinsic synaptic membrane component. Exogenously directed activity (toward added gangliosides) parallels endogenously directed activity; the preferred substrate appears to be the higher gangliosides (23) in the membranes. The possibility that these gangliosides may be specifically associated with the synaptic area has been presented above. The intrinsic sialidase responds to the presence of the neurotoxic cations Hg^{++} and Cu^{++} (27).

TABLE 3
DISTRIBUTION OF LABEL UPON TREATMENT
OF MEMBRANES WITH NaB[^{3}H]$_4$ ALONE
(% CPM)

Phase*	-Ca^{++}	+Ca^{++}
Upper	53	13
Lower	47	87

*after extraction of membranes according to Suzuki (17)

Synaptosomal plasma membranes were incubated at pH_{++}3.9 for$_+$90 minutes in the presence of various concentrations of Ca^{++} or Na$^+$. Assaying for released, soluble sialic acid (Figure 2) shows that activity is inhibited by calcium ion, beginning at approximately 5 x 10^{-3} M.

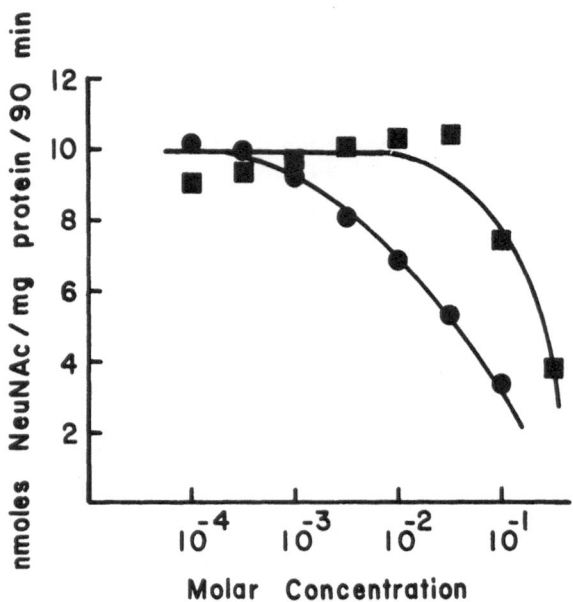

Figure 2. Release of sialic acid after incubation for 90 min. at pH 3.9. Circles, calcium ion; squares, sodium ion. Soluble sialic acid was measured by the thiobarbituric acid assay (28).

Samples were extracted with phosphate-buffered tetrahydrofuran and partitioned with ethyl ether. This procedure was adopted for these studies since it has been shown to more completely separate glycoprotein and glycolipid (29). Sialic acid remaining bound to protein after 90 min. incubation is shown in Figure 3A. Unhydrolyzed protein-bound sialic acid was seen to increase beginning at about 1 mM Ca^{++}, demonstrating an inhibition toward this particular substrate class. Na$^+$ had no observable effect. Unhydrolyzed lipid-bound substrate (Figure 3B) was seen to increase only after 10 mM Ca^{++} had been reached. Very high Na$^+$ concentrations also caused unhydrolyzed lipid-bound sialic acid to increase. The inhibitory action of Ca^{++} was confirmed in a study of the time course of release of sialic acid from protein and lipid substrate (Figure 4). Release from protein was initially very fast, and later exhibited a biphasic nature. This biphasic release of sialic acid was

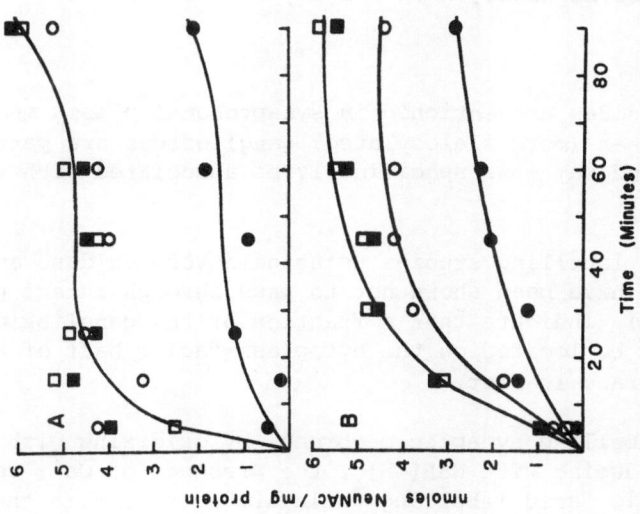

Figure 4. Time course of release of sialic acid from endogenous protein (A) and lipid (B). Circles, calcium ion; squares, sodium ion. Open figures are 5mM; closed figures are 50 mM.

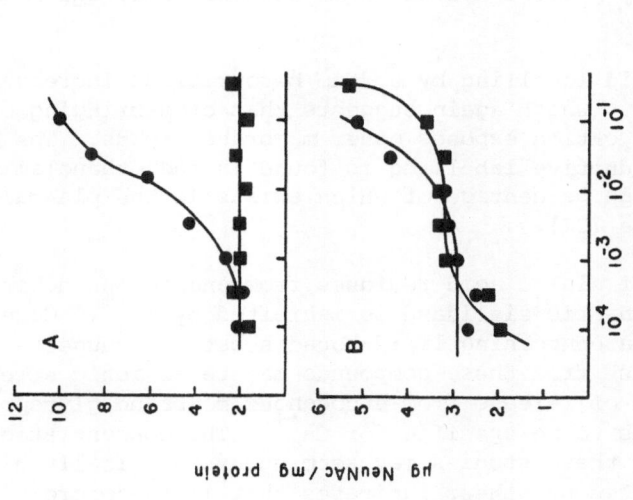

Figure 3. Sialic acid remaining bound to protein (A) and lipid (B) after incubation for 90 min at pH 3.9. Circles, calcium ion; squares, sodium ion. Bound sialic acid was measured by the periodate-resorcinol method (10).

observed previously in our laboratory with synaptosomes from immature rat brain (30). Cleavage of sialic acid from lipid was not observed to be biphasic in this study (Figure 4B).

CONCLUSIONS

1. Gangliosides are enriched in synaptosomal plasma membranes (8,23). The higher (more sialosylated) gangliosides are particularly abundant and may more specifically be associated with the synaptic region.

2. Surface labelling studies using galactose oxidase and periodate, which have been shown not to pass through intact plasma membranes (17,18), indicate that a fraction of the gangliosides G_{M1} and G_{D1b} may be located on the cytoplasm-facing half of the presynaptic membrane leaflet.

3. Upon labelling synaptic membranes by oxidizing with periodate and reducing with $NaB[^3H]_4$, the presence of Ca^{++} decreases ganglioside sialic acid labelling while at the same time the labelling of phospholipids and cerebroside increases. This effect was not observed when oxidizing with galactose oxidase, suggesting a lateral separation or steric modification of the ganglioside sialosyl residues by Ca^{++} so as to expose those groups lying closer to the bimolecular lipid membrane core. Recent carbon-13 and proton nuclear magnetic resonance studies of gangliosides suggest the sialic acid structure depicted in Figure 5, illustrating the possible oxygen cage (31). Crosslinking of the carboxyl groups of two such residues by Ca^{++} may protect hydroxyls on carbons 7, 8, and 9 from periodate oxidation.

4. Phospholipid labelling by sodium borotritiide increased in the presence of Ca^{++}, which again suggests that crossbridging of sialic acid by this cation exposes other membrane lipids. The majority of this reductive labelling is found in the ethanolamine phosphatides, a large percentage of which exists in the plasmalogen form in brain tissue (22).

5. Cleavage of sialic acid residues from endogenous substrate by the intrinsic synaptic sialidase is inhibited by Ca^{++}. Careful scrutiny of the data concerning lipid-bound substrate suggests that inhibition of release from these compounds may be an ionic strength effect. Inhibition of release from endogenous membrane glycoprotein, however, seems to be specific for Ca^{++}. The concentrations of Ca^{++} employed in these studies may seem unphysiologically high. However, work reported by others indicates that in the microenvironment of the lipid membrane the concentration of divalent cations may be considerably higher than in bulk solution under physiological conditions (32,33).

The concept has been advanced many times that a key event during nerve cell stimulation is a displacement of Ca^{++} from certain negatively charged membrane sites, resulting in a change in the polarity, and subsequent permeability, of the synaptic membrane. Sialic acid residues bound to both protein and lipid in the synaptic area can be viewed as representing a dense negatively charged screen. The action of the intrinsic synaptic sialidase represents a means for modifying this charged screen. The results presented here suggest that a dynamic role for calcium ion may be to modulate the availability of protein-bound sialic acid substrate to the enzyme.

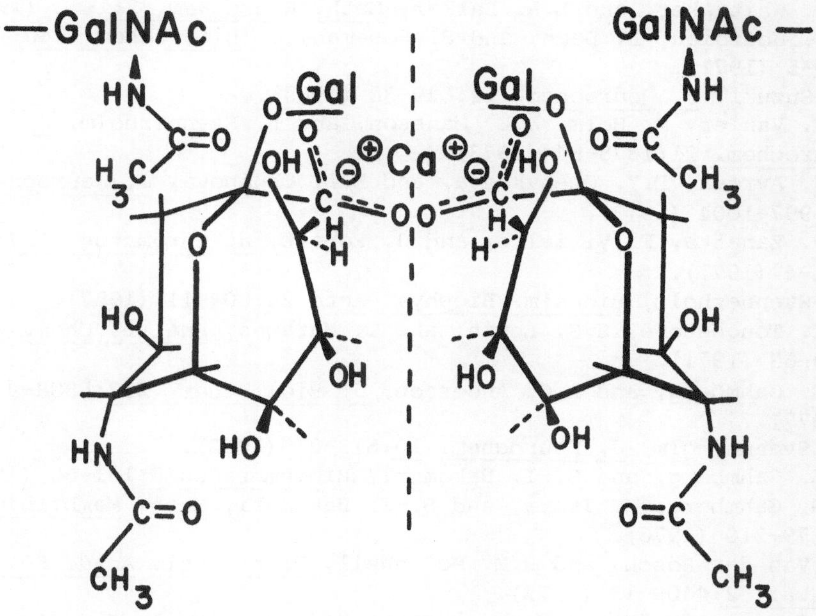

Figure 5. Structure of sialic acid as suggested by NMR studies (31) (31). Two N-acetylneuraminic acid molecules are shown (each side of dashed line), as linked to the ganglioside oligosaccharide sequence. Cross-bridging by calcium ion (center) may protect vicinal hydroxyls on carbons 7, 8, and 9 from periodate oxidation.

REFERENCES

1. E.G. Lapetina, E.F. Soto, and E. DeRobertis, Biochim. Biophys. Acta 135:33-43 (1967).
2. D.W. Wooley, and B.W. Gommi, Nature 202:1074 (1964).
3. G.A. Dette, and W. Wesemann Hoppe-Seyler's Z. Physiol. Chem. 359:399-406 (1978).
4. T.B. Helting, O. Zwister, and H. Wiegandt, J. Biol. Chem. 252: 194-8 (1977).
5. R.W. Ledeen, and J. Mellanby, in "Perspectives in Toxicology", (A.W. Bernheimer, ed.) New York:Wiley (1977).
6. R.W. Ledeen, J. Supramol. Struct. 8:1-17 (1978).
7. K.C. Leskawa, H.C. Yohe, M. Matsumoto, and A. Rosenberg, Trans. Am. Soc. Neurochem. 10:141 (1979).
8. K.C. Leskawa, H.C. Yohe, M. Matsumoto, and A. Rosenberg, Neurochem. Res., in press (1979).
9. V.P. Whittaker, and L.A. Barker, Meth. Neurochem. 2:1-52 (1972).
10. G.W. Jourdian, L. Dean, and S. Roseman, J. Biol. Chem. 246: 430-5 (1971).
11. K. Suzuki, J. Neurochem. 12:629-38 (1965).
12. M.T. Vanier, M. Holm, J.E. Mansson, and L. Svennerholm, J. Neurochem. 21:1375-84 (1973).
13. N.F. Avrova, E.Y. Chenykaeva, and E.L. Obukhova, J. Neurochem. 20:997-1004 (1973).
14. J.P. Zanetta, F. Vitiello, and J. Robert, J. Chromatog. 137: 481-4 (1977).
15. L. Svennerholm, Biochim. Biophys. Acta 24:604-11 (1957).
16. J.C. Touchstone, S.S. Levin, and T. Murawec, Analyt. Chem. 43: 858-63 (1971).
17. C.G. Gahmberg, and L.C. Anderson, J. Biol. Chem. 252:5888-94 (1977).
18. L. Svennerholm, J. Neurochem. 10:613-23 (1963).
19. C.G. Gahmberg, and S.-I. Hakomori, Biomembranes 8:131-65 (1977).
20. C.G. Gahmberg, K. Itaya, and S.-I. Hakomori, Meth. Memb. Biol. 7:179-210 (1976).
21. J. Van der Bosch, and H.M. McConnell, Proc. Natl. Acad. Sci., U.S.A. 72:4409-13 (1975).
22. K. Suzuki, in "Basic Neurochemistry" (G.J. Siegel, R.W. Albers, R. Katzman, B.W. Agranoff, eds.) Boston:Little, Brown & Co. (1976).
23. C.-L. Schengrund, and A. Rosenberg, J. Biol. Chem. 245:1196-2000 (1970).
24. G. Tettamanti, I.G. Morgan, G. Gombos, G. Vincendon, and P. Mandel, Brain Res. 47:515-8 (1972).
25. P.A. Roukema, D.H. Van den Eijnden, J. Heijlman, and G. Van der Berg, FEBS Lett. 9:267-70 (1970).
26. C.-L. Schengrund, and A. Rosenberg, Biochem. 10:2424-8 (1971).
27. H.C. Yohe, and A. Rosenberg, Neurochem. Res. 3:101-13 (1976).
28. L. Warren, J. Biol. Chem. 234:1971-5 (1959).

29. G. Tettamanti, F. Bonali, S. Marchesini, and V. Zambotti,
    Biochim. Biophys. Acta 296:160-170 (1973).
30. H.C. Yohe, and A. Rosenberg, J. Biol. Chem. 252:2412-8 (1977).
31. P.L. Harris, and E.R. Thornton, J. Am. Chem. Soc. 100:6738-45
    (1978).
32. H. Hauser, A. Drake, and M.C. Phillips, Eur. J. Biochem. 62:
    335-44 (1976).
33. S.G.A. McLaughlin, G. Szabo, and G. Eisenman, J. Gen. Physiol.
    58:667-89 (1971).

HYDROLYSIS OF GANGLIOSIDES IN MICELLAR AND LIPOSOMAL

DISPERSION BY BACTERIAL NEURAMINIDASES

S. Gatt, B. Gazit, B. Cestaro[1] and Y. Barenholz

Laboratory of Neurochemistry, Dept. of Biochemistry
Hebrew University Hadassah Medical School
Jerusalem, Israel

In aqueous media gangliosides form aggregates whose micellar weights are about 250,000 daltons (GAMMACK, 1963; YOHE, ROARK and ROSENBERG, 1976). In these micelles the hydrophylic portion, which contains the carbohydrate and neuraminyl residues is directed to the bulk water-phase and thereby interacts with the soluble bacterial neuraminidases. This paper presents a brief discussion of studies, done in our laboratory on the hydrolysis of mono-, di- and trisialogangliosides by the neuraminidases of *Cl. perfringens* and *V. cholerae*. These studies suggested that in micellar dispersion the gangliosides are poor substrates for these enzymes. Enzymatic reaction rates increased considerably when the gangliosides were mixed with a lipid or detergent. The effect of adding a bile salt was absolute in the case of GM1 which was not hydrolyzed at all unless a bile salt was added (see also references of WENGER and WARDELL, 1972, 1973), but was also observed using GM3, which is hydrolyzed by these neuraminidases (WIEGANDT, 1966; BURTON, 1963). The effect of an additive was very striking when di- and trisialogangliosides were used. These compounds are hydrolyzed by both bacterial neuraminidases, yielding GM1 as the end product (WIEGANDT, 1966; BURTON 1963). Rates of hydrolysis increased somewhat when certain lipids were added to a dispersion of mixed brain gangliosides (LIPOVAC, BIGALLI and ROSENBERG, 1971). We have now shown that rates of hydrolysis increased by 1-2 orders of magnitude when these gangliosides were incorporated in a unilamellar, vesicular phospholipid bilayer composed of phosphatidylcholine or sphingomyelin. These studies suggested that in the micelles of pure ganglioside, factors such as steric hindrance or interactions between

---

[1]Permanent address : Institute of Biological Chemistry, Faculty of Medicine and Surgery, University of Milano, Italy.

the carbohydrate chains of neighboring molecules interfere with
the utilization, by the bacterial enzymes, of the neuraminyl residues of these compounds. In the micelle of pure ganglioside these
interactions are maximal, they decrease in mixed aggregates with
other lipids and are minimal in unilamellar vesicles.

HYDROLYSIS OF DI- AND TRISIALOGANGLIOSIDES BY BACTERIAL NEURAMINIDASES

Di- and trisialogangliosides are substrates for the neuraminidases of *V. cholerae* and *Cl. perfringens*. LIPOVAC et al. (1971)
showed that a lipid dispersion increased the rate of hydrolysis of
a preparation of mixed gangliosides. We have studied, in detail the
effect of phosphatidylcholine (PC) and sphingomyelin (SM) on the
hydrolysis of pure GD1a, GD1b and GT1 by these two neuraminidases.

Fig. 1 shows the rate of hydrolysis of GD1a by itself or when
mixed with egg phosphatidylcholine dispersed as multilayered liposomes (▲) or as ultrasonically-irradiated, small unilamellar vesicles
(●). Relative to micelles of pure GD1a (■), the rates were up to 50
times greater using a mixed dispersion of GD1a and 90 mole % PC.
Fig. 1 describes, for each of the two enzymes and three dispersion
states, the rate of hydrolysis of GD1a as a function of time, enzyme
or substrate concentration.

Fig. 2 shows the rate of hydrolysis of the sialyl residues of
GD1a, GD1b or GT1 in mixed dispersion with phosphatidylcholine or
sphingomyelin, as a function of the relative content of the ganglioside in the mixture. In all the experiments shown in the figure the
greatest reaction rates were obtained when the ganglioside constituted
5-10 mole % of the entire mixture. The figure also shows that the
rates of hydrolysis were practically the same with each of the three
gangliosides and the two phospholipids.

Fig. 3 relates the reaction rate to the integrity of the vesicular structure of a GD1a-phospholipid dispersion. In this case the
phospholipids were composed of a mixture of phosphatidylcholine
(PC) and phosphatidylethanolamine (PE), 8:1 (M/M). Inclusion of PE
permitted its interaction with trinitrobenzenesulfonic acid, resulting in formation of a yellow adduct whose content could be determined spectrophotometrically (LITMAN, 1973). Since this reagent does
not penetrate into intact, sealed vesicles this permits determining
the percent PE in the outer layer of unilamellar vesicles; this
value was found to be about 60 % using PC to PE, 9/1 (M/M) (LITMAN,
1973). Fig. 3 shows that the ultrasonically-irradiated mixed dispersion of GD1a, PC and PE consisted entirely of sealed vesicles
when the ganglioside constituted 5-10 mole %. With increasing GD1a,
the proportion of the sealed vesicles decreased and that of mixed
micelles increased. The figure shows that with each dispersion the

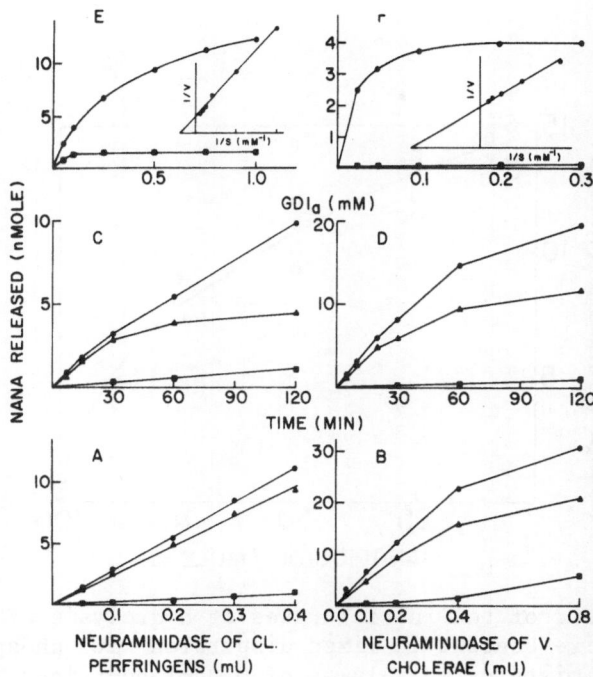

Fig. 1. Hydrolysis of GD1a dispersions by two bacterial neuraminidases. ■, micelles of GD1a; ▲, multilamellar liposomes having 10 %
GD1a and 90 % egg phosphatidylcholine; ●, small, unilamellar liposomes having 10 % GD1a and 90 % egg phosphatidylcholine.

rate of hydrolysis of GD1a paralleled the percentage of sealed
vesicles present.

These and similar experiments in which GM1 or cholesterol were
also included (CESTARO, BARENHOLZ and GATT, submitted) suggested
the following:

1. Di- and trisialogangliosides are poor substrates for the
bacterial neuraminidases when dispersed as micelles of the pure
ganglioside.

2. Rates increase considerably (20-50 fold) when these gangliosides are incorporated into small, unilamellar vesicles of phosphatidylcholine or sphingomyelin. The maximal content of GD1a in these
vesicles is about 10 mole %.

3. The above results as well as studies using nuclear magnetic
resonance (LICHTENBERG, BARENHOLZ and GATT, submitted; see also
this book, paper by BARENHOLZ et al. and the paper by SHAROM and
GRANT, 1978; who used spin-labelled gangliosides) suggest that in-

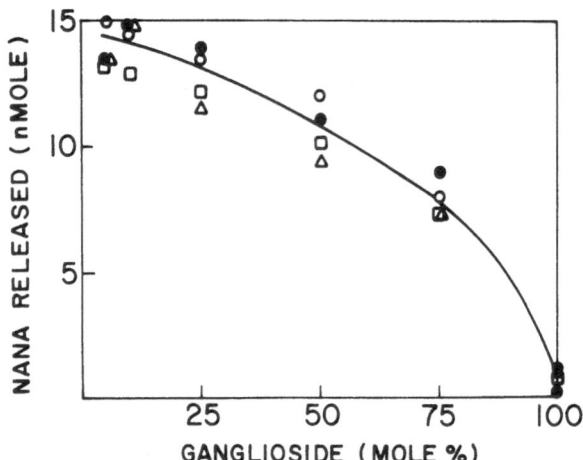

Fig.2. Dependence of the initial rates of hydrolysis of gangliosides
on their relative content in mixed dispersion with phospholipids.
The incubation mixtures in volumes of 0.2 ml contained 20 μmole of
acetate buffer, pH 5.6; 0.2 m units of *V. cholerae* neuraminidase;
0.125 μmole CaCl$_2$; 75 μmole of phospholipid.
● GD1a, SM; O GD1a, egg PC; □ GD1b, PC; Δ GT1, PC.

teractions between carbohydrate chains of adjacent ganglioside
molecules interfere with the utilization, by the neuraminidase of
the neuraminyl residues of the gangliosides. In the micelle of pure
ganglioside these interactions are maximal while in the unilamellar
vesicles they are probably minimal.

HYDROLYSIS OF GM1 BY THE NEURAMINIDASE OF *CL. PERFRINGENS*

This work was undertaken to gain further insight into the observations of WENGER and WARDELL (1972, 1973) that the neuraminyl linkages of GM1 and GM2 which are resistant to the action of the neuraminidase of *Cl. perfringens*, were hydrolyzed when bile salts were
included in the reaction medium. For this purpose, GM1 was mixed
with sodium taurocholate or taurodeoxycholate and the mixed dispersions were incubated with a partially purified preparation of the
above enzyme (Sigma, type VI). We confirmed the findings of WENGER
and WARDELL and added further data which are summarized as follows:

1. The curves which describe the rate of hydrolysis as a function
of increasing concentrations of bile salt were biphasic. The rates

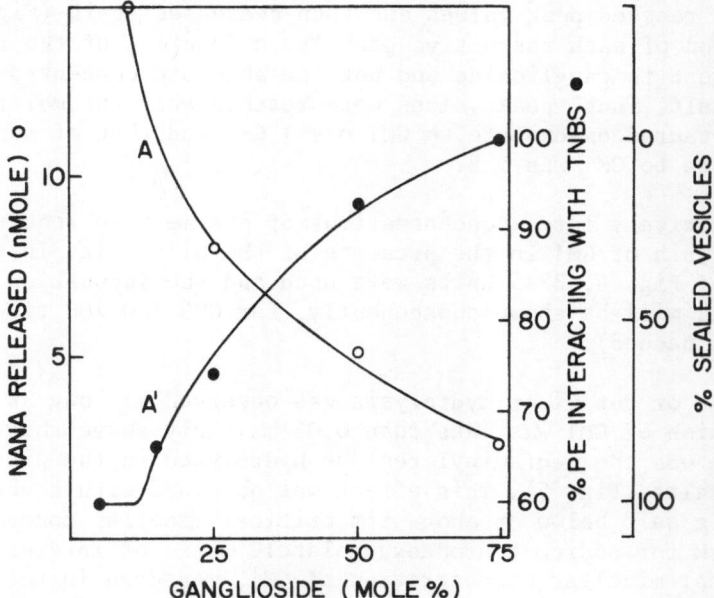

Fig. 3. Correlation of the relative content of sealed vesicles and the rate of hydrolysis of GD1a in mixed dispersion with phospholipids. Curve A describes the NANA released and curve A' the percent PE which interacted with trinitrobenzenesulfonic acid.

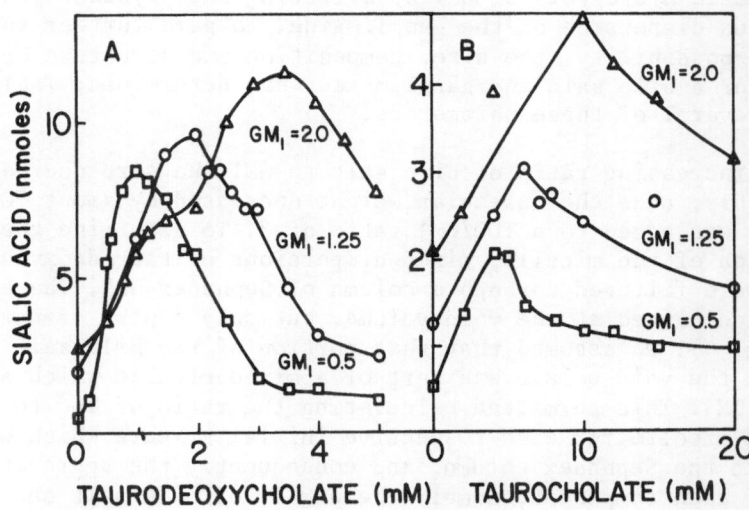

Fig. 4. Dependence of the rate of hydrolysis of GM1 on bile salts. 32.5 m units of neuraminidase of *Cl. perfringens* were used in this experiment; the GM1 concentrations were 0.5 (□), 1.25 (O) and 2.0 (Δ) mM.

increased, reached peak values and then decreased again (Fig. 4). The position of each respective peak was a function of the ratio of the detergent to ganglioside and not the absolute concentration of the bile salt. Thus, peak values were reached when the molar ratio of sodium taurodeoxycholate to GM1 was 1.6-2 and that of sodium taurocholate to GM1 was 5-8.

2. Relatively large concentrations of enzyme were required for the hydrolysis of GM1 in the presence of the bile salt. In the experiment of Fig. 4, 32.5 units were used and the incubation time was 2 h (as will be shown subsequently, for GM3 100-200 times less enzyme was needed).

3. None or but minor hydrolysis was observed as long as the concentration of GM1 was less than 0.03 mM; only above this concentration was the neuraminyl residue hydrolyzed in the presence of the bile salts (Fig. 5). This effect was obtained with concentrations of the bile salt below or above its critical micellar concentration (about 2 mM for sodium taurodeoxycholate). It is of interest that the critical micellar concentration of GM1, measured in the presence of taurodeoxycholate using a fluorescent probe was about 0.03 mM (BARENHOLZ, GAZIT and GATT, in preparation).

4. The carbohydrate residue of GM1, devoid of the ceramide moiety was not hydrolyzed, in the absence or presence of bile salts.

The above observations suggested that the bile salts influence the enzymatic hydrolysis of GM1 by affecting the physical state of the aqueous dispersion of the ganglioside. To gain further insight into this possibility, the size, composition and structure of mixed micelles of a bile salt and ganglioside were determined; Table 1 records several of these parameters.

With increasing ratio of bile salt to GM1 the size decreased considerably; thus the molecular weight decreased by about 70 % when bile salt was added to a TDC/GM1 ratio of 3. To determine the actual composition of the micelle, mixed dispersions of taurodeoxycholate and GM1 were filtered through a column of Sephadex G25. The entire ganglioside eluted at the void volume, but only a part of the bile salt (Fig. 6). We assumed that that portion of the bile salt which eluted at the void volume was part of a mixed micelle which also included GM1. This permitted calculating the ratio of TDC to GM1 in the mixed micelle for each respective initial mixture which was applied to the Sephadex column, and consequently the aggregation number of each component, namely the number of bile salt and ganglioside molecules per micelle. The values shown in Table 1 suggest that with increasing ratio of TDC to GM1 the micellar size and aggregation numbers decreased, and the degree of asymmetry (as measured by the ratio of r_s to r_{eq}) increased.

Table 1. Physical Properties of Mixed Dispersions of GM1
and Sodium Taurodeoxycholate

Ratio TDC/GM1	$S^o_{20,w}$	$D^o_{20,w}$ x10^7 (cm^2/sec)	r_s	M.W. x10^{-3}	N_{GM1}	N_{TDC}	N	r_s/r_{eq}
0	9.77	2.16	99.2	484	307	0	307	1.875
1.75	3.6	1.85	115	246	130	90	220	2.67
3.0	2.6	2.14	100	140	71	57	128	2.84

Explanation of symbols: TDC-taurodeoxycholate; $S^o_{20,w}$-sedimentation coefficient; $D^o_{20,w}$-diffusion coefficient; r_s-stokes radius; r_{eq}-equivalent radius; M.W.-molecular weight; N-aggregation number; N_{GM1}, N_{TDC}-number of molecules of GM1 and TDC, respectively in one mixed-micelle.

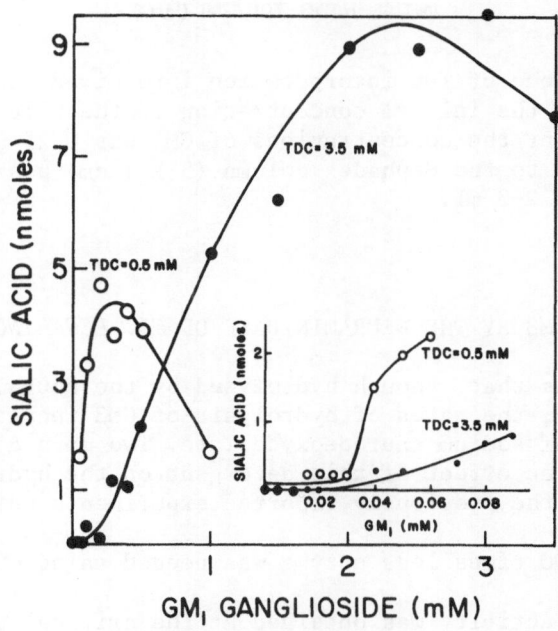

Fig. 5. Dependence of GM1 hydrolysis on substrate concentration at 2 fixed concentrations of sodium taurodeoxycholate and 32.5 m units of the clostridial neuraminidase.

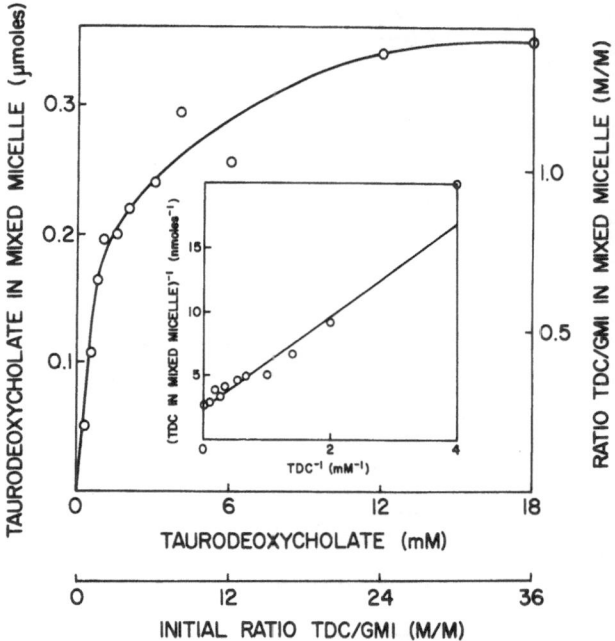

Fig. 6. Dependence of TDC incorporation into mixed micelles of bile salt and GM1 on the initial concentration of the bile salt. In the entire experiment the concentrations of GM1 was 0.25 mM; the initial volume, applied to the Sephadex column (5 x 1 cm) was 0.5 ml. The void volume was 2-3 ml.

HYDROLYSIS OF GM3 BY THE NEURAMINIDASE OF *CL. PERFRINGENS*

Fig. 7 shows that, though hydrolyzed by the neuraminidase of *Cl. perfringens*, the rates of hydrolysis of GM3 increased 4-10 fold upon addition of sodium taurodeoxycholate. Two main differences were noted between the effect of this detergent on the hydrolysis of GM3 as compared to the previously reported experiments using GM1.

1. About 200 times less enzyme was needed using GM3.

2. Optimal activity was obtained at the critical micellar concentration of the bile salt, irrespective of ganglioside concentration. In parallel experiments (GATT and GAZIT, submitted), similar observations were made using glycophorin, orosomucoid and even the low-molecular-weight substrate neuraminlactose. This suggests that with GM3 and the three non glycosidic substrates, the bile salt influences the enzymatic reaction by activating the enzyme rather than affecting the dispersion state of the substrate.

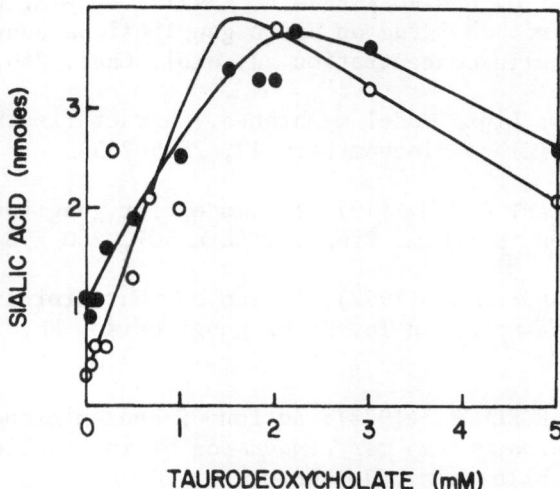

Fig. 7. Dependence of hydrolysis of GM3 on increasing concentrations of taurodeoxycholate: 0.2 m unit of the clostridial enzyme and two concentrations of GM3 (0.5 and 1.5 mM) were used.

IN SUMMARY:

Aqueous dispersions of pure gangliosides contain micelles of these compounds. In this dispersion state, the rates of hydrolysis of the neuraminyl residues by bacterial neuraminidases are slowest. Incorporation of gangliosides into mixed dispersion with other lipids or into mixed micelles with bile salts considerably increases the reaction rates. The greatest reaction rates are obtained when di- or trisialogangliosides are incorporated into unilamellar vesicles of lecithin or sphingomyelin.

Acknowledgements: This work was supported in part by grants from NIH NSO2967 and from DGRST N°78.7.0367). The following generous gifts are acknowledged: GM1 was obtained from Dr. L. Svennerholm; GD1a, GD1b and GT1 from Dr. G. Tettamanti; GM3 from Dr. G. Schwarzmann; the carbohydrate residue of GM1 from Dr. H. Wiegandt and synthetic sodium taurodeoxycholate from Drs. L. Krabisch and B. Borgstrom.

REFERENCES

BURTON R.M. (1963): The action of neuraminidase from *Clostridium perfringens* on gangliosides. J. Neurochem. 10, 503-512.

GAMMACK D.B. (1963): Physicochemical properties of ox-brain gangliosides. Biochem. J. 88, 373-383.

LIPOVAC V., BIGALLI G. and ROSENBERG A. (1971): Enzymatic action of
sialidase of *Vibrio cholerae* on brain gangliosides above and below
the critical micelle concentration. J. biol. Chem. 246, 7642-7648.

LITMAN B.J. (1973): Lipid model membranes. Characterization of mixed
phospholipid vesicles. Biochemistry 12, 2545-2554.

SHAROM F.J. and GRANT C.W.M. (1978): A model for ganglioside behaviour
in cell membranes. Biochim. Biophys. Acta 507, 280-293.

WENGER D.A. and WARDELL S. (1972): Action of neuraminidase from
Clostridium perfringens on Tay-Sachs ganglioside. Physiol. Chem.
Phys. 4, 224-228.

WENGER D.A. and WARDELL S. (1973): Action of neuraminidase (EC
3.2.1.18) from *Clostridium perfringens* on brain gangliosides in
the presence of bile salts. J. Neurochem. 20, 607-612.

WIEGANDT H. (1966): Ganglioside. Rev. Physiol. Biochem. Exptl.
Pharmacol. 57, 190-222.

YOHE H.C., ROARK D.E. and ROSENBERG A. (1976): C_{20}-sphingosine as a
determining factor in aggregation of gangliosides. J. biol. Chem.
251, 7083-7087.

DISTRIBUTION OF GANGLIOSIDES IN TISSUES, SUBCELLULAR FRACTIONS, CELLS AND FLUIDS

BRAIN AND RETINAL GANGLIOSIDE COMPOSITION FROM DIFFERENT SPECIES

DETERMINED BY TLC AND HPTLC

P.F. Urban[1], S. Harth[1], L. Freysz[1] and H. Dreyfus[2]

Centre de Neurochimie du CNRS, and Unité 44 de l'INSERM, 11 rue Humann, 67085 Strasbourg Cedex, France.

Gangliosides are compounds found in the CNS at relatively high concentrations (MORGAN *et al.*, 1971, 1973; TETTAMANTI *et al.*, 1972; BRECKENRIDGE, GOMBOS and MORGAN, 1972). In establishing the physiological role of gangliosides, their distribution in different brain areas, cell types, and subcellular fractions should be useful. The presence of a high amount of gangliosides in plasma membranes and in synaptic junctions has been established (HAMBERGER and SVENNERHOLM, 1971; WIEGANDT, 1967).

In this paper, we report our investigations on the ganglioside patterns in the brain and in retinas of different species. In view of the necessity of an analytical method which can be applied to discrete brain areas or retinas without a loss of glycolipids, we developed a rapid sensitive technique for ganglioside isolation and pattern determination.

MATERIALS AND METHODS

Chemicals were of highest available purity and organic solvents were saturated with nitrogen before use. Gangliosides were fractionated either on 20 x 20 cm precoated silica gel plates DC-Fertig platten, Kieselgel 60 ref. 5721, or on 10 x 20 cm HPTLC Fertig platten Kieselgel 60 ref. 5641 (Merck).

Calf, lamb, hamster, rabbit, chick, mouse and frog retinas were taken as quickly as possible after death.

---

[1]Chargés de Recherche au CNRS; [2]Chargé de Recherche à l'INSERM.

149

Whole brains of adult chick, mouse, frog, were obtained after
decapitation. Brain areas of adult rat and mouse were dissected and
weighed at 4°C (KEMPF, personal communication).

Neuronal and glial cells from rabbit brain were isolated by the
following procedure. Adult rabbits of 2 to 2.5 kg were used for the
preparation of different cell types. Neurons and astroglial cells
were isolated from the grey matter and oligodendroglial cells from
the white matter of the cerebral hemispheres by the method of IQBAL
and TELLEZ (1972) with slight modifications as reported elsewhere
(FREYSZ *et al.*, 1979). The morphological integrity of the isolated
cells was examined by phase contrast and electron microscopy. The
neuronal fraction contained mainly perikarya with few processes,
whereas the astroglial cells retained arborized processes. Oligo-
dendrocytes were round cells with a large nucleus surrounded by a
small cytoplasmic layer as reported by FEWSTER *et al.* (1973) and
PODUSLO and NORTON (1975). The use of enzyme markers (acetylcholine
transferase, dopamine-β-hydroxylase, UDP-galactosylceramide transferase, 2'3'-cyclic nucleotide 3'-phosphohydrolase, β-galactosidase
(FREYSZ, unpublished results) and phase contrast microscopic observation indicated that the purity of each fraction was of 85 to 90 %
(contamination was mainly due to capillaries). Cross contamination
of neurons by glial membrane fractions and of glial cells by neuronal membranes was relatively low.

Tissues or cells were homogenized in 20 vol (w/v) of chloroform-
methanol (C-M) (2:1 v/v). After centrifugation, the pellet was resuspended in the same volume of C-M (1:2 v/v) containing 5 % water.
A final extraction was performed with an equal volume of C-M (2:1
v/v) and the lipid extracts were pooled.

Ganglioside analyses were performed as described elsewhere
(HARTH *et al.*, 1978). This method is rapid, requires less material,
and reduces the loss of lipids.

Typical HPTLC of gangliosides from a total lipid extract of
chick, rabbit and frog retinas are shown in Fig. 1. The chromatogram
on the left concerns a total lipid extract from retinas without
further purification. The chromatogram on the right side concerns
rabbit and frog retinal purified gangliosides, the HPTLC shows a
cleaner profile. The relative distribution of gangliosides was
evaluated by densitometric transmission scanning on a Vernon densitometer TRI 5.

RESULTS AND DISCUSSION

Previously, using the technique of SUZUKI (1964) with slight
modifications by DREYFUS *et al.* (1976), we reported the ganglioside
patterns in retinas from various species. Mammalian retinas are

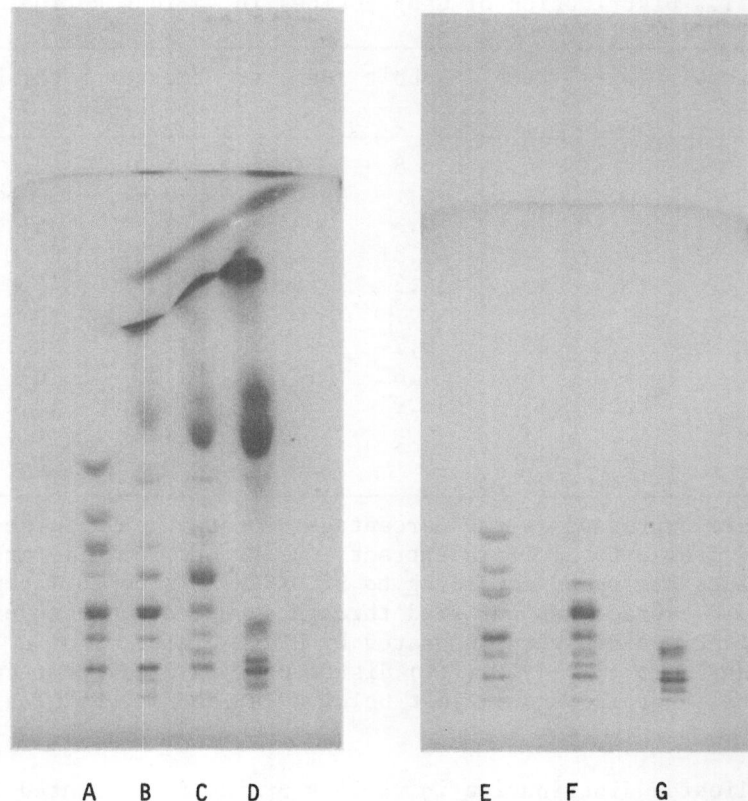

Fig. 1. Thin layer chromatography of retinal gangliosides of a total
lipid extract. Left side: (A) marker line; (B) chick; (C) rabbit;
(D) frog. Right side: before TLC the extract is purified through a
small Sephadex G25 superfine column; (E) marker line; (F) rabbit;
(G) frog. HARTH *et al*. (1978).

characterized by high level of GD3. Those from avian species contain
high amounts of GD1a, while amphibian retinas show high levels of
polysialogangliosides with relatively low levels of monosialogangliosides.

Using our recent technique, which requires less material and is
less time consuming, we could confirm the high amount of GD3 in mammalian retinas (Table 1). Furthermore, we detected a spot migrating
between GD1b and GT1b which was tentatively identified as GT1L, probably a lactonic form of GT1b (SVENNERHOLM, personal communication).
In avian retina, we found high amounts of GD1a as in the previous
technique. GT1L was not detected in chick retina. The species specificity in ganglioside profiles for retinas could be confirmed with
the new technique.

Table 1. Distribution of Gangliosides in Various Retinas

Gangliosides[a]	Calf	Lamb	Rabbit	Hamster	Mouse C57BL/6J	Chick	Frog
GM3	2.8	4.1	3.8	4.2*	5.0	3.6	0.6
GM2	0.6	1.6	0	0.4	0.6	0	0.3
GM1	2.1*	4.9*	7.3*	3.5	3.8*	5.4*	0
GD3	37.3*	32.3*	41.8*	30.2*	34.2*	15.0	1.8
GD1a	13.2	13.2	16.5	6.0	11.7	43.9	0
<GD1a	0	0	0	0	0	0	24.1*
GD1b	15.0	19.3	8.2	11.8*	12.1	9.5	12.6*
GT1L	4.3	5.6	7.6	12.2	11.0	0	22.0*
GT1b	22.3	17.9	13.9	24.2	19.8	18.2	26.0
GQ1	2.4	1.1	0.9	7.5	1.8	4.4	4.0
GQ'	0	0	0	0	0	0	8.6

Results are expressed as the percentage of total ganglioside-NeuAc
recovered from a total lipid extract. (a) Except for GT1L and GQ',
gangliosides are named according to SVENNERHOLM's nomenclature (1963).
Total lipid extract was purified through a Sephadex G25 superfine
column, gangliosides were separated by HPTLC plates using a 3 solvent
system (HARTH et al., 1978). (*) distributed in 2 adjacent spots.
<GD1a: this spot is located just below GD1a, and may be GT1a.

        Ganglioside distribution in various brains is presented in Table
2. In mammalian brain, GD1a is the most abundant ganglioside. GD3
represents about 3-6 % of the total gangliosides and GM1 accounts
for 13 to 14 %. The amount of these three gangliosides differs from
those observed in retinas of the same mammals. In frog brain, we
found high levels of GD1b, GT, GQ and more complex ganglioside
species. Frog retinas contain appreciable amounts of polysialogangliosides (GD1b→GQ'), while in the brain tetra- and probably pentasialogangliosides (GP) were strikingly high. On the other hand, in chick
brain, we noted for the adult animal 30 % of GD1a, but also 15 % of
GD3. These two characteristics including the presence of about 6 %
of GM4 indicate a species specificity in the ganglioside pattern in
the retina. GM4 is considered as an index for myelination and its
presence in the CNS has been reported by COCHRAN, YU and LEDEEN
(1977). GT1L was not detected in chick brain. In addition, GM3 and
GD3 decreased during chick brain ontogenesis (Table 2) whereas GM4,
GM1, GD1a, GT1b and GQ1 increased. These properties were already
described (DREYFUS et al., 1975).

        Considering various brain areas in mammalian species we found
in rat as well as in mouse (Tables 3 and 4) approximately 30 % of
GT1b + GT1L. GD1a is the highest ganglioside in mouse hippocampus and
in rat and mouse hypothalamus GD1a and GT1 are present in the same
range.

Table 2. Distribution of Gangliosides in Various Brains

Gangliosides[a]	Chick		Frog	Mouse C57BL/6J
	Embryo 11 days	Adult		
GM4	–	5.8	0.5	0.7
GM3	7.3	4.4	1.3	1.9
GM2	0.7	0.4	1.0	0.7
GM1	2.0	13.3	1.1	12.8
GD3	40.5	15.3	1.2	3.3
GD1a	24.0	29.2	8.7	33.6
GD1b	10.4	8.6	19.5	11.4
GT1L	–	–	6.0	7.6
GT1b	12.2	17.6	37.4	23.9
GQ1	2.9	5.3	3.3	4.1
GQ'	–	–	17.9	–
GP	–	–	23.4	–

Results are expressed as the percentage of total ganglioside–NeuAc recovered from a total lipid extract. (a) Except for GM4, GT1L, GQ' and GP, gangliosides were named according to SVENNERHOLM's nomenclature. TLC with 3 solvent systems (HARTH *et al.*, 1978). GQ' and GP migrate below GQ1 and could be isomeric forms of tetra- or penta-sialogangliosides.

Table 3. Distribution of Gangliosides in Various Areas of Rat Brain

Gangliosides[a]	Cortex	Pons-Medulla	Hypothalamus	Striatum caudate nucleus
GM4	–	2	–	–
GM3	3.2	5.1	5.5	3.5
GM2	3.2	0.3	1.9	1.8
GM1	14.1	13.0	13.6	17.5
GD3	3.2	2.4	4.0	2.9
GD1a	32.1	16.3	27.1	40.6
GD1b	13.2	19.2	16.6	13.4
GT1L	3.2	13.1	5.3	2.9
GT1b	22.7	24.7	20.9	14.8
GQ1	5.1	3.9	5.1	2.6

Results are expressed as the percentage of total ganglioside–NeuAc recovered from a total lipid extract. (a) Except for GM4 and GT1L, gangliosides are named according to SVENNERHOLM's nomenclature. Direct TLC technique (HARTH *et al.*, 1978).

Table 4.   Distribution of Gangliosides in Various Areas
of Mouse Brain (C57BL/6J Strain)

Gangliosides[a]	Cerebellum	Cortex	Hippocampus	Hypothalamus	Pons-Medulla
(GM4)[b]	0.8	−	0.5	0.5	1.4
GM3	3.2	0.5	1.9	1.9	1.9
GM2	1.0	0.5	0.5	0.9	1.2
GM1	11.0	11.8	12.7	15.4	22.4
GD3	3.6	6.3	2.4	2.5	2.7
GD1a	18.3	32.4	31.0	25.5	15.6
GD1b	9.0	11.7	11.3	12.9	11.7
GT1L	14.4	8.7	7.6	7.4	11.7
GT1b	33.0	26.1	27.0	28.6	26.1
GQ1	5.7	2.0	5.1	4.4	5.3

Results are expressed as the percentage of total ganglioside-NeuAc
recovered from a total lipid extract. (a) Except for GT1L and GM4,
gangliosides are named according to SVENNERHOLM's nomenclature.
(b) This compound migrates over GM3, gives with resorcinol spray
the same color as the other gangliosides, it may be GM4; it has not
been identified. Direct TLC technique (HARTH *et al.*, 1978).

Table 5.   Distribution of Gangliosides in White and Grey Matter
of Human and Rabbit Brains

Gangliosides[a]	Human brain		Rabbit brain	
	White matter	Grey matter	White matter	Grey matter
GM4	8.4	0.7	−	−
GM3	4.1	3.8	0.8	2.4
GM2	1.7	2.6	Traces	0.6
GM1	19.5	19.3	23.3	18.2
GD3	7.6	10.0	4.2	5.5
GD1a	16.0	24.8	36.6	37.9
GD1b	17.2	17.0	11.7	9.4
GT1L	2.5	3.7	5.5	4.6
GT1b	20.0	16.7	16.7	19.6
GQ1	3.0	1.4	1.2	1.8

Results are expressed as the percentage of total ganglioside-NeuAc
recovered from a total lipid extract. (a) Except for GT1L and GM4,
gangliosides were named according to SVENNERHOLM's nomenclature.
Direct TLC technique (HARTH *et al.*, 1978). Rabbit tissue was submitted to Sephadex G25 purification before TLC separation.

Table 6. Distribution of Gangliosides in Neuronal, Oligodendroglial
and Astroglial Cells Isolated from Adult Rabbit Brain

Gangliosides[a]	Astrocytes	Oligodendrocytes	Neurons
GM3	1.7	1.8	1.7
GM2	1.1	1.8	1.2
GM1	15.3	16.1	15.6
GD3	11.0	15.0	7.6
GD1a	35.8	40.3	37.3
GD1b	13.3	11.4	13.2
GT1L	4.1	3.2	4.2
GT1b	17.7	10.4	18.0
GQ1	–	–	1.2

Results are expressed as the percentage of total ganglioside-NeuAc
recovered from a total lipid extract. (a) Except for GT1L, gangliosides are named according to SVENNERHOM's nomenclature. Direct technique followed by HPTLC with 3 solvent systems.

Recently, ANDO, CHANG and YU (1978) reported characteristic
ganglioside patterns for white and grey matter from brains of various
species. Results concerning white and grey matter of human and rabbit
brains are reported in Table 5. White matter of human brain contains
8 % of GM4 reflecting the high amount of myelin as reported by these
authors. The major ganglioside in these tissues was GD1a, followed
by GM1 and GT1.

In order to analyse the ganglioside patterns of well characterized
cells we prepared relatively pure fractions of astrocytes, oligo-
dendrocytes and neuronal cells isolated from adult rabbit brain. The
analysis of the ganglioside distribution is reported in Table 6, which
shows that GD1a is the major ganglioside for all cell types as it was
for white and grey matter from rabbit brain. Neurons contain somewhat less GD3 and are the only cell type where GQ1 could be detected.

It might be useful to consider the differences in functional activity of retinas in different species in order to find out if there
are correlations between a qualitative or quantitative aspect of
retinal cell activity and specific gangliosides.

There is also a striking difference in ganglioside patterns between chick, frog and mouse brain. The functional significance of
these differences merits further study for understanding the role of
gangliosides. In order to explore possible correlations between the
differences in patterns of ganglioside and the functional activity,
ganglioside patterns in different brain areas were investigated.

In white matter as well as in grey matter values rather close to
one another were found for GM1 and GT1 in human as well as in rabbit
brain, while for GD1a and GD1b there are striking differences.

There was good agreement of ganglioside analyses of the same
brain areas in rodent populations, but notable differences were
found between regions in the brain. This suggests a correlation
between ganglioside patterns and functional activity.

ACKNOWLEDGEMENTS —This work was supported by a grant from the
INSERM (Contract N° 78-4-0176), and the DGRST (Contract N° 78.7.0367).

REFERENCES

ANDO S., CHANG N. and YU R.K. (1978): High-performance thin-layer
    chromatography and densitometric determination of brain ganglios-
    ide compositions of several species. Anal. Biochem. 89, 437-450.

BRECKENRIDGE W.C., GOMBOS G. and MORGAN I.G. (1972): The lipid comp-
    osition of adult rat brain synaptosomal plasma membranes. Biochim.
    Biophys. Acta 266, 695-707.

COCHRAN F.B., YU R.K. and LEDEEN R.W. (1977): Comparison of CNS
    myelin gangliosides in several vertebrate species. 6th Intern.
    Meeting of Neurochemistry, Copenhagen, Denmark, Vol. 6, Abstr.
    p. 540, n° 496.

DREYFUS H., URBAN P.F., EDEL-HARTH S. and MANDEL P. (1975): Develop-
    mental patterns of gangliosides and of phospholipids in chick
    retina and brain. J. Neurochem. 25, 245-250.

DREYFUS H., URBAN P.F., HARTH S., PRETI A. and MANDEL P. (1976):
    Retinal gangliosides: composition, evolution with age. Biosynthetic
    and metabolic approaches, in "Ganglioside Function", PORCELLATI G.,
    CECCARELLI B. and TETTAMANTI G., Eds., Advances in Experimental
    Biology, Vol. 71, Plenum Press (New York and London), pp. 163-188.

FEWSTER M.E., BLACKSLONE S.C. and IHRIQ T.Y. (1973): The preparation
    and characterization of isolated oligodendroglia from bovine white
    matter. Brain Res. 63, 263-271.

FREYSZ L., FAROOQUI A.A., ADAMCZEWSKA-GONCERZEWICZ Z. and MANDEL P.
    (1979): Lysosomal hydrolases in neuronal, astroglial and oligo-
    dendroglial enriched fractions of rabbit and beef brain. J. Lipid
    Res. 20, 503-508.

HAMBERGER A. and SVENNERHOLM L. (1971): Composition of gangliosides
    and phospholipids of neuronal and glial cell enriched fractions.
    J. Neurochem. 18, 1821-1829.

HARTH S., DREYFUS H., URBAN P.F. and MANDEL P. (1978): Direct thin layer chromatography of gangliosides of a total lipid extract. Anal. Biochem. 86, 543-551.

IQBAL K. and TELLEZ-NAGEL I. (1972): Isolation of neurons and glial cells from normal and pathological human brains. Brain Res. 45, 296-301.

LEDEEN R.W., YU R.K. and ENG L.F. (1973): Gangliosides of human myelin: sialosyl-galactosylceramide (G7) as a major component. J. Neurochem. 21, 829-839.

MORGAN I.G., ZANETTA J.P., BRECKENRIDGE W.C., VINCENDON G. and GOMBOS G. (1973): The chemical structure of synaptic membranes. Brain Res. 62, 405-411.

MORGAN I.G., WOLFE L.S., MANDEL P. and GOMBOS G. (1971): Isolation of plasma membranes from rat brain. Biochim. Biophys. Acta 241, 737-751.

PODUSLO S.E. and NORTON W.T. (1975): Isolation of specific brain cells. Methods in Enzymol. 35, 561-576.

SUZUKI K. (1964): A simple and accurate micromethod for quantitative determination of ganglioside patterns. Life Sci. 3, 1227-1233.

SVENNERHOLM L. (1963): Chromatographic separation of human brain gangliosides. J. Neurochem. 10, 613-623.

TETTAMANTI G., MORGAN I.G., GOMBOS G., VINCENDON G. and MANDEL P. (1972): Sub-synaptosomal localization of brain particulate neuraminidase. Brain Res. 47, 515-518.

WIEGANDT H. (1967): The subcellular localization of gangliosides in the brain. J. Neurochem. 14, 671-674.

GLYCOLIPIDS AND CELL DIFFERENTIATION

G. Rebel[1], J. Robert and P. Mandel

Centre de Neurochimie du CNRS, 11 Rue Humann,
67085 Strasbourg Cedex, France

Differentiation of a cultured cell refers generally to a change in the expression of some peculiar functions specific to this cell. Generally differentiation involves only a modification of some biochemical properties of the cell such as synthesis of melanin by melanocytes (KREIDER *et al.*, 1975) or synthesis of collagen by bone fibroblasts (MANNER and KULEBA, 1974) or tendon cells (SCHWARTZ *et al.*, 1976). However, differentiation is sometimes characterized by a net change in the cell morphology which accompanies the biochemical changes as for example with myoblasts.

Few works have been devoted to the changes that occur in lipid during cell differentiation. Most of them concern gangliosides and other glycolipids, localized mainly in the plasma membrane, probably on its outer part (see for review CRITCHLEY and VICKER, 1977), analysing the possible relations between these lipids and the observed changes of the cell morphology.

EPITHELIOID CELLS

When treated with butyrate some epithelioid cells undergo a striking morphological differentiation (GINSBURG *et al.*, 1973). HeLa cells are normally round or polygonal. When short chain fatty acids (C3-C5) are added to the culture medium, the cells stop growing and become more fibroblastic, developing extended processes. Concomitant with this change, a specific increase of the GM3 ganglioside is observed. Extensive studies of this type of differentiation have been performed on HeLa cells by the group of FISHMAN and BRADY (FISHMAN

---

[1]Chargé de Recherche au CNRS.

et al., 1974, 1976; SIMMONS *et al.*, 1975) and on KB cells by MACHER
et al., 1978). The main properties of the short chain fatty acids
induced differentiation are summarized below;

(a) The best results are obtained with butyrate. Propionate
and valerate are also active.

(b) Only GM3 increases. No changes occur in the other glycolipids.

(c) Specific INDUCTION of the CMP-NeuAc lactosylceramide neuraminyltransferase is observed. No changes in the activity of the other
glycolipid synthesizing enzymes are reported.

(d) After withdrawal of the inducer from the culture medium, the
lactosylceramide neuraminyltransferase decreases and the normal
morphology of the cells reappears.

(e) The morphological differentiation and the increase of the
GM3 are not consequences of the stop of the cell growth occasioned
by butyrate, as thymidine does not induce any differentiation. However, butyrate added to cells arrested by thymidine, induces a
morphological change and an increase in the synthesis of GM3.

(f) Inhibitors of the cytoskeleton polymerisation like colcemid
inhibit, in the presence of butyrate, the morphological change but
are without effects on the increase of the GM3 synthesis.

(g) Addition of GM3 to the medium of undifferentiated cells does
not induce a morphological change.

NEUROBLASTOMA CELLS

A comparison of neuroblastoma cells grown in either suspension
or in monolayer, in the same medium, has been done by YOGEESWARAN *et
al.* (1973). When grown in suspension these cells are round. After
they are plated on a solid support, they flatten and become polygonal.
Some cells show small processes. Despite the great disparity in their
morphology, no differences in the ganglioside were found between the
two kinds of cells. However, we now know that this morphological
change does not correspond to a differentiation. Furthermore, the
morphology of neuroblastoma cells growing in monolayer in normal
medium, remains very different from that of normal neurons. A study
of neuroblastoma cells after differentiation induced, either by with-
drawing the serum from the culture medium or by adding to this medium
inducers such as dibutyryl cyclic AMP (dbcAMP) or bromodeoxyuridine
(BrdU) was performed in our laboratory (REBEL *et al.*, 1973; CIESIEL-
SKI *et al.*, 1977). After such treatments cells developed numerous
long processes and seemed similar to neuronal cells.

Table 1. Total Lipid Sialic Acid of Control and Differentiated
                    Neuroblastoma Cells

Clones	Control cells	Sialic acid (μg/g dry weight) Differentiated cells		
		Serum withdrawal	dbcAMP	BrdU
NlE 19	301 ±20	396 ±22 (+32 %)[a]	–	–
NS 20	459 ±17	510 ±25 (+11 %)	–	–
NlE 115	282 ±18	413 ±18 (+46 %)	595 ±35 (+211 %)	–
Neuro 2A	240 ±20	303 ±19 (+26 %)	399 ±26 (+66 %)	
M1	498 ±18	630 ±21 (+26 %)	736 ±21 (+47 %)	564 ±23 (+13 %)

[a]Increase in percent compared to control cells.

Regardless of the method used to differentiate neuroblastoma
cells and of the method used to recover the cells (scraping or treatment with EDTA), the differentiated cells always exhibit an increased
amount of total gangliosides, as compared with the non differentiated
cells. However, this increase differs notably with respect to the
clone tested and the method used to differentiate the cells (Table 1).
It should be noted that dibutyryl cyclic AMP, which produces a more
extensive biochemical and morphological differentiation than serum
withdrawal or BrdU, also increases the amount of gangliosides to a
greater extent.

A change in the ganglioside pattern was observed after differentiation of all the clones tested. This change was more or less extensive and varied according to the clone tested (Table 2).

Differentiation of clone Neuro 2A induced a net decrease in GDla
and a corresponding increase in the monosialogangliosides GM3 and
GM2. Better was the morphological differentiation, more the GDla
decreased (compare serum deprivation vs. dbcAMP). An opposite result
was obtained with NlE 115 where GDla increased at the expense of GM2
and GM3 during differentiation. We must also point out that, regardless of the clone tested or the method used to differentiate the
cells, we have never observed the synthesis of tri- and tetrasialogangliosides which account for up to 15 % of the total brain gangliosides (NORTON et al., 1975).

Table 2.   Ganglioside Pattern of Some Neuroblastoma Clones

Clones	Percent of total gangliosides					
	GM3	GM2	GM1	GD3	GD1a	GD1b
Neuro 2A						
Control	21.3	44.9	Traces	–	33.8	–
–Serum	33.6	39.9	–	5.3	21.2	–
+dbcAMP	24.4	49.9	4.9	7.6	13.2	–
N1E 115						
Control	4.7	63.9	8.7	Traces	22.7	–
–Serum	–	23.6	14.2	15.9	46.3	–
+dbcAMP	Traces	22.6	18.3	Traces	58.9	0.2
M1						
Control	5.7	23.9	12.6	–	53.1	4.7
+BrdU	1.0	17.3	13.7	2.8	57.4	7.8
+dbcAMP	20.5	26.3	13.4	Traces	39.1	0.5

CULVER *et al.* (1977) have reported that treatment of N1E 115 cells with dimethyl-triazeno-imidazole-carboxamide (NSC 45388) induces a biochemical differentiation (measured by the activity of some enzymes involved in neurotransmitter metabolism) and a cessation of cell growth. However, these events were not accompanied by a change in the cell morphology. Analysis of the gangliosides of control and NSC 45388-treated cells shows no qualitative nor quantitative differences (REBEL and PRASAD, unpublished results).

MOSKAL *et al.* (1974) then DUFFARD *et al.* (1977) reported that neuroblastoma cells have the enzyme activities involved in the synthesis of GM3, GM2, GD3 and GD1a. A low incorporation of CMP-NeuAc into GD1b of N1E 115 cells has been observed by MOSKAL *et al.* (1974). By growing N1E 115 cells in roller tubes, MOSKAL *et al.* (1974) have shown also that addition of dibutyryl cyclic AMP to the medium does not change the incorporation of neuraminic acid into lactosylceramide, GM3 or GD1b. A doubling of the incorporation into GM1 was observed.

PRESPER *et al.* (1978) have studied the synthesis of fucoglycolipids in neuroblastoma cells. Of the two clones studied (N18 and IRM 32), only IRM 32 cells showed a high fucosyltransferase activity, leading to the synthesis of fucolipids of the blood group B1 and H1. After differentiation of these cells by BrdU, a strong inhibition of the synthesis of these fucolipids occurred. By contrast, no changes in the incorporation of fucose were observed when the cells were treated with 6-mercaptoguanosine which produces also a morphological differentiation of the cells.

Table 3. Gangliosides NN Cells

	GM3	GM2	GM1	GD3	GD1a	GD1b
Control	66.8[a]	–	1.3	30.6	2.3	–
Differentiated	30.9	2.8	23.5	21.4	20.6	1.6

[a]In percent of total gangliosides. Total lipid neuraminic acid was 3.9 and 4.0 nmol/mg protein in control and differentiated cells

GLIAL CELLS

The two clones of glial cells used (NN and C6 cells) do not undergo a morphological differentiation when treated with dibutyryl cyclic AMP under the same conditions as neuroblastoma. Significant but transitory morphological changes have been obtained when C6 cells were grown in presence of norepinephrine (OEY, 1975). Under the same conditions, we observed a parallel change of the ganglioside amounts and of the cell morphology. When the morphological change reached its maximum, a 65 % increase in the amount of total gangliosides was found. Twenty four hours after addition of norepinephrine to the culture medium, cells have recovered their normal morphology and the ganglioside levels have practically returned to the basal value (REBEL and OEY, unpublished results). GM3 is practically the only ganglioside present in C6 cells (ROBERT et al., 1977). Thin-layer chromatography revealed that no other gangliosides could be found in the differentiated C6 cells.

When grown in presence of BrdU for longer than 12 days, NN cells assumed the morphology of mature astroblasts. A complete study of the different membrane lipids has been conducted on normal and differentiated NN cells (ROBERT, MANDEL and REBEL, submitted). No significant changes in the phospholipids, cholesterol or neutral glycolipids were observed between the two types of cells. On the contrary, a notable change was observed with the gangliosides. If the amount of total gangliosides did not vary, a strong increase of GM1 and GD1a was observed in the differentiated cells (Table 3).

The change in the morphology of the NN cells appears only after 10 to 12 days of culture in the presence of BrdU. Using the percentage of GM1 as a marker to follow the change in the gangliosides, we have observed that this ganglioside does not account for more than 3 % of the total gangliosides up to the 10th day. It increased to nearly 8 % at the 12th day, and then reached 20 % on the 14th day.

DUFFARD et al. (1978) have shown that the GM3-N-acetylgalactosaminyltransferase activity is very low in undifferentiated NN cells. However, these cells possess high GM1 and GD1a synthesizing activity.

Therefore, our results suggest an induction of the GM2 synthesizing enzyme during growth in presence of BrdU. However, this increase of galactosaminyltransferase activity is likely to be limited, as the amount of GM2 ganglioside in differentiated NN cells remains very low. This contrasts with neuroblastoma cells, which are characterized by a high GM3-N-acetylgalactosaminyltransferase activity (DUFFARD *et al.*, 1977) and a high GM2 level (see Table 2, and YOGEESWARAN *et al.*, 1973).

As occurs with some normal cells, the morphology of NN cells, growing in normal culture medium, changes with the cell density. This change is probably related to the "cell density inhibition of growth" which alters the cytoskeleton (FOLKMAN and MOSCONA, 1978) and is not considered to be a differentiation. Analysis of cells recovered at low or high density revealed that both types of cells had the same ganglioside pattern. GM3 and GD3, accounting for more than 95 % of the total gangliosides.

OTHER CELLS

Few other systems of differentiation have been studied. The most physiological system is probably the intestinal mucosa. This tissue is characterized by the differentiation of crypt cells into villus cells. GM3 is the main ganglioside found in intestinal mucosa. GLICK-MAN and BOUNHOURS (1976) have reported a ten-fold increase in the amount of gangliosides during the differentiation of crypt cells. This change is the consequence of a strong increase of the lactosyl-ceramide N-acetyl-neuraminyltransferase.

Myoblast differentiation is accompanied by a strong morphological change. Myoblast cells contain besides GM3 which predominates, GM2, GM1 and GD1a. Quantitative or qualitative changes could not be detected when myoblasts and myotubes were compared (WHATLEY *et al.*, 1976). However, a net transitory increase of the incorporation of ^{14}C-galactose into all the neutral glycolipids and gangliosides occurred during the fusion period (McEVOY and ELLIS, 1977). Simultaneously, an increase in the amount of GD1a was observed (WHATLEY *et al.*, 1976). Mutant myoblasts, which are unable to differentiate, have a simplified ganglioside pattern, containing rather exclusively GM3 (WHATLEY *et al.*, 1976).

Lymphocyte is a cell which in tissue culture grows only in suspension. Lymphoblast cells undergo a biochemical differentiation giving rise to different kinds of lymphocytes. Comparing lymphoblastoid cells with peripheral lymphocytes, LEVIS *et al.* (1976) found that the amoung of GM3 (the only ganglioside of lymphoblastoid cells) markedly decreases in peripheral lymphocytes. In contrast, a notable increase of the neutral glycolipids is observed.

GENERAL CONCLUSION

Morphological differentiation of cells grown in monolayer is accompanied by a change in their gangliosides. We generally observe an increase in the amount of these glycolipids. The phenomenon is probably associated with the induction of enzymes involved in gangliosides synthesis. Such an induction has been proved in the case of HeLa cells treated with butyrate. Similarly, the changes observed with glial cells, intestinal mucosa cells and myoblasts could be explained by the same phenomenon.

The time-course of the changes in gangliosides in norepinephrine treated C6 cells or BrdU treated NN cells, as well as the results obtained with neuroblastoma differentiated with NSC 45388, indicate a good correlation between the changes in the cell morphology and the change in the gangliosides. However, experiments performed with HeLa cells treated with 'utyrate in presence of colcemid, show that the ganglioside changes could be obtained in absence of the morphological differentiation. More experiments are needed to understand the exact mechanism of the observed changes.

REFERENCES

CIESIELSKI-TRESKA J., ROBERT J., REBEL G. and MANDEL P. (1977): Gangliosides of active and inactive neuroblastoma clones. Differentiation 8, 31-37.

CRITCHLEY D.R. and VICKER M.G. (1977): Glycolipids as membrane receptors important in growth regulation and cell-cell interactions, in "Dynamic Aspect of Cell Surface Organisation", POSTE G. and NICOLSON G.L., Eds., Elsevier (Amsterdam), pp. 307-370.

CULVER B., SAHU S.K., VERNADAKIS A. and PRASAD K.N. (1977): Effect of 5-(3,3-dimethyl-1-triazeno) imidazole-4-carboxamide (NEC 45388, DTIC) on neuroblastoma cells in culture. Biochem. Biophys. Res. Commun. 76, 778-783.

DUFFARD R.O., FISHMAN P.H., BRADLEY R.M., LAUTER C.J., BRADY R.O. and TRAMS E.G. (1977): Ganglioside composition and biosynthesis in cultured cells derived from CNS. J. Neurochem. 28, 1161-1166.

FISHMAN P.H., BRADLEY R.M. and HENNEBERRY R.C. (1976): Butyrate induced glycolipid biosynthesis in HeLa cells: properties of the induced sialyltransferase. Arch. Biochem. Biophys. 172, 618-626.

FISHMAN P.H., SIMMONS J.L., BRADY R.O. and FREESE E. (1974): Induction of glycolipid biosynthesis by sodium butyrate in the HeLa cells. Biochem. Biophys. Res. Commun. 59, 292-299.

FOLKMAN J. and MOSCONA A. (1978): Role of cell shape in growth control. Nature 273, 345-348.

GINSBURG E., SALOMON D., SREEVALSAN T. and FREESE E. (1973): Growth inhibition and morphological changes caused by lipophylic acids in mammalian cells. Proc. Nat. Acad. Sci. 70, 2457-2461.

GLICKMAN R.M. and BOUHOURS J.F. (1976): Characterization, distribution and biosynthesis of the major ganglioside of rat intestinal
mucosa. Biochim. Biophys. Acta 424, 17-25.

KREIDER J.W., WADE D.R., ROSENTHAL M. and DENSLEY T. (1975): Maturation and differentiation of B16 Melanoma cells induced by theophyl-
line treatment. J. Nat. Cancer Inst. 54, 1457-1467.

LEVIS G.M., KARLI J.N..and CRUMPTON M.J. (1976): Plasma membrane glycosphingolipids (GSLs) of the human lymphoblastoid cell line BRI 8
and differences between the GLSs of BRI 8 cells and those of peripheral lymphocytes. Biochem. Biophys. Res. Commun. 68, 336-342.

MACHER B.A., LOCKNEY M., MOSKAL J.R., FUNG Y.K. and SWEELEY C.C.
(1978): Studies on the mechanism of butyrate-induced morphological
changes in KB cells. Exptl. Cell Res. 117, 95-102.

MANNER G. and KULEBA M. (1974): Effect of dibutyryl-cyclic AMP on
collagen and non-collagen protein synthesis in cultured human
cells. Connective Tissue Res. 2, 167-176.

McEVOY F.A. and ELLIS D.E. (1977): Glycolipids and myoblast differentiation. Biochem. Soc. Trans. 5, 1719-1721.

MOSKAL J.R., GARDNER D.A. and BASU S. (1974): Changes in glycolipid
glycosyltransferase and glutamate decarboxylase and their relationship to differentiation in neuroblastoma cells. Biochem. Biophys.
Res. Commun. 61, 751-758.

NORTON W.T., ABE T., PODUSLO S.E. and DE VRIES G.H. (1975): The lipid
composition of isolated brain cells and axon. J. Neurosci. Res. 1,
57-75.

OEY J. (1975): Noradrenaline induces morphological alterations in
nucleated and enucleated C6 glioma cells. Nature 257, 317-319.

PRESPER K.A., BASU M. and BASU S. (1978): Biosynthesis *in vitro* of
fucose-containing glycosphingolipids in human neuroblastoma IMR-32
cells. Proc. Nat. Acad. Sci. 75, 289-293.

REBEL G., CIESIELSKI-TRESKA J. and MANDEL P. (1973): Etude des gangliosides d'un clone de cellules de neuroblastome. C. R. Acad. Sci.
(Paris) série D 277, 1193-1195.

ROBERT J., REBEL G. and MANDEL P. (1977): Glycosphingolipids from
cultured astroblasts. J. Lipid Res. 18, 517-522.

SCHWARZ R., COLARUSSO L. and DOTY P. (1976): Maintenance of differ-
rentiation in primary cultures of avian tendon cells. Exptl. Cell
Res. 102, 63-71.

SIMMONS J.L., FISHMAN P.H., FREESE E. and BRADY R.O. (1975): Morphological alterations and ganglioside sialyltransferase activity induced by small fatty acids in HeLa cells. J. Cell Biol. 66, 414-424.

WHATLEY R., NG S.K.C., ROGERS J., McMURRAY W.C. and SANWAL B.D.
(1976): Developmental changes in gangliosides during myogenesis of
a rat myoblast cell line and its drug resistant variants. Biochem.
Biophys. Res. Commun. 70, 180-185.

YOGEESWARAN G., MURRAY R.K., PEARSON M.L., SANWAL B.D., McMORRIS F.
A. and RUDDLE F.H. (1973): Glycosphingolipids of clonal lines of
mouse neuroblastoma and neuroblastoma X L cell hybrids. J. Biol.
Chem. 248, 1231-1239.

GANGLIOSIDES OF THE CNS MYELIN MEMBRANE

R.W. Ledeen, F.B. Cochran, R.K. Yu, F.G.
Samuels and J.E. Haley

Depts. of Neurology and Biochemistry, Albert
Einstein Coll. of Med. and Dept. of Neurology,
Yale Univ.

BACKGROUND

Myelin gangliosides have a number of unusual
properties that set them apart from gangliosides of
most other nervous system membranes. One is their low
concentration, which in the case of rat amounts to
about one-tenth that in synaptic plasma membranes. This
may account for the fact that when first detected in
isolated myelin[1,2] they were believed to represent
neuronal contamination. Subsequent studies of Suzuki
and coworkers[3,4,5] provided strong if not conclusive
evidence that they are intrinsic to myelin itself. A
key finding in those studies was the distinctive pattern
of molecular distribution, the main feature of which
was a high proportion of G_{M1}. The same phenomenon was
later observed in myelin from man[6] and mouse[7]. We have
undertaken a survey of several other vertebrates to
assess possible variations in pattern and concentration.

Another distinctive feature is the presence of
sialosylgalactosyl ceramide ($G_{M4} = G_7$) in at least some
species. In mature human myelin it accounts for 15-20%
of total myelin ganglioside. This unusual ganglioside,
shown to be closely related structurally[6] and probably

167

metabolically[8] to myelin galactosyl ceramide, was barely
detectable in gray matter[6] and completely undetectable
in peripheral nerve[9] and extraneural tissues[10]. The
recent demonstration[11] of G_{M4} in isolated human oligodendroglia indicated its probable biosynthetic origin;
that study indicated a limited pool size in these cells
with over 85% of white matter G_{M4} being localized in
myelin. Study of mouse brain[7] myelin revealed considerably lower levels of G_{M4} while data for other species
showed further wide variations in regard to this
component (see below).

Finally, unusual developmental changes constitute
yet another distinctive feature of myelin gangliosides.
Thus, myelin from young rats, while still possessing
G_{M1} as the major component, had relatively more di-
and trisialogangliosides than the adult pattern[3]. G_{M4}
also shows an increase in development but on a different
time-scale then G_{M1}; in the mouse it was undetectable
until 35 days of age, after which it slowly increased
over several months[7]. Human myelin also showed increasing G_{M4} with age[6]. An additional aspect of development
was revealed in the fact that total ganglioside concentration in mouse myelin doubled between the youngest
(23 days) and oldest (490 days) animals[7]. This is very
likely a species-related phenomenon.

ORIGIN OF MYELIN GANGLIOSIDES

While the studies cited above support the concept
of gangliosides as true myelin constituents, the
evidence in toto does not rigorously exclude the possibility that they might belong to contaminating structures. Residual quantities of tightly adhering membranes
could, depending on their composition, account for a
greater or lesser portion of the observed ganglioside.
The axolemma membrane is of particular concern in this
regard because of indications that it adheres tightly
to myelin during the isolation process[12] and that it
has a relatively high ganglioside concentration[13].

To study this question we exploited the finding of
DeVries[14] that EGTA is an effective agent for stripping
axolemma from myelin. This was analogous to the report

by Elam[15] that another chelating agent removed radiolabeled substances from myelin which originated in the
axon. We studied this question by comparing myelin
isolated by the conventional procedure of Norton and
Poduslo[16] with that obtained by a modification of the
DeVries procedure in which the myelin was treated 4
times with 10 mM EGTA to remove as much of the axolemma
as possible. Myelin protein patterns determined by
SDS polyacrylamide gel electrophoresis showed progressive loss of high molecular weight components with each
EGTA treatment, the final product having fewer of these
bands than conventional myelin. The integrity of the
myelin did not appear damaged by the EGTA.

Myelin was then isolated from rabbit optic tracts
following intraocular injection of $[3-^{14}C]$serine as
precursor for labeling lipids and proteins, some of
which migrated into the optic tract via axonal transport.
Seven days after injection the bulk of label in the
optic tract was in the axolemma and other axonal elements, thus providing a marker for these potential
contaminants. Expressing the results as contralateral-
ipsilateral differences (since only one eye was injected)
myelin isolated by the conventional method was found to
have approximately 3 times as much radioactivity in
total lipid and 5 times as much in total protein as
myelin isolated by the EGTA procedure. When the
experiment was repeated with $[^{3}H]$N-acetylmannosamine,
gangliosides in conventional myelin had 40% more
radioactivity than those in the EGTA preparation*. The
amount of radioactivity recovered in the isolated
myelin was approximately 6% of that present in the
original homogenate.

---

*The greater difference found in the first experiment
is believed due to the fact that the dissected optic
tracts were stored overnight in the cold prior to myelin
isolation, whereas in the $[^{3}H]$N-acetylmannosamine
experiment the dissected tissues were processed immediately. Studies with human brain[6] and other tissues
previously indicated the importance of isolating myelin
from fresh tissues to minimize contamination.

To compare ganglioside concentrations, myelin was obtained from brainstems of rats and rabbits by the 2 above procedures and the gangliosides isolated[6] and quantified[17] as previously described. The results indicated virtually no difference in ganglioside concentration or TLC pattern. These experiments showed that while EGTA effectively removes radiolabeled gangliosides believed to originate in axolemma remnants, the quantity of ganglioside thus removed is very small. This provides additional evidence that the majority and possibly all of the gangliosides present in isolated myelin are intrinsic to that membrane.

SPECIES COMPARISON

Fully mature animals were used in this study to enable species comparison without the additional complication of developmental changes. Myelin was isolated from white matter when the brain was large enough (human, monkey, goat, bull and rabbit) and from whole brain for the remainder. The procedure was that of Norton and Poduslo[16] with an additional purification step consisting of a third discontinuous sucrose gradient; this gave preparations equal in purity to those of EGTA-treated myelin when fresh brain was employed. Gangliosides were isolated by the DEAE-Sephadex method[6] in order to maximize recovery and avoid selective loss of the less polar gangliosides. Thin-layer chromatography was carried out with HP-TLC plates (Silica gel 60) using chloroform-methanol-water (50:40: 10) with 0.02% $CaCl_2$ $2H_2O$. Plates were sprayed with resorcinol[18] and patterns quantified by densitometric scanning[19].

The species surveyed showed a considerable range of concentrations and pattern distributions (Table 1). Rhesus monkey and human were similar, both having appreciable G_{M4} in addition to G_{M1}. The cebus monkey and chimpanzee (not shown) had similar concentrations but somewhat less G_{M4}. At the other end of the evolutionary scale, codfish and frog had the lowest overall concentrations and markedly different patterns that were devoid of G_{M4} (Fig. 1). Polysialoganglio-sides predominated and G_{M1} was no longer the major

TABLE 1

Ganglioside Concentrations and Patterns in Vertebrate Brain Myelin.

Species	Sialic Acid μg/100 mg	Pattern Distribution[a]									
		G_{M4}	G_{M3}	G_{M2}	G_{M1}	G_{D3}	G_{D1a}	G_{D2}	G_{D1b}	G_{T1}	G_{Q1}
Human	62	20	4	3	32	2	8	1	20	9	2
Monkey[b]	69	22	-	5	45	-	5	1	10	8	3
Bull	56	1	-	5	48	-	8	-	21	9	5
Goat	53	8	1	5	45	-	7	2	19	9	4
Rabbit	76	0	1	2	59	1	6	1	16	10	5
Guinea Pig[c]	91	5	1	3	52	1	8	1	14	12	5
Frog[d]	44	0	-	5	20	3	5	25	3	13	13
Codfish[d]	31	0	0	6	4	2	6	4	6	17	29
Chicken	229	30	2	3	29	9	10	1	11	6	3
Pigeon	246	24	2	1	27	7	14	-	11	11	3

[a]Distribution expressed as % of lipid-bound sialic acid. Dashes indicate less than 0.5%. Symbols are those of Svennerholm[20].

[b]These data pertain to Rhesus only.

[c]G_{M1} in this sample and others includes the minor band running just ahead on TLC.

[d]Some band assignments are tentative. In addition to those shown here, frog and codfish had approximately 10% and 13% Gp1, respectively; codfish had another 14% in other (unlisted) components.

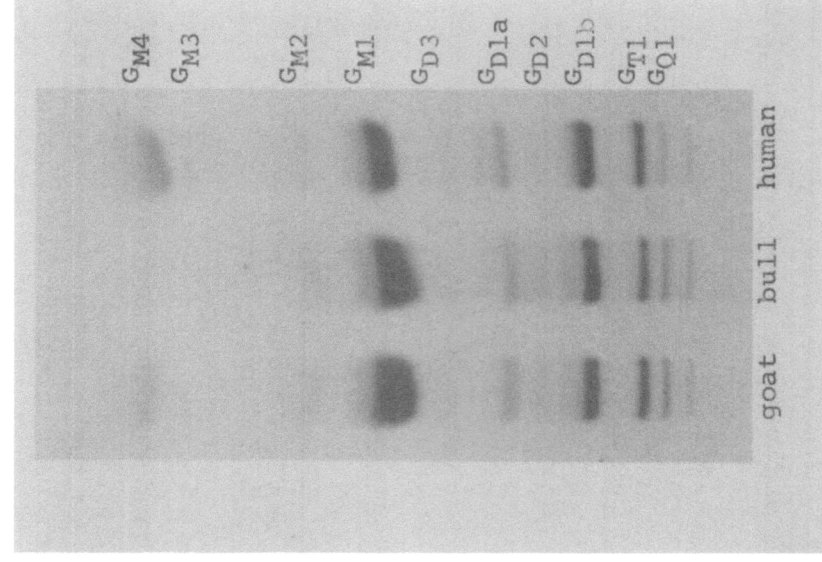

Fig. 2

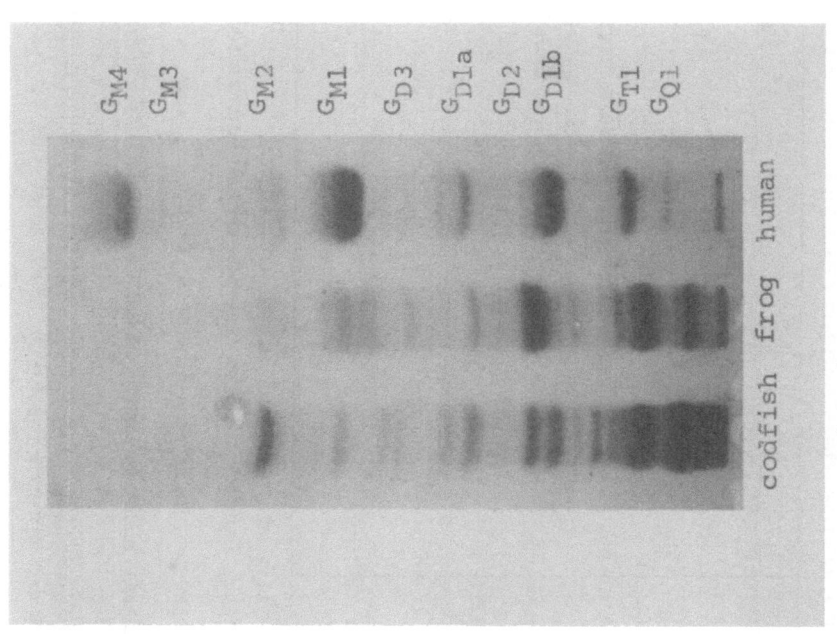

Fig. 1

fraction. The other mammals had concentrations similar to or somewhat higher than the primates with patterns resembling those of rat[3] and mouse[7] in the predominance of G_{M1} (Figs. 2 & 3). The amount of G_{M4} in these mammals was variable; it was generally below the human level and in one species (rabbit) was undetectable.

Avian myelin proved exceptional in having total ganglioside concentrations 2- to 4-fold higher than mammalian myelin and an unusual pattern (Fig. 4) in which G_{M4} was approximately equal to G_{M1}. The small difference in TLC migratory rates of G_{M1} reflected fatty acid differences. Developmentally, avian myelin was unusual in having a relatively constant ganglioside level and pattern from hatching.

CONCLUSIONS

Evidence is presented that an insignificant portion of the gangliosides present in purified myelin derive from adhering axolemma, increasing the likelihood they are intrinsic to myelin itself. Species comparisons revealed higher myelin ganglioside concentrations and higher proportions of monosialogangliosides in mammals than in fish or amphibia, analogous to the trend previously demonstrated for whole brain gangliosides[21]. Human and rhesus monkey had the highest G_{M4} concentrations among mammals while fish and amphibia had none of this component. Avian myelin gangliosides exhibited unique characteristics which included high overall concentration and equivalent proportions of G_{M1} and G_{M4}. It may be noted that myelin from spinal cord[22] had reduced ganglioside concentration but similar pattern as brain myelin, while that from peripheral nerve[9] differed in both respects.

The function of gangliosides in myelin, as in other membranes, is still a matter of speculation. Although they are minor components, their molar concentration was shown[9] to approximate that of the P_1 basic protein in both CNS and PNS myelin, leading to the suggestion of an in situ association. The possibility that myelin gangliosides can influence the conformation of this protein is under investigation.

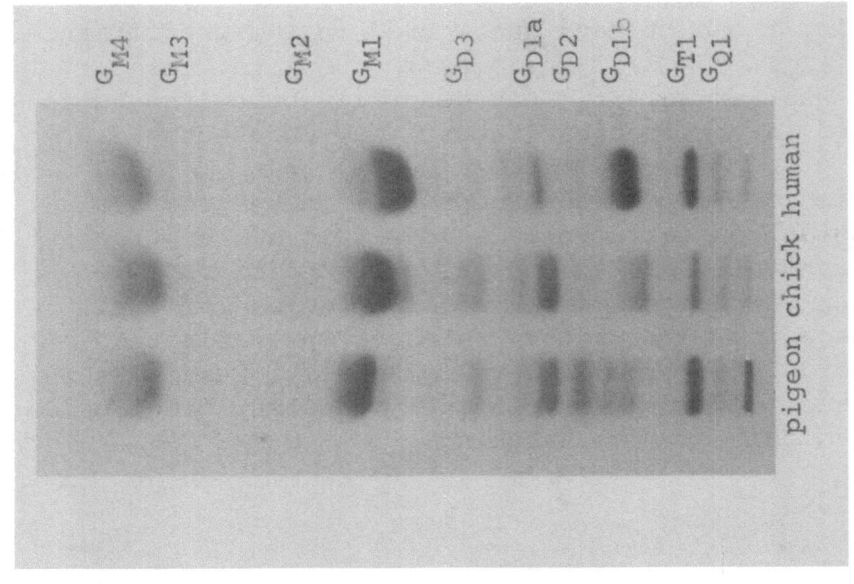

pigeon chick human

Fig. 4

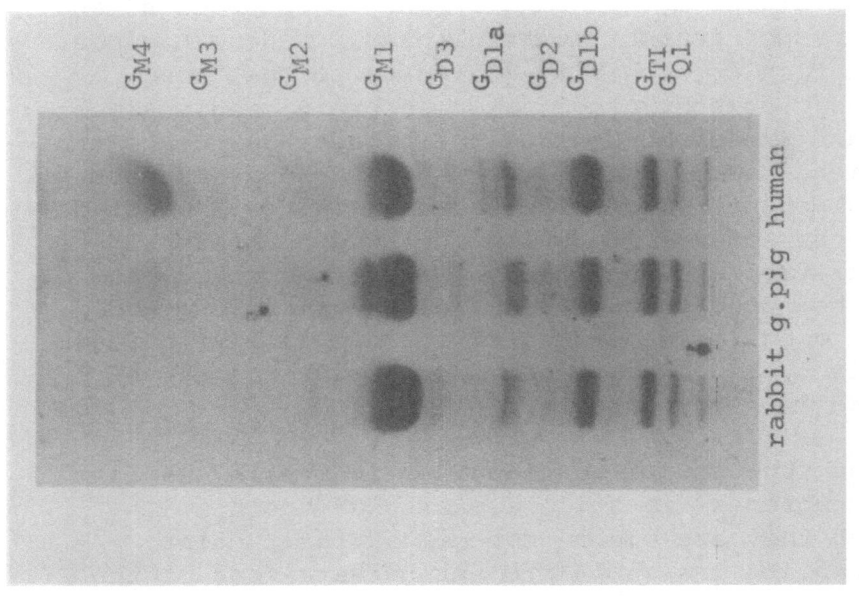

rabbit g.pig human

Fig. 3

ACKNOWLEDGEMENTS

This work was supported by grants NS 04834, NS 03356, NS 10931, and NS 11853 from the National Institutes of Health, U.S. Public Health Service.

REFERENCES

1. W. T. Norton and L. A. Autilio, The lipid composition of purified bovine brain myelin, J. Neurochem. 13:213 (1966).
2. E. F. Soto, L. S. de Bohner and M. D. C. Calvino, Chemical composition of myelin and other sub-cellular fractions isolated from bovine white matter, J. Neurochem. 13:989 (1966).
3. K. Suzuki, S. E. Poduslo and W. T. Norton, Gangliosides in the myelin fraction of developing rats, Biochim. Biophys. Acta 144:375 (1967).
4. K. Suzuki, J. F. Poduslo and S. E. Poduslo, Further evidence for a specific ganglioside fraction closely associated with myelin, Biochim. Biophys. Acta 152:576 (1968).
5. K. Suzuki, Formation and turnover of myelin ganglioside, J. Neurochem. 17:209 (1970).
6. R. W. Ledeen, R. K. Yu and L. F. Eng, Gangliosides of human myelin: sialosylgalactosylceramide (G_7) as a major component, J. Neurochem. 21:829 (1973).
7. R. K. Yu and S. I. Yen, Gangliosides in developing mouse brain myelin, J. Neurochem. 25:229 (1975).
8. R. K. Yu and S. H. Lee, In vitro biosynthesis of sialosylgalactosylceramide (G_7) by mouse brain microsomes, J. Biol. Chem. 251:198 (1976).
9. J. W. Fong, R. W. Ledeen, S. K. Kundu and S. W. Brostoff, Gangliosides of peripheral nerve myelin, J. Neurochem. 26:157 (1976).
10. T. N. Seyfried, S. Ando and R. K. Yu, Isolation and characterization of human liver hematoside, J. Lipid Res. 19:538 (1978).

11. R. K. Yu and K. Iqbal, Sialosylgalactosyl ceramide
     as a specific marker for human myelin and
     oligodendroglial perikarya: gangliosides of
     human myelin, oligodendroglia and neurones,
     J. Neurochem. 32:293 (1979).

12. G. H. DeVries, W. T. Norton and C. S. Raine, Axons:
     Isolation from mammalian central nervous
     system, Science 172:1370 (1972).

13. G. H. DeVries, W. Payne and C. Zmachinski,
     Molecular composition of a rat CNS axolemma-
     enriched fraction, Trans. Am. Soc. Neurochem.
     10:159 (1979).

14. G. H. DeVries, Isolation of axolemma-enriched
     fractions from bovine central nervous system,
     Neurosci. Letts. 3:117 (1976).

15. J. S. Elam, Dissociation of axonally transported
     proteins from myelin with EDTA, J. Neurochem.
     31:351 (1978).

16. W. T. Norton and S. E. Poduslo, Myelination in rat
     brain: method of myelin isolation, J. Neurochem.
     21:749 (1973).

17. R. K. Yu and R. W. Ledeen, Gas-liquid chromato-
     graphic assay of lipid-bound sialic acid:
     measurement of gangliosides in brain of several
     species, J. Lipid Res. 11:506 (1970).

18. L. Svennerholm, Quantitative estimation of sialic
     acids. II. A colorimetric resorcinol-hydro-
     chloric acid method, Biochim. Biophys. Acta
     24:604 (1957).

19. S. Ando, N,-C. Chang and R. K. Yu, High-performance
     thin-layer chromatography and densitometric
     determination of brain ganglioside compositions
     of several species, Analyt. Biochem. 89:437
     (1978).

20. L. Svennerholm, Chromatographic separation of
     human brain gangliosides, J. Neurochem. 10:613
     (1963).

21. N. F. Avrova, Brain ganglioside patterns of
     vertebrates, J. Neurochem. 18:667 (1971).

22. K. Ueno, S. Ando and R. K. Yu, Gangliosides of
     human, cat, and rabbit spinal cords and cord
     myelin, J. Lipid Res. 19:863 (1978).

GANGLIOSIDES IN FISH BRAIN

Natalia F. Avrova

Sechenov Institute of Evolutionary Physiology and
Biochemistry, U.S.S.R. Academy of Sciences,
194223 Leningrad, U.S.S.R.

Gangliosides are characteristic components of plasma membranes
of the nervous cells. The complexity of ganglioside molecule determines the wide range of its variability, as all the components may
vary due to the systematic position of species, the age of animals
and the environmental factors, the organ specificity of composition
being pronounced as well. Comparative biochemical studies are of
importance for better understanding of evolutionary interrelationships of various animal groups and may throw light on functional
roles of the substances studied. Progressive development of the
central nervous system is characteristic of the process of phylogen-
etic development of vertebrates. In this connection brain appears
to be interesting object especially taking into account ganglioside
localization. In our studies of vertebrate brain gangliosides, the
most detailed study was done on fish brain. The fishes are extremely
numerous group diverse in their habitats, mode of life and level of
development of the nervous system.

The fishes were caught in the expeditions on research ships.
Brain gangliosides were extracted, their sialic acid and sphingosine
content was determined. Gangliosides were separated by thin-layer
chromatography on silica gel. Composition of hexoses, hexosamines,
sialic acids, sphingoids and fatty acids was determined by the
methods of gas-liquid chromatography. The method of chromato-mass-
spectrometry was also used for their identification. Ganglioside
structure was studied by the methods of partial acid and enzymatic
(by neuraminidase) hydrolysis. In some cases the methods of sequential enzymatic hydrolysis and the method of permethylation were used.

Gangliosides are present in the brain of representatives of all
vertebrate classes. The higher the level of development of the

nervous system, the higher the brain ganglioside concentration. In
progressive evolution of vertebrates the relative content of brain
polysialogangliosides decreased, the degree of ganglioside fatty
acid saturation and the relative content of sphingoids with 20
carbon atoms increased etc. From this standpoint it is possible
to distinguish between the progressive and primitive features of
ganglioside composition and structure.

Investigation of great number of fish species (up to 60) provides
evidence that 3 large taxons (cartilaginous fishes, teleosts and
ganoids) possess distinct and pronounced differences in the brain
ganglioside composition and content. In teleost brain, gangliosides
with gangliotetraosyl chain predominate; tetra- and pentasialoganglioside content is high (AVROVA, 1971; KREPS et al., 1975). Large
amounts of polyenoic fatty acids (mainly of C22:6ω3) were shown in
brain gangliosides of teleosts dwelling in cold water. Up to the
present polyenoic acids have not been found in brain gangliosides.

In brain gangliosides of all the fish species studied, sphingosine predominates (80 to 99 % of total sphingoids). Only N-acetyl
derivatives of neuraminic acid were found in brain gangliosides of
representatives of various fish taxons, NeuGc being practically
absent.

In the brain of cartilaginous fishes (elasmobranches) ganglioside content is higher than in teleosts (Table 1). Relatively low
content of tetra- and pentasialogangliosides may also be considered
a progressive feature of chemical organization of the brain in
cartilaginous fish. The predominance of gangliosides with lactosyl
and gangliotriaosyl chain is characteristic of elasmobranch brain.
The major gangliosides were shown to be $II^3(NeuAc)_2$-LacCer,
$II^3(NeuAc)_2$-GgOse$_3$Cer, $II^3(NeuAc)_3$LacCer, II^3NeuAc-LacCer etc.
(AVROVA et al., 1979, in press).

Ganglioside composition in the brain of elasmobranches varies
from one species to another and depends on taxonomic position of
animals. The brain of some elasmobranches has high ganglioside
content, high degree of fatty acid saturation, high content of
ganglioside stearic acid, low content of tetra- and pentasialogangliosides and other characters typical to brain gangliosides of
higher vertebrates. These species may be characterized as
"progressive" elasmobranches, while the species with low brain
ganglioside content, high content of polysialogangliosides etc.
may be referred to as the "primitive" forms. From this point of
view the elasmobranches studied form the following 3 groups:
"primitive" (sharks *Heptranchias*, *Squalus*, *Isistius*, *Dalatias* and
rays *Raja*, *Rhinobatos*), "intermediate" (sharks *Squatina*), and
"progressive" (sharks *Carcharhinus*, *Orectolobus* and rays *Dasyatis*,
Myliobatis).

Table 1. Gangliosides in Brain of Cartilaginous Fishes, Teleosts and Ganoids (Means ± s.e.m.)

Ganglioside characteristics, units of measurement	Taxon		
	Cartilaginous fishes, 14 species	Teleosts, up to 37 species	Ganoids 4 species
Content, mcg of sialic acid per 1 g of wet brain weight	High 385 ±38	Low 257 ±17	Average 300 ±17
Tetra- and pentasialoganglioside content; % of ganglioside sialic acid	Low 23.5 ±3.4	High 42.4 ±4.8	Very low 11.5 ±0.7
The content of gangliosides with lactosyl and gangliotriaosyl chain; % of sialic acid	Very high 60–80 %	Low 10–20 %	Low 15–20 %
Content of acids C22:1, C24:1; % of total fatty acids	Average 22.0 ±3.3	Low 13.5 ±1.2	High 35.9 ±2.8
Content of polyenoic acids – C22:6 (n-3) and other; % of total fatty acids	Not found or minute amounts	High for cold water species	Not found

This biochemical approach is consistent with comparative morphological data on elasmobranch brain (NORTHCUTT, 1977). In certain aspects elasmobranch brain appears to develop in parallel with the brain of the main line of vertebrates. Classification of elasmobranches encounters numerous difficulties, many classification schemes have been suggested. Our data are in agreement with evolutionary interrelationships postulated in classification of COMPAGNO (1973).

In brain of ganoids (great sturgeon, Russian sturgeon etc.) gangliosides with gangliotetraosyl chain predominate (in contrast to elasmobranch brain), the content of tetra- and pentasialogangliosides is low (in contast to teleost brain). Fatty acid composition of ganoid brain gangliosides has unic features as well. It confirms the conception that ganoids should be considered not as a superorder of bony fishes (*Osteichthyes*), but as a taxon of a higher range.

Table 2. Fatty Acid Composition of Brain Gangliosides in
Fishes from Various Thermal Habitats

Species, temperatures of dwelling	Fatty acids			
	Satu-rated	C18:0	Mono-enoic	Poly-enoic
ELASMOBRANCHES				
From cold water				
Raja whitleyi (1–2°C)	27.8	11.2	72.2	Traces
Dalatias philippsi (2–4°C)	38.2	17.9	58.1	3.7
Orectolobus ornatus (16°C)	42.9	28.6	44.3	12.8
From warm water				
Raja clavata (16°C)	45.9	30.0	54.1	–
Squalus acanthias (16°C)	46.6	24.9	53.4	–
Carcharhinus longimanus (25°C)	88.4	72.5	11.6	–
TELEOSTS				
From cold water				
Antimora rostrata (2–6°C)	30.7	18.5	48.0	21.3
Comephorus dybowski (0–7°C)	47.3	28.4	35.4	17.3
Cottocomephorus inermis (0–9°C)	45.0	30.7	25.3	29.7
From warm water				
Lethrinus chrysostomus (23°C)	79.8	52.7	16.4	3.8
Lactophris tricornis (25°C)	93.0	75.7	7.0	–
Coryphaena hippurus (29°C)	79.6	57.6	20.4	–

Fatty acid composition of gangliosides from fish brain depends
on the ecology as well. Positive correlation between the degree of
saturation, the relative content of stearic acid and the temperature of environment was found, when 45 species were studied. At low
temperature these values are low, especially for deep sea fishes.
Both the ambient temperature and hydrostatic pressure change with
depth, the effect of their change on fatty acid composition being
synergistic. Thus, in brain gangliosides of the ray _Raja whitleyi_
(2000 m) saturated fatty acids constitute 28, stearic acid – 11 %,
while the average values for cartilaginous fishes are 60 and 41 %,
respectively. In brain gangliosides of the teleost _Antimora rostrata_
(1200–2300 m) saturated fatty acids constitute approximately 31,
monoenoic – 48, polyenoic – 21 % of total fatty acids, while in
brain gangliosides of tropical species living in upper strata of the
sea polyenoic acids are absent and saturated fatty acids constitute
more than 80 % of the total (Table 2).

In the evolution of vertebrates the degree of ganglioside fatty acid saturation increases (AVROVA and ZABELINSKII, 1971). It may be also considered as thermal adaptation since in mammals and birds cell membranes function at relatively high temperatures (37-40°C). Thus gangliosides appear to take part in regulation of physical chemical characteristics of the membrane, in particular of its "fluidity".

Another hydrophobic component of ganglioside molecule is sphingoid, it forms another carbon chain also situated in the membrane lipid bilayer. The ecological conditions studied appear not to influence the sphingoid composition. For example in brain gangliosides of sea elasmobranches *Raja whitleyi* and *Dalatias philippsi* saturated sphingoids constitute from 1 to 7 %, while in thermophilic species their content is 11-12 %. Composition of sphingoids from brain gangliosides of teleosts does not depend on the ambient temperature as well, sphingosine being predominant in all the species studied. The relative content of saturated sphingoids (sphinganine and icosasphinganine) in brain gangliosides of mammals and birds does not show any difference from that of cyclostomes, fishes, amphibia and reptiles. It constitutes 3-15 % of total long chain base content. Thus ganglioside sphingoids appear not to be involved in regulation of membrane fluidity.

High content of tetra- and pentasialogangliosides is characteristic of the brain of cold-blooded animals, especially of teleosts and amphibia, while in the brain of warm-blooded animals tetrasialoganglioside content is very low and pentasialoganglioside is absent. It is tempting to suggest that these peculiarities of brain ganglioside composition are due to differences in the body temperature. However, in the brain of reptiles polysialoganglioside content is very low (AVROVA, 1971), though they belong to ectothermic animals.

BREER (1975) and RAHMANN (1976) studying 4 species of teleosts suggested that the high content of tetra- and pentasialogangliosides is characteristic of species dwelling in cold water. However our study of brain gangliosides in 20 species of teleosts and elasmobranches did not confirm their data. According to our data tetra- and pentasialoganglioside content of the perch *Perca fluviatilis* living in the lakes of Leningrad district and caught in Spring (13°C) does not differ from that in number of tropical species (23-25°C) from the same superorder. Especially low content of these polysialogangliosides (23 and 26 % of ganglioside sialic acid) was found in the brain of fresh water percomorphs *Comephorus dybowski* and *Cottocomephorus grewingki*, dwelling at 0-8°C in the Baikal lake. The studies of elasmobranches do not confirm the conclusions of BREER and RAHMANN as well. For example in the brain of deep sea ray *Raja whitleyi* tetra- and pentasialogangliosides constitute 26 % of ganglioside sialic acid, and in Black Sea ray dwelling in warm water (*Raja clavata*) 33 % (Table 3).

Table 3. Brain Ganglioside Patterns in Fishes from Various
Thermal Habitats

Species, temperatures of dwelling	Ganglioside fraction		
	Tetra- and penta-sialo	Tri-sialo-	Di- and mono-sialo-
	g a n g l i o s i d e s		
ELASMOBRANCHES			
<u>From cold water</u>			
Raja whitleyi (1–2°C)	26.0	14.4	59.6
Heptranchias dakini (9–10°C)	20.2	4.9	74.9
<u>From warm water</u>			
Raja clavata (15°C)	33.5	22.2	44.3
Squalus acanthias (15°C)	29.0	25.6	45.4
Trigonorrhina fasciata (20°C)	18.0	16.9	65.1
TELEOSTS			
<u>From cold water</u>			
Comephorus dybowski (0–7°C)	22.6	15.3	62.1
Cottocomephorus grewingki (1–8°C)	26.1	19.2	54.7
Perca fluviatilis (12°C)	47.5	29.2	23.3
<u>From warm water</u>			
Malacanthus plumieri (25°C)	49.4	17.9	32.7
Haemulon aurolineatus (25°C)	56.9	20.1	23.0
Coryphaena hippurus (29°C)	37.9	8.9	53.2

The absence of correlation is presumably due to the important
role which the carbohydrate component and sphingoid residue play in
the determination of ganglioside specific functions (reception of
toxins and hormones, immunological properties, participation in the
process of cell to cell interaction etc.). HOLMGREN, MANSSON and
SVENNERHOLM (1974) found that the removal of fatty acid from NeuAc-
GgOse$_4$Cer does not cause the loss of its ability to be the receptor
of the cholera toxin: the sialosylgangliotetraosylsphingosine formed
may act as the receptor after incorporation in cell membrane. At
the same time any change in the composition or structure of carbohydrate component of the molecule or removal of sphingosine residue
results in the loss of receptor properties. The antigenic specificity
of ganglioside molecule is as well determined by the composition and
structure of these components of ganglioside molecule, but not by
the nature of fatty acid residue (KARLSSON, 1970).

Thus the composition and structure of carbohydrate components and sphingoids of brain gangliosides in fish species from the same taxons (superorders etc.) do not correlate with such environmental factors as temperature and the depth of dwelling. They depend mainly on the systematic position of the species. The relatively constant composition of sialoglycolipids, its relative independence from the environmental factors appear to be necessary for realization of ganglioside specific function. On the other hand, ganglioside fatty acid composition was found to be affected by the temperature and the depth of fish habitats. Changes in fatty acid composition of gangliosides appear to be involved in adaptive mechanisms which account for operation of nervous cell membranes under varying environmental conditions.

REFERENCES

AVROVA N.F. (1971): Ganglioside patterns of vertebrate brain. J. Neurochem. 18, 667-674.

AVROVA N.F. and ZABELINSKII S.A. (1971): Fatty acid and long chain base composition of vertebrate brain gangliosides. J. Neurochem. 18, 675-681.

BREER H. (1975): Ganglioside pattern and thermal tolerance of fish species. Life Sci. 16, 1459-1463.

COMPAGNO L.J.V. (1973): Interrelationships of living elasmobranches, in "Interrelationships of Fishes", GREENWOOD P.N., MILES R.C. and PATTERSON C., Eds., Academic Press (New York), pp. 15-61.

HOLMGREN J., MANSSON J.E. and SVENNERHOLM L. (1974): Tissue receptor for cholera exotoxin: structural requirement of GM1-ganglioside in toxin binding and inactivation. Med. Biol. 52, 229-233.

ISHIZUKA J. and WIEGANDT H. (1972): An isomer of trisialoganglioside and the structure of tetra- and pentasialogangliosides from fish brain. Biochim. Biophys. Acta 260, 279-289.

KARLSSON K.A. (1970): Sphingolipid long chain bases. Lipids 5, 878-893.

KREPS E.M., AVROVA N.F., CHEBOTARËVA M.A., CHIRKOVSKAYA E.V., KRASILNIKOVA V.I., KRUGLOVA E.E., LEVITINA M.V., OBUKHOVA E.L., POMAZANSKAYA L.F., PRAVDINA N.I. and ZABELINSKII S.A. (1975): Phospholipids and glycolipids in the brain of marine fish. Comp. Biochem. Physiol. 52B, 283-292.

NORTHCUTT R.G. (1977): Elasmobranch central nervous system organization and its possible evolutionary significance. Amer. Zoologist 17, 411-429.

RAHMANN H. (1976): Possible functional role of gangliosides, in "Ganglioside Function", PORCELLATI G., CECCARELLI B. and TETTAMANTI G., Eds. Advances in Experimental Biology, Vol. 71, Plenum Press (New York and London), pp. 151-161.

GANGLIOSIDES OF BRAIN AND OF EXTRANEURAL TISSUES:

STRUCTURAL RELATIONSHIP TO PROTEIN-LINKED GLYCANS

Heikki Rauvala and Jukka Finne
Department of Medical Chemistry
University of Helsinki,Siltavuorenpenger
10A, SF-00170 Helsinki 17, Finland

INTRODUCTION

Several membrane-associated functions involve an interaction between a biologically active agent, such as antibody (Hakomori and Kobata, 1974; Talmadge and Burger, 1975), protein hormone (Fishman and Brady,1976) or toxin (Craig and Cuatrecasas, 1975), and a membrane carbohydrate chain. The carbohydrate chains of both membrane glycolipids and glycoproteins are in principle suited for such specific interactions, as they both contain a wide diversity of specific saccharide structures projecting to the external milieu from the cell.

In studies on the membrane receptor sites of antibodies, hormones or toxins, the responsible structure has often been variously identified as glycolipid or glycoprotein. A classical example is the controversy concerning the chemical nature of the blood group active substances of the erythrocyte membrane. We therefore found it necessary to investigate, whether this kind of controversy might, at least in some cases, be only apparent, and could be explained by the occurrence of similar carbohydrate determinants in glycolipids and glycoproteins.

In the following we will present evidence that similarities between the carbohydrate chains of the sialic

acid-containing glycolipids, the gangliosides, and the
carbohydrate chains of glycoproteins exist. The structural similarities mainly occur at the terminal sugar
sequences of the carbohydrate chains, whereas the parts
near the protein or lipid are often different. The analogies are most evident in the structures of large carbohydrate chains, like the poly(glycosyl) chains of the
erythrocyte membrane, which contain both sialic acid
and blood group determinants in their complex structures.

"CORE STRUCTURES" AND "TERMINAL STRUCTURES" OF THE CARBOHYDRATE CHAINS OF GANGLIOSIDES AND GLYCOPROTEINS

In order to facilitate the description of the structures of gangliosides and glycoproteins, the carbohydrate
chains are divided into core and terminal parts. Most
glycolipids contain a lactose unit, Gal(β1-4)Glc, as
their core, which is linked to ceramide (Hakomori, 1976).
In the alkali-stable N-glycosidically-linked carbohydrate units of glycoproteins a rather invariable core
structure is also found, which consists of mannose and
N-acetylglucosamine (Montreuil, 1975). In the alkali-
labile O-glycosidically-linked glycoprotein saccharides
an invariable core cannot be discerned, but the different
carbohydrate sequences are directly linked to the peptide.

Most of the structural variation of glycolipids and
glycoproteins is due to their terminal sugar sequences.
These can be often described as derivatives of disaccharides, which are linked to the core structures. In the
following we will give a major emphasis to the description of these terminal sugar sequences. In order to
shorten the presentation of the different structures,
we do not specify the different types of sialic acids.

BRAIN GANGLIOSIDES

N-acetylgalactosamine-containing sugar sequences.
Carbohydrate structures which are related to the disaccharide Gal(β1-3)GalNAc typically occur in brain
gangliosides as the terminal sugar moieties, which are
linked to a sialosylated lactose core. This type of
gangliosides also occurs in lower amounts in several
extraneural tissues (Wiegandt, 1973; Rauvala,1976a).

Table 1. Terminal sugar sequences of galactosamine-containing gangliosides and glycoproteins.

		Gangliosides	Glycoproteins
A[1]	GalNAc-	Klenk et al.,1963	Carlson,1968
B[1]	Gal(β1-3)GalNAc-	Kuhn and Wiegandt, 1963a	Thomas and Winzler, 1969
C[1]	$\overset{\displaystyle \text{Sia}\alpha 2}{\underset{\displaystyle 3}{\mid}}$ Gal(β1-3)GalNAc-	Kuhn and Wiegandt 1963b	Thomas and Winzler 1969 Spiro and Bhoyroo,1974
D	$\overset{\displaystyle \text{Sia}\alpha 2}{\underset{\displaystyle 3}{\mid}}$ GalNAc(β1-4)Gal(β1-3)GalNAc-	Svennerholm et al., 1973	
E[1]	$\overset{\displaystyle \text{Fuc}\alpha 1}{\underset{\displaystyle 2}{\mid}}$ Gal(β1-3)GalNAc-	Wiegandt, 1973; Sonnino et al.,1978	Carlson,1968; Wold et al.,1975
F[1]	$\overset{\displaystyle \text{Fuc}\alpha 1}{\underset{\displaystyle 2}{\mid}}$ GalNAc(α1-3)Gal(β1-3)GalNAc-		Carlson, 1968
G	$\overset{\displaystyle \text{Fuc}\alpha 1}{\underset{\displaystyle 2}{\mid}}$ Gal(α1-3)Gal(β1-3)GalNAc-		Newman and Kabat,1976
H	$\overset{\displaystyle \text{Sia}(\alpha 2-8)\text{Sia}\alpha 2}{\underset{\displaystyle 3}{\mid}}$ Gal(β1-3)GalNAc-	Kuhn and Wiegandt, 1963b	Finne et al.,1977b,1977c

[1] Similar structures with an additional sialic acid at C-6 of GalNAc occur in glycoproteins.

The most simple structure in the ganglioside series containing N-acetylgalactosamine is the GM_2 ganglioside* (Table 1), which occurs in high amounts in Tay-Sachs brain (Klenk et al., 1963). The basic disaccharide (Table 1) is found in the GM_1 and GD_{1B} gangliosides, and the sialosylated derivative (Table 1,C) occurs in the GD_{1A} and GT_{1B} gangliosides. An H active fucosyl derivative (Table 1,E) fucosyl-GM_1 (Wiegandt, 1973), occurs in liver and in other extraneural tissues. A similar structure, fucosyl-GD_{1B}, was recently found in brain (Sonnino et al., 1978).

The majority of the alkali-labile O-glycosidic glycoprotein saccharides are structurally related to Gal(β1-

*The shorthand nomenclature of Svennerhom (1963) is used for gangliosides.

3)GalNAc. These structures occur in a relatively high
content in brain (Finne,1975), and they also commonly
occur in glycoproteins of extraneural tissues and in
plasma glycoproteins (Finne and Krusius,1976; Vaith and
Uhlenbruck, 1978). As in gangliosides, the basic disaccharide and its sialosyl derivative are commonly found
Table 1,B and C). The A and B blood group active O-glycosidic structures (Table 1,F and G) have not been described in comparable gangliosides, although the H active
precursor (Table 1,E) has been bound in both gangliosides
and in glycoproteins. An α-linked isomer of the basic
disaccharide, Gal(α1-3)GalNAc, was found to occur in
glycoproteins but not in gangliosides (Finne et al.,
1977a).

Although the terminal sugar sequences of the galactosamine series (Table 1) are generally similar in gangliosides and in glycoproteins it should be noted that
the gangliosides contain a β-N-acetylgalactosaminidic
linkage to the lactose core, whereas the O-glycosidic
glycoprotein structures are linked α-glycosidically to
the peptide through their N-acetylgalactosamine residue.
This difference could be an explanation for the fact
that most of the glycoprotein structures shown in Table 1
can contain a sialic acid substituent at the C-6 of the
N-acetylgalactosamine residue. This type of sialic acid
substitution has not been found in gangliosides.

Oligosialosyl sequences. The occurrence of disialosyl and trisialosyl sequences was first shown for brain
gangliosides. In these glycolipids the disialosyl
sequence is linked to the C-3 of the galactose residue
of the lactosylceramide core in GT_{1B} and GQ gangliosides
(Kuhn and Wiegandt,1963a,1963b; Ishizuka and Wiegandt,
1972), and it is also found as linked to the galactosamine-containing terminal structure (Table 1,H) in GQ
and GP gangliosides. The ganglioside GT_{1A}, which was
recently characterized from human brain (Ando and Yu,
1977) also contains a disialosyl group linked to the
terminal sugar sequence (Table 1,H). A trisialosyl group
is linked to the galactose residue of the lactosylceramide core in the GP ganglioside (Ishizuka and Wiegandt,
1972). Gangliosides containing sialosyl-sialosyl sequences have been suggested to be involved in the receptor
structures of protein hormones (Fishman and Brady,1976).

Through the use of methylation technique for the
analysis of sialic acids it became possible to reliably
study the substitutions of these sugar residues (Rauvala
and Kärkäinen, 1977; Finne et al, 1977b; Haverkamp et
al., 1977). When performed before and after neuraminidase
treatment, methylation analysis reveals the occurrence
of the sialosyl(α2-8)sialosyl sequences. Analysis of
several rat tissues showed that the sialosyl-sialosyl
sequences also occur in glycoproteins, especially in
brain tissue (Finne et al, 1977b). The highest amounts
were found in glycopeptides from the plasma membrane
fraction of the young rat brain, which contained equal
amounts of the 8-0-substituted sialic acid in glycoproteins and gangliosides (Finne et al.,1977c). More recently, these sugar sequences were also found in a glycoprotein isolated from trout eggs (Inoue and Iwasaki,
1978) and in glycoproteins of salivary glands (Slomiany
et al.,1978).

The sialosyl(α2-8)sialosyl groups of glycoproteins
were found both in the alkali-labile O-glycosidic chains
and in the alkali-stable N-glycosidic chains. The
galactosamine-containing structures (Table 1,H) are
probably similar to those of brain gangliosides. In the
N-glycosidic chains the sialosyl-sialosyl sequences are
linked to glucosamine-containing structures (see below).

EXTRANEURAL GANGLIOSIDES

Carbohydrate chains of hematosides. Simple gangliosides containing lactose as their neutral carbohydrate chain, the hematosides (Yamakawa and Suzuki, 1951)
prevail in most extraneural tissues, such as liver,
spleen and kidney (Puro et al.,1969; Wiegandt, 1973;
Rauvala,1976a). Depending on the tissue and the animal
species the hematosides contain different proportions
of N-acetyl- and N-glycolylneuraminic acid (Iwamori and
Nagai,1978a). In addition, the disialosyl hematosides,
in which two sialic acid residues are linked together by
an α2-8 bond (Puro, 1969; Iwamori and Nagai,1978b) are
of common occurrence. The monosialosyl hematoside is
usually found in a higher amount than the disialosyl
hematoside. Comparison of the hematoside-type structures
to the saccharide moieties of glycoproteins may not be
relevant, although some structural specificity might be

shared, e.g. in the structure of the disialosyl hematosides and the disialosyl sequences of glycoproteins.

N-acetylglucosamine-containing sugar sequences. Besides the hematosides, several more complex gangliosides
are found in extraneural tissues. These include galactosamine-containing gangliosides similar to those found in
brain, but the major structures generally contain N-
acetylglucosamine instead of N-acetylgalactosamine (Wiegandt,1973; Li et al., 1973; Siddiqui and Hakomori,1973;
Rauvala,1976b).

The terminal sugar sequences of the glucosamine-
containing gangliosides (Table 2) are derivatives of
lactosamine, Gal(β1-4)GlcNAc, which is linked to the
lactosylceramide core. Most of the complex N-glycosidic
glycoprotein chains also contain terminal sugar sequences of this structural series. The carbohydrate chains
of these glycoproteins, which contain mannose and N-acetylglucosamine in their core portions, have been characterized in detail from many soluble glycoproteins
(Montreuil, 1975). In general, similar structures seem
to be present in the N-glycosidic carbohydrates of several
tissues (Krusius et al.,1976; Krusius and Finne, 1977;
Järnefelt et al.,1978a). As shown in Table 2, the same
sialosyl derivatives of the lactosamine series have been
found in gangliosides and in glycoproteins. The sialic
acid- and fucose-containing ganglioside (Table 2,C) has
been found in human kidney(Rauvala,1976b). This ganglioside is probably of limited occurrence, and it could not
be detected, for instance, in the erythrocyte membrane
(Rauvala,unpublished). The same terminal sugar sequence
has recently been found in glycoproteins of brain and
other tissues (Krusius and Finne, 1978).

The disialosyl derivative of lactosamine (Table 2,D)
occurs in glycoproteins of brain and extraneural tissues
(Finne et al., 1977b, 1977c). This sugar sequence was
recently also detected in a ganglioside of human kidney
(Rauvala et al.,1978). The sialosyl galactose derivative
of lactosamine (Table 2,E) has been only described in a
glycolipid of human erythrocyte membrane (Stellner and
Hakomori,1974), but the neutral structure lacking the
sialic acid residue has also been found in N-glycosidic
glycans of calf thymocyte glycoproteins (Kornfeld,1978).

Table 2. Terminal sugar sequence of glucosamine-containing gangliosides and glycoproteins.

		Gangliosides	Glycoproteins
A	Siaα2 ↓ 3 Gal(β1-4)GlcNAc-	Siddiqui and Hakomori 1973; Li et al.,1973	Spiro, 1964
B	Siaα2 ↓ 6 Gal(β1-4)GlcNAc-	Wiegandt, 1973	Baenziger and Kornfeld, 1974
C	Siaα2 Fucα1 ↓ ↓ 3 3 Gal(β1-4)GlcNAc-	Rauvala, 1976b	Krusius and Finne, 1978
D[1]	Sia(α2-8)Siaα2 ↓ 3 Gal(β1-4)GlcNAc-	Rauvala et al.,1978	Finne et al.,1977b,1977c
E	Siaα2 ↓ 3 Gal(β1-3)Gal(β1-4)GlcNAc-	Stellner and Hakomori, 1974	

[1] The position of the linkage between the disialosyl group and the core disaccharides of glycoproteins has not yet been chemically confirmed.

Terminal sugar sequences, which are structurally related to Gal(β1-3)GlcNAc (type I chains) instead of Gal(β1-4)GlcNAc (type II chains), are found both in glycoproteins (Montreuil,1975) and glycolipids (Hakomori, 1976). The possible occurrence of sialic acid derivatives in this type of structures has not been established for either glycolipids or glycoproteins.

Occurrence of poly(glycosyl) chains in glycolipids and glycoproteins. It has been shown that the glyco-lipid paragloboside, Gal(β1-4)GlcNAc→Gal-Glc-Cer (Sid-diqui and Hakomori,1973), and the N-glycosidic saccharide Gal(β1-4)GlcNAc →(Man)$_3$-(GlcNAc)$_2$-Asn (Kornfeld and Kornfeld,1970; Thomas and Winzler,1971), occur in human erythrocyte membrane (up to four terminal glucosamine-containing branches are found in the glycopeptides). These chains are the precursors of the sialic acid-con-taining structures (Table 2).

It has been shown that in blood group active glyco-lipids the Gal-GlcNAc units can form repeating sequen-ces (Watanabe et al.,1975; Gardas,1976;Kościelak et al., 1976;Dejter-Juszynski et al.,1978), which contain tens

of sugar residues. As it seems evident that the shorter
sialic acid-containing sugar chains and their neutral
precursors are largely analogous in the glycolipids and
in the N-glycosidic glycoprotein saccharides of the
erythrocyte membrane, we decided to study, whether this
kind of analogy might also apply to the blood group active chains. As these glycolipid antigens were shown
to contain extended sugar chains formed of Gal-GlcNAc
repeating units, it was to be expected that similar carbohydrate structures could also be found in the glycoproteins. It was found that in fact the major part of
the N-glycosidic carbohydrate in glycoproteins of the
human erythrocyte membrane occurs in high-molecular
weight glycopeptides (Finne et al.,1978). Structural
studies showed that these poly(glycosyl) peptides contain 20-70 sugar residues in an alkali-stable chain
linked through N-acetylglucosamine to the peptide. The
repeating saccharide structure of these chains is the
-3)Gal(β1-4)GlcNAc(β1- unit, and the branching points
are at the C-6 of the galactose residues (Krusius et al.,
1978; Järnefelt et al.,1978b). Thus, the repeating
carbohydrate structure is essentially the same as that
found in the poly(glycosyl) ceramides.

The terminal sugar sequences of the poly(glycosyl)
peptides contain similar sialic acid-terminated structures (Fig.1,A and B) as those found in shorter gangliosides and glycoprotein chains (Table 2). In addition,
the precursors of these chains (Fig.1,C and D) and the
blood group H, A and B determinants (Fig.1,E, F and G)
are also found. The neutral terminal sugar sequences,
including the blood group determinants, co-exist with
the sialic acid-containing sequences in the same complex
molecule (Finne et al.,1978). The chains are heterogeneous with respect to the number of sugar residues
and the structure of the terminal determinants (Krusius
et al., 1978). It seems evident that the terminal sugar
sequences are very similar to those found in the corresponding glycolipids, which have been shown to contain
blood group determinants (Watanabe et al.,1975; Gardas,
1976; Kościelak et al., 1976) and also sialic acid
(Dejter-Juszynski et al., 1978). The positions of the
sialic acid linkages in the poly(glycosyl) ceramides
have however not yet been determined.

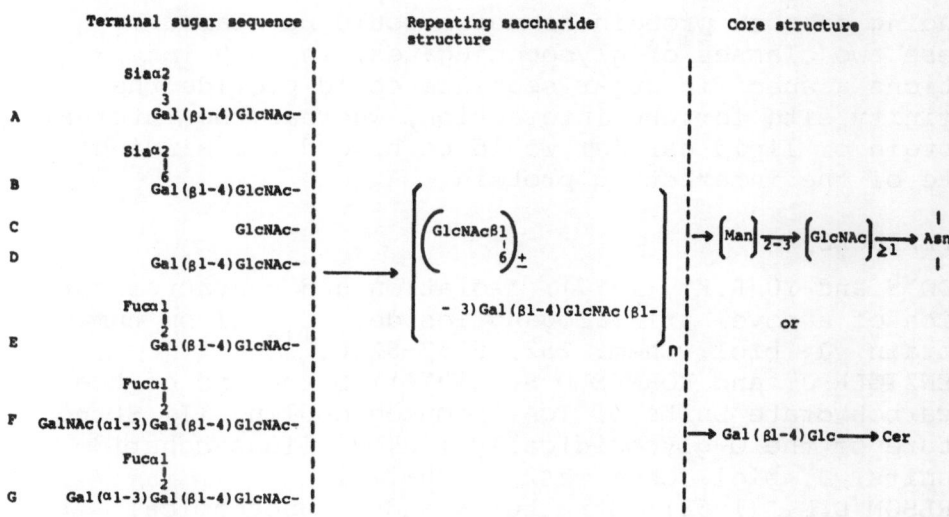

Fig.1. Proposed general structure of the poly(glycosyl) chains of human erythrocyte membrane.

CONCLUSIONS

Comparable carbohydrate structures to those of brain and extraneural gangliosides occur in the N-glycosidic and O-glycosidic carbohydrate chains of glycoproteins. The similarities in the ganglioside and glycoprotein structures mainly occur at the non-reducing terminal sugar sequences. The degree of the structural similarity is a function of the chain length of the carbohydrate, and it becomes especially evident in the structure of the poly(glycosyl)-type chains of glycolipids and glycoproteins.

The analogies in the structure of the ganglioside and glycoprotein glycans probably reflect the inability of some glycosyl transferases and hydrolases to distinguish between the lipid- and protein-linked membrane carbohydrates. It is possible that also in other protein-sugar interactions the carbohydrate chain can be either lipid- or protein-linked. This seems evident for the reaction of blood group agglutinins with the ABH antigens of the erythrocyte membrane, which occur as linked both to lipid and to protein. The finding of similar

sugar sequences in glycoproteins and glycolipids may
also indicate that the structural specificity for the
binding of some protein hormones could be shared by
these two classes of glycoconjugates. In such interactions a specific sugar sequence could provide the
affinity site for the interaction, whereas the internal
protein or lipid carrier would be beyond the binding
site of the interacting protein.

REFERENCES

ANDO S.and YU R.K. (1977): Isolation and characterization of a novel trisialoganglioside, G_{TIa}, from human
brain. J. biol. Chem. 252, 6247-6250.

BAENZIGER J. and KORNFELD S. (1974): Structure of the
carbohydrate units of IgA_1 immunoglobulin. II. Structure of the O-glycosidically linked oligosaccharide
units. J. biol. Chem. 242, 7270-7281.

CARLSON D.M. (1968): Structures and immunochemical
properties of oligosaccharides from pig submaxillary
mucins. J. biol. Chem. 243, 616-626.

CRAIG S.W. and CUATRECASAS P. (1975): Mobility of cholera
toxin receptors on rat lymphocyte membranes. Proc.
natn. Acad. Sci. U.S.A. 72, 3844-3848.

DEJTER-JUSZYNSKI M., HARPAZ N., FLOWERS H.M. and SHARON
N. (1978): Blood-group ABH-specific macroglycolipids
of human erythrocytes: isolation in a high yield
from a crude membrane glycoprotein fraction. Eur. J.
Biochem. 83, 363-373.

FINNE J. (1975): Structure of the O-glycosidically
linked carbohydrate units of rat brain glycoproteins.
Biochem. biophys. Acta 412, 317-325.

FINNE J. and KRUSIUS T.(1976): O-glycosidic carbohydrate
units from glycoproteins of different tissues:
demonstration of a brain-specific disaccharide, α-
galactosyl-(1-3)-N-acetylgalactosamine. FEBS Lett. 66,
94-97.

FINNE J, MONONEN I. and KÄRKKÄINEN,J.,(1977a): Analysis
of hexosaminitol-containing disaccharide alditols from
rat brain glycoproteins and gangliosides as O-trimethyl-
silyl derivatives by gas chromatography mass spectrometry. Biomed. Mass Spectrom. 4, 281-283.

FINNE J.,KRUSIUS T. and RAUVALA H.(1977b): Occurrence of
disialosyl groups in glycoproteins. Biochem. biophys.
Res. Commun. 74, 405-410.

FINNE J.,KRUSIUS T.,RAUVALA H. and HEMMINKI K. (1977c):
The disialosyl group of glycoproteins. Occurrence in
different tissues and cellular membranes. Eur. J.
Biochem. 77, 319-323.

FINNE J.,KRUSIUS T.,RAUVALA H.,KEKOMÄKI R. and MYLLYLÄ
G. (1978): Alkali-stable blood group A- and B-active
poly(glycosyl) peptides from human erythrocyte membrane. FEBS Lett. 89, 111-115.

FISHMAN P.H. and BRADY R.O. (1976): Biosynthesis and
function of gangliosides. Science, N.Y. 194, 906-915.

GARDAS A. (1976): A structural study on a macro-glyco-
lipid containing 22 sugars isolated from human erythrocytes. Eur. J. Biochem. 68, 177-183.

HAKOMORI S. (1976): Glycolipids of animal cell membranes,
in "MTP International Review of Science", Organic
Chemistry, Series Two, Aspinall, G.O., Ed.,Vol. 7,pp.
223-249.

HAKOMORI S. and KOBATA A. (1974): Blood group antigens,
in "The Antigens", Sela M.,Ed., Vol.2, Academic Press
(New York), pp. 79-140.

HAVERKAMP J., KAMERLING J.P., VLIEGENTHART J.F.G., VEH
R.W. and SCHAUER R. (1977): Methylation analysis determination of acylneuraminic acid residue type $2 \rightarrow 8$
glycosidic linkage, application to GT_{1b} ganglioside
and colominic acid. FEBS Lett. 73, 215-219.

INOUE S. and IWASAKI M. (1978): Isolation of a novel
glycoprotein from the eggs of trout: occurrence of
disialosyl groups on all carbohydrate chains. Biochem.
biophys. Res. Commun. 83, 1018-1023.

ISHIZUKA I. and WIEGANDT H. (1972):An isomer of trisialoganglioside and the structure of tetra- and pentasialogangliosides from fish brain. Biochim. biophys.
Acta 260, 279-289.

IWAMORI M. and NAGAI Y. (1978a): GM_3 ganglioside in
various tissues of rabbit. Tissue-specific distribution of N-glycolylneuraminic acid-containing GM_3. J.
Biochem. Tokyo 84, 1609-1615.

IWAMORI M. and NAGAI Y.,(1978b): Isolation and characterization of GD_3 ganglioside having a novel disialosyl residue from rabbit thymus. J. biol. Chem. 253,
8328-8331.

JARNEFELT J.,FINNE J., KRUSIUS T. and RAUVALA H. (1978a):
Protein-bound oligosaccharides of cell membranes.
Trends Biochem. Sci. 3, 110-114.

JÄRNEFELT J., RUSH J., LI Y.-T. and LAINE R.A. (1978b):
Erythroglycan, a high molecular weight glycopeptide

with the repeating structure (galactosyl-(1→4)-2-deoxy-
2-acetamido-glucosyl(1→3)) comprising more than one-
third of the protein-bound carbohydrate of human erythrocyte stroma. J. biol. Chem. 253, 8006-8009.

KLENK E.,LIEDKE U., and GIELEN W. (1963): Das Gangliosid
des Gehirns bei der infantilen amaurotischen Idiotie vom
Typ Tay-Sachs. Hoppe-Seyler's Z. physiol. Chem. 334,
186-192.

KORNFELD R.and KORNFELD S.(1970): The structure of a
phytohemagglutin receptor site from human erythrocytes.
J. biol. Chem. 245, 2536-2545.

KORNFELD R.(1978): Structure of the oligosaccharides of
three glycopeptides from calf thymocyte plasma membranes.
Biochemistry 17, 1415-1423.

KOSCIELAK J., MILLER-PODRAZA H., KRAUZE R.and PIASEK A.
(1976): Isolation and characterization of poly(glycosyl)
ceramides (megaglycolipids) with A, H and I blood-
group activities. Eur. J. Biochem. 71, 9-18.

KRUSIUS T.and FINNE J. (1977): Structural features of
tissue glycoproteins. Fractionation and methylation
analysis of glycopeptides derived from rat brain, kidney and liver. Eur. J. Biochem. 78, 369-379.

KRUSIUS T. and FINNE J. (1978): Characterization of a
novel sugar sequence from rat-brain glycoproteins containing fucose and sialic acid. Eur. J. Biochem. 84,
395-403.

KRUSIUS T.,FINNE J. and RAUVALA H. (1976): The structural
basis of different affinities of two types of acidic
N-glycosidic glycopeptides for concanavalin A-Sephar-
ose. FEBS Lett. 71, 117-120.

KRUSIUS T.,FINNE J. and RAUVALA H. (1978): The poly(glycosyl) chains of glycoproteins. Characterization of a
novel type of glycoprotein saccharides from human
erythrocyte membrane. Eur. J. Biochem. 92, 289-300.

KUHN R. and WIEGANDT H. (1963a): Die Konstitution der
Ganglio-N-tetraose und des Gangliosids G_I. Chem. Ber.
96, 866-880.

KUHN R. and WIEGANDT H. (1963b): Die Konstitution der
Ganglioside G_{II}, G_{III} und G_{IV}. Z. Naturforsch. 18b,
541-543.

LI Y.-T., MÅNSSON J.-E., VANIER M.-T. and SVENNERHOLM
L. (1973): Structure of the major glucosamine-containing ganglioside of human tissues. J. biol. Chem.
248, 2634-2636.

MONTREUIL J. (1975): Recent data on the structure of
the carbohydrate moiety of glycoproteins. Metabolic
and biological implications. Pure appl. Chem. 42,431-

477.

NEWMAN W. and KABAT E.A. (1976): Immunochemical studies on blood groups. Structures and immunochemical properties of nine oligosaccharides from B-active and non-B-active blood group substances of horse gastric mucosae. Archs Biochem. Biophys. 172, 535-550.

PURO K. (1969): Carbohydrate components of bovine-kidney gangliosides. Biochim. biophys. Acta 187, 401-413.

PURO K.,MAURY P. and HUTTUNEN J.K. (1969): Qualitative and quantitative patterns of gangliosides in extraneural tissues. Biochim. biophys. Acta 187, 230-235.

RAUVALA H. (1976a): Isolation and partial characterization of human kidney gangliosides. Biochim. biophys. Acta 424, 284-295.

RAUVALA H. (1976b): Gangliosides of human kidney. J. biol. Chem. 251, 7517-7520.

RAUVALA H. and KÄRKKÄINEN J. (1977): Methylation analysis of neuraminic acids by gas chromatography-mass spectrometry. Carbohydr. Res. 56, 1-9.

RAUVALA H., KRUSIUS T. and FINNE J. (1978): Disialosyl paragloboside, a novel ganglioside isolated from human kidney. Biochim. biophys. Acta 531, 266-274.

SIDDIQUI B. and HAKOMORI S. (1973): A ceramide tetrasaccharide of human erythrocyte membrane reacting with anti-type XIV pneumococcal polysaccharide antiserum. Biochim. biophys. Acta 330, 147-155.

SLOMIAMY B.L., SLOMIANY A. and HERP A. (1978): Studies on the occurrence of disialosyl groups in glycoproteins of salivary glands. Eur. J. Biochem. 90, 255-260.

SONNINO S., GHIDONI R., GALLI G. and TETTAMANTI G. (1978): On the structure of a new, fucose containing ganglioside from pig cerebellum. J. Neurochem. 31, 947-956.

SPIRO R.G. (1964): Periodate oxidation of the glycoprotein fetuin. J. biol. Chem. 239, 567-573.

SPIRO R.G. and BHOYROO V.D. (1974): Structure of the O-glycosidically linked carbohydrate units of fetuin. J. biol. Chem. 249, 5704-5717.

STELLNER K. and HAKOMORI S. (1974): A ceramide pentasaccharide of human erythrocyte membrane. J. biol. Chem. 249, 1022-1025.

SVENNERHOLM L. (1963): Chromatographic separation of human brain gangliosides. J. Neurochem. 10, 613-623.

SVENNERHOLM L, MÅNSSON J.-E. and LI Y.-T. (1973): Isolation and structural determination of a novel ganglioside, a disialosylpentahexosylceramide from human brain. J. biol. Chem. 248, 740-742.

TALMADGE K.W. and BURGER M.M. (1975): Carbohydrates and cell-surface phenomena, in "MTP International Review of Science", Biochemistry, Series One, Whelan W.J., Ed., Vol.5, Butterworths (London), pp.43-93.

THOMAS D.B. and WINZLER R.J. (1969): Structural studies of human erythrocyte glycoproteins. Alkali-labile oligosaccharides. J. biol. Chem. 244, 5943-5946.

THOMAS D.B., and WINZLER R.J. (1971): Structure of glycoproteins of human erythrocytes. Alkali-stable oligosaccharides. Biochem. J. 124, 55-59.

VAITH P. and UHLENBRUCK G. (1978): The Thomsen agglutination phenomenon: a discovery revisited 50 years later. Z. Immun.-Forsch. 154, 1-14.

WATANABE K., LAINE R.A. and HAKOMORI S. (1975): On neutral fucoglycolipids having long, branched carbohydrate chains: H-active and I-active glycosphingolipids of human erythrocyte membranes. Biochemistry 14, 2725-2733.

WIEGANDT H. (1973): Gangliosides of extraneural organs. Hoppe-Seyler's Z. physiol. Chem. 354, 1049-1056.

WOLD J.K., SMESTAD B. and MIDTVEDT T. (1975): Intestinal glycoproteins of germfree rats.IV. Oligosaccharides obtained by chemical degradation of a water-soluble glycoprotein fraction. Acta chem. scand. B 29, 703-709.

YAMAKAWA T. and SUZUKI S. (1951): The chemistry of the lipids of posthemolytic residue or stroma of erythrocytes. I. Concerning the ether-insoluble lipids of lyophilized horse blood stroma. J. Biochem.,Tokyo 38, 199-212.

A SENSITIVE ASSAY FOR GANGLIOSIDES

IN THE SUBNANOMOLE RANGE

Günter Schwarzmann , Ullrich Schlemmer
and Herbert Wiegandt

Physiologisch-Chemisches Institut I
der Universität Marburg
D-3550 Marburg, GFR

INTRODUCTION

The analysis of the chemical quantity of glycosphingolipids during ontogenic development or transformation of cells is as important as the characterization of their glycosphingolipid pattern and metabolism.

In cases where cells and in particular cultured
cells may not become available in larger quantities it
is a necessity to have a very sensitive procedure for
quantification of glycosphingolipids at hand.

For the analysis of the chemical quantity of glycosphingolipids, in particular if their exact composition
is not known, the estimation of the sphingoid base content is a suitable measure (Oulevey et al., 1977). The
sphingoid base, containing only one amino group, is a
characteristic substituent of glycosphingolipids, one molecule is found in every molecule of glycosphingolipids.
However, for this procedure the separation of glycosphingolipids from sphingomyeline is obligatory and is
achieved by the method of Saito and Hakomori (1971).

The most sensitive assays for sphingoid bases reported so far are based on the reaction of these bases
with either 1-naphthylamino-4-sulfonate (Coles and
Gray, 1970) or fluorescamine (Kisic and Rapport, 1974;
Naoi et al., 1974).

In a different approach promising a much higher
sensitivity for the spingoïd base estimation a method
was elaborated based on selective N-(1-^{14}C)-acetylation
of the sphingoid bases of glycosphingolipids following
acid hydrolysis of the latter.

PROCEDURE

For quantification of glycosphingolipids their
sphingoid bases are released by acid hydrolysis as described by Gaver and Sweeley (1965). For measurement of
tritium labelled fatty acids the latter are extracted
from the hydrolysate into hexane. However, this extraction may be omitted, if fatty acids are not to be
analyzed.

The hydrolysate is made alcaline by the addition
of sodium hydroxide. Water and chloroform are then
added to an appropriate ratio for Folch partition
(Folch et al., 1957) of the free sphingoid bases.
Following partition the lower phase is immediately
transferred to a clean tube, neutralized with methano-
lic HCl and evaporated to dryness. Selective N-acetyl-
ation of the bases is performed in either methanol or
tetrahydrofurane in the presence of a ten-fold molar
excess of sodium hydrogen carbonate. A hundred-fold
molar excess of freshly distilled $(1-^{14}C)$ acetic anhydride in toluene is added at once. Under these conditions labelling is achieved quantitatively within
thirty to sixty minutes, with no O-acetylation of the
free bases taking place.

The reaction mixture is repeatedly dried in a
dessicator under high vacuum following several addi-
tions of mixtures of toluene-acetic anhydride and
toluene-methanol. This procedure ensures complete removal of any of the unreacted $(1-^{14}C)$acetic anhydride
and free $(1-^{14})$acetic acid, which otherwise will give
rise to unduly high background counts, thus impairing
correct measurement of the $N-(1-^{14}C)$acetyl sphingoid
bases.

The labelled sphingoid bases are then extracted
into ethyl acetate in the precence of an equal volume
of water to separate any residual labelled contaminant
(acetates and or hexosamines) which also might interfere with the sphingoid estimation.

For determination of sphingoid bases in the sub-
nanomole range acetic anhydride with a radioactivity of
about 25 Ci/mol is employed, whereas 1 Ci/mol is sufficient for detection of sphingoid bases in the lower
nanomole range.

RESULTS AND DISCUSSION

A standard curve was obtained with pure sphingosine or
N-acetyl sphingosine (Fig. 1) and served as a basis for
determining the sphingoid base content of gangliosides.

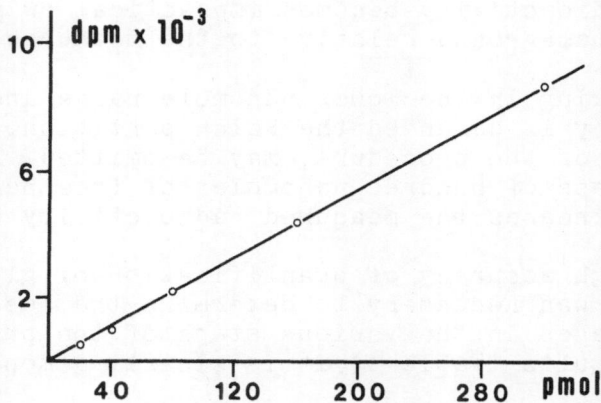

Fig. 1. Radioactivity as a function of sphingoid concentration. The procedure is given in the text.

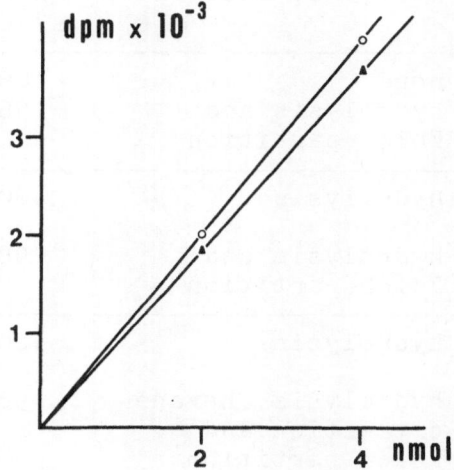

Fig. 2. Effect of free galactosamine on the yield of radioactivity; procedure without Folch partition (▲) and in the presence of 100 nanomoles of galactosamine (o).

A consistent relationship was found between sphingosine concentration down to 40 picomoles and radioactivity in the ethyl acetate phase. Measuring at lower concentrations by employing acetic anhydride of higher specific radioactivity becomes impractical owing to an increasing background relative to the measurable radioactivity.

If working in the lower nanomole range and if no high accuracy is demanded the Folch partition, for simplification of the procedure, may be omitted. In this case an excess of hundred nanomoles of free hexosamines (Fig. 2) increases the measured radioactivity by about 9 percent.

For high accuracy of quantification of glycosphingolipids it was necessary to determine the losses of sphingoid bases in the various steps of the procedure. From the results (Table 1) it is clearly demonstrated

Table 1. Comparison of the Recoveries of Radioactivity after Different Steps of Sphingoid Base Determination

Sphingoid	Treatment prior to N-(1-^{14}C)-acetylation	Recovery of Radioactivity[a]	
		^{14}C (%)	^{3}H (%)
sphingosine	none	100	–
sphingosine	hydrolysis and Folch partition	96.5	–
N-acetyl-sphingosine	hydrolysis	100	–
N-acetyl-sphingosine	hydrolysis and Folch partition	98	–
N-acetyl-^{3}H-sphinganine[c]	hydrolysis	100	98.5
N-acetyl-^{3}H-sphinganine	hydrolysis, hexane-extraction and Folch partition	96	95[b]

[a]Values are the mean of six to twelve experiments; prior to measurement N-(1-^{14}C)acetyl-sphingosine and N-(1-^{14}C)acetyl-^{3}H-spinganine were extracted into ethyl acetate.
[b]Two percent of the tritium label was found in the hexane.
[c]N-acetyl-sphingosine was tritiated according to Schwarzmann (1978).

that small losses occur. The combination of these results show that sphingoid bases are lost to the extent of 1.5% , 2% and 1.5% in hydrolysis, hexane extraction and Folch partition, respectively. As could be shown with the aid of N-acetyl[^{3}H]sphinganine, the N-acetyl derivatives are completely extracted into ethyl acetate.

Table 2. Determination by their Content of
Sphingoid Bases of Pure Tritium
Labelled Gangliosides

Sample[a]	nmol[b]	nmol[c]	deviation (%)
1	3.80	4.0	5 (3)
2	1.90	2.0	5 (3)
3	0.96	1.0	4 (2)
4	3.80	4.0	5 (3)
5	1.90	2.0	5 (3)
6	0.94	1.0	6 (4)

Values in brackets are corrected for losses into the hexane phase.

[a]Sample 1 to 3: II^{3}NeuAc-GgOse$_4$-[^{3}H]Cer;
sample 4 to 6: II^{3}NeuAc-Lac-[^{3}H]Cer.
[b]Values are the mean of four experiments.
[c]Values are expected from the estimation of sialic acids according to Svennerholm (1957), performed on larger quantities of the same stock.

Table 3. Distribution of Tritium Labelled
Gangliosides Subjected to Sphingoid
Base Determination

	Percentage of tritium label	
	in hexane[a]	in ethyl acetate[a]
II^{3}NeuAc-GgOse$_4$-[^{3}H]Cer	1.8	98.2
II^{3}NeuAc-Lac-[^{3}H]Cer	42	58

[a]Values are the mean of sixteen determinations and are corrected for partition into hexane of tritium labelled sphingoid base.

To test the applicability of this procedure tritium labelled gangliosides from known stocks were analyzed. The results obtained (Table 2) show a deviation of 2 to 4 percent from sialic acid estimations performed on larger quantities of the same stock. This is very probably due to analytical errors.

Tritium labelled glycosphingolipids offer the possibility to evaluate the distribution of carbon-carbon double bonds in the ceramide portion in its native state. This is clearly demonstrated here for the two gangliosides, II^3NeuAc-GgOse$_4$-Cer and II^3NeuAc-Lac-Cer (Table 3). Whereas almost all double bonds residue in the sphingoid of II^3NeuAc-GgOse$_4$-Cer, a considerable amount of double bonds is also found in the fatty acids of the hematoside of human spleen (Table 3).

In very small amounts of cellular material or tissue samples the individual glycosphingolipids are present in quantities below the limit of detection by conventional techniques. In such cases they can be detected by autoradiography following labelling and thin-layer chromatographic separation. For quantification of such small amounts of glycosphingolipids their extractability from plates is one important prerequisite. A recovery by extraction with chloroform-methanol-water mixtures of labelled gangliosides exceeding 90 percent was obtained (Table 4). In this case 50 and 120 picomoles of II^3NeuAc-Lac-$[^3H]$Cer and II^3NeuAc-GgOse$_4$-$[^3H]$Cer, respectively, were spotted onto plates (silica gel G, 0.2 mm; Merck AG, Darmstadt, GFR) on areas 60 mm^2 in size and eluted.

In those instances where only picomole quantities of material are available, this technique, employing two different radioactive labels, briefly illustrated here for the two gangliosides, may show to be of considerable help in quantification of glycosphingolipids and in characterization of their ceramide portion.

Table 4. Recovery of Tritium Labelled Gangliosides from Thin-Layer Plates

	DPM[a] applied	DPM[a] recovered	Recovery (%)
II^3NeuAc-GgOse$_4$-$[^3H]$Cer	18034	16750	93
II^3NeuAc-Lac-$[^3H]$Cer	34405	32950	)6

[a]Values are the mean of four experiments.

REFERENCES

COLES, L. and GRAY, G.M. (1970): Fluorimetric determination of sphingosine and its application to natural mixtures of glycosphingolipids. J. Lipid. Res. 11, 164-166

FOLCH, J., LEES, M. and SLOANE-STANLEY, G.H. (1957): A simple method for the isolation and purification of total lipids from animal tissues. J. Biol. Chem. 226, 497-509

GAVER, R.C. and SWEELEY, C.C. (1965): Methods for methanolysis of sphingolipids and direct determination of long-chain bases by gas chromatography. J. Am. Oil Chemists' Soc. 42, 294-298

KISIC, A. and RAPPORT, M.M. (1974): Determination of long-chain base in glycosphingolipids with Fluorescamine. J. Lipid Res. 15, 179-180

NAOI, M., LEE, Y.C. and ROSEMAN, S. (1974): Rapid and sensitive determination of sphingosine bases and sphingolipids with Fluorescamine. Anal. Biochem. 58, 571-577

OULEVEY, J., BODDEN, E. and THIELE, O.W. (1977): Quantitative determination of glycosphingolipids illustrated by using erythrocyte membranes of various mammalian species. Eur. J. Biochem. 79, 265-267

SAITO, T. and HAKOMORI, S.-I. (1971): Quantitative isolation of total glycosphingolipids from animal cells. J. Lipid Res. 12, 257-259

SCHWARZMANN, G. (1978): A simple and novel method for tritium labeling of gangliosides and other sphingolipids. Biochim. Biophys. Acta 529, 106-114

SVENNERHOLM, L. (1957): Quantitative estimation of sialic acids. Biochim. Biophys. Acta 24, 604-611

METABOLISM OF GANGLIOSIDES

INTRODUCTORY REMARKS ON GANGLIOSIDE METABOLISM

Shimon Gatt

Laboratory of Neurochemistry, Dept. of Biochemistry,
Hebrew University - Hadassah Medical School,
Jerusalem, Israel

Most of sialylated glycolipids occur in nervous tissue. In extraneural tissues which have mostly non-sialylated glycolipids, the gangliosides constitute a relatively small fraction. Only little information is available on the metabolic sequences which result in biosynthesis, deposition and subsequent degradation of the ganglioside in neural membranes. Some attempts were made to measure ganglioside biosynthesis by injecting labelled precursor into animals. However, these have not provided sufficient information on the metabolic events occurring *in vivo*. This short review of ganglioside metabolism will be confined to the reactions which synthesise or degrade these compounds and the enzyme which catalyze them, namely the glycolipid transferases and hydrolases. Though strictly-defined as sialic acid-containing glycolipids, a discussion of the metabolism of gangliosides must also include metabolic sequences of ceramide and its mono- or oligohexosyl derivatives. Studies on these enzymes were begun in the 1960's and in the last 15 years a considerable number of publications appeared. In this discussion the earlier work on the transferases and hydrolases will be reviewed and the reader will be directed to monographs which treat the more recent aspects of ganglioside metabolism in a comprehensive manner. The most updated discussion of enzymes of glycolipid metabolism is the book which summarizes the proceedings of the Meeting on Enzymes of Lipid Metabolism held at Mont Ste Odile exactly two years ago, only a short distance from the present meeting place (GATT, FREYSZ and MANDEL, 1978).

ENZYMATIC REACTIONS OF GANGLIOSIDE BIOSYNTHESIS

The pioneering work on the transferases was done by ROSEMAN (1970) who used mostly embryonic chick brain.

They described several enzymatic reactions and suggested a sequential transfer of carbohydrate and sialyl residues from their linkages with the respective nucleotides to the lipid backbone. The first lipid acceptor probably is ceramide, though a transfer of glucose for UDP-Glc to sphingosine to form glucosylsphingosine ("psychosine") cannot be entirely excluded. Ceramide is formed by the condensation of sphingosine with a free fatty acid or with fatty acyl coenzyme A. The series of steps resulting in the biosynthesis of a ganglioside, was suggested by ROSEMAN (1970).

Similar experiments were done in numerous other laboratories, notably those of BASU, BRADY, BURTON, CAPUTTO, DAIN, HAUSER, KANFER, KEENAN, RADIN, SANDHOFF and JATZKEWITZ, SHAH, and the Strasbourg lipid group. In the recent few years an increasing number of papers appeared in which glycolipid biosynthesis was investigated using cells in tissue culture. Considerable work was also done on the alteration of ganglioside pattern and biosynthesis in transformed cells but this is out of the scope of this presentation.

In general, work with the biosynthetic enzymes ("transferases") is complicated by the fact that they are intrinsic membraneous proteins which cannot be solubilized. NESKOVIC et al. (see GATT, FREYSZ and MANDEL, 1978), succeeded in solubilizing one such enzyme (a galactosyltransferase) and purifying it partially. This purification was done in the presence of a nonionic detergent, during purification, a requirement for external lipids was observed. From an enzymological standpoint, the reactions catalyzed by the transferases represent a heterogeneous system in which the enzyme and one substrate (the lipid) are membranous while the second substrate (the nucleotide) is water-soluble. Further information on these reactions will probably have to await a thorough analysis of enzyme-substrate interactions in intact membranes.

ENZYMATIC REACTIONS OF GANGLIOSIDE DEGRADATION

Enzymes which hydrolyze the gangliosides were identified and characterized in the 1960's. We extracted from brain tissue and partially purified the following enzymes: ceramidase, β-glucosidase, β-galactosidase, β-N-acetylhexosaminidase and neuraminidase. All these enzymes have acidic pH optima and are enriched in a lysosome-rich particle of rat brain. Potentially they can degrade the entire ganglioside molecule to fatty acid and sphingosine which is further degraded to a long-chain aldehyde. While these studies were done in our laboratory, BRADY and his coworkers studied several of these enzymes in other tissues. Special attention was directed, by several investigators to the mammalian sialidases. In the 1970's interest in the glycolipid hydrolases was renewed as a consequence of their implications in the lipid storage diseases. It is beyond the scope of this discussion, which is confined to sequences of ganglioside metabolism, to describe the considerable work done on enzyme detection, assay procedures and purification, as well as the existence of isoenzymes. For a comprehensive discussion on these aspects in normal and pathological tissue one should refer to several books (HERS and VAN HOOF, 1973; VOLK and SCHNECK, 1975, 1976; GLEW and PETERS, 1977; STANBURY, WYNGAARDEN and FREDRICKSON, 1978) and many reviews, e.g. a chapter in the last Annual Reviews of Biochemistry (BRADY, 1978). These as well as the proceeding of two symposia (PORCELLATI, CECCARELLI and TETTAMANTI, 1976; GATT, FREYSZ and MANDEL, 1978) have full coverage and extensive quotation of the entire subject. Interesting modern aspects, such as activators and inhibitors of glycolipid hydrolases can also be found in these publications.

The extensive work on glycolipid hydrolases was done using enzymes whose acidic pH optima suggested a lysosomal origin. In brain, lysosomes have not been clearly-defined, subcellular fractionation results in a low yield of enzymes in the "lysosomal fraction" and the latter exhibits a considerable heterogeneity. Some of the glycolipid hydrolases can be easily solubilized while other require detergents (e.g. glucosidase and cerebroside galactosidase). It is of interest that even sphingomyelinase, which could be extracted with aqueous media devoid of detergents was probably still a multiprotein aggregate with a molecular weight of about 300,000 daltons (GATT and GOTTESDINER, 1976).

Recently, a sphingomyelinase was described in brain tissue which probably is not lysosomal. It requires magnesium ions and its optimal pH is between 7 and 8. The fact that this enzyme was present in brain of a patient with Niemann Pick disease defines it as an entity separate from the corresponding lysosomal enzyme. The presence of a non-lysosomal sphingomyelinase raises the possibility that enzymes may also exist which degrade the glycolipids. The presence of such enzymes would permit selective turnover of gangliosides in membranes without the necessity to resort to the total-degradation process of the lysosomal apparatus.

Acknowledgement: The experimental work which is described in this paper and which was done in the laboratory of S. Gatt was supported in part by N.I.H. grant NS 02967.

REFERENCES

BRADY R.O. (1978): Sphingolipidoses. Ann. Rev. Biochem. 47, 687-713.

GATT S., FREYSZ L. and MANDEL P., Eds. (1978): Enzymes of lipid metabolism. Plenum Press (New York), 791 pp.

GATT S. and GOTTESDINER T. (1976): Solubilization of sphingomyelin-ase by isotonic extraction of rat brain lysosomes. J. Neurochem. 26, 421-422.

GLEW R.H. and PETERS S.P., Eds. (1977): Practical enzymology of the sphingolipidoses. A.R. Liss (New York), 305 pp.

HERS H.G. and VAN HOOF J.F., Eds. (1973): Lysosomes and storage diseases. Academic Press (New York).

PORCELLATI G., CECCARELLI B. and TETTAMANTI G., Eds. (1976): Ganglioside function: biochemical and pharmacological implications. Plenum Press (New York), 306 pp.

ROSEMAN S. (1970): The synthesis of complex carbohydrates by multi-glycosyltransferase systems and their potential function in inter-cellular adhesion. Chem. Phys. Lipids 5, 270-297.

STANBURY J.B., WYNGAARDEN J.B. and FREDRICKSON D.S., Eds. (1978): The metabolic basis of inherited disease. McGraw Hill (New York), 4th Ed.

VOLK B.W. and SCHNECK L., Eds. (1975): The gangliosidoses. Plenum Press (New York), 277 pp.

VOLK B.W. and SCHNECK L., Eds. (1976): Current trends in sphingo-lipidosis and allied disorders. Plenum Press (New York), 612 pp.

BIOSYNTHESIS OF GANGLIOSIDES IN TISSUES

Subhash Basu, Manju Basu, Jao-Long Chien[+] and
Kathleen A. Presper

Department of Chemistry, Biochemistry and Biophysics
Program, Univ. of Notre Dame, Notre Dame, IN 46556.

The presence of neuraminic acid-containing glycosphingolipids
(gangliosides) in mammalian (1-5) and avian (6) tissues has been
known for the last 40 years. Recently the chemical structures of 20
different gangliosides isolated from neural (7-9) and non-neural
(10-12) tissues have been elucidated. In addition to the hexosamine-
free and N-acetylgalactosamine-containing gangliosides, the structures of at least 3 different gangliosides (13-18) containing
nLcOse$_4$Cer (Gal(β1-4)GlcNAc(β1-3)Gal(β1-4)Glc-Ceramide) as the core
structure have been reported by several laboratories.

On the basis of our previous studies with an embryonic chicken
brain membrane system (19-24), we have proposed the stepwise biosynthesis of GD1a ganglioside, starting from lactosylceramide (Fig. 1).
Cumar et al. (25) have reported the biosynthesis _in vitro_ of GD1b
starting from GD3 using a double labeling technique. Several of
these glycolipid glycosyltransferases of different tissues such as
rat brain (26-30), mouse brain (31), rat liver Golgi apparatus (32-
34), and bovine thyroid Golgi apparatus (35) which are involved in
the biosynthesis of GD1a and GT1 have been reported by various laboratories. We now report our studies of the biosynthesis _in vitro_ of
AcNeu-nLcOse$_4$Cer by an enzyme preparation isolated from embryonic
chicken brain, bovine spleen and cells of human neural tissue origin.
Recently Schneck and his coworkers (36) have established a permanent
line of glial cells (TSD) from fetal Tay-Sachs-diseased cerebrum for
the study of this disease. In addition to GM2 a high level of N-
acetylglucosamine-containing ganglioside has been found in these
TSD cells.

[+]Present address: Department of Neurology, Medical University of
South Carolina, Charleston, S.C. 29403.

MATERIALS AND METHODS

Preparation of the Golgi-Rich Membranes from Bovine Spleen.

A membrane fraction rich in Golgi bodies was prepared from fresh bovine spleen tissue according to a modification of a method published previously (37,38). Within 10 hours after dissection, the fresh bovine spleen tissue (100 g) was homogenized with a Polytron 20ST (Kinematica, Lucerne, Switzerland) homogenizer in 6 volumes of 0.5 M sucrose, 1% dextran (average M_r 264,000; Sigma), 5 mM mercapto-ethanol, 1 mM $MgCl_2$ and 0.04 M Tris-maleate buffer, pH 6.4. All steps were carried out between 4° and 6°. The homogenate (600 ml) was centrifuged for 15 min at 5,000 x g. The supernatant fluid was centrifuged for 45 min at 30,000 x g. The reddish brown pellet (BSR-2; 40 ml) was rehomogenized in 2 volumes of 0.32 M sucrose, layered on 2 vol of 1.25 M sucrose, centrifuged for 30 min at 83,000 x g in an SW 27 swinging bucket rotor. A pinkish white layer (BSG; 10 ml) was collected from the junction of 0.32 M and 1.25 M sucrose. The pellet that sedimented at the bottom of the tube and the buffy coat on top of the pellet were designated BSP (12.6 ml) and BSUP (5.3 ml), respectively. The BSG fraction was resuspended in 3 volumes of distilled water and centrifuged at 9,000 x g for 20 to 30 min to remove plasma membranes. The pellet (BSWG) was resuspended in 0.32 M sucrose and used as enzyme source.

Isolation of a Membrane Fraction Rich in Glycosyltransferases from Embryonic Chicken Brain.

Sixty fresh frozen 11-day-old embryonic chicken brains (10 g) were homogenized in 3 volumes of 0.32 M sucrose containing 0.1% 2-mercaptoethanol. The total homogenate was layered on top of a discontinuous sucrose gradient (0.75 M and 1.2 M) and centrifuged at 25,000 rpm for 2 hours in a SW-27 swinging bucket rotor. Three visible membrane bands were observed: Fraction CBM-1, (5.5 ml; 4.5 mg/ml) at the junction between 0.32 M and 0.75 M sucrose; Fraction CBM-2 (10 ml; 3.7 mg/ml) between 0.75 M and 1.25 M; and Fraction CBM-3 at the bottom of 1.25 M sucrose. Fraction CBM-2 was used as the enzyme source.

Paper Chromatographic Assay of [^{14}C]-Labeled Products

After the required incubation period the mixtures were spotted on Whatman No. 3MM paper and developed in descending fashion with 1% sodium tetraborate (pH 9.0). This process (39, 40) removed CMP, CMP-[^{14}C]AcNeu and [^{14}C]-AcNeu from the enzymatic product, which remained at the origin. After air-drying, the paper was cut 4 cm below the origin, and ascending chromatography was performed in the reverse direction with chloroform-methanol-water (60:35:8;

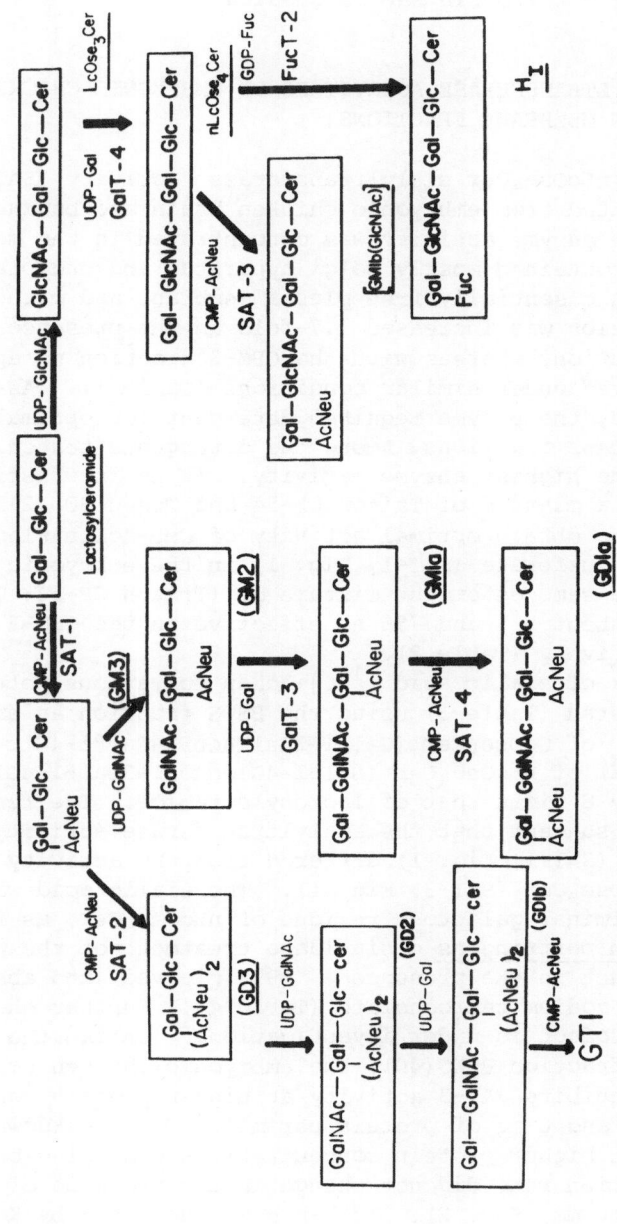

Fig 1. Proposed pathways for biosynthesis of gangliosides and blood group related glycosphingolipids in animal tissues.

v/v/v). The appropriate areas of each chromatogram were cut out, and radioactivity was determined by liquid scintillation counting.

RESULTS AND DISCUSSION

GLYCOLIPID: SIALYLTRANSFERASE ACTIVITIES OF EMBRYONIC CHICKEN BRAIN AND BOVINE SPLEEN MEMBRANE FRACTIONS.

A CMP-AcNeu:nLcOse$_4$Cer sialyltransferase activity (SAT-3; Fig. 1) was isolated from embryonic chicken brain and bovine spleen. In the spleen the enzyme activity was concentrated in the membrane fraction, which contained mostly Golgi apparatus and endoplasmic reticulum and was essentially free plasma membrane and mitochondria. The rate of reaction was increased 1.7-fold in the presence of MgCl$_2$ and the BSWG fraction, whereas with the CBM-2 fraction no appreciable change was observed under similar conditions (Table 1). As indicated in Tables 1 and 2, the enzyme required detergent for optimal activity in both membrane fractions. Among the detergents tested, Triton CF-54 produced the highest enzyme activity. It is interesting that Triton CF-54 and a mixture of Triton CF-54 and Tween 80 (2:1) were first used (20) to obtain optimal activity of CMP-AcNeu: lactosyl-ceramide sialyltransferase (SAT-1, Fig. 1) in the embryonic chicken brain whereas the same detergent mixture of (Triton CF-54: Tween 80 (2:1)) was only about 53% and 75% as effective in the CBM-2 and BSWG fraction, respectively (Table 2).

The transfer of sialic acid [^{14}C]-AcNeu to various potential acceptors was tested (Table 3) using the BSWG fraction as enzyme source. The V$_{max}$ of GgOse$_4$Cer (Galβ1-3GalNAcβ1-4Galβ1-4Glc-Cer) was about 3 times that of nLcOse$_4$Cer (Galβ1-4GlcNAcβ1-3Galβ1-4Glc-Cer) and approximately 8 times that of lactosylceramide. The results shown in Table 3 suggest that the sialyltransferase activity with lactosylceramide (SAT-1, Fig. 1) differed from the activity with GgOse$_4$Cer or nLcOse$_4$Cer (SAT-3, Fig. 1). The sialic acid was tranferred to the terminal galactose residue of nLcOse$_4$Cer, as found after <u>Clostridium perfringens</u> α-sialidase treatment of the purified radioactive product in the presence (99% cleaved) and absence (cleaved 94%) of sodium taurocholate (Table 4).[i] Further characterization of the product is under investigation.[i] Unlike the bovine spleen membrane fraction BSWG(40), the embryonic chicken brain fraction CBM-2 inhibits SAT-3 activity at higher protein concentrations (between 2 and 6 mg of protein per ml). The breakdown of CMP-[^{14}C]AcNeu at higher protein concentrations was ruled out because the reaction rate did not change in the presence of 5'-AMP (5 mM) or EDTA (10 mM, Fig. 2). It has been suggested by Kean (41)

---

[i] Manuscript in preparation-Chien, J.L., Basu, M., Basu, S., and Stoffyn, P.

Table 1

Requirements for CMP-AcNeu: nLcOse$_4$Cer Sialyltransferases

Incubation Mixture	[^{14}C]AcNeu-nLcOse$_4$Cer formed	
	CBM-2 Frac.	BSWG Frac.
	pmol/mg/hr	
	(a)	(b)
Complete	1230	2260
" Minus nLcOse$_4$Cer	390	330
" Minus Triton CF-54	440	140
" Minus Mg^{+2} Plus EDTA	1160	1380

(a) The complete incubation mixture contained the following components (in micromoles) in final volumes of 0.04 ml: nLcOse$_4$Cer, 0.0125; cacodylate-HCl buffer, pH 6.5, 10; MgCl$_2$, 0.25; Triton CF-54, 50 µg; CMP-[^{14}C] AcNeu, 26,000 cpm (6 x 10^{6}cpm per µmole) and enzyme fraction CBM-2, 0.1 mg of protein. After 2 hr at 37°, the mixtures were assayed by a double chromatographic technique as described in the text. Under these conditions, the rate of reaction remained constant for at least 2 hr and was proportional to protein concentration.

(b) The complete incubation mixture contained the following components (in micromoles) in final volumes of 0.04 ml: nLcOse$_4$Cer, 0.05; HEPES buffer, pH 6.45, 10; MgCl$_2$, 0.25; Triton CF-54, 50 µg; CMP-[^{14}C] AcNeu, 0.026 (1.0x 10^6 cpm per µmole); enzyme fraction BSWG, 0.14 mg of protein. After 2 hr at 37°, the incubation mixtures were assayed by the double chromatographic method described in the text. Under these conditions the rate of reaction remained constant with time of incubation for 3 hr and was proportional to protein concentration.

that the CMP-AcNeu hydrolase activity is also inhibited by 0.1% 2-mercaptoethanol. The enzyme preparation used for the present study also contained 0.1% 2-mercaptoethanol. The inhibition of SAT-3 activity at higher protein concentrations may have been due to the presence of a specific inhibitor (43) and is under investigation in embryonic chicken brain and cultured cells (TSD) established from a Tay-Sachs diseased fetal brain (42).

Table 2

Effects of Different Detergents on CMP-AcNeu: nLcOse$_4$Cer
Sialyltransferase Activity

Detergent added	Concentration	[^{14}C]AcNeu-nLcOse$_4$Cer formed	
		CBM-2 Frac.	BSWG Frac.
	µg/40µl	pmol/mg/hr	
		(a)	(b)
None	-	130	170
Triton CF-54	50	530	1770
Triton CF-54: Tween 80 (2:1)	50	280	1320
Triton X-100	50	260	1180
Taurocholate(Na$^+$)	100	100	350
Tween 20	100	-	390
Tween 65	100	120	-

(a) The incubation mixtures contained the following components (in micromoles) in final volumes of 0.04 ml: nLcOse$_4$Cer, 0.0125; Triton CF-54, 50 µg or other detergents as indicated; HEPES buffer, pH 6.5, 10; MgCl$_2$, 0.25; CMP-[^{14}C]AcNeu, 0.015 (1.8 x 10^6 cpm per micromole) and enzyme fraction CBM-2, 74 µg of protein. After 1 hr at 37°, the incubation mixtures were assayed by a double chromatographic technique as described in the text.

(b) The complete incubation mixture contained the following components (in micromoles) in final volumes of 0.04 ml: nLcOse$_4$Cer, 0.05; detergents as indicated; cacodylate-HCl buffer, pH 6.45, 10; MgCl$_2$ 0.25; CMP-[^{14}C]AcNeu, 0.026 (1.15 x 10^6 cpm per µmole); enzyme fraction BSWG, 0.19 mg of protein. After 2 hrs at 37°, the incubation mixtures were assayed by the double chromatographic method described in the text.

Table 3

Glycolipid: Sialyltransferase Activities - Substrate Competition
Experiment with BSWG Fraction

Substrate	[^{14}C]AcNeu incorporated		
	Found	Theoretical for	
		One enzyme[a]	Two enzymes[b]
	nmol/mg/hr		
Lactosylceramide.......	2.5		
nLcOse$_4$Cer............	5.8		
GgOse$_4$Cer............	19.4		
nLcOse$_4$Cer + Lactosylceramide.......	8.4	4.1	8.4
nLcOse$_4$Cer + GgOse$_4$Cer............	16.5	10.0	25.3

Conditions were the same as described in Table 1(b), except that
different substrates (1.13 mM) or substrate mixtures and 0.05 μmole
of CMP-[U-^{14}C]AcNeu were used in final volumes of 0.045 ml. The
incubation mixtures were assayed by a double chromatographic method
described in the text.

[a]The following equation was used to calcuate the values for one-
enzyme theory

$$v_t = \frac{V_a(a/K_a) + V_b(b/K_b)}{1 + (a/K_a) + (b/K_b)}$$

[b]Calculated value for two enzymes $\qquad v_t = v_a + v_b$

GLYCOLIPID GALACTOSYLTRANSFERASE ACTIVITIES OF BSWG AND CBM-2 FRACTIONS.

At least two different glycolipid galactosyltransferase activities that catalyze the sequential transfer of galactose from UDP-[^{14}C]galactose to LcOse$_3$Cer (GlcNAcβ1-3Galβ1-4Glc-Cer) to form a pentaglycosylceramide (Galα-Galβ-GlcNAcβ1-3Galβ1-4Glc-Cer; nLcOse$_5$Cer) were also detected in the Golgi-rich membrane fraction isolated from bovine spleen. As shown in Table 4, the activity of UDP-Gal: LcOse$_3$Cer (β1-4)galactosyltransferase (GalT-4, Fig. 1; EC 2.4.1.86 (44-46)) was 5-fold higher than the activity of EC 2.4.1.87 (UDP-Gal: nLcOse$_4$Cer(α1-3)galactosyltransferase)(47). However, the different kinetic parameters (40)[i] and the substrate competition experiment (Table 4) suggest that the two enzymatic activities in fraction BSWG are due to two different proteins. Similar experiments were performed to distinguish GalT-3 (UDP-Gal: GM2 (β1-3)galactosyltransferase; EC 2.4.1.62; Fig. 1) and GalT-4 (UDP-Gal: LcOse$_3$Cer (β1-4)galactosyltransferase) activities in the CBM-2 fraction (Table 5)[ii]. The activity (V_{max}) of GalT-4 was 1.5 times higher than that of GalT-3 and was not inhibited in an experiment with mixed substrates (Table 5).

The purified ^{14}C-labeled product obtained from exogenous LcOse$_3$Cer and UDP-[^{14}C]galactose co-chromatographed with authentic nLcOse$_4$Cer (Galβ1-4GlcNAcβ1-3Galβ1-4Glc-Cer) and the presence of a terminal β-galactose residue in the radioactive product was made evident by its cleavage by purified papaya β-galactosidase (38).

GalT-3 (EC 2.4.1.62) and GalT-4 (EC 2.4.1.86) activities have recently been distinguished from each other in the CBM-2 fraction on the basis of their inhibition by p-nitrophenyl-2-acetamido-2-deoxy-β-D-glucopyranoside (Fig.3). However, p-nitrophenyl-β-D-GalNAc did not inhibit GalT-3 activity and the mechanism is under investigation with solubilized enzymes (48,49).

BIOSYNTHESIS OF GANGLIOSIDES IN CULTURED CELLS (TSD) FROM TAY-SACHS DISEASED CEREBRUM.

Different glycolipid glycosyltransferase activities (Table 6) and their inhibition (43) have been tested in an SV-40 transformed glial cell culture derived from the cerebrum of a 20-week-old Tay-Sachs diseased (TSD) fetus.[iii] As shown in Table 6, the TSD cells contained very little activity of GalT-3 (UDP-Gal: GM2 (β1-3) galactosyltransferase), which catalyzes the conversion of GM2 to GM1a. There was also no inhibition of GalT-3 activity in these cell lines. Strong inhibition of the GDP-Fuc: nLcOse$_4$Cer (α1-2)fucosyl-

---

[ii] Manuscript in preparation-Basu,M., Basu,S., and Stoffyn,P.
[iii] The work is in progress in collaboration with Drs. Linda M. Hoffman, Steven E. Brooks and Larry Schneck of Kingsbrook Jewish Medical Center.

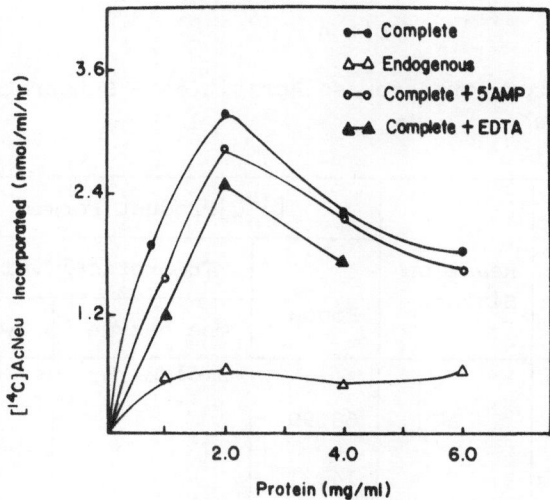

Fig. 2.   Effect of protein concentration on the rate of the
          sialyltransferase-catalyzed reaction. Conditions were
          the same as described in Table 1a except that the indicated
          quantities of CBM-2 fraction were used in the presence of
          5.0 mM 5'-AMP and 10.0 mM EDTA (pH 7.2).

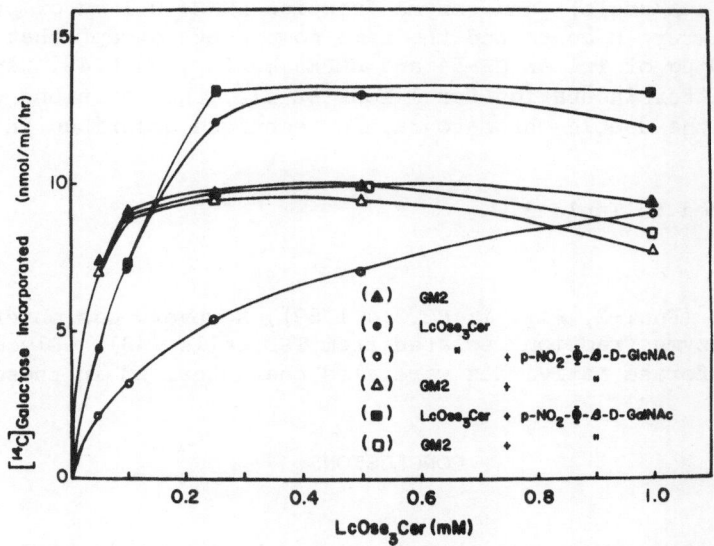

Fig. 3.   Inhibition of glycolipid galactosyltransferase activities
          of CBM-2 by p-nitrophenyl-β-D-glycosides. Incubation cond-
          itions were the same as described in Table-5, except that
          0.9 mM p-nitrophenyl-β-D-glycosides were used with varying
          concentrations of LcOse₃Cer and GM2.

Table 4

Glycolipid: Galactosyltransferase Activities - Substrate Competition
with BSWG Fraction

| Substrate | Reaction Mixture | [^{14}C]Product Formed | | |
| | | Found | Theoretical values[a] for | |
			One enzyme	Two enzymes
		nmol/ml/hr		
0.25 mM LcOse$_3$Cer	A	68.99		
0.75 mM nLcOse$_4$Cer	B	14.31		
0.25 mM LcOse$_3$Cer + 0.75 mM nLcOse$_4$Cer	A	83.06	17.95	83.3

Reaction mixture A contained the following components (in
micromoles) in final volumes of 0.04 ml: Triton CF-54, 50 µg;
MnCl$_2$, 0.5; MES buffer, pH 7.25, 10; UDP-[^{14}C]galactose, 0.075
(1.4 x 10^6 cpm/µmole); and enzyme fraction BSWG, 0.1 mg of protein.
Reaction mixture B contained the same components except that 30 µg
replaced 50 µg of Triton CF-54 and HEPES buffer, pH 6.45, replaced
the MES. After incubation for 1 hour at 37°, the reactions were
assayed by the double chromatographic technique described in the
text.

[a]As described in Table 3.

transferase (FucT-2, Fig. 1; EC 2.4.1.89), however, was observed in
the same enzyme fraction isolated from TSD cells (43). Glycolipid
sialyltransferase activities were also characterized in these cells.

CONCLUSIONS

These studies suggest that in vitro biosynthesis of GM1b
(GlcNAc) from LcOse$_3$Cer proceeds (Fig. 1) by the sequential transfer of galactose and the terminal sialic acid by GalT-4 and SAT-3
glycosyltransferases, respectively. The activities of these two
enzymes are present in bovine spleen Golgi-rich membrane and the
membrane fraction isolated from embryonic chicken brain. Hoffman

Table 5

Glycolipid:Galactosyltransferase Activities - Substrate Competition Experiment with CBM-2 Fraction.

Substrate	Concentration	[^{14}C] product		
		Found	Theoretical value for	
			One enzyme [a]	Two enzymes [b]
	mM	nmol/ml/2 hr		
LcOse$_3$Cer	0.25	27.8		
GM2	0.1	17.8		
LcOse$_3$Cer+	0.25			
GM2	+ 0.1	42.3	17.3	45.6

The complete incubation mixture contained either one or a mixture of LcOse$_3$Cer and GM2 in the indicated amounts (µmoles) in final volumes of 0.05 ml: detergent Triton CF-54: Tween 80 (2:1) 100 µg; cacodylate-HCl buffer, pH 7.3, 10 µmoles; MnCl$_2$, 0.25 µmoles; UDP-[^{14}C]galactose, 0.026 µmoles (2.65 x 10^6 cpm per µmole) and enzyme protein (CBM-2 Fraction, isolated from 14-day-old embryonic chicken brain), 210 µg. After 2 hr at 37°, the incubation mixtures were assayed by a double chromatographic technique as described in the text.

[a,b,] As described in Table 3. K_m values of LcOse$_3$Cer and GM2 were 0.16 mM and 0.06 mM, respectively.

et al. (36) have found a high level of N-acetylglucosamine-containing ganglioside in TSD transformed cells and our present experiments (43) suggest that the synthesis in vitro of GM1b(GlcNAc) also occur in these cells. The activity of GalT-4 was 33-fold higher than the activity of GalT-3 in TSD cells;the significance of the low activity of GalT-3 is yet to be determined.

Table 6

Glycolipid: Glycosyltransferase Activities in TSD Cells.

Acceptor	Donor		
	UDP-[^{14}C]-Gal	CMP-[^{14}C]-AcNeu	GDP-[^{14}C]-Fuc
	pmol/mg protein/hr		
	(a)	(b)	(c)
Lactosylceramide		190	
LcOse$_3$Cer	2980		0
nLcOse$_4$Cer	540	150	220*
nLcOse$_5$Cer			350
GM3		40	
GM2	90		
GM1		180	
GgOse$_4$Cer		440	

Complete incubation mixtures for the glycosyltransferase assays contained the following components (in micromoles) in final volumes of 0.1 ml: acceptor glycosphingolipids, 0.05; and enzyme fractions consisting of homogenates of TSD cells, 0.25 mg of protein (a) Galactosyltransferase assay- Triton CF-54: Tween 80 (2:1), 0.3 mg; MnCl$_2$, 0.25; cacodylate-HCl buffer (pH 7.3), 10.0; UDP-[^{14}C]galactose, 0.04 (1.85 x 10^6 cpm/μmole), and was incubated at 37° for 2 hours.   (b)Sialyltransferase assay- Triton CF-54: Tween 80 (2:1), 0.3 mg; MgCl$_2$, 0.125; cacodylate-HCl buffer (pH 6.4), 10.0; CMP-[^{14}C]AcNeu,0.027 (2.47 x 10^6 cpm/μmole), and was incubated at 37° for 2 hr.   (c) Fucosyltransferase assay- detergent G-3634-A (Atlas Chemical), 0.2 mg; MgCl$_2$, 0.125; cacodylate-HCl buffer (pH 6.4), 10.0; GDP-L-[^{14}C]Fuc, 0.015 (2.7 x 10^6 cpm/μmole), and was incubated for 45 min at 37°. The reactions were stopped by the addition of 2.5 μmoles of EDTA and 10 μl of chloroform: methanol (2:1) and then assayed by a double chromatographic method described previously (39).

*Under similar condition the fucosyltransferase activity using bovine spleen Golgi-rich membrane fraction was 5670 pmoles per mg of protein per hour.

ACKNOWLEDGEMENT

This work was supported by U.S. Public Health Service Grants NS-09541 and CA-14764 and a Grant -in Aid from Miles Laboratories, Inc. (Elkhart, IN) to S.B. The authors are grateful to Dr. Prabir Bhattacharya for his help in the preparation of this manuscript.

REFERENCES

1. Klenk,E.(1942)Z.Physiol.Chem., 273, 76.
2. Yamakawa,T. and Suzuki,S.(1953)J.Biochem., 40, 7.
3. Svennerholm,L.(1962)Biochem.Biophys.Res.Commun., 9, 436.
4. Kuhn,R. and Weigandt,H.(1963)Chem.Ber., 96, 866.
5. Ledeen,R., and Salsman,K.(1965)Biochemistry, 4, 2225.
6. Garrigan,O.W. and Chargaff,E.(1963)Biochim.Biophys,Acta, 70, 452.
7. Svennerholm,L.(1972)Methods Carbohydrate Chem., 6, 464.
8. Iwamori,M. and Nagai,Y.(1978)Biochim.Biophys.Acta, 528,257.
9. Ledeen,R.W.(1978)J.Supramol.Struct., 8,1.
10. Puro,K.(1969)Biochim,Biophys.Acta, 187, 401.
11. Wiegandt,H. and Bucking,H.W.(1970)Eur.J.Biochem., 15, 287.
12. Rauvala,H.(1976)J.Biol.Chem., 251, 7517.
13. Kuhn,R. and Wiegandt,H.(1964)Z. Naturforsch., 19b, 80.
14. Wiegandt,H. and Schulze,B.(1969)Z. Naturforsch., 24b, 945.
15. Svennerholm,L., Bruce,A., Mansson, J-E., Rynmark,B.M. and Vanier,M.-T.(1972)Biochim.Biophys.Acta, 280, 626.
16. Ohashi,M. and Yamakawa,T.(1977)J.Biochem., 81, 1675.
17. Chien,J.-L.,Li,S-C., Laine,R.A. and Li,Y-T.(1978)J.Biol.Chem., 253, 4031.
18. Uemura,K., Yuzawa,M. and Taketomi,T.(1978)J.Biochem., 83, 463.
19. Basu,S., Kaufman,B. and Roseman,S.(1965)J.Biol.Chem., 240, 4115.
20. Basu,S.(1966)Ph.D. Thesis, Univ. of Michigan.
21. Basu,S., Kaufman,B. and Roseman,S.(1968)J.Biol.Chem., 243, 5802.
22. Kaufman,B., Basu,S. and Roseman,S.(1968)J.Biol.Chem., 243, 5804.
23. Kaufman,B., Basu,S. and Roseman,S.(1966)in Proceedings of the Third International Symposium on the Cerebral Sphingolipidoses (Aronson,S.M. and Volk,B.W., eds.)p. 193, Pergamon Press, New York.
24. Steigerwald,J.C., Basu,S., Kaufman,B. and Roseman,S.(1975) J.Biol.Chem., 250, 6727.
25. Cumar,F.A., Tallman,J.F., and Brady,R.O.(1972)J.Biol.Chem., 247, 2322.
26. Arce,A., Maccioni,H.F. and Caputto,R.(1966)Arch.Biochem.Biophys., 116, 52.
27. Arce,A., Maccioni,H.F. and Caputto,R.(1971)Biochem.J., 121, 483.
28. Yip,G.B. and Dain,J.A.(1970)Biochim.Biophys.Acta, 206, 252.
29. Ng,S.S. and Dain,J.A.(1977)J.Neurochem., 29, 1075.
30. Ng,S.S. and Dain,J.A.(1977)J.Neurochem., 29, 1085.

31. Yu,R.K. and Lee,S.H.(1976)J.Biol.Chem., 251,
32. Keenan,T.W., Morre,D.J. and Basu,S.(1974)J.Biol.Chem.,249, 310.
33. Richardson,C.L., Keenan,T.W. and Morre,D.J.(1977)Biochim. Biophys.Acta, 488, 88-96.
34. Richardson,C.L., Merritt,W.D., Keenan,T.W. and Morre,D.J.(1978) Experientia, 34, 571.
35. Pacuszka,T., Duffard,R.O., Nishimura,R.N., Brady,R.O. and Fishman, P.H.(1978)J.Biol.Chem., 253, 5839-5846.
36. Hoffman,L.M., Brooks,S.E., Amsterdam,D. and Schneck,L.(1978) Trans.Amer.Soc.Neurochem., 9, 206.
37. Basu,S., Basu,M. and Chien,J.-L.(1975)J.Biol.Chem., 250, 2956-2962.
38. Basu,S., Basu,M., Moskal,J.R., and Chien,J.-L.(1976) in Glycolipid Methodology, ed. Witting,L.A. (American Oil Chemist's Society Press, Champaign, Ill), 123-138.
39. Chien,J.-L., Williams,T. and Basu,S.(1973)J.Biol.Chem., 249, 310-315.
40. Chien,J.-L.(1975)Ph.D. Thesis, University of Notre Dame, Notre Dame, Indiana.
41. Kean,E.L.(1972) in Methods in Enzymology, ed. Ginsburg,V. (Academic Press, New York), 28, pp. 983-990.
42. Hoffman,L.M., Amsterdam,D. and Schneck,L.(1976)Brain Res., 111, 109-117.
43. Basu,M., Presper, K.A., Basu,S., Hoffman,L.M. and Brooks,S.E. (1979)Proc.Natl.Acad,Sci. U.S.A., 76, (in press).
44. IUPAC-IUB(1977)Commission on Biochemical Nomenclature "The Nomen-clature of Lipids", Lipids, 12, 455-468.
45. IUPAC-IUB(1976)Commission on Biochemical Nomenclature "Enzyme Nomenclature"(and amendments thereto)Biochim,Biophys.Acta, 429, 1-45.
46. Basu,M. and Basu,S.(1972)J.Biol.Chem., 247, 1489-1495.
47. Basu,M. and Basu,S.(1973)J.Biol.Chem., 248, 1700-1706.
48. Moskal,J.R.(1977)Ph.D. Thesis,University of Notre Dame.
49. Berger,E.G., Kozdrowski,I., Weiser,M. M., van den Eijnden,D.H. and Schiphorst,E.C.M.(1978)Eur.J.Biochem., 90, 213-222.

SIALYLTRANSFERASE ACTIVITIES IN TWO NEURONAL MODELS : RETINA AND CULTURES OF ISOLATED NEURONS

H. Dreyfus[1], S. Harth[2], A.N.K. Yusufi, P.F. Urban[2] and
P. Mandel

Unité 44 de l'INSERM, and Centre de Neurochimie du CNRS,
11 rue Humann, 67085 Strasbourg Cedex, France

INTRODUCTION

Gangliosides have been found in various organs and tissues, however they are localized in a rather high amount and with a great diversity in the central nervous system (see for review LEDEEN and YU, 1976). The synthesis of gangliosides occurs by different pathways, requiring glycosyltransferases and sialyltransferases. Activities of these enzymes were studied both in non-nervous and in nervous tissues, with respect to their subcellular localization and kinetic properties. It was postulated that sialic acid residues on cell surfaces are involved in membrane-related cellular phenomena such as cellular recognition and adhesion, malignant transformation, contact inhibition and cellular migration (see for review SCHAUER, 1973). Further sialosyl groups seem to be involved in the receptor function of gangliosides (YAMAKAWA and NAGAI, 1978). Moreover, it was hypothesized that the formation of an enzyme-substrate complex between a glycosyltransferase from one cell and one acceptor from another could be responsible for cellular recognition (ROSEMAN, 1970).

Besides the knowledge of glycolipid composition of membranes their synthesis and integration would be of great importance. The present paper deals with studies of sialyltransferase activities in two neuronal systems: the retina and cultured neuronal cells devoid of glial cells. The retina presents a rather simpler morphological organization of neuronal and glial cells than that of brain, and therefore facilitates the investigation of functional activity.

Tissue culture of pure neurons offers the possibility to study

---

[1]Chargé de Recherche à l'INSERM; [2]Chargés de Recherche au CNRS.

the metabolism of gangliosides at the level of intact neurons. For the retina, the activities were studied during ontogenesis so that they could be correlated with the accretion of gangliosides and with the morphological development of both glial and neuronal cells. For the neuronal cells, we investigated the possibility of the "ecto" synthesis of gangliosides in parallel with the total activity.

MATERIALS AND METHODS

Retinas and Neuronal Cultured Cells

Chick retinas were treated as previously described (DREYFUS et al., 1976). Cultures of isolated neurons were obtained as previously described (PETTMANN, LOUIS and SENSENBRENNER, 1979). The cell suspension was prepared from cerebral hemispheres of 8-day-old chick embryo and seeded on polylysine-coated Petri dishes. These cultures remained healthy until the 8th day of culture.

Glycolipid and CMP-NeuAc Substrates

Lactosylceramide and GM3 were isolated from bovine spleen and purified. GM1, GD1a and GT1b were purchased from Supelco. GD1b was kindly provided by Dr. A. PRETI (Institute of Biochemistry, Milan). CMP-NeuAc of different specific activities were used: (a) CMP-(sialic-4-^{14}C) acid, 1.68 mCi/mmol from NEN; (b) and (c) CMP-N-acetyl (4,5,6,7,8,9-^{14}C) neuraminic acid, 4.1 mCi/mmol and 256-304 mCi/mmol respectively from Radiochemical Center.

Preparation of Antisera and Purification of Antibodies against GM3 and GM1 Gangliosides

Antibodies against GM1 and GM3 were produced by two intradermal injections of 5 mg of each ganglioside in total at a 10-day interval using methylated bovine serum albumin (BSA) and complete Freund's adjuvant as carrier. An immunoglobulin fraction (50 % ammonium sulphate saturation) passed through a BSA/methylated BSA affinity column (CUATRECASAS and ANFINSEN, 1971) was used, and normal immunoglobulins prepared similarly for use as controls.

Sialyltransferase Assays and Ganglioside Analysis

Sialyltransferases were assayed with the endogenous substrates as well as by adding exogenous substrates. Thus lactosylceramide (Lac-Cer) was used for the assays of CMP-NeuAc: lactosylceramide sialyltransferase (ST1); GM3 for CMP-NeuAc: GM3 sialyltransferase (ST2); GM1 for CMP-NeuAc: GM1 sialyltransferase (ST3); GD1a for CMP-NeuAc: GD1a sialyltransferase (ST4); GD1b for CMP-NeuAc: GD1b sialyltransferase (ST5); GT1b for CMP-NeuAc: GT1b sialyltransferase (ST6).

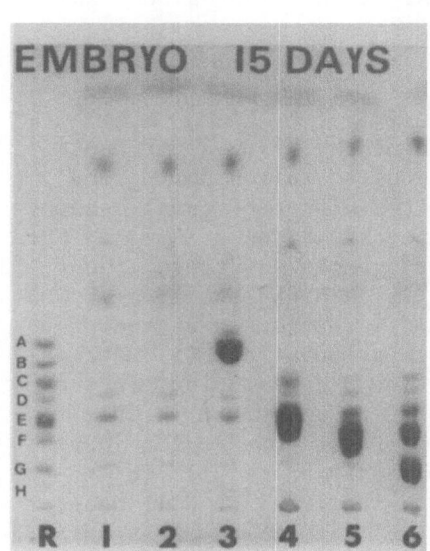
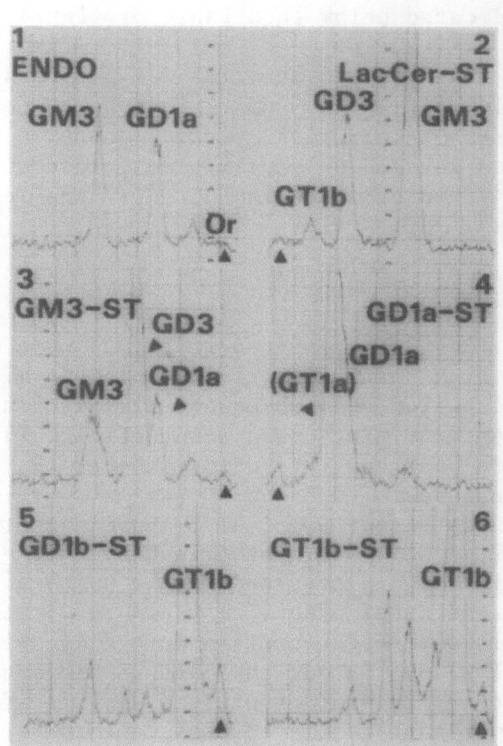

Fig. 1. Radioactivity scans (right side) and TLC (left side) of the
total retinal lipid extract obtained after incubation of substrates
with 15-day-old embryo retinal homogenate. Incubation conditions:
Triton CF-54 (300 μg); (1) without exogenous substrate; (2) Lac-Cer
25 nmol; (3) GM3 50 nmol; (4) GM1 50 nmol; (5) GD1a 50 nmol; (6) GD1b
25 nmol; sodium cacodylate buffer 500 mM pH 6.35 (1,2,4,5,6) and pH
7.0 (3); MgCl$_2$ 5 mM; CMP-NeuAc 247 pmol (304 mCi/mmol); retinal homogenate 300-500 μg prot. Final volume 0.1 ml 37°C, 90 min incubation
(1,2,4,5,6) and 60 min (3). Total lipids were purified through a
Sephadex G-25 superfine column, and TLC of gangliosides was performed
as described elsewhere (HARTH *et al.*, 1978). Reference gangliosides
were obtained from a mucopolysaccharidosis diseased brain; (A) GM3;
(B) GM2; (C) GM1; (D) GD3; (E) GD1a; (F) GD1b; (G) GT1b; (H) GQ1.
Radioactivity scans were obtained using the Thin Layer Scanner II,
LB 2723, Berthold.

After incubation, the reaction was stopped by addition of 20 vol
chloroform-methanol (C-M) 2:1 (v/v). For neuronal cells we tested
also some sialyltransferase activities determined directly in the
dish (in monolayer). The reaction was stopped by freezing, and the

incubation medium was decanted, cells were washed with NaCl (0.154 M) and extracted by C-M mixtures. The different assay systems used are indicated below each table or figure.

The lipidic material was purified using Sephadex G-25 superfine columns (1 g; Ø 1 cm). Elution was performed first with 2 x 2.5 ml C-M-H_2O (60:30:4.5 by vol.) and then with 2 ml C-M 2:1 (v/v).Methods used for TLC of gangliosides, radioactivity scan and radioactivity determination are indicated in the legends of the figures and tables.

RESULTS

Results presented in Fig. 1 were obtained by using a CMP-NeuAc of high specific activity which permitted us to follow during one incubation the subsequent sialylation of the different gangliosides. When examining endogenous activity (1), the syntheses of GM3 and GD1a were observed. By adding Lac-Cer (2) as exogenous substrate, the synthesis of GM3 was increased. With GM3 (3) a high labelling of GD3 was obtained; GD1a, GT1b and GM3 contained lower levels of radioactivity. GT1b was synthesized by adding GD1b as substrate (5) whereas GD1a (4) did not lead to the synthesis of this ganglioside. In the latter case, practically all radioactivity was recovered in GD1a; this peak might contain some GT1a which migrated to a region just below GD1a in our chromatographic system. Addition of GT1b (6) to the incubation did not produce any significant labelling of GQ1; however lower ganglioside species contained radioactivity which might be explained by degradation of GT1b during the incubation due to the presence of neuraminidases.

Table 1 summarizes the sialyltransferase activities during retinal ontogenesis. For endogenous activity the highest synthesis was observed for GM3 from 8 to 15-day-old embryos. A parallel labelling of GD3 was observed which decreased with age, whereas the synthesis of GD1a increased until 18 days of embryonic age. The percentage of NeuAc incorporation showed that at hatching GD1a contained more radioactivity (48 %) than GM3 (32 %). CMP-NeuAc:Lac-Cer sialyltransferase (ST1) showed its highest activity between 8 to 15-day-old embryos (about 3 pmol NeuAc incorporated/h/mg prot. in GM3). Labelling of GD1a was also increased until 15-18 days. CMP-NeuAc:GM3 sialyltransferase (ST2) and :GM1 sialyltransferase (ST3) exhibited their highest specific activities very early in the embryonic period. At 8 days, similar activities were found for GD3 and GD1a (15-17 pmol/ h/prot.). The activities decreased more rapidly for ST2 than for ST3. The synthesis of GD1a in adult retina remained higher 4.4 pmol). Sialyltransferases using disialogangliosides (GD1a, GD1b) as substrates possessed highest activities in 15-day-old embryos. The percentages of incorporation of NeuAc into GD1a (+ GT1a) and GT1b did not change during retinal ontogenesis.

Table 1. Sialyltransferase Activities in Chick Retina

Substrate	Compound labelled by NeuAc	E 8 days		E 15 days		P 1 day		Adulthood	
		Act.^{a}	% Inc.	Act.^{a}	% Inc.	Act.^{a}	% Inc.	Act.^{a}	% Inc.
(1) Endogenous activity	GM3	0.82	71	1.00	53	0.25	32	0.33	72
	GD3	0.22	19	0.06	3	0.01	1	0.01	2
	GD1a	0.07	6	0.71	37	0.38	48	0.06	13
(2) Lac-Cer	GM3	2.88	83	2.53	70	1.18	68	0.77	70
	GD1a	0.08	2	0.89	25	0.40	23	0.16	15
(3) GM3	GD3	15.15	95	2.89	57	2.30	62	1.72	70
	GD1a	0.17	1	0.95	19	0.64	17	0.24	10
(4) GM1	GD1a + (GD3)	16.96	96	7.93	99	3.97	97	4.36	95
(5) GD1a	GD1a^b		86	2.73	86	0.79	80	0.53	77
(6) GD1b	GT1b	11.45	94	12.80	91	4.11	87	2.19	87

(a) Activities are expressed as pmol NeuAc incorp./h/mg prot.;
(b) this spot may contain GT1a. (E) Embryonic period; (P) post-hatching period. Incubation and TLC conditions as in legend of Fig. 1. Radioactivity contained in each spot was determined using a Packard-PRIAS liquid scintillation spectrometer with the Kieselgel dispersed in 50 µl H_2O, 1 ml ethanol and 5 ml of 0.4 % scintimix 3 in toluol.

Fig. 2 represents the decrease in the synthesis of GD1a produced by adding increasing quantities of GM1-antibodies to the incubation mixture. For retina and brain of 12-day-old chick embryo, 50 % inhibition was observed by using 100 µl of this antibody preparation. Under the same incubation conditions, with higher quantities of antibodies, the synthesis of GD1a in brain homogenates decreased more rapidly than in retinal homogenates.

Table 2 shows that when a total brain ganglioside mixture was used as substrate (first column), syntheses of GD1a, GT1b and GM3 were observed. These syntheses are explained by the presence of GM1, GD1a and GD1b in the brain mixture, GM3 may reflect the endogenous activity. The three compounds showed a decrease in their labelling after the addition of GM1-antibodies which shows that GM1-antibodies may not be specific for only GM1. The second and third columns of Table 2 show inhibition of the syntheses of GD1a and GD3 produced by addition of GM1- and GM3-antibodies, respectively. When GM3 was utilized as exogenous substrate, a synthesis of GD3 was observed; but we noted also a high labelling of GM3. This might be explained by a rapid degradation of GM3 (and/or GD3) in lactosylceramide and a

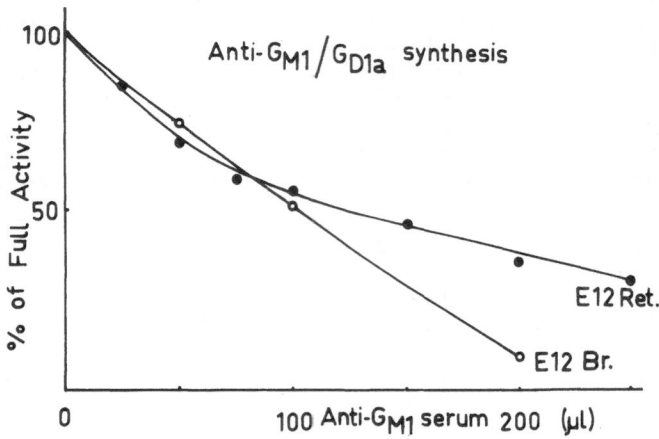

Fig. 2. Effect of the addition of increasing volumes of a preparation of GM1-antibodies (see methods) on the CMP-NeuAc:GM1 sialyltransferase in 12-day-old embryo retinal (E12 Ret) and brain (E12 Br) homogenate. Incubation conditions for the control: GM1 50 nmol; Triton CF-54 (300 µg); sodium cacodylate buffer pH 6.35 500 mM; CMP-NeuAc 60 nmol (1.68 mCi/mmol); retinal or brain homogenate 400 µg prot. Total volume 0.32 ml. 37°C, 90 min incubation. TLC and radioactivity determination as in legend of Table 1.

Table 2. Effect of GM1- and GM3-Antibodies on the Synthesis of Gangliosides

Gangliosides labelled by NeuAc	Substrates									
	Brain ganglioside mixture		GM1		GM3		GM1 + GM3			
	Cont.	+Anti-GM1 150 µl	Cont.	+Anti-GM1 150 µl	Cont.	+Anti-GM3 80 µl	Cont.	+Anti-GM1 150 µl	+Anti-GM3 80 µl	+Anti-GM3 200 µl
GM3	114	25	-	-	3716	3245	3590	2989	3276	1981
GD3	-	-	118	100	323	141	180	160	141	-
GD1a	343	280	970	780	-	-	1100	-	-	-
GT1b	260	136	-	-	-	-	-	-	-	-

Results expressed as cpm/mg prot./60 min of incubation, 50 nmol GM3 and pig brain gangliosides were also added. For incubation conditions, TLC of lipids and radioactivity determination, see legend of Fig. 2.

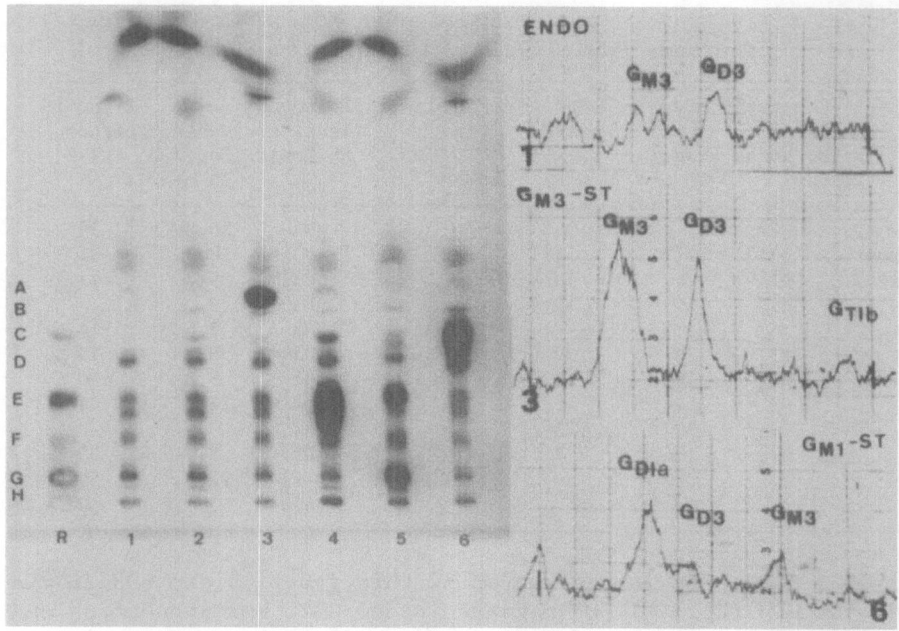

Fig. 3. Radioactivity scans (right side) of three lanes (1,3,6) of
TLC (left side) of total lipid extract of neuronal cells obtained
after incubation in the dish of the different substrates and neuronal
cells (7 days of culture). Incubation conditions: in one dish (approx.
200 μg prot.); (1) without exogenous substrate; (2) Lac-Cer 250 nmol;
(3) GM3 250 nmol; (4) GD1a 250 nmol; (5) GT1b 250 nmol; (6) GM1 250
nmol; sodium cacodylate buffer pH 7.0 500 mM; $MnCl_2$ 5 mM; CMP-NeuAc
294 nmol (4 mCi/mmol). Final volume 1.5 ml, 37°C, 90 min incubation.
TLC of lipids, reference gangliosides (R) and radioactivity scans,
see legend of Fig. 1.

re-synthesis of GM3. Addition of GM3 antibodies did not produce a
significant decrease in this labelling. When GM1 and GM3 were added
together as substrates, we noted a decrease in the labelling of GD1a
(15 %) and a decrease in that of GD3 (40 %). The labelling of GM3
remained quite stable. Addition of GM1- as well as GM3-antibodies,
inhibited the synthesis of GD1a, whereas the level of incorporation
of NeuAc into GD3 was decreased more by adding GM3-antibodies. The
last line of column 4 (Table 2) shows that higher amounts of GM3-
antibodies led to a total inhibition of the synthesis of GD1a and
GD3.

     For cells directly incubated in the dish (in monolayer system),
slight synthesizing activities for GM3 and GD3 were observed when no
exogenous glycolipid substrate was added (Fig. 3). GD3 and GM3 were

Table 3.   Sialyltransferase Activities in Monolayer Cultured
Neuronal Cells (7 Days of Culture)

Substrate	Compound labelled by NeuAc	pmol NeuAc inc./ 90 min	% Incorporated	Substrate	Compound labelled by NeuAc	pmol NeuAc inc./ 90 min	% Incorporated
(1) Endogenous activity	GM3	7.4	35	(3) GM3	GM3	20.3	39
	GD3(a)	2.2	11		GD3(a)	18.2	35
	GD3(b)	7.6	36		GD3(b)	5.7	11
	GD1a	3.1	15				
				(4) GT1b	GM3	7.6	14
(2) Lac-Cer	GM3	8.1	39		GD3(a)	8.5	15
	GD3(a)	3.2	15		GQ1	5.3	10
	GD3(b)	5.4	26		GQ'(c)	22.7	41
	GD1a	2.7	13				
	GD1b	1.6	7	(5) GM1	GD3	8.1	19
					GD1a	22.8	53
					GD1b	3.1	7

(a) (b) GD3: two spots in the region of this ganglioside. GD3 (a) is
located over GD3 (b). (c) GQ' migrated below GQ1. For incubation and
TLC conditions, see legend of Fig. 3. For radioactivity determination, see legend of Table 1.

labelled when using GM3 as exogenous substrate and GD1a was synthesized from GM1 (Table 3). The corresponding specific activities and
the relative percentages of incorporation are given in Table 3. GD3
could be divided into two spots as shown in Fig. 3. Synthesis of different gangliosides was obtained: mainly GD3 (endogenous activity), GM3
and GD3 from lactosylceramide; GM3, GQ (2 compounds) from GT1b; GD1a
from GM1. When similar quantities of gangliosides (250 nmol) were introduced in the incubation mixture, 24-28 pmol of NeuAc were found in
GD3, GQ and GD1a that were synthesized from GM3, GT1b and GM1, respectively.

DISCUSSION

     We reported previously that glycosyltransferases and sialyltransf-
erases are involved in the first steps of ganglioside biosynthesis in
retina (DREYFUS et al., 1978). In this study, we have compared the
different sialyltransferase (ST) activities at different stages of
chick retinal ontogenesis. The presence of sialyltransferases in retina
can be shown using endogenous as well as exogenous substrates. The
endogenous activities permitted the synthesis of GM3, GD3 and GD1a.
Until hatching, the synthesis of GD1a increased following the increasing amount of GD1a during ontogenesis and its in $vivo$ labelling (DREYFUS et al., 1975, 1976). The high specific activities

determined for CMP-NeuAc:GM3 ST fit with the high amount of GD3 in
the embryonic chick retina. CMP-NeuAc:GM1 ST (ST3) presented its
highest activity in 8-day-old embryos, while the level of GD1a was
low (DREYFUS et al., 1975): this suggests that GM1-ST is not the
only determinant factor for GD1a synthesis. CMP-NeuAc:GD1a ST (ST4)
and :GD1b ST (ST5) presented their highest activities later than
CMP-NeuAc:Lac-Cer ST (ST1), ST2 and ST3. We believe that GT1a is
synthesized from GD1a; however it is difficult to separate GT1a from
GD1a in our chromatographic systems. The differences observed between
the activities and the evolutions of the different sialyltransferases
may parallel the appearance of new structures in the retina (HUGHES
and La VELLE, 1974). It is difficult to correlate the evolution of
sialyltransferase activities with all morphological events of developing chick retina; although sialyltransferase presented high activities before the appearance of the first synapses in the retina (13
to 15-day-old embryo) and only ST4 and ST5 activities increased in
parallel with synaptogenesis. In fact, all changes in ganglioside
patterns and sialyltransferase activities were seen before the
period at which the retina acquires its functional activity.

We have used antibodies against GM1 and GM3 in order to explore
the substrate specificities of the sialyltransferases and to study
the effects of the ganglioside environment on their activities. The
synthesis of GD1a was inhibited by addition of anti-GM1 to the incubation medium, which indicated that the presence of GM1 was essential for the synthesis of GD1a.

When a whole brain ganglioside mixture was used as exogenous
substrates (Table 2) we found that the addition of GM1-antibodies
produced a decrease in the synthesis of all gangliosides. Thus, it
seems probable that GM1-antibodies may not be specific only for GM1,
but that it also interacts with other gangliosides. This finding is
in agreement with that of Dr. NAGAI (personal communication) about
the non-specificity up to a certain dilution of GM1-antibodies. It
appears, however (Table 2), that the specificity of certain sialyltransferases could be demonstrated; indeed the addition of antibodies
against one specific ganglioside did not favour the synthesis of another ganglioside. Thus inhibition of GD1a synthesis when GM1-anti-
bodies are used, is in favour of a certain specificity of GM1-anti-
bodies and/or GM1-sialyltransferase. However the rate of synthesis
differs depending on whether one single or a mixed population of
gangliosides are given as exogenous substrates. Thus, when more than
one ganglioside is present as substrate, the activity of some sialyltransferases could be influenced, resulting from a competition between the different ganglioside substrates. Moreover, it has been
demonstrated that GM1-antibodies had an effect on CMP-NeuAc:GM3 ST,
but that this was smaller than the corresponding effect of GM3-anti-
bodies against CMP-NeuAc:GM1 ST. The level of incorporation of NeuAc
into one ganglioside depends on the concentration of the substrates
and also on the presence of other gangliosides. However, we should

keep in mind that the activities measured *in vitro* permit the
evaluation of the enzyme to synthesize preferentially one compound,
and cannot be extrapolated to the *in vivo* activities.

We have determined that part of the sialyltransferase activities
in pure neuronal cells is located on the external surface of the
plasma membranes of neuronal cells (Fig. 3, Table 3). These findings
are significant with regard to the hypotheses of ROSEMAN (1970) who
proposed that ectoglycosyltransferases are implicated in cellular
adhesion and recognition. On the other hand, for these neuronal cells,
we can assume that the gangliosides present at the outer site of the
membrane are derived in part from the integration of the Golgi apparatus and in part from a local synthesis. We still need more controls
(KEENAN and MORRE, 1975) in order to demonstrate with certainty the
occurrence of ectoenzyme systems that also act on glycoprotein
substrates.

It is obvious that ganglioside synthesis requires the intervent-
ion of several transferases and that their parallel activities determine the ultimate synthesis of the gangliosides. Sialyltransferases
are involved in some specific steps which seem to be similar in retina
and in neuronal cells and which follow the sequences: Lac-Cer→GM3→
GD3, GM1→GD1a→GT1a, GD1b→GT1b→GQ1b. GM1 and GD1a seem not to be
respective precursors of GD1b and GT1b.

ACKNOWLEDGEMENTS — This investigation was supported by a grant
from the Institut National de la Santé et de la Recherche Médicale
(Contract N° 78-4-017-6). The skillful technical assistance of Mrs.
A. GOMEZ DE GRACIA was greatly appreciated.

REFERENCES

CUATRECASAS P. and ANFINSEN C.B. (1971): Affinity chromatography,
   in "Methods in Enzymology", JACOBY W.B., Ed., Vol. 22, Academic
   Press (New York and London), pp. 345-378.

DREYFUS H., HARTH S., PRETI A., URBAN P.F. and MANDEL P. (1978):
   Studies on retinal ganglioside metabolism, in "Enzymes of Lipid
   Metabolism", GATT S., FREYSZ L. and MANDEL P., Eds., Advances in
   Experimental Biology, Vol. 101, Plenum Press (New York), pp. 655-
   665.

DREYFUS H., URBAN P.F., EDEL-HARTH S. and MANDEL P. (1975): Developmental patterns of gangliosides and phospholipids in chick retina
   and brain. J. Neurochem. 25, 245-250.

DREYFUS H., URBAN P.F., HARTH S., PRETI A. and MANDEL P. (1976):
   Retinal gangliosides: composition, evolution with age. Biosynthetic

and metabolic approaches, in "Ganglioside Function", PORCELLATI G., CECCARELLI B. and TETTAMANTI G., Eds., Advances in Experimental Biology, Vol. 71, Plenum Press (New York and London), pp. 163-188.

HARTH S., DREYFUS H., URBAN P.F. and MANDEL P. (1978): Direct thin layer chromatography of gangliosides of a total lipid extract. Analyt. Biochem. 86, 543-551.

HUGHES W.F. and La VELLE A. (1974): On the synaptogenic sequence in the chicken retina. Anat. Rec. 179, 297-302.

KEENAN T.W. and MORRE D.J. (1975): Glycosyltransferases: do they exist on the surface membrane of mammalian cells? FEBS Lett. 55, 8-43.

LEDEEN R.W. and YU R.K. (1976): Gangliosides of the nervous system, in "Glycolipid Methodology", WITTING L.A., Ed., American Oil Chemist's Society (Champaign, Illinois), pp. 187-214.

ROSEMAN S. (1970): The synthesis of complex carbohydrates by multi-glycosyltransferase systems and their potential function in inter-cellular adhesion. Chem. Phys. Lip. 5, 270-297.

SCHAUER R. (1973): Chemistry and biology of the acetylneuraminic acids. Angew. Chem. Internat. Edit. 12, 127-138.

YAMAKAWA T. and NAGAI Y. (1978): Glycolipids at the cell surface and their biological functions. Trends Biochem. Sci. 3, 128-131.

SIALYLTRANSFERASES IN YOUNG RAT BRAIN

Joel A. Dain and Sai-Sun Ng

Department of Biochemistry and Biophysics
University of Rhode Island
Kingston, Rhode Island 02881 USA

INTRODUCTION

Brain sialyltransferases catalize the transfer of sialic acid (NeuNAc) from CMP-NeuNAc to glycoprotein and glycolipid acceptors. In this study sialyltransferase activities were simultaneously examined utilizing both endogenous and exogenous glycoprotein and glycolipid acceptors. The assumptions were made that the activities obtained with exogenously added substrates are a measure of the amount of sialyltransferases present and that the endogenous activities give additional information on the identity of the endogenous glycoprotein and glycolipid acceptors. Unless otherwise indicated, experiments were performed with a total particulate preparation obtained by centrifuging the homogenate of 11-15 day rat cerebra at 105,000 x g for 1 hr. Four types of sialyltransferase reactions were investigated (A, B, C and D below). Each were assayed by measuring the incorporation (^{14}C)-NeuNAc from (^{14}C)-CMP-NeuNAc into a glycolipid or glycoprotein acceptor (NG & DAIN, 1977a,b). The abbreviations for individual gangliosides are those proposed by SVENNERHOLM (1964).

(A) Endogenous glycolipids

$$\text{Cer-Glc-Gal} \longrightarrow \underset{\underset{\text{NeuNAc}}{|}}{\text{Cer-Glc-Gal}} \text{ (GM}_3\text{)}$$

$$\underset{\underset{\text{NeuNAc}}{|}}{\text{Cer-Glc-Gal-GalNAc-Gal}} \text{(GM}_1\text{)} \longrightarrow \underset{\underset{\text{NeuNAc}}{|} \qquad \underset{\text{NeuNAc}}{|}}{\text{Cer-Glc-Gal-GalNAc-Gal}}\text{(GD}_{1a}\text{)}$$

239

(B)   Exogenous glycolipid

Cer-Glc-Gal-GalNAc-Gal(GM$_1$) $\longrightarrow$ Cer-Glc-Gal-GalNAc-Gal(GD$_{1a}$)
       | | |
      NeuNAc   NeuNAc   NeuNAc

(C)   Endogenous glycoproteins

Glycoproteins $\longrightarrow$ Glycoproteins-NeuNAc

(D)   Exogenous glycoprotein

Desialated (DS) fetuin $\longrightarrow$ DS-fetuin-NeuNAc

RESULTS AND DISCUSSION

The sialyltransferase activities with the endogenous glycoprotein and glycolipid acceptors in the standard assays were linear with time for at least 60 min, while those with the exogenously added GM$_1$ (B) and DS-fetuin (D) were linear with time only for
about 30 min. Activities were directly proportional to the amount
of enzyme used in the range of up to 0.75 mg protein.

The activities of reactions A, B, C and D expressed as nmoles
NeuNAc incorporated per 0.5 mg protein per 30 min at 37°C and
pH 6.3 were 0.094, 0.17, 0.039 and 0.64, respectively. Incorporation into the endogenous glycolipids(A) was always higher than incorporation into endogenous glycoproteins (C) indicating the
greater availability of glycolipid acceptor sites for sialic acid.

The Endogenous Glycolipids

Radioactivity scanning of a thin layer chromatogram of reaction (A) (Fig 1) indicated that GM$_3$ and GD$_{1a}$ were labeled.This suggested that the endogenous substrates were Cer-Glc-Gal and GM$_1$.
In reaction (A) 60% of the incorporated NeuNAc was identified as
ganglioside GM$_3$ and 40% as GD$_{1a}$ (Fig 1 (A in Fig 1)). The incorporation into other gangliosides was not significant indicating
that at this early stage in the development of the rat cerebrum
the glycolipid acceptor sites for sialic acid appear to be almost
exclusively Cer-Glc-Gal and GM$_1$ ganglioside.

Distinguishing the Glycolipid Sialyltransferases

Exogenously added GM$_1$ in reaction (B) was converted to GD$_{1a}$
without any apparent effects on endogenous GM$_3$ formation (Fig 1
(B in Fig 1)) suggesting that the sialyltransferase acting on
Cer-Glc-Gal is different from the one acting on GM$_1$. This conclusion was further substantiated by the finding that endogenous
(A) and the exogenous (B) glycolipid activities could be differentiated in a heat inactivation study (Fig 2) with the exogenous
(B) activity being more heat labile than the endogenous (A) activity. Since more than 60% of the (A) activity was in the conversion of Cer-Glc-Gal to GM$_3$ this further substantiated the con-

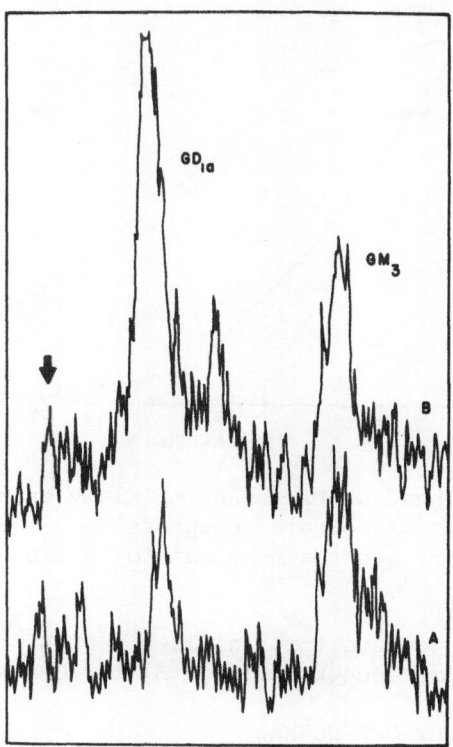

Fig. 1 TLC radiochromatogram of endogenous (A) and the combined
endogenous and exogenous (B) glycolipid sialyltransferase reaction
products. The arrow indicates the origin. The radioactivity peaks
were identified with crude beef brain ganglioside carrier. The
silica gel TLC was developed with chloroform-methanol-water (0.2%
$CaCl_2$), 60/35/8, by vol.

clusion that the sialyltransferase that acts on Cer-Glc-Gal is
different from the one that acts on the GM_1 ganglioside. KAUFMAN
et al. (1968) has also differentiated these two sialyltransferase
activities by heat inactivation studies in embryonic chick brain.

The Endogenous Glycoproteins

     A detailed study on the nature of the endogenous glycoprotein
acceptors has not been reported. SDS-polyacrylamide gel electrophoresis of the endogenous labeled glycoproteins (C) yielded about
20 species of labeled sialoglycopeptides (NG & DAIN, 1977b) indicating the great complexity of the products in this reaction and
the probability that we are dealing with a number of different
glycoprotein sialyltransferases(SCHACHTER & RODEN, 1973). The
sialyltransferase using glycoprotein and glycoprotein acceptors
appear to be two distinct classes of enzymes. Exogenously added

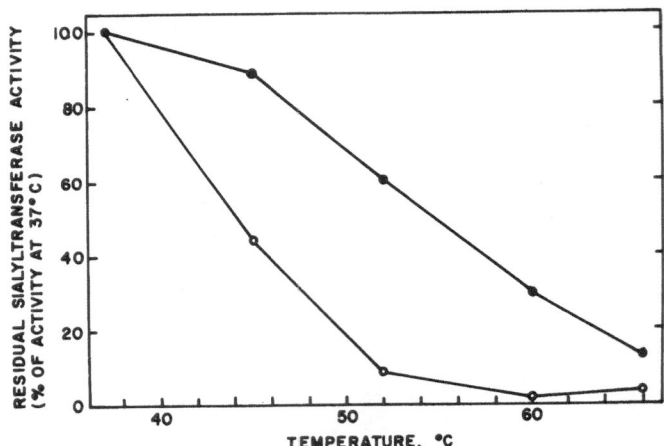

Fig. 2 Effect of heat pretreatment on sialyltransferase activities with endogenous glycolipid acceptors (●————●) and exogenous
GM_1 (○————○). Heat pretreatment was for 5 min at the temperature indicated.

DS-fetuin did not compete for sialylation with both endogenous and
exogenous glycolipid acceptors (Fig. 3) for CMP-NeuNAc.

The Binding Site for CMP-NeuNAc

The apparent Km values for CMP-NeuNAc for each of the four
reactions (A, B, C and D) was 0.13 mM. This value is comparable
to the Km of 0.15 mM in cultured mouse neuroblastoma cell with
Cer-Glc-Gal as an acceptor (KEMP & STOOLMILLER, 1976) and the Km
of 0.14 mM in embryonic chick brain with either GM_1 or GD_{1b} at the
acceptor (MESTRALLET et al., 1977). These similarities in the Km
in diverse species suggest that the binding sites on the neural
tissue sialyltransferases for the CMP-NeuNAc are similar in structure and affinity properties and have a common conservative evo-
utionary origin. On the other hand, the Km values for the glycoprotein and glycolipid acceptors are quite varied and thus indicates a divergence in evolutionary modifications for glycolipid
and glycoprotein substrate acceptor sites.

Subcellular Localization of Sialyltransferase Activities

The subcellular localization of glycosyltransferases in brain
tissue has been a subject of still unsettled controversy (NG &
DAIN, 1977b). The disagreements revolve around the fractionation
techniques used. In all the reported studies on brain subcellular
fractionations, differential centrifugation were made followed by
discontinuous sucrose density gradients. These methods all involved pelleting and resuspension procedures which may introduce
unnecessary artifacts in assessing the subcellular localization

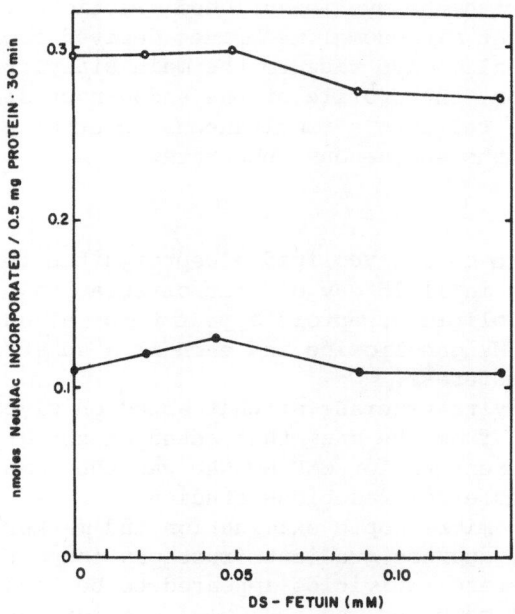

Fig. 3 Effect of exogenous DS-fetuin on the incorporation of NeuNAc into endogenous glycolipids (●————●) and endogenous glycolipids + exogenous GM₁ (o————o).

of the glycosyl transferases. In order to avoid pelleting and resuspension procedures, the rat cerebra homogenate with nuclei and large debris removed, was applied directly onto a continuous sucrose density gradient (0.4-1.6M) 60,000G (NG & DAIN, 1977b). All four sialyltransferase activities showed a similar distribution pattern with peak activities around 0.7M sucrose. Significantly, these peak activities did not overlap with the protein peak fractions. Electron microscopic examination of the gradient fractions with the peak sialyltransferase activity was associated with an enrichment in smooth membrane fragments and vesicles, some of which resembled structures of the Golgi complexes. Mitochondria, synaptosomes and myelin were absent, but present in other fractions. The well defined sialyltransferase peaks in the sucrose gradient suggest that the membranes bearing the transferases are not derived from mitochondria, synaptosomes or myelin, or a much broader enzymes activity peak overlapping with one or more of the protein peaks would have been expected. From these observations, the conclusion can be drawn that sialyltransferases are concentrated in membrane structures which are not related to synaptosomes and mitochondria. These sialyltransferase-enriched structures are presumably derived from the endoplasmic reticulum, the Golgi complexes and the plasma membrane. A lesser sialyltransferase acti-

vity associated with the myelin-enriched fractions is believed to
be due to the light microsomal membranes derived from the same
structures which also gave rise to the main sialyltransferase
peak. In addition, the profile of the endogenous sialyltransferase activities reflects a simultaneous occurrence of the sialyltransferases and the endogenous substrates.

SUMMARY

1.   There are more glycolipid acceptor sites for NeuNAc than
for glycoproteins in 11-15 day old rat cerebra.

2.   The glycolipid acceptors appear to be almost exclusively
Cer-Glc-Gal and GM_1 ganglioside and each is a substrate for a different sialyltransferase.

3.   The sialyltransferase(s) that acted on glycoprotein could
be differentiated from the ones that acted on the glycolipids.

4.   The apparent Km for CMP-NeuNAc was the same for all four
of the sialyltransferase reactions studied.

5.   Electron microscopic examination and marker enzyme studies on continuous sucrose gradient fractions found that most of
the sialyltransferase activities appeared to be localized in
smooth microsomal membrane and the Golgi complex derivatives and
not associated with the synaptosomes.

ACKNOWLEDGEMENT

The authors wish to thank the Journal of Neurochemistry for
permission to reproduce some of the figures and graphs from NG &
DAIN, 1977a,b. This investigation was supported by NIH Grant
NS05104.

REFERENCES

KAUFMAN B., BASU S. & ROSEMAN S. (1968):Enzymatic synthesis of
     disialogangliosides from monosialogangliosides by sialyl-
     transferases from embryonic chicken brain. J. Biol. Chem.
     243, 5804-5807.
KEMP S.F. & STOOLMILLER A.C. (1976):Studies on the biosynthesis of
     glycosphingolipids in cultured mouse neuroblastoma cells.
     Characterization and acceptor specificities of N-acetyl-
     neuraminyl- and N-acetylgalactosaminyltransferases. J.
     Neurochem. 27, 723-732.
MESTRALLET M., CUMAR F.A. & CAPUTTO R. (1977):Trisialoganglioside
     synthesis by a chicken brain sialyltransferase. Comparative
     study with the similar reaction for the synthesis of disialo-
     ganglioside. Mol. Cell. Biochem. 16, 63-70.
NG S.S. & DAIN J.A. (1977a):Sialyltransferases in rat brain:Reac-
     tion kinetics, product analyses, and multiplicities of enzyme
     species. J. Neurochem. 29, 1075-1083.

NG S.S. & DAIN J.A. (1977b): Sialyltransferase in rat brain:Intra-
    cellular localization and some membrane properties. J.
    Neurochem. 29, 1085-1093.
SCHACHTER H. & RODEN L. (1973): The biosyntheis of animal glyco-
    proteins, in "Metabolic Conjugation and Metabolic Hydrolysis",
    FISHMAN W.H., Eds., Vol. 3, Academic Press (New York and
    London), pp. 2-149.
SVENNERHOLM L. (1964):The gangliosides. J. Lipid Res. 5, 145-155.

CELL BIOLOGICAL AND IMMUNOLOGICAL SIGNIFICANCE OF GANGLIOSIDE CHANGES ASSOCIATED WITH TRANSFORMATION[*]

Sen-itiroh Hakomori, William W. Young, Jr.[†],
Leonard M. Patt, Teruo Yoshino[‡], Laurel Halfpap and
Clifford A. Lingwood[§]
Biochemical Oncology, Fred Hutchinson Cancer Research
Center and Departments of Pathobiology and Microbiology,
University of Washington, Seattle, Washington 98104.

INTRODUCTION

Glycolipid changes associated with oncogenic transformation have been extensively studied by a number of investigators on chemical and enzymatic basis. An incomplete synthesis of complex glycolipids and an associated accumulation of its precursor glycolipids are the common change observable in many transformed cells irrespective of the transforming agents (see for reviews, Hakomori 1973; 1975; Brady and Fishman 1974; Richardson et al 1976). However, the other type of glycolipid change, i.e. a synthesis of a new glycolipid absent in normal tissue, or foreign to the host may occur in some experimental tumour and in human tumour; a novel fucolipid in rat hepatoma (Baumann et al 1978), A-like antigen in tumours of blood group O or B individuals (Hakomori et al 1967; Häkkinen 1970), blood group P_1 and P antigen in a tumour of the rare pp individual (Levine et al 1951), and Forssman antigen in tumours of F$^-$ individuals (Hakomori et al 1977) are typical examples. These components are normally absent in progenitor cells or in host's tissues and therefore could be a tumour-associated antigen. All of these antigens are glycosphingolipid and may be synthesized through an activation of a new glycosyltransferase which is foreign

---

[*]This investigation has been supported by grants from the National Institutes of Health CA20026 and GM23100, and a grant from the Cancer Research Institute, New York. [†]W.W.Y is supported by a Fellowship from the Cancer Research Institute, New York. [‡]Present address: Department of Chemistry, International Christian University, Tokyo. [§]Present address: Hospital for Sick Children, Toronto, Ontario Canada.

to the host (see for reviews Hakomori and Young 1978; Young and
Hakomori 1978).

Although glycolipid changes are, in general, closely
associated with oncogenic transformation and is regarded to be a
change caused by activation of transforming gene ("src" gene) in
cells transformed by avian sarcoma virus (Hakomori et al 1977),
very little is known of its biological implication and significance.
This paper is aimed at synthesizing some ideas as to how the glycolipid changs is related to cell biological and immunological
significance based mainly on our recent observations.

CELL BIOLOGICAL SIGNIFICANCE OF GANGLIOSIDE CHANGES

A possible role of glycolipids and gangliosides in cell growth
regulation has been suggested by several phenomena described in the
following subsections. Incomplete synthesis of higher ganglioside
or glycolipid may well be related to a loss of growth regulation
demonstrated in transformed cells as will be discussed below.
a) Exogenous addition of ganglioside and its analogues modify
cell growth behavior. As we have originally described (Laine et al
1973); exogenous addition of globoside in culture medium resulted
in accumulation of globoside in cell surface membrane; as a
consequence, an extension of pre-replicative phase (G_1 to S), a
reduction of cell growth and cell saturation density, an increase
of adhesive property of cells to substratum, and a change of
morphology have been observed. Similarly, Keenan et al (1975)
described exogenous addition of various gangliosides resulted in
a similar morphological and growth behavioral change of cells.
Although the mechanism of this phenomena is not exactly known,
glycolipid metabolism, particularly their catabolism, could be
greatly inhibited. Kirsten tumour cells (3T3KiMSV) when incubated
with culture medium containing 120-130 µg/ml of hematoside reduces
membrane sialidase activity to a great extent (Halfpap and Hakomori
unpublished observation). Membrane sialidase has been implicated
as a parameter in cell growth and transformation (Schengrund et al
1973; Yogeeswaran and Hakomori 1975).

We have synthesized ganglioside analogue containing N-tri-
fluoroacetylneuraminic acid (abbreviated TFA-hematoside) (Yoshino,
Halfpap and Hakomori unpublished observation). The synthesis was
successfully made starting fully acetylated N-glycolylhematoside
reacted trimethyl oxonium salt of antimonium chloride ($Me_3SO^+SbCl_4^-$)
followed by reaction with trifluoroacetic anhydride in piridine
and methylene dichloride followed by treating with sodium methoxide
as follows.

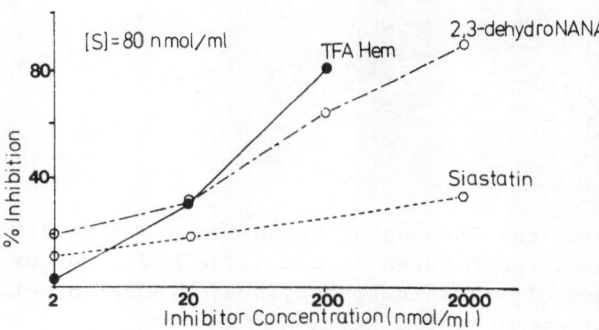

Membrane fraction was prepared by hypotonic lysis of confluent
3T3KiMSV cells grown in 10 cm plates according to the method of
Schengrund et al (1972). The membrane was suspended in 0.5 ml of
1 mM sodium acetate buffer pH 4.5, and 50 µl aliquots were used for
assay. To the membrane suspension is added a given amount of
inhibitors dissolved in the same acetate-Triton X solution and the
[3H]-labeled substrates are added followed by incubation for 60 min
at 37°C. The incubated solutions were lyophilized in a conical
microtube and the residue was extracted with 20-30 µl of chloroform-
methanol (2:1). The extract was developed on TLC plate, and the
ratio of the radioactivity of lactosylceramide (CDH) and hematoside
was determined, and expressed as % inhibition in the ordinate.

Fig. 1. Inhibition of Membrane-Bound Sialidase by Trifluoroacetyl
    Hematoside (TFA-Hematoside) and Other Inhibitors

As compared with sialidase inhibitor such as Umezawa's
"siastatin" (2-acetamido-3,4-dihydroxy-5-carboxypiperidine)
(Umezawa et al 1974) or Meindl's transitional sialidase inhibitor
(2,3-dehydro-2-deoxy-N-acetylneuraminic acid) (Meindl et al 1974),
hematoside containing N-trifluoroacetyl neuraminic acid inhibit
membrane sialidase to a greater extent (see Fig. 1) and consequently,
cell growth behavior and morphology of 3T3KiMSV cells have been
greatly modified (see Fig. 2).

b) Cell contact-dependent glycolipid synthesis or "glycolipid
response". The synthesis of a particular glycolipid increases upon
cell to cell contact. This phenomenum is called "glycolipid
response" and was considered to be related to contact inhibition
of cell growth since contact inhibitability and contact-dependent
glycolipid synthesis showed a parallelism in various cells (Hakomori

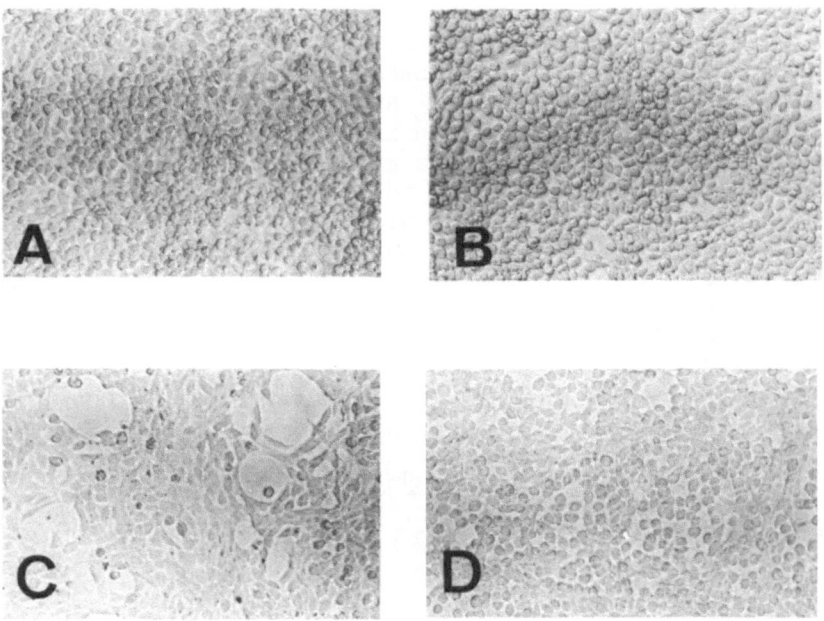

Fig. 2.   Morphological Changes of Transformed 3T3 Cells with Murine
          Sarcoma Virus Kirsten Strain (3T3KiMSV), cultured in the
          Presence of N-Trifluoroacetylneuraminosyl Hematoside.
     A) Cells grown in normal medium as control.
     B) Cells grown in the medium containing 125 nmole/ml of
        N-immunoethylsialylhematoside.
     C) Cells grown in medium containing 125 nmole/ml N-trifluoro-
        acetylhematoside.
     D) Cells grown in medium containing N-glycolylsialylhemato-
        side.

1970; Sakiyama et al 1972; Critchley and MacPherson 1973; Kijimoto
and Hakomori 1971; Chandrabose et al 1975), although the degree of
glycolipid contact response may not be correlated with the tumor-
genicity (Sakiyama et al 1973). Highly contact inhibitable cells
showed contact response of glycolipid at the early stage (touching
stage) rather than at the later stage (confluency) of cell contact
(Yogeeswaran and Hakomori 1975). The concentration of GDla ganglioside of 3T3 cells increases significantly at the touching phase
and decreases to a normal level when cells were contact inhibited
at confluency. The increase of GDla was ascribed to an inhibited
sialidase activity (Yogeeswaran and Hakomori 1975). Similarly,
GDla ganglioside of muscle cells increases several fold at the
touching stage before cell fusion initiated (Whatley et al 1976).
The cell contact response of neutral glycolipid such as ceramide
trihexoside was correlated with the enhancement of α-galactosyl-
transferase activity. Only the activity of UDP-galactose:lacto-
sylceramide-α-galactosyltransferase, but not UDP-galactose:gluco-
sylceramide-β-galactosyltransferase of both BHK and NIL cells
increases several fold at the contact inhibited stage as compared
to the sparse growing stage. A similar enzyme response on cell
contact was not observable on polyoma-transformed cells (Kijimoto
and Hakomori 1973). Based on these observations, a model was
presented (Hakomori et al 1972) in which a surface ektoprotein that
recognizes glycolipid was postulated (see Fig. 3).

   c) A mimic contact response of glycolipid and growth inhibition induced by anti-glycolipid Fab. If the model presented in
Figure 3 is true, anti-glycolipid antibodies, particularly directed
to a "contact-sensitive glycolipid", applied on cell surface should
induce growth inhibition and mimic contact response of glycolipid
synthesis.

   The affinity purified anti-GM_3 ganglioside Fab inhibited cell
growth and reduced cell saturation density of NIL and 3T3 cells to
a great extent, however, cell growth of transformed NIL and 3T3
cells was not inhibited and saturation density was not reduced in
the presence of anti-GM_3 antibodies Fab. The affinity purified
anti-globoside Fab did not induce cell growth inhibition and did
not reduce cell saturation density for both NIL and 3T3 cells
(Lingwood and Hakomori 1977).

   The enhanced synthesis of GM_3 ganglioside, but not GDla, or
$GDlb$ and GM_1 gangliosides, was induced by anti-GM_3 antibodies Fab
(see Fig. 4). This enhancement of GM_3 synthesis was most
remarkable when sparse cells were treated with anti-GM_3 antibodies
Fab, but to a lesser extent when crowded cells were treated
(Lingwood and Hakomori 1977). Therefore, the effect of antiganglioside antibodies to ganglioside synthesis varied depending
on the cell saturation density.

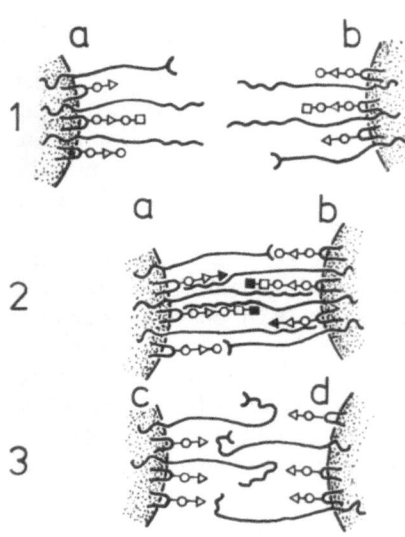

Fig. 3. A model for ekto-
protein that reorganizes
"contact sensitive" glycolipid (Hakomori et al 1972).
1. growing cells: glyco-
lipids and ektoproteins
arranged on cell surfaces
in a certain order.
2. Confluent cells: when
cell a and b meet, glyco-
lipid and ektoprotein linked
together through a complementary structure. Some
glycolipid carbohydrates
can extend their chains for
better linkages. 3. Transformed malignant cells;
carbohydrate chains are
incomplete, consequently
no complementary structures
were found between glyco-
lipid and ektoprotein.

Fig. 4. Effect of anti-GM$_3$
Fab on ganglioside synthesis
in 3T3 cells.
3T3 cell culture (10 cm plate;
4×10^4 cells/cm^2) were incubated in the presence of [^{14}C]-
Gal (1 μCi/ml) for a total of
12 hrs.
Fab anti-GM$_3$ was added (3 μg/
ml) to each culture at 0, 4,
and 8 hrs of incubation. At
12th hrs of incubation, cells
were washed, harvested and
radiolabeled gangliosides
were extracted and analyzed
on TLC. The activities of
each ganglioside were determined (Lingwood and Hakomori
1977).

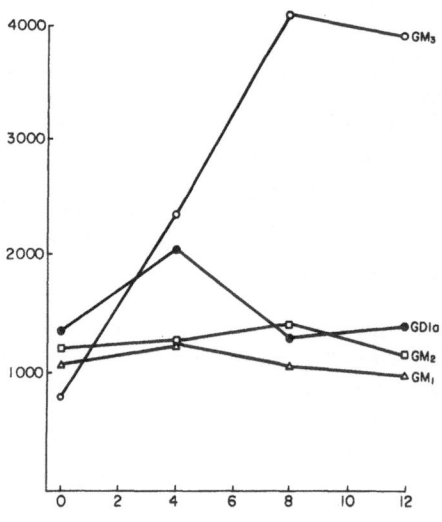

These observations suggested further that contact-dependent glycolipid response as well as the process of "contact inhibition" could be based on a possible interaction between antibody-like ektoprotein and contact-sensitive glycolipid. This assumption has been supported by that 1) anti-glycolipid antibodies Fab enhance synthesis of contact-sensitive glycolipid, and 2) the antibody Fab induces a similar state of growth inhibition as seen in contact inhibited cells but failed to induce growth inhibition for transformed cells. The putative ektoprotein has not been isolated nor characterized, so far.

d) <u>A reduction of cell saturation density induced by cell surface proteins</u>. Cell surface proteins (CSP), metabolically labeled with $[^{35}S]$-methionine was released on incubation of cell sheet with phosphate buffered saline containing ethidium bromide (10 μg/ml, 37°C, 30 min). The protein fraction will bind to trypsinized cells and the binding of released CSP fraction from 3T3 cells was reduced on incubating CSP with GM_3 prior to incubation of CSP with 3T3 cells. $[^{35}S]$-CSP bound to 3T3 cells was 280 cpm/10^6 cells, whereas that preincubated with GM_3 was 150 cpm/10^6 cells by mean values of triplicate experiment (Lingwood and Hakomori unpublished data). In a previous study, various glycolipids covalently linked to amino propyl glass column were used to study which component of CSP will bind to glycolipid-glass columns. A specific elimination of two components with molecular weight 200,000 and 150,000 was observed when CSP was past through GM_3 column. These components were not eliminated by lactosylceramide-, Forssman-, and globoside-column. However, those proteins which bound to a specific glycolipid column were difficult to elute by various chaotropic reagents and detergents (Carter et al 1978). Although these results are preliminary, some of the ethidium bromide released protein has a specific affinity to GM_3 ganglioside and some other glycolipids as well.

Addition of such cellular proteins to 3T3 cell culture induced a remarkable reduction of cell saturation density for 3T3 cells but not for Kirsten virus transformed 3T3 cells (see next page, Fig. 5; Lingwood and Hakomori unpublished observation). This indicates that the CSP present in ethidium bromide released fraction could recognize GM_3 and some other cell surface glycolipids, then induce the enhanced susceptibility of contact inhibition. This result is compatible with that anti-GM_3 Fab induces enhanced susceptibility to contact inhibition.

e) <u>Glycolipid carbohydrate chain as a serum growth factor receptor</u>. Serum growth factors are essential to maintain cell growth <u>in vitro</u>, although the significance of such factors <u>in vivo</u> is not known. We found that the major activity in cell growth stimulation of fetal calf serum was specifically eliminated through affinity columns composed of a specific glycolipid covalently linked to Sepharose (Young et al 1978) or a glycolipid-liposomes entrapped in polyacrylamide (Marcus 1976). The growth stimulatory activity

was trapped most strongly and specifically by ganglio-N-tetraosylceramide (asialo GM_1)-column, but was not eliminated by lactosylceramide, α-galactosyllactosylceramide, globoside, ganglio-N-triosylceramide, hematoside, hamatosidol sialic acid carboxyl of
hematoside was reduced to alcohol) (see next page, Fig. 6). The
column containing GM_1 ganglioside, and GM_1 gangliosidol (sialic
acid carboxyl of GM_1 ganglioside was reduced to alcohol) reduced
the growth stimulating activity. Both the increase of cell numbers
and [^{3}H]-thymidineincorporation into cells were inhibited by
culturing cells in the media containing fetal calf serum which was
treated with ganglio-N-tetraosylceramide- or GM_1 ganglioside-poly-
acrylamide column.

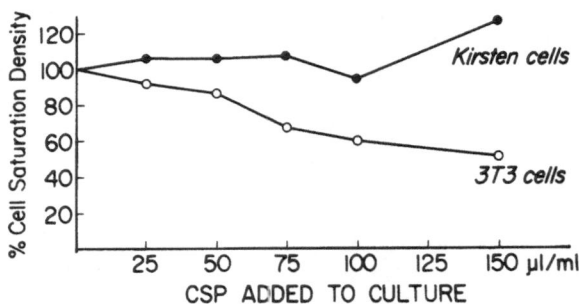

Fig. 5. Effect of cellular proteins ("CSP") released by ethidium
bromide on cell growth.

Cell surface proteins (CSP) released by incubating 3T3 cell sheets
in a glucose containing basic salt solution containing 10 µg/ml of
ethidium bromide at 37°C for 30 minutes. Some cells detached from
the culture dishes but are viable. The solution was separated by
centrifugation and extensively dialyzed and concentrated through
membrane filtration. Cells were cultured in medium containing
25-150 µg/ml of cell surface protein. Cell saturation density was
determined after 3 days for 3T3 cells and after 4 days for 3T3 cells
transformed by murine sarcoma virus Kirsten strain ("Kirsten cells").
Note that cell saturation density was reduced as much as 40-50% in
a medium containing 150 µg/ml of "CSP". Kirsten cell growth was
unaltered. Each point is the mean value of triplicate.

The active factor was recovered by eluting the column with 2 M potassium thiocyanate or with 8 M urea followed by dialysis and concentration through membrane filtration (Table I). However, the activity was not fully recovered from the column; some activity must be lost during the elution, dialysis and concentration of the eluate in most of the experiment.

Furthermore, pre-treatment of 3T3 cells with monovalent anti-bodies to ganglio-N-tetraosylceramide or to GM_1-ganglioside inhibit growth stimulation by serum (Table II). These findings suggest that a specific carbohydrate chain $Gal\beta1\rightarrow3GalNAc\beta1\rightarrow4Gal\beta1\rightarrow R$ may function recepting serum growth stimulation. Cell growth inhibition induced by monovalent antibodies to GM_3 and GM_1 as described in the previous section, could be due to the block of recepting serum growth factor. Growth inhibition on cell contact may be induced by masking of receptor glycolipid through a complementary ektoprotein as shown in Figure 3.

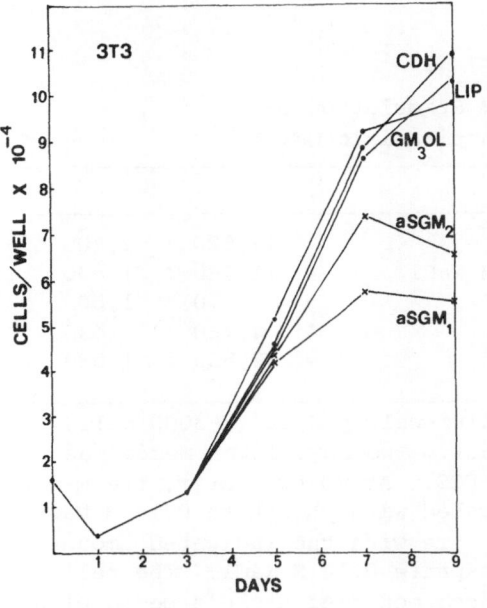

Fig. 6. Growth curve of 3T3 cells grown in media containing fetal calf serum which was passed through glycolipid affinity columns. Balb 3T3 cells were plated in 24 well plates (each well diameter 1.5 cm) at the cell number indicated for day 0. On day 2, the media was removed, cells were washed with Dulbecco-modified Eagle's (DME) medium without serum, and 1 ml of DME containing 0.4% fetal calf serum was added. On day 3, the medium was removed and 1 ml of the indicated media was added. Values are the mean of triplicate measurement of trypsinized cells from wells and counted by a Coulter counter at the indicated time. CDH, GM_3OL, $aSGM_2$, $aSGM_1$, LIP are columns prepared from ceramide dihexoside, GM_3-ganglioside having sialic acid whose carboxyl group was reduced to primary alcohol, asialo-GM_2, asialo-GM_1, and cholesterol lecithin.

TABLE I.   Stimulation of 3T3 Cell Growth by Eluate from
                  Ganglio-N-Tetraosylceramide-Sepharose 4B Column.

	Cells/Well	
	I	II
DME	44,640 ± 2,700	35,290 ± 1,540
FCS	60,456 ± 15,950	110,520 ± 28,890
LACTOSE (.2M)	36,170 ± 2,230	42,234 ± 4,310
KSCN (2M)	42,340 ± 5,456	42,880 ± 1,640
UREA (6M)	64,636 ± 12,150	70,500 ± 3,080

30 ml of fetal calf serum diluted with the same volume of
phosphate-0.14 M NaCl was passed slowly through a 10 ml column
volume of ganglio-N-tetraosylceramide covalently bound to Sepharose
4B. The column was washed with phosphate-0.14 M NaCl until normal
ultra-violet absorption material was washed off. Then the column
was sequentially eluted with the indicated solutions. The eluates
(50 ml) were dialyzed extensively against 0.15 M saline and then
phosphate buffered-0.14 M NaCl, and concentrated to 5 ml. Cells
were seeded in Falcon 24 well plates (1.5 cm diameter for each well)
in 1% fetal calf serum plus 50 μl eluates. Values are 2 separate
experiments, I and II. In both experiments, triplicate wells were
counted in a Coulter cell counter after 5 days incubation at 37°C.

TABLE II.   Inhibition of Serum Stimulation by
                   Pre-Incubation with Anti-Asialo GM_1-Fab

Condition	
Control	13,620 ± 2,400
Pre-incubated with phosphate-0.14 M NaCl	11,420 ± 1,400
" with 20 1 Fab*	7,270 ± 1,800
" with 50 1 Fab	4,760 ± 430
Unstimulated	2,530 ± 1,030

*Balb/c 3T3 cells were seeded in multi-well plates (2-3000 cells
per well, each had a diameter 0.9 cm). Two days later media was
changed to the one containing C.4% FCS. At 40 hrs later, the media
was aspirated and the wells were washed with phosphate 0.14 M NaCl.
The wells were then incubated for 6 hrs with the indicated amount of
purified anti-asialo GM_1-Fab or phosphate 0.14 M NaCl. The cells
were then washed once in warm Dulbecco modified Eagle's media without serum and then a fresh 1% FCS was added. Sixteen hrs later,
1 μCi of [3H]-Thy was added and the incubation continued for 5-6 hrs.
The cells were then processed to determine [3H]-Thy incorporation
into trichloroacetic acid insoluble material.

IMMUNOLOGICAL SIGNIFICANCE OF GLYCOLIPID CHANGES ASSOCIATED WITH
TRANSFORMATION.

This subject has been reviewed repeatedly in the past (Hakomori
and Young 1978; Young and Hakomori 1978), therefore we will discuss
exclusively on one glycolipid accumulated in one type of transformed
cell rather than general changes of glycolipid antigen associated
with oncogenic transformation.

Ganglio-N-triosylceramide (asialo GM_2) is detectable at the
surface of KiMSV tumour cells in Balb/c mice but not in progenitor
3T3 cells nor in various organs of host mice by immunological and
chemical methods, including cell surface labeling with galactose
oxidase-NaB $^3H_4$, therefore this glycolipid is regarded as a
tumour-associated antigen (Rosenfelder 1977). In these tumour
cells as compared to their progenitor 3T3 cells a blocked synthesis
of higher ganglioside (GM_1, GD1a) was obvious (Mora et al 1973)
and accumulation of GM_2 and ganglio-N-triosylceramide has been
clear (Rosenfelder et al 1977).

Antibodies directed to ganglio-N-triosylceramide produced in
rabbit was identified as IgG and demonstrated an effective killing
of 3T3KiMSV cells but not 3T3 cells (see Fig. 7). In the presence
of human lymphocytes and antibodies only 3T3KiMSV cells were killed
and 3T3 cells were not killed (antibody-dependent K cell toxicity)
(see Fig. 8). Passive immunization of Balb/c mice with anti-
ganglio-N-triosylceramide inhibited tumour cell growth if not
completely (see Fig. 9). However, active immunization of Balb/c
mice with ganglio-N-triosylceramide failed to protect KiMSV tumour
growth. Various methods with KiMSV immunization resulted in no
antibody formation or only a weak antibody production directed to
ganglio-N-triosylceramide. Cellular immune response to this glycolipid has not been observed so far. These results suggest that
ganglio-N-triosylceramide is not a typical tumour-associated antigen since KiMSV tumour is not immunogenic and animals immunized
with ganglio-N-triosylceramide do not reject KiMSV tumour. However,
ganglio-N-triosylceramide may function as a specific cell surface
marker which could be distinguishable from cell surface structures
of various other cells.

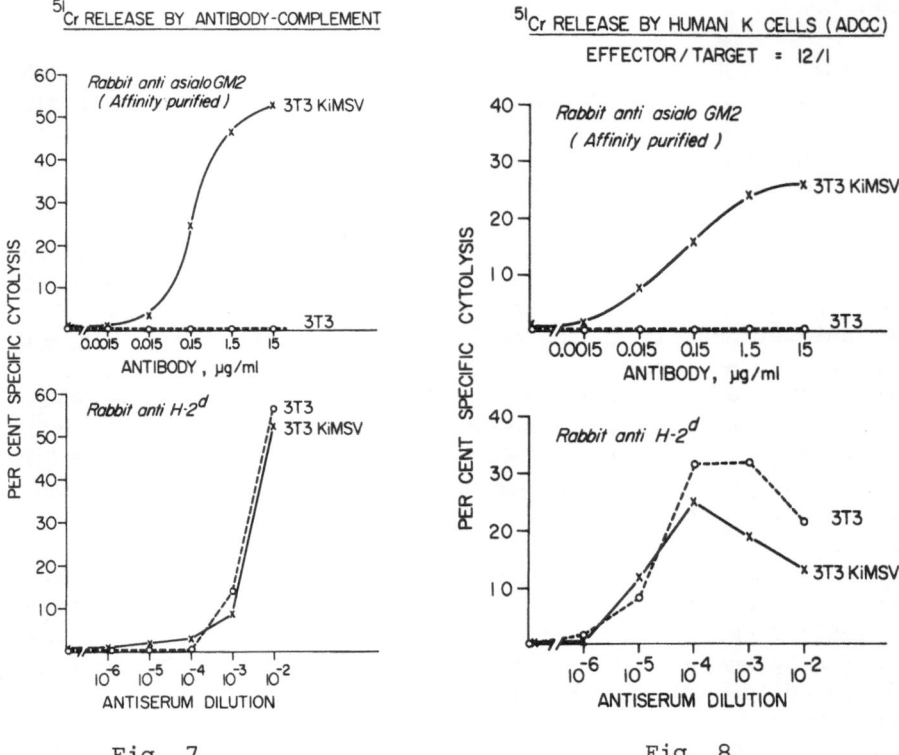

Fig. 7

Fig. 8

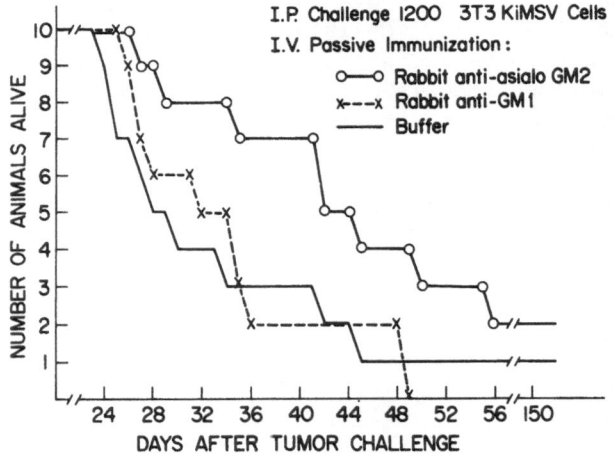

Fig. 9

Fig. 7. Cytotoxicity of Balb/c 3T3 and 3T3KiMSV Cells by Antibody-
         Complement.

     The cytotoxicity assay was performed in V-bottom microtiter
plates, and all dilutions were made with MEM plus 10% fetal calf
serum. Each well received the appropriate antibody dilution plus
5×10^3 labeled target cells in a total volume of 0.15 ml.
Following incubation at room temperature for 30 minutes, 0.05 ml
of a 1:10 dilution of guinea pig serum was added, and incubation
continued for 45 minutes at $37^\circ C$. The plates were centrifuged
at 300 xg for 1 minute, and 0.1 ml of the cell-free supernatant
was assayed for ^{51}Cr content.

Fig. 8. Antibody Dependent Cell-Mediated Cytotoxicity of Balb/c
         3T3 and 3T3KiMSV Cells.

     Human peripheral blood lymphocytes were purified by Ficoll-
Hypaque separation followed by incubation at $37^\circ C$ for 1 hr to
remove adherent cells. The cytotoxicity assay was performed in a
microtiter plate, each well containing 6×10^4 lymphocytes, 5×10^3
^{51}Cr-labeled 3T3 or 3T3KiMSV cells and appropriate dilutions of
antibody in a final volume of 0.2 ml. The plates were centrifuged
at 300 xg for 1 minute, then incubated at $37^\circ C$ for 4 hours. After
incubation 0.1 ml of the cell-free supernatant was assayed for
^{51}Cr content.

Fig. 9. Survival of Mice Injected with 3T3KiMSV Tumour Cells.

     Groups of 10 mice were challenged intraperitoneally with
1200 3T3KiMSV cells. After 2 hours each group received a single
intravenous injection of either rabbit anti-asialo GM_2, rabbit
anti-GM_1, or buffer alone.

REFERENCES

Baumann, H., and Hakomori, S. (see the last reference)

Brady, R.O., and Fishman, P., 1974, Biosynthesis of glycolipids in virus-transformed cells, Biochim. Biophys. Acta 335:121.

Carter, W.G., Fukuda, M., Lingwood, C., and Hakomori, S., 1978, Chemical composition, gross structure and organization of transformation-sensitive glycoproteins, Ann. N.Y. Acad. Sci. 32:160.

Chandrabose, K.A., Graham, J.M., and Macpherson, I., 1976, Glycolipid glycosyltransferases of a hamster cell line in culture. II. Subcellular distribution and the effect of culture age and density, Biochim. Biophys. Acta 429:112.

Critchley, D.R., and Macpherson, I., 1973, Cell density-dependent glycolipids in NIL_2 hamster cells derived from malignant and transformed cell lines, Biochim. Biophys. Acta 246:145.

Hakomori, S., 1970, Cell density-dependent changes of glycolipid concentrations in fibroblasts, and loss of this response in virus-transformed cells, Proc. Natl. Acad. Sci. U.S.A. 67:1741.

Hakomori, S., 1973, Glycolipids of tumor cell membrane, Adv. in Cancer Res. 18:265.

Hakomori,S., 1975, Structures and organization of cell surface glycolipids and glycoproteins, dependency on cell growth and malignant transformation, Biochim. Biophys. Acta 417:55.

Hakomori, S., Kijimoto, S., and Siddiqui, B., 1972, Glycolipids of normal and transformed cells - a difference in structure and dynamic behavior, in: "Membrane Research," C. Fred Fox, ed., Academic Press, New York.

Hakomori, S., and Young, W.W., Jr., 1978, Tumor associated glycolipid antigens and modified blood group antigens, Scand. J. Immunol. 6:97.

Keenan, T.W., Schmid, E., Franke, W.W., and Wiegandt, H., 1975, Exogenous ganglioside suppress growth rate of transformed and untransformed 3T3 mouse cells, Exp. Cell Res. 92:259.

Kijimoto, S., and Hakomori, S., 1971, Enhanced glycolipid: alpha-galactosyltransferase activity in contact-inhibited hamster cells and loss of this response in polyoma transformants, Biochem. Biophys. Res. Commun. 44:557.

Laine, R.A., and Hakomori, S., 1973, Incorporation of exogenous glycosphingolipids in plasma membranes of cultured hamster cells and concurrent change of growth behavior, Biochem. Biophys. Res. Commun. 54:1039.

Lingwood, C., and Hakomori, S., 1977, Selective inhibition of cell growth and associated changes in glycolipid metabolism induced by monovalent antibody to glycolipids, Exp. Cell Res. 108:385.

Meindl, P., Bodo, G., Palese, P., Schulman, J., and Tuppy, H., 1974, Inhibition of neuraminidase activity by derivatives of 2-deoxy-2,3-dehydro-N-acetylneuraminic acid, Virology 58:457.

Mora, P.T., Fishman, P.H., Bassin, R.H., Brady, R.O., and McFarland, V.W., 1973, Transformation of Swiss 3T3 cells by murine sarcoma virus if fillowed by decrease in a glycolipid glycosyl transferase, Nature (New Biol) 245:226.

Richardson, C.L., Keenan, T.W., and Morré, D.J., 1976, Ganglioside biosynthesis, characterization of CMP-N-acetylneuraminic acid: lactosylceramide sialyltransferase in Golgi apparatus from rat liver, Biochim. Biophys. Acta 488:88.

Sakiyama, H., Gross, S.K., and Robbins, P.W., 1972, Glycolipid synthesis in normal and virus-transformed hamster cell lines, Proc. Natl. Acad. Sci. U.S.A. 69:872. .

Sakiyama, H., and Robbins, P.W., 1973, Glycolipid synthesis and tumorigenicity of clones isolated from the NIL_2 line of hamster embryo fibroblasts, Fed. Proc. 32:86.

Schengrund, C-L., Lausch, R.M., and Rosenberg, A., 1973, Sialidase activity in transformed cells, J. Biol. Chem. 248:4424.

Umazawa, A., Aoyagi, T., Komiyama, T., Morishima, H., Hamada, M., and Takeuchi, T., 1974, Purification and characterization of sialidase inhibitor, siastatin, produced by streptomyces.

Whatley, R., Ng, S.K.-C., Roger, J., McMurray, W.C., and Sanwal, B.D., 1976, Developmental changes in gangliosides during myogenesis of a rat myoblast cell line and its drug resistant variants, Biochem. Biophys. Res. Commun., 70:180.

Yogeeswaran, G., and Hakomori, S., 1975, Cell contact-dependent ganglioside changes in mouse 3T3 fibroblasts and a suppressed sialidase activity on cell contact, Biochemistry 14:2151.

Young, W.W., Jr., and Hakomori, S., 1978, Status of blood group carbohydrate chains in human tumors, in: "Glycoproteins and Glycolipids in Disease Processes", E.F. Walborg, Jr., ed., American Chemical Society Symposium Monograph, Washington, D.C.

Young, W.W., Jr., Laine, R.A., and Hakomori, S., 1978, Covalent attachment of glycolipids to solid supports and macromolecules, Methods in Enzymol. 50:137.

Baumann, H., and Hakomori, S., 1978, Neutral fucolipids and fucogangliosides of hepatoma cells, Proc. Amer. Assn. Cancer Res., p. 227.

GANGLIOSIDES, NEURAMINIDASE AND SIALYLTRANSFERASE AT THE NERVE ENDINGS.

Tettamanti G., Preti A., Cestaro B., Venerando B.,
Lombardo A., Ghidoni R. and Sonnino S.

Department of Biological Chemistry, Medical School,
University of Milan, Milan, Italy

INTRODUCTION

Gangliosides are characteristic glycolipid components of the plasma membranes of mammalian cells. They are particularly abundant in the nervous tissue, specially the grey matter, where their concentration is about one tenth that of total phospholipids. The high content of gangliosides in the neuronal membranes, the great variety in the composition of their oligosaccharide chains, and their peculiar location in the outer membrane surface are enough evidence to stimulate research and speculation on the possible involvement of gangliosides in brain specific functions. As a matter of fact, gangliosides are just located- the synaptic junctions- where a specialized physiological event takes place, and definitely synaptic membranes would be and would behave differently without gangliosides. However, in order to provide a plausible working hypothesis for any specific roles of gangliosides in brain function, a more precise knowledge on the contribution given by gangliosides to the local environment of the membrane, in terms of capability and quality of interactions with both the lipid and protein components of the membrane itself, is required.

In accordance with this line, the following situations will be reviewed and discussed in the present report: (a) content of gangliosides in the plasma membranes surrounding nerve terminals ("synaptosomal membranes"); (b) presence and content in the same membranes of the enzymes, neuraminidase and sialyltransferase, causing modifications in the chemical structure of gangliosides; (c) synaptosomal membranes–cytosol relationships at the nerve ending as for gangliosides and associated enzymes; (d) basic physico-chemical aspects of phospholipid–ganglioside interactions in artificial membrane systems; (e) ganglioside contribution to the supramolecular organization of the lipid matrix and glycocalix of synaptosomal membranes.

Content of gangliosides, neuraminidase and sialyltransferase in the synaptosomal membranes.

The occurrence of gangliosides and neuraminidase in the membranes surrounding nerve endings was ascertained by several investigators (1–9). Quite recently (10) definite evidence, as to the presence in the same membranes of a sialyltransferase, has been provided. Conversely, what aliquot of the total brain content of gangliosides, neuraminidase and sialyltransferase is carried by nerve ending surfaces is still an open and controversial matter. We tried to give a reasonable solution to both the problem and the controversy. Nerve endings and nerve ending plasma membranes (synaptosomal membranes) were prepared from calf brain cortex by the procedure schematically represented in fig. 1 and described in details elsewhere (10). For obtaining pure preparations we adopted the following criteria: (a) to use as the nerve ending source the 1000–11.500 g sediment from brain homogenate ("P_2 fraction"), thus avoiding massive contamination by microsomal membranes and particles; (b) to wash several times the preparation for getting it rid of contaminating membranes; (c) to remove myelin fragments and light membranes by a first centrifugation on 9% Ficoll, leading nerve endings to precipitate in the pellet; then to purify nerve endings by a second centrifugation on a 9% / 16% Ficoll gradient; (d) to isolate the membranes deriving from hypoosmotically ruptured nerve endings and to characterize them, as plasma membranes, by the use of either positive (ATP-ase, 5'-nucleotidase) and negative (NADH-citochrome C and NADPH-citochrome C reductases, rotenone insensitive) markers.(°)

––––––––––––––––––––––––––––––

(°) Upon morphologiacal examination, by electron microscopy, the final nerve ending preparation appeared to be constituted over 80% by well preserved nerve endings.

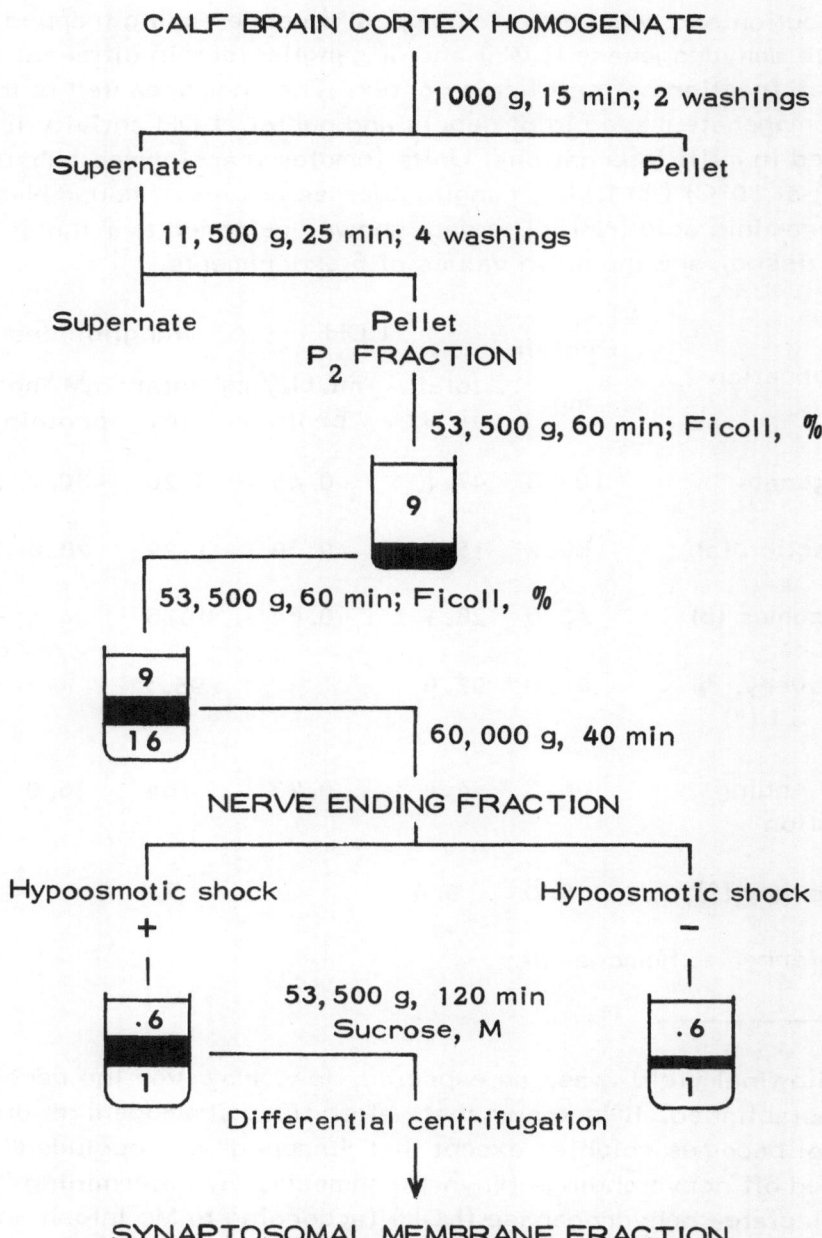

Figure 1. Flowsheet of the procedure for isolating synaptosomal plasma membranes from calf brain.

TABLE I

Distribution and recovery of "trapped" (nerve ending trapped)
Lactate dehydrogenase (LDH) and of gangliosides in different subcellular fractions of calf brain cortex. The "homogenate" is the tissue homogenate made rid of debris and nuclei. LDH activity is expressed in milli International Units (nmoles transformed substrate
min $^{-1}$ at 30°C) (m I.U.); gangliosides as nmoles of bound N-acetylneuraminic acid (nM). The data shown, referred to 1 g starting
fresh tissue, are the mean values of 6 experiments.

Preparation	Protein mg	LDH total ml.U.	LDH ml.U./mg protein	Gangliosides total nM	Gangliosides nM/mg protein
Homogenate	104.8	47.1	0.45	3220	30.7
P$_2$ fraction (a)	39.2	15.5	0.39	1129	28.8
Microsomes (b)	43.6	28.1	0.64	1936	44.4
Recovery, % (a + b) (°)	80.0	92.6		95.2	
Nerve ending fraction	6.3	4.4	0.73	164	26.0
Recovery (°), %	5.0	9.4		5.1	

(°) Referred to homogenate.

The final yield was, as expected, low. However the recovery
was established. It is known that, after tissue homogenization, the
cytosol becomes soluble, except that "trapped" or "occluded" into
pinched off nerve endings ("synaptosomes"). By determining "trapped" lactate- dehydrogenase (LDH) (according to Mc Intosh and
Plummer, 11) we can have a good record for "trapped" cytosol, in
other words of nerve endings. As shown in Table I the "P$_2$ fraction" and the "nerve ending fraction" when analyzed for trapped
LDH contained, respectively, 30% and 9.4% of the nerve endings
originally present in the homogenate. If we assume that LDH is

TABLE II

Gangliosides, neuraminidase, sialyltransferase and some marker enzymes in the "Nerve ending fraction" and in the "Synaptosomal membrane fraction", obtained from the nerve ending fraction by hypoosmotic treatment. Calf brain. Gangliosides are expressed as nmoles bound N-acetylneuraminic acid; enzyme activities in milli International Units (1 nmole transformed substrate min^{-1}, at 37°C or 30°C). The data shown, referred to 1 g starting fresh tissue, are the mean values of 6 experiments.

Parameter	Nerve ending fraction		Synaptosomal membrane fraction	
	Activity (or concentration)		Activity (or concentration)	
	total	specific	total	specific
ATP-ase(°)	1422.0	226.0	153.5	830.0
5'-nucleotidase(°)	101.7	15.3	14.2	76.2
Ach-esterase(°)	34.2	5.4	3.4	18.4
Gangliosides	163.8	26.0	16.7	90.5
Neuraminidase(–)	4.4	0.7	0.5	2.73
Sialyltransferase	119.9	19.0	14.1	76.3
NADH-Cyt. C reductase(")	165.0	26.0	3.5	19.0
NADPH-Cyt. C reductase(")	14.8	2.3	0.3	1.7

(°) Assayed at 37°C; (") Assayed at 30°C; (–) Assayed at 37°C using GD1a ganglioside as substrate and an enzyme preparation made rid of endogenous substrates. Sialyltransferase activity expressed as c.p.m. min^{-1} of incubation using ^{14}C-NeuAc-CMP and lactosylceramide as substrates.

---

evenly distributed in the different nerve endings, which is likely, the recovery of nerve endings in the "nerve ending fraction" was 9.4%.

When hypoosmotically treated, the "nerve ending fraction" gave origin to synaptosomal membranes, part of which were collected in

the "synaptosomal membrane fraction" (see, for details, Venerando et al.,12 and Preti et al.,10). These membranes displayed
(Table II) : (a) no "trapped" LDH activity, thus excluding the presence of unruptured nerve endings; (b) markedly enhanced specific activity of authentic plasma membrane markers (ATP-ase, Ach-
-esterase, 5' nucleotidase), thus qualifying them as plasma membranes; (c) very low absolute content and lowered specific activities of intracellular membrane markers (NADH-citochrome C reductase, NADPH-citochrome C reductase), thus proving a low
contamination of membranes of other origin. The recovery of authentic plasma membranes markers from the "nerve ending fraction" to the "synaptosomal fraction" ranged (see Table II) from
10 to 12%, thus stating the recovery of synaptosomal membranes.
In conclusion, the final recovery of synaptosomal membranes
through the all procedure (from homogenate to the "synaptosomal
membrane fraction") was 10-12% of the 9.4%, that is about 1%.

The behaviour of gangliosides, neuraminidase and sialyltransferase in the procedure leading to synaptosomal membranes (see
Table I, II and III) closely resembled that of authentic plasma membrane markers. For all of them the specific activity (or concentration) ratio, referred to the "nerve ending fraction" ranged from
3.5 to 5.0. Noteworthy, about 1/6 (as protein) of the material present in the "synaptosomal membrane fraction" did not derive from
nerve ending rupture by hypoosmotic shock since it was obtained
in the absence of hypoosmotic treatment (Table III). However this
membranous material displayed the features of plasma membranes:
thus it can be considered as made up by synaptosomal plasma membranes present, as contaminant, in the "nerve ending fraction".

It should be emphasized that, while the data concerning gangliosides and neuraminidase are just confirmatory of previous reports (4,6,8) those regarding sialyltransferase are quite original.

Now, the contents of gangliosides, neuraminidase and sialyltransferase in the "synaptosomal membrane fraction", when divided by the recovery factor (1%), give the figures of the "total"
content of these parameters in the synaptosomal membranes present in brain cortex. As shown in Table IV the content of gangliosides, neuraminidase and sialyltransferase in the synaptosomal
membranes covers, respectively, 52%, 65% and 40% of the total
brain content of each parameter. If we consider, according to

TABLE III

Gangliosides, neuraminidase, sialyltransferase, and some marker enzymes in the "Synaptosomal membrane fraction" (band floating over 1.0 M sucrose), obtained from the nerve ending fraction submitted, or not, to hypoosmotic shock. Calf brain. Of each parameter the specific activity (or concentration) ratio (specific activity -or concentration- in the synaptosomal membrane fraction / specific activity-or concentration- in the nerve ending fraction) is given.

Parameter	Synaptosomal membrane fraction	
	with Hypoosmotic shock	without Hypoosmotic shock
Protein, (mg) (°)	0.185	0.030
Gangliosides	3.48	3.02
ATP-ase	3.83	3.45
5'-nucleotidase	4.97	4.80
Ach-esterase	3.55	3.41
Neuraminidase	3.90	3.35
Sialyltransferase	4.13	4.02
NADH-Cit. C reductase	0.75	0.72
NADPH-Cit. C reductase	0.79	0.75

(°) Referred to 1 g starting fresh tissue. See also legend to Table II.

Ledeen (13) that the nerve endings cover about 10% of the "volume" of brain tissue, and that the remainder "volume", due to the presence of long processes, has a much higher membrane/volume ratio, we conclude that the membranes wrapping nerve endings carry much more gangliosides (over 5-fold) and neuraminidase (over 6.5-fold) than the other brain plasma membranes. As to sialyltransferase, due to the presence of this enzyme in the Golgi apparatus too, the evaluation is less precise: likely the enrichment is of the same order of magnitude. The data concerning gangliosides reported here, perfectly fit to and strengthen, the elegant evi-

TABLE IV

Gangliosides, neuraminidase, and sialyltransferase in the total
membranes of calf brain cortex ("total particulate", 0-105,000 g
pellet of brain homogenate) and in the total synaptosomal plasma
membranes. The content of each parameter in the total synaptosomal membranes has been calculated on the basis of the established
1 % recovery of these membranes in the "Synaptosomal membrane
fraction". The data are referred to 1 g of starting fresh tissue.

Parameter	Total membranes (a)	Total synaptosomal membranes (b)	(b) / (a) %
Gangliosides (°)	3220	1674	52.0
Neuraminidase (+)	765	500	65.3
Sialyltransferase (−)	3500	1410	40.3

(°) As nmoles bound N-acetylneuraminic acid
(+) As milli I.Units ; see also legend to Table II.
(−) As c.p.m., min^{-1} of incubation; see also legend to Table II.

dence recently provided by Hanson et al. (9).

In conclusion, the synaptosomal membranes functionally differentiate from the other plasma membranes of brain tissue for ena-
bling neurotransmission. A biochemical correlation to this functional specialization is the striking enrichment in gangliosides and
in the enzymes capable to modify the sialic acid/ganglioside ratio. The fact that neuraminidase works best at acidic pHs (4.0)
and sialyltransferase at neutral pHs, may link the optimal function-
ality of these enzymes (hence the sialylation-desialylation cycle
of gangliosides) to local fluctuations of pH value.

Presence and content of gangliosides and neuraminidase in the cytosol of nerve endings.

Venerando et al.(12) deviced a simple procedure for separating the cytosol contained in neuronal bodies and glial cells from
that contained into nerve endings. In terms of protein (see Table

TABLE V

Gangliosides, neuraminidase , and Lactate dehydrogenase (LDH) in the cytosol derived from neuronal bodies and glial cells and in the cytosol derived from nerve endings. Calf Brain. Enzyme activities are expressed as milli International Units (m I.U.) (nmoles transformed substrate min^{-1} at 30°C for LDH, at 37°C for neuraminidase); gangliosides as nmoles bound N-acetylneuraminic acid (n M) . The data shown, referred to 1 g starting fresh tissue, are the mean values of 6 experiments.

Parameter	Cytosol	
	Neuronal bodies and glial cells	Nerve endings
Protein, mg	30. 30	5. 60
LDH		
total m I.U.	87. 30	15. 80
m I.U. /mg protein	2. 86	2. 82
Neuraminidase (°)		
total m I.U.	0. 60	1. 23
m I.U. /mg protein	0. 025	0. 22
Gangliosides		
total n M	94	26. 50
nM /mg protein	3. 39	4. 73

(°) Assayed using α(2→3) sialyllactose as substrate.

V) the latter cytosol was about 1/6 of the former. As shown in the same Table, while the specific activity of LDH in the two cytosols was practically the same, the specific concentration of gangliosides was 1.4 times higher in the cytosol from nerve endings and the specific activity of neuraminidase ("soluble" neuraminidase) still greater (9 fold). In other words soluble gangliosides and, particularly, soluble neuraminidase, tend to accumulate in the cytosol of nerve endings. The pattern of the different gangliosides in the cytosol from nerve endings and in the cytosol from neuronal bodies and glial cells was described to be similar to each other

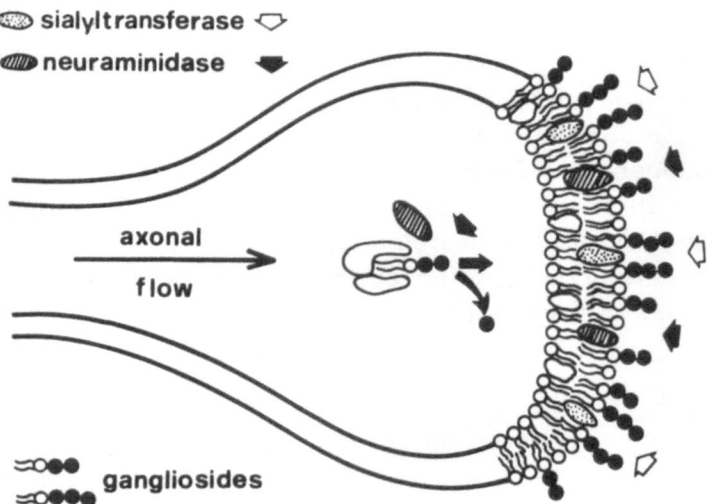

Figure 2. Location of gangliosides, neuraminidase, and sialyl-transferase at the nerve ending.

and to resemble that of membrane bound gangliosides(14). In addition, the gangliosides from both cytosols precipitated with proteins upon stepwise addition of ammonium sulphate and behaved as ganglioside-protein complexes on gel permeation chromatography and on hydroxylapatite column chromatography.

The presence at the level of nerve ending cytosol of gangliosides having higher specific concentration than in the perikaryon cytosol, together with their occurrence as lipoprotein complexes, makes plausible the hypothesis that gangliosides, synthesized in the neuronal body, are carried to the terminals by axonal flow. The form of lipoprotein complexes may greatly facilitate the transport with the rapid phase of the flow (Morgan et al., 15; Forman and Ledeen, 16).

A schematic picture of the location of gangliosides, neuraminidase and sialyltransferase in the nerve ending is shown in Fig 2. The location of gangliosides and neuraminidase in the outer plasma membrane surface has a consistent support (9, 17, 18). The sidedness of sialyltransferase at this level has not yet been ascertained. However, the reported occurrence of sialyltransferase at

the external surface of a number of cells(19, 20, 21) makes this assignment very probable.

Interactions of gangliosides with phospholipids in artificial mem-
brane systems

Gangliosides are soluble amphiphiles. In water solutions they
undergo micellization, above a certain concentration (critical micellar concentration, CMC). Light scattering measurements gave,
for gangliosides GM1, GD1a, GD1b and GT1b (°) values of CMC
about 10^{-6} M. This value is significantly lower than that reported
in previous studies (23-26) but in the line with what is being presently delivered by different laboratories. The mixing of different
gangliosides and the presence of ions, Ca^{2+} included, in concentrations within the physiological range, did not appreciably affect
the CMC while influencing the aggregation number.

Conversely phospholipids, for instance phosphatidylcholine
and phosphatidylethanolamine form, when properly treated in water solution, bilayered aggregates, namely multilamellar vesicles
(liposomes).

When phosphatidylcholine and phosphatidylethanolamine (used
as a surface sided marker to be revealed with 2, 4, 6-trinitroben-
zene sulfonic acid, TNBS, 27) and gangliosides are mixed together, the solvent removed, and the residue dissolved with a proper buffer, mixed aggregates are formed. The type of aggregate
depends upon the molar ratio gangliosides/phospholipids.
As shown in Fig 3a till a certain value of the ganglioside/phospho-
lipid molar ratio, the process of aggregation led to liposomes.
In fact the level of turbidity remained the same as in the absence
of gangliosides and the % of outer surface sided phosphatidylethanolamine remained unchanged at about 60%. Over that value micelles were formed as shown by gradual decrease of turbidity and
increase of aminogroups available for TNBS. The critical value
of the ganglioside/phospholipids molar ratio for transition, or
"transition ratio", varied with the different gangliosides : from
0.25 with GM1 ganglioside to 0.09 for GQ ganglioside. In other
words it rose by decreasing the sialic acid content per ganglioside. In addition, and as shown in Fig 3b, while monovalent cations

_____

(°) The ganglioside nomenclature of Svennerholm (22) is followed.

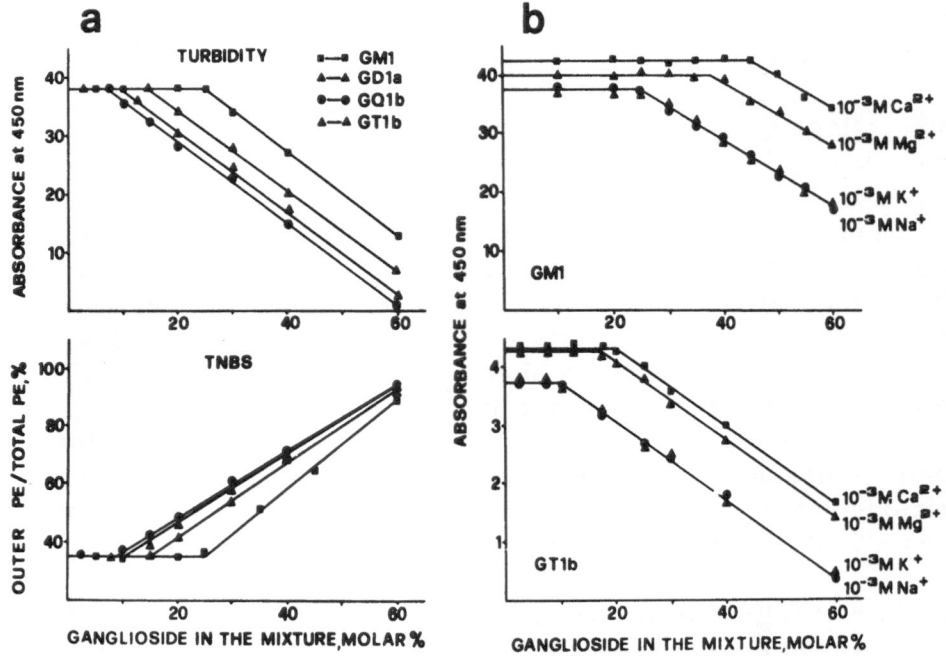

Figure 3. (a) Physicochemical features of mixed aggregates of phosphatidylcholine, phosphatidylethanolamine (PE, used as surface marker) and different gangliosides (GM1, GD1a, GT1b, GQ1b) at increasing proportions of ganglioside. The highest value of turbidity and lowest value of outer PE/total PE ratio corresponds to liposomes. The lowering of turbidity and concurrent enahncement of the outer PE/total PE ratio indicates the presence of micelles. Both liposomes and micelles are of the mixed species (made up by phospholipids and gangliosides). The "break" point is indicated as the "transition ganglioside:phospholipid molar ratio". (b) Effect of different monovalent and divalent cations on the "transition ganglioside/phospholipid molar ratio".

(K^+, Na^+) did not influence the "transition ratio", divalent cations (Mg^{2+}, Ca^{2+}), particularly Ca^{2+}, at concentrations within the physiological range, almost doubled it: from 0.25 to 0.45 for GM1 ganglioside; from 0.1 to 0.2 for GT1b ganglioside. This means that Ca^{2+} ions determine a liposomal supramolecular organization in the presence of such amounts of gangliosides which, in the absence of calcium, would give a micellar organization.

Vesicles (liposomes) of phosphatidylcholine (containing ^{14}C--choline) upon incubation in the presence of ganglioside micelles

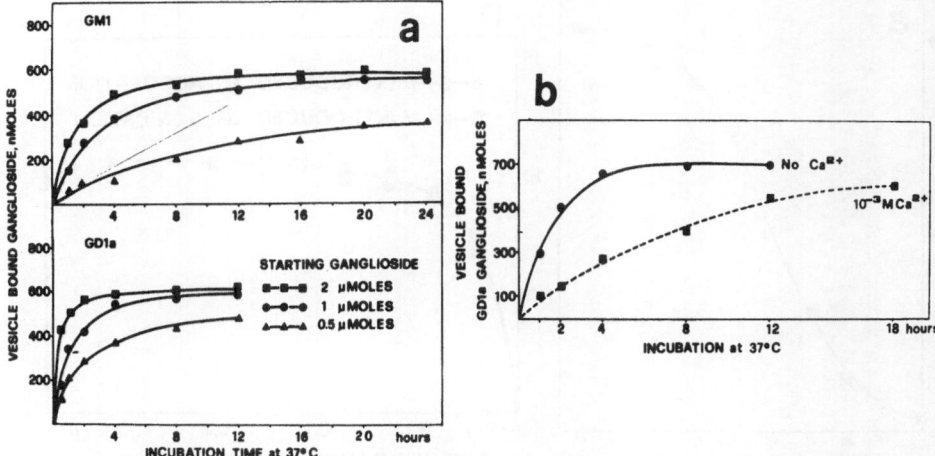

Figure 4. (a) Effect of incubation time (at 27°C) and of ganglioside concentration on the incorporation of gangliosides (GM1, GD1a) into phosphatidylcholine monolamellar vesicles (liposomes). Phosphatidylcholine (in liposomes): 9 μmoles. Gangliosides: from O.5 to 2 u moles. The mixtures (O.5 ml), after incubation, were passed through a 1 X 2O cm Sepharose 4B column in order to separate vesicles from ganglioside micelles. (b) Effect of Ca^{2+} ions on the incorporation of GD1a ganglioside into phosphatidylcholine mono - lamellar vesicles. Phosphatidylcholine (in liposomes): 9 μmoles. Starting GD1a ganglioside: 1 μ mole.

(containing tritium labeled gangliosides), do incorporate gangliosides; concurrently no incorporation of phospholipid into ganglioside micelle occurs (28). The process of ganglioside incorporation into phospholipid vesicles is a time, ionic strength, pH, ganglioside concentration, temperature dependent phenomenon (28). For instance, starting from 9 μmoles of phospholipid (as monolamellar vesicles) and from 0.5–2 μmoles of ganglioside (GM1, GD1a), the incorporation of ganglioside into vesicles preceeded proportionately with time and ganglioside concentration (see Fig 4a). The incorporation rate was higher with GD1a than with GM1 ganglioside. Interestingly, in both cases, a maximum and equal level of incorporation was reached (0.6 μmoles), as displayed by a saturation process dependent upon the ganglioside lipid moiety, not the quality of the oligosaccharide chain. This saturation level corresponds to a ganglioside/phospholipid molar ratio of 0.07, which is much lower than the transition ratio discussed above. The presence of Ca^{2+} ions (see Fig 4b) caused the maximum level of ganglioside incorporation to be rea-

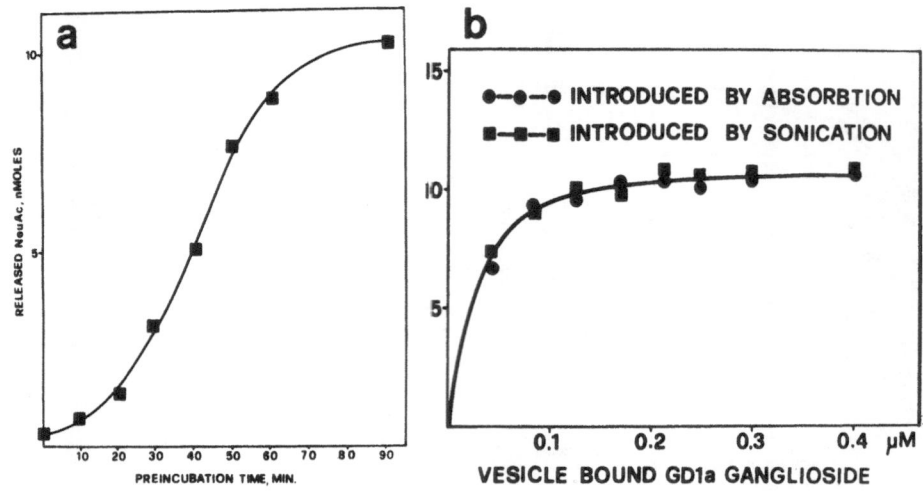

Figure 5. (a) Release of N-acetylneuraminic acid (NeuAc) by Vibrio cholerae neuraminidase from GD1a ganglioside after preincubation in the presence of phosphatidylcholine monolamellar vesicles(liposomes). GD1a ganglioside (1 μmole) and phosphatidylcholine-as liposomes- (9 μmoles) were preincubated at 37°C for the given time, in a volume of O.5 ml (pH 6.8), then 2 Units of Vibrio cholerae neuraminidase were added and the incubation carried out for 5 min. Released NeuAc was assayed according to Warren (33). (b) Release of NeuAc from GD1a ganglioside introduced into phosphatidylcholine (PC) liposomes by: (i) absorbtion, and (ii) sonication.

(i) Phosphatidylcholine vesicles were incubated at 37°C for 4 hours with GD1a ganglioside (PC/GD1a molar ratio = 9/2), then the formed mixed vesicles (carrying a PC/GD1a molar ratio = 9/O.6) were separated by Sepharose 4B column chromatography and employed for neuraminidase assay.

(ii) Phosphatidylcholine and GD1a ganglioside (molar ratio =9/O.6) were dried, suspended in proper buffer solution, submitted to sonication, then employed for neuraminidase assay.

---

ched markedly later.

    The interaction of ganglioside micelles with phospholipid vesicles led to insertion of ganglioside molecules into the lipid matrix of the vesicle, likely by a fusion process. In fact, when mixtures of phospholipid vesicles and ganglioside micelles were incubated at 37°C for different times, then treated with Vibrio cholerae neuraminidase, a release of N-acetylneuraminic acid (NeuAc) was recorded,

which followed the sigmoidal kinetics reported in Fig 5a. Since Vibrio cholerae neuraminidase was found (29) to display on GD1a ganglioside-phospholipid mixed liposomes Vmax values more than 50--fold higher than on GD1a ganglioside micelles, the sigmoidal kinetics are likely the expression of the following phenomenon. Initially all ganglioside is present in micellar form, thus yielding the lowest record of neuraminidase activity. By allowing interaction with phospholipid vesicles gangliosides become inserted into the vesicles, thus yielding a mixed vesicle, which is a much better substrate for the enzyme: therefore the rate of NeuAc release increased.
As a further proof, when phospholipid vesicles, which incorporated after incubation a certain amount of GD1a ganglioside (7%, in molar terms), were submitted to the action of Vibrio cholerae neuraminidase in parallel with mixed GD1a ganglioside-phospholipid vesicles (carrying the same proportion of ganglioside) prepared by sonication, the resulting V/S relationshipswere exactly the same (Fig 5b). This means that, in both cases, the vesicles were the same (mixed vesicles carrying the same molar content of ganglioside) and, also, that the distribution of ganglioside in the inner-outer lipid layer was the same: 40% inside, 60% outside, as reported for sonicated mixed vesicles (29).

All this experimental evidence suggests the following picture of the process of ganglioside micelle-phospholipid vesicle interaction. Initially a contact, or adhesion, occurs, mainly determined by the ganglioside carbohydrate chains, which is followed by insertion of the ganglioside lipid moiety into the lipid matrix of the vesicle. This latter process, which leads to distribution of gangliosides on the inner and outer lipid layer, is a fusion phenomenon. After a certain amount of ganglioside has been incorporated into the vesicle, hydrgen bonds (and other weak bonds) are being formed within the oligosaccharide chains, thus resulting in a stabilization of the vesicular structure. On the other hand, the acquired surface charge prevent further adhesion (and incorporation)of ganglioside micelle, thus mimicing a "saturation" process. It should be stressed, in particular, that gangliosides, unlike artificial micellar detergents (30, 31), and probably because of the presence of their carbohydrate moiety, do not succeed lysing phospholipid vesicles.

Considerations on the structural location of gangliosides in the
synaptosomal membranes.

As determined by Breckenridge et al.(4), the content of gangliosides in purified rat brain synaptosomal membranes is, in molar
terms, about 1/8-1/10 that of total phospholipids. This figure can
be accepted as valid for the same membranes from other animals.
If we consider (17,18) the largely asymmetrical distribution of gangliosides toward the outer membrane surface, the ganglioside/phos-
pholipids molar ratio can be estimated to be, here, 0.25-0.2.
Due to the large preponderance of oligosialogangliosides in the synaptosomal membranes, this ratio falls in the range of the "transition ratio" observed in simple ganglioside-phospholipid systems.
The presence of proteins and cholesterol may be expected to enhance
the transition ratio, thus allowing the maintenance of the bilayered
structure even in the presence of such high amounts of gangliosides.

However the distribution of gangliosides at the surface can hardly be even, since interactions within the oligosaccharide chains will
necessarily lead gangliosides to get together and to form "patches"
(32). Likely this tendency is greater for oligosialogangliosides
than for monosialogangliosides. "Patches" in fixed positions
of the membrane may be obtained due to the presence of outer sided
glycoproteins, the oligosaccharide chains of which interacting with
the correspondent chains of gangliosides. In correspondence to
"patches", the ganglioside/phospholipid ratio necessarily increases, this inducing a local tendency to drastic change in the supramolecular organization of the membrane.

The uneven distribution of gangliosides in the membrane determines very peculiar situations. First, the fluidity of the membrane would be different in the various membrane areas: much lower
in correspondence to ganglioside "patches", higher elsewhere.
Second, in the "patch" area the organization of the membrane would
reach a maximum degree of instability, and the membrane potential,
due to the local density of sialic acid residues, a maximum value.
In addition, the "patches" will pilot and facilitate interactions with
some external ligands (hydrophilic or carbohydrate interactions).
Conversely, the ganglioside free areas will enable hydrophobic
interactions with ligands of other nature (see Fig 6). Finally, this
organization displays great flexibility and disposition to regulation.
In fact changes of the sialic acid content of gangliosides, which are

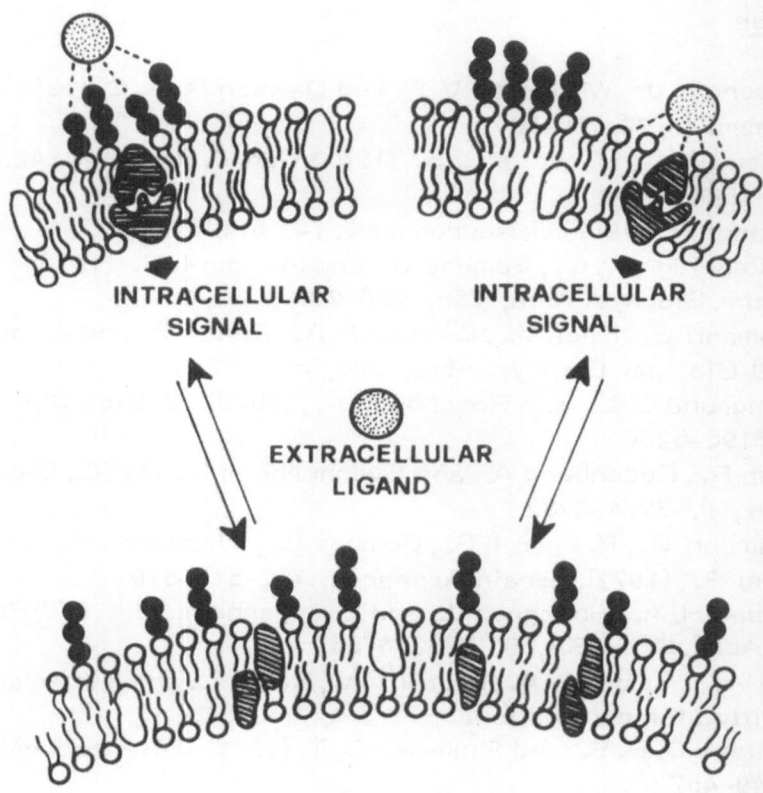

Figure 6. "Patch" distribution of gangliosides in the outer surface of synaptosomal membranes. Importance of the "patch" organization of the glycocalix in orienting the binding of extracellular ligands to the membrane surface and the following induction of intracellular responses.

---

possible at this level, will cause diffusion of gangliosides through the membrane, and change either the fluidity of the membrane lipid matrix and the features of the glycocalix. Of course the role of neuraminidase and sialyltransferase in this system is fundamental.

It is reasonable to expect that, in the near future, integrated progress of research in different fields–physicochemistry of ganglioside contribution to membrane organization; enzyme events occurring at the membrane surface; ligands–glycocalix interactions–will provide enough information to figure out the role played by gangliosides in synaptosomal membranes.

References

1) Eichberg J. Jr, Whittaker V. P. and Dawson R. W. C. (1954)
    Biochem. J. , 92, 91-100

2) Lowden J. A. and Wolfe L. S. (1964) Can. J. Biochem. , 42,
    1587-1594

3) Wiegandt H. (1967) J. Neurochem. , 14, 671-674

4) Breckenridge W. C. , Gombos G. and Morgan I. G. (1972)
    Biochim. Biophys. Acta, 266, 695-707

5) Tettamanti G. , Preti A. , Lombardo A. , Bonali F. and Zambotti V.
    (1973) Biochim. Biophys. Acta, 306, 466-477

6) Schengrund C. L. and Rosenberg A. , (1970), J. Biol. Chem. ,
    254, 6196-6200

7) Ohman R. , Rosenberg A. and Svennerholm L. (1970), Bioche-
    mistry, 9, 3774-3782

8) Tettamanti G. , Morgan I. G. , Gombos G. , Vincendon G. and
    Mandel P. , (1972), Brain Research, 47, 515-518

9) Hansson H. A. , Holmgren J. and Svennerholm L. (1977)Proc.
    Natl. Acad. Sci. USA, 9, 3782-3786

10) Preti A. , Lombardo A. , Fiorilli A. , Caimi L. and Tettamanti G.
    submitted for publication.

11) Mc Intosh C. H. S. and Plummer D. T. (1976) J. Neurochem. ,
    27, 449-457

12) Venerando B. , Preti A. , Lombardo A. , Cestaro B. and Tetta-
    manti G. (1978) Biochim. Biophys. Acta, 527, 17-30

13) Ledeen R. W. (1978) J. Supram. Structure, 8, 1-17

14) Sonnino S. , Ghidoni R. , Marchesini S. and Tettamanti G.
    (1979) J. Neurochem. , 32, 312-316

15) Morgan I. G. , Tettamanti G. and Gombos G. (1976) Adv. Exptl.
    Med. Biol. , 71, 137-150

16) Forman D. S. and Ledeen R. W. (1972)Science, 177, 1630-1633

17) Rosenberg A. (1978) Adv. Exptl. Med. Biol. , 101, 439-446

18) Sandhoff K. , Pallman P. , Wiegandt H. and Ziegler W. (1975)
    Adv. Exptl. Med. Biol. , 101, 463-474

19) Shur B. D. and Roth S. (1975) Biochim. Biophys. Acta, 415,
    473-512

20) Porter C. W. and Bernacki R. J. (1975) Nature, 256, 648-650

21) Colombino L. F. , Bosmann H. B. and Mc Lean R. J. (1978),
    Experimental Cell Research, 112, 25-30

22) Svennerholm L. (1970) in "Comprehensive Biochemistry" (Florkin M. and Stotz E. H., eds), Elsevier Amsterdam, vol 18, pp 201-227

23) Gammack D. B. (1963) Biochem. J., 88, 373-383

24) Howard R. E. and Burton R. (1964) Biochim. Biophys. Acta, 84, 435-440

25) Yohe H. and Rosenberg A. (1972), Chem. Phys. Lipids, 9, 279-285

26) Yohe H., Roark D. E. and Rosenberg A. (1976) J. Biol. Chem. 251, 7083-7089

27) Barenholz Y., Gibbs D., Sitman B. J., Goll J., Thompson T. E. and Carlson F. D. (1977) Biochemistry, 16, 2806-2810

28) Cestaro B., Ippolito R., Ghidoni R., Orlando P. and Tettamanti G. (1979), submitted for publication in the Bull. Mol. Biol. Med.

29) Cestaro B., Barenholz Y. and Gatt S. (1979) submitted for publication in the J. Biol. Chem.

30) Inoue K. and Kitagawa T. (1976) Biochim. Biophys. Acta, 426, 1-16

31) Richards M. H. and Gardner C. R. (1978) Biochim. Biophys. Acta, 543, 508-522

32) Sharom F. J. and Grant C. W. M. (1978) Biochim. Biophys. Acta 507, 280-293

33) Warren L. (1959), J. Biol. Chem. 234, 1971-1974

"NEURAMINIDASE-RESISTANT" SIALIC ACID

RESIDUES OF GANGLIOSIDES

Roland Schauer[a], Rüdiger W. Veh[b], M. Sander[b],
Anthony P. Corfield[a], and Herbert Wiegandt[c]

[a]Biochemisches Institut, Universität Kiel, FRG
[b]Institut für Anatomie, Universität Bochum, FRG
[c]Institut für Physiol. Chemie, Universität
Marburg, FRG

Gangliosides have been found to contain up to five
sialic acid moieties bound to galactose residues in α,
2->3 linkages either at the end of the oligosaccharide
chain or within the chain, and may form di- and trisia-
lyl groups with α,2->8 linkages (1,2). N-Acetylneuraminic
acid (NeuAc) and N-glycolylneuraminic acid (NeuGl) have
been identified as the commonly occurring sialic acids
in gangliosides, while the O-acetylated sialic acids occur less frequently. The 9-O-acetyl-N-acetylneuraminic
acid has been found in gangliosides from a variety of
vertebrates including man (3), and 4-O-acetyl-N-glycolyl-
neuraminic acid (4-OAc-NeuGl) has been isolated from horse erythrocyte hematoside (4).

Studies with viral, bacterial and mammalian neuraminidases have shown that the sialic acids can be removed
from the different gangliosides with exception of the "internal" sialic acid residue (5-7). In all these gangliosides the resistant neuraminic acid is linked to cis-3,4
substituted galactose with sialic acid in 3- and N-acetylgalactosamine in 4-position. The resistance of the sialic
acid is believed to be due to a steric hindrance caused
by this structural feature. After enzymic release of the
peripheral galactose and N-acetylgalactosamine from GM1
(II^{3}NeuAc-GgOse$_4$-Cer) or N-acetylgalactosamine from GM2
(II^{3}NeuAc-GgOse$_3$-Cer) to form GM3 (II^{3}NeuAc-Lac-Cer) the
sialic acid is in terminal position and can therefore be
removed by the action of neuraminidase. This shows that
the position and not the nature of sialic acid is responsible for neuraminidase resistance.

While the internal sialic acid residue of GM1 and GM2 is completely resistant towards the action of Vibrio cholerae neuraminidase, it can be cleaved by the neuraminidases from Clostridium perfringens (8,9) and Arthrobacter ureafaciens (10) below the critical micelle concentration or after the addition of bile salts. This auxiliary role of the amphiphilic bile salts on enzyme activity may be explained by an influence on the enzyme protein, or by interaction with the ceramide moiety of gangliosides thus influencing the size of micelles and mixed micelle formation.

In order to obtain more insight into this problem the monosialogangliotetraose (Des-GM1; $II^3NeuAc-GgOse_4$) Gal(ß,1-> 3)GalNAc(ß,1->4)Gal(ß,1->4)Glc, the free redu-

$$(3)$$
$$\uparrow$$
$$(\alpha,2)$$
$$NeuAc$$

cing carbohydrate moiety of GM1, was prepared from the parent ganglioside and studied with bacterial and viral neuraminidases.

It is known from glycoprotein metabolism that an acetyl group at C_4 of NeuAc or NeuGl renders their glycosidic linkages resistant towards the action of bacterial neuraminidases (11). We were therefore interested in studying the behaviour of 4-O-acetylated sialic acids as components of gangliosides towards neuraminidases. As a suitable substrate GM3 from horse erythrocyte membranes was chosen, containing terminal 4-OAc-NeuGl.

ISOLATION OF 4-OAc-NeuGl-GM3 (II^3[4-OAc-NeuGl]-Lac-Cer)

GM3 was isolated from horse erythrocytes similar to the method described by Hakomori and Saito (12). The procedure applied is shown in Fig. 1. The sphingolipid fraction from ethanol precipitation was passed over two silica gel columns with the solvents indicated in Fig. 1. Thus, 62 mg pure 4-OAc-NeuGl-GM3 in addition to 312 mg NeuGl-GM3 was obtained from 10 l packed erythrocytes. The occurrence of an ester group in 4-OAc-NeuGl-GM3 could be demonstrated by a two-dimensional thinlayer chromatographic procedure on silica gel HPTLC-plates in chloroform/methanol/water (65/25/4, by vol.) with an intermediate 12 h treatment in ammonia vapour. After this saponification the R_f value of GM3 is appreciably reduced. The ester group is localized on the sialic acid moiety and not on the lactosyl-ceramide

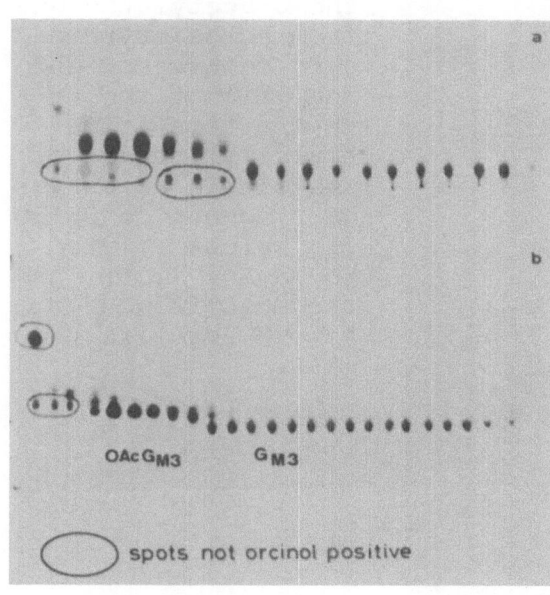

packed horse erythrocytes

$\downarrow$

saponine hemolysis

$\downarrow$

centrifugation for 30 min
at 70,000 g

$\downarrow$

extraction of the pellet
with 96% ethanol

$\downarrow$

precipitation of the
sphingolipid fraction at
-10°C
$\qquad\downarrow$

first (a) silica gel
chromatography,$CHCl_3$:
CH_3OH = 6 : 4
$\qquad\downarrow$

second (b) silica gel
chromatography, $CHCl_3$:
CH_3OH : H_2O = 65 : 25 : 4
$\qquad\downarrow$

pure O-acetylated GM3

Fig. 1. Preparation of 4-OAc-NeuGl-GM3 from horse erythrocytes. TLC of glycolipids from the first (a) and
second (b) chromatography on silica gel using
$CHCl_3/CH_3OH/H_2O$ (65/25/4, by vol.) as solvent.
The sialic acid-containing spots were stained
with the orcinol/HCl/Fe^{3+} spray reagent.

part of GM3, as was shown by two-dimensional thinlayer
chromatography of the compound before and after release of
its sialic acid moiety by mild acid hydrolysis (pH 2, for-
mic acid, 70°C, 1 h, (13)). The structure of the sialic
acid released from GM3 and purified by ion-exchange chromatography was elucidated mainly by a combination of gas-
liquid chromatography and mass spectrometry. The retention
time of the methyl ester pertrimethylsilyl ether derivative
on SE-30 was 2.1 relative to NeuAc (R_t 1.0), similar to the
value reported in the literature (14). The fragmentation
pattern obtained by mass spectrometry corresponds to that
of 4-OAc-NeuGl (Fig. 2). Thus, Hakomori's GM3 was proved
to be 4-OAc-NeuGl-GM3 (Fig. 3) and therefore appeared to
be a suitable substrate for the investigation of mammalian
ganglioside-specific neuraminidases. The sialic acid moiety
of this hematoside was radioactively labelled by periodate
cleavage of its C_7-C_9 side chain and reduction with tritiated borohydride (15). Conditions were chosen to yield
mainly the C_8-analogue of 4-OAc-NeuGl which in the non-O-

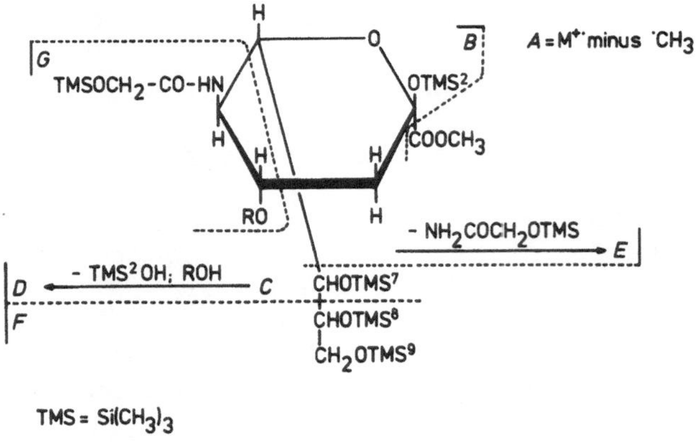

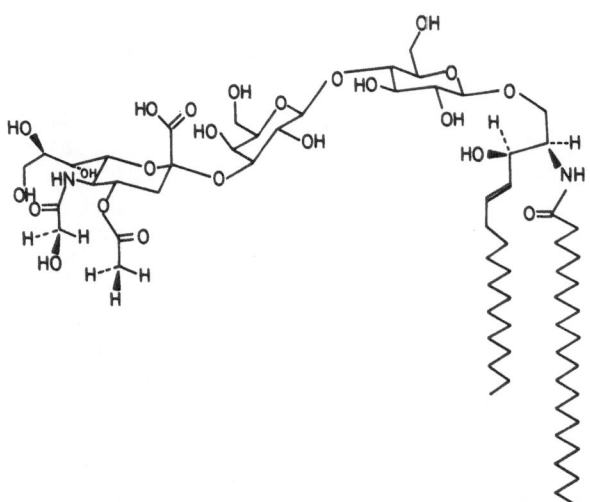

TMS = Si(CH$_3$)$_3$

compound	R	R$_t$	A	B	C	D	E	F	G
NeuGl	TMS	1.81	756	712	566	386	317	205	261
4-O-Ac-NeuGl	Ac	2.02	726	682	536	386	–	205	231

Fig. 2. Mass spectrometry of 4-OAc-NeuGl isolated from the O-acetylated GM3 from horse erythrocyte membranes. GLC analysis of the methyl esters trimethylsilyl ethers of the sialic acids was carried out on SE-30 (R$_t$ values related to NeuAc, 1.00), and the mass fragments A to G identified (14).

Fig. 3. Structure of 4-OAc-NeuGl-GM3 from horse erythrocyte membranes.

acetylated form is a substrate for a very sensitive neuraminidase assay (16). No loss of O-acetyl groups could be detected after this modification procedure. NeuGl-GM3 was tritiated in a corresponding way.

PREPARATION OF GM1 and DES-GM1

The ganglioside was prepared from bovine brain and checked for purity using a radioactive method described in ref. (15). Des-GM1 was prepared by ozonolysis and alkaline fragmentation of GM1 according to ref. (17). Its purity was proved by paper and thinlayer chromatography and paper electrophoresis at pH 1.9 and 6.5. The preparation was resistant to Vibrio cholerae neuraminidase indicating the absence of terminal sialic acid residues.

NEURAMINIDASES

Neuraminidase from Vibrio cholerae was purchased from Behring-Werke, Marburg. Neuraminidase from Clostridium perfringens was purified according to Nees et al. (18). The enzyme, which was finally purified by preparative polyacrylamide gel electrophoresis, had a specific activity of 600 U/mg protein when tested with sialyllactose. It was not contaminated with other enzyme activities and appeared to be a homogeneous protein using several analytical methods. Viral neuraminidases (Newcastle Disease virus and Fowl Plague virus)were a gift from Prof. Rott, University Giessen. Particulate or solubilized neuraminidases from human heart are described in ref. 19.

Due to our interest in the metabolism of O-acetylated sialic acids, horse liver has been chosen as a further enzyme source, since the glycoconjugates from horse tissues contain considerable amounts of 4-O-acetyl-N-acylneuraminic acids (11,13). Two membrane-bound neuraminidase preparations, one glycoprotein-specific and the other glycolipid-specific, were obtained from horse liver homogenates. A crude mitochondrial-lysosomal fraction (20) was further fractionated with the aid of a Ludox-polyvinylpyrrolidone gradient (Fig. 4). The glycoprotein-specific neuraminidase activity was found to be 20-fold enriched in the lysosomal fraction. In contrast, the ganglioside-specific neuraminidase followed the microsomal markers. The two neuraminidase activities show remarkable differences. While the glycoprotein neuraminidase hydrolyzes both C_7- and C_8-analogues of NeuAc at similar rates (collocalia glycopeptides as substrate), the ganglioside neuraminidase shows a strong preference for the C_8-analogue. Furthermore, the glycoprotein neuraminidase is not affected by 2-deoxy-2,3-dehydro-N-acetylneuraminic acid, while the ganglioside neuraminidase which is strongly inhibited by this compound underlines a clear distinction between these two neuraminidases, to our knowledge not previously demonstrated in other tissues. In addition, the glycoprotein-specific horse liver neuraminidase is the only neuraminidase known not to be inhibited by 2-deoxy-2,3-dehydro-NeuAc.

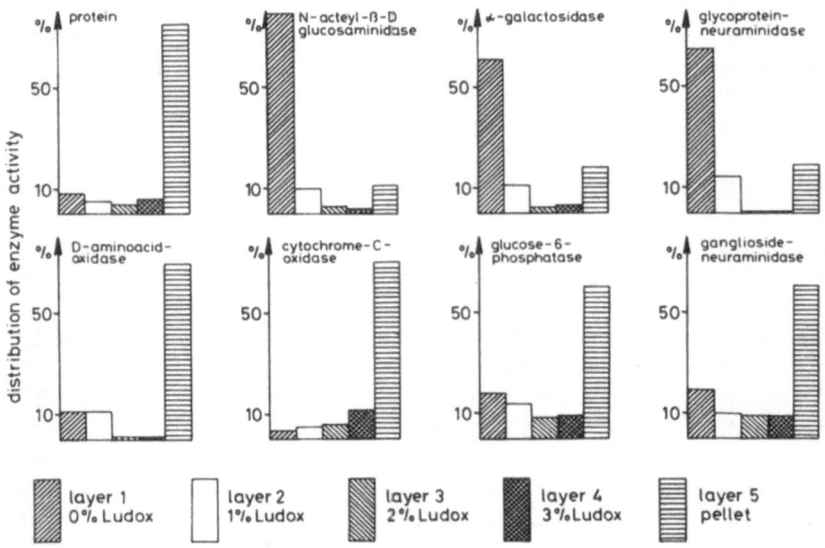

Fig. 4. Subcellular distribution of glycoprotein- and glycolipid-specific neuraminidases in horse liver. A mitochondrial-lysosomal fraction was subfractionated on a Ludox-Polyvinylpyrrolidone gradient. The relative distribution of protein and enzyme activities in the different fractions are shown.

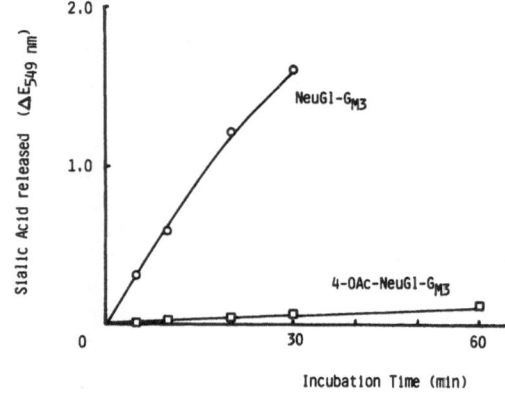

Fig. 5. Cleavage of NeuGl-GM3 (O) and 4-OAc-NeuGl-GM3 (□) by Newcastle Disease virus neuraminidase. Incubations contained 10^{-3}M substrate and 0.15 M phosphate buffer, pH 7.0. Released sialic acid was measured using the periodic acid/thiobarbituric acid colorimetric method.

EXPERIMENTS WITH O-ACETYLATED GM3

NeuGl-GM3 and 4-OAc-NeuGl-GM3 at 10^{-3}M concentration were incubated in 0.1 ml (100 nmoles sialic acid) for 2 and 5 h at 37°C with neuraminidases from Clostridium perfringens, human brain, human heart and horse liver. The incubations were carried out in 0.1 M acetate buffer at the pH-optima of these enzymes: pH 5.0, 4.1, 4.4 and 4.3, respectively. After incubation the unreacted substrate was precipitated by the addition of ovalbumin and tri-chloroacetic acid, and the radioactivity of the superna-tant representing the liberated sialic acid was counted (16). The results are shown in Table 1. In all experi-ments no radioactivity or only low amounts were released from 4-OAc-NeuGl-GM3 in contrast to NeuGl-GM3 which ser-ves as a good substrate for the neuraminidases tested. The hydrolysis of 6 to 15 % of 4-OAc-NeuGl-GM3 observed in the assays with the solubilized or particulate brain neurami-nidase may be due to de-O-acetylation of the substrate during the relatively long incubation, under the influence of the slightly acidic buffer or the presence of esterases in the brain preparation.

Table 1. Effect of 4-O-acetyl substitution of the neurami-
         nic acid in GM3 on the activity of different
         neuraminidases. For incubation conditions see the
         text. The values represent nmoles sialic acid re-
         leased during 2 h and 5 h. n.t., not determined.

substrate	NeuGl-GM3		4-OAc-NeuGl-GM3	
neuramini- dase source	2 h	5 h	2 h	5 h
Clostridium perfringens	5.5	n.t.	o.o	n.t.
Human brain (particulate)	1o.6	25.2	o.8	1.5
Human brain (solubilized)	1o.6	n.t.	1.4	n.t.
Human heart (particulate)	2.3	6.1	o.o	o.o
Horse liver (particulate)	2.1	5.2	o.o	o.5

Incubation of 4-OAc-NeuGl-GM3 and NeuGl-GM3 with New-castle Disease and Fowl Plague virus neuraminidases also demonstrated very low release of 4-OAc-NeuGl relative to NeuGl (Fig. 5).

Saponification of the O-acetyl groups of GM3 by mild alkali before incubation with the neuraminidases results in cleavage rates corresponding to that of NeuGl-GM3. This shows that the resistance of 4-OAc-NeuGl-GM3 is due to the presence of an ester group at C_4 of neuraminic acid. The experiments furthermore clearly demonstrate the strong in-fluence of this ester group not only on bacterial, but also on mammalian and viral neuraminidases.

EXPERIMENTS WITH GM1 AND DES-GM1

There are conflicting reports in the literature concerning the susceptibility of GM1 towards neuraminidases.
We have therefore reinvestigated this problem using pure
GM1 and pure Clostridium perfringens neuraminidase to avoid
errors which might be due to impurities of the substrate or
the enzyme. Furthermore, the enzymic behaviour of GM1 was
compared to that of the corresponding Des-GM1. It is shown
in Fig. 6 that GM1 is substrate for Clostridium perfringens neuraminidase also in our hands, especially at concentrations below the critical micelle concentration. Addition of cholate markedly increases enzyme activity. This
activation by cholate is optimal at pH 5 (Fig. 7). In contrast to this behaviour of GM1, its free, reducing oligosaccharide moiety is completely resistant towards the action of Clostridium perfringens neuraminidase, whether cholate is present or not. It is furthermore remarkable that
both GM1 and Des-GM1 are inactive with neuraminidases from
Vibrio cholerae and from Newcastle Disease or Fowl Plague
viruses, either in the presence or absence of cholate.

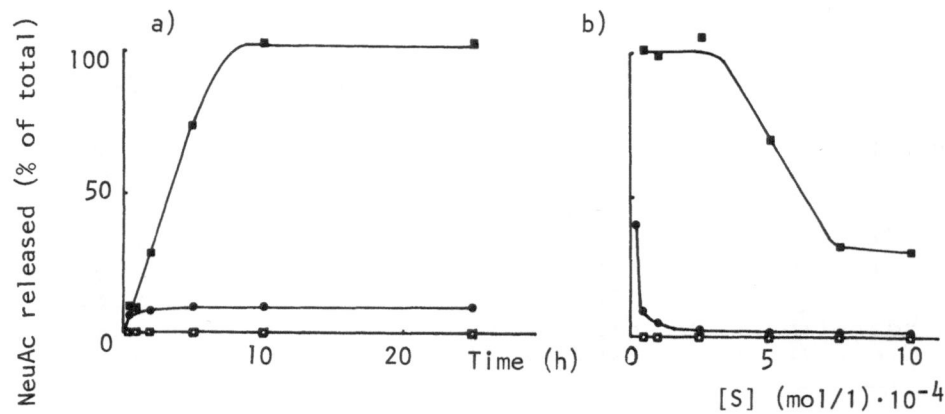

Fig. 6. Cleavage of GM1 and Des-GM1 by Clostridium perfringens neuraminidase. a) Influence of time on
cleavage at a substrate concentration of $5 \times 10^{-5}M$
in 0.05 M sodium acetate, pH 5.1, at 37°C. ■ ,GM1
with 1 mg/ml sodium cholate; ● ,GM1 without cholate;
□ ,Des-GM1 with 1 mg/ml sodium cholate; O,Des-GM1
without cholate. b) Influence of substrate concentration after 25 h incubation in 0.05 M sodium acetate, pH 5.1, at 37°C. Symbols as in a).

The reactivity of GM1 with Clostridium perfringens
neuraminidase in the presence of detergents agrees with the
observations made by Wenger and Wardell (8) and by Rauvala
(9). In a recent report (10) Des-GM1 and GM1 have been

described to be good substrates for neuraminidase from
Arthrobacter ureafaciens, GM1 being cleaved about 3 times
faster than its isolated oligosaccharide.

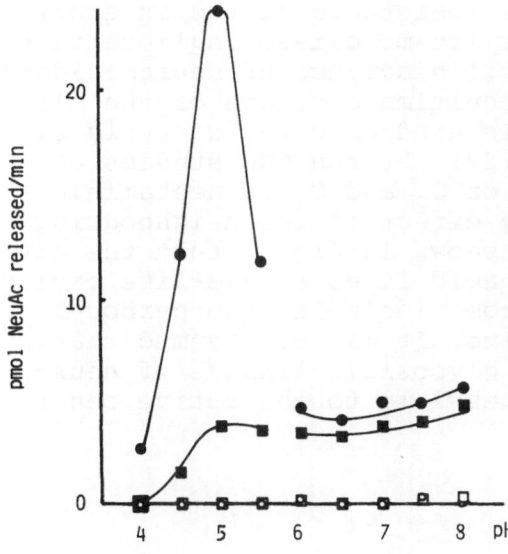

Fig. 7. Influence of pH on
the cleavage rate of GM1 and
Des-GM1. Incubations from pH
4.0-5.5 in 0.05 M sodium acetate buffer and from pH 6.o
-8.o in 0.05 M sodium phosphate buffer. Incubations
were for 5 h at 37°C. ●,GM1
with 1 mg/ml sodium cholate;
■ ,GM1 without cholate; O,
Des-GM1 with 1 mg/ml sodium
cholate;□ ,Des-GM1 without
cholate.

CONCLUDING REMARKS

In this report two ganglioside sialic acid residues
have been described which are considered to be "resistant"
towards the action of neuraminidases. The resistance of
both glycoprotein- and glycolipid-bound 4-O-acetylated
sialic acids is not yet understood. It may be assumed that
a substituent at C_4 of neuraminic acid sterically hinders
the action of neuraminidases. Carbon-13 NMR spin-lattice
relaxation studies of NeuAc carried out by Czarniecki and
Thornton (21) have supported this opinion, since a hydro-
gen-bond has been shown to exist between the hydroxyl
group at C_4 of neuraminic acid and the carbonyl group of
the N-acyl residue (Fig. 8). The action of neuraminidases
seems to depend on the conformation of this part of glycosidically bound N-acylneuraminic acid molecules, which

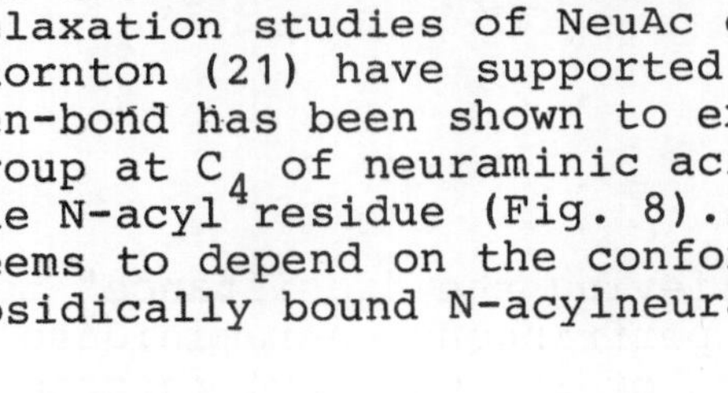

Fig. 8. Schematic re-
presentation of NeuAc
conformation based on
C-13 NMR spin-lattice
relaxation studies of
Czarniecki and Thorn-
ton (21).

could be altered by substitution of the 4-hydroxyl group.
In contrast to 4-O-acetyl groups, 9-O-acetyl residues of
sialic acids have only a small effect on neuraminidase
activity (11,13).

The relative or complete resistance of the internal
sialic acid residue of GM1 or its monosialogangliotetraose
is believed to be due to steric hindrance of neuraminidase
action by the ultimate and penultimate sugars of the oligosaccharide chain (5-9). This hindrance was recently explained by Thornton's group (22). Proton NMR studies of
GD1a and GM1 revealed shifts of C_2 and C_3 of neuraminic
acid suggesting an anisotropy effect of the neighbouring
N-acetylgalactosamine. It is shown in Fig. 9 that the glycosidic oxygen of neuraminic acid lines a cage-like cavity
together with other oxygen atoms including the carbonyl
oxygen of N-acetylgalactosamine. It may be assumed that
this conformation around the glycosidic linkage of neuraminic acid is not easily accessible to the active centre
of neuraminidases.

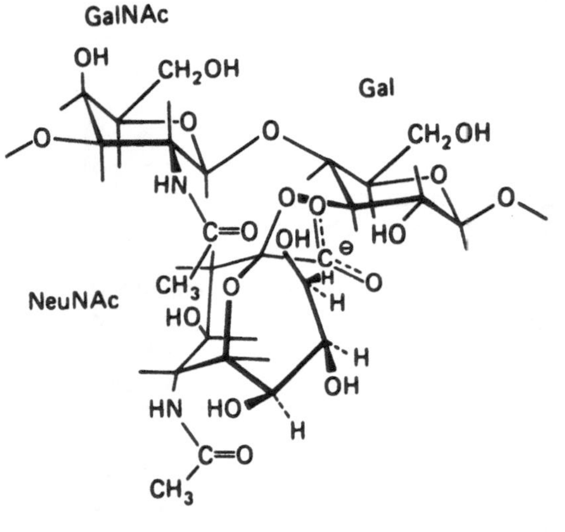

Fig. 9. Schematic
diagram how GalNAc,
NeuAc and Gal-1 resi-
dues of GM1 and Des-
GM1 lead to an "oxy-
gen cage" in which the
glycosidic linkage of
NeuAc may be hidden.
Based on C-13 NMR data
of Harris and Thornton
(22).

The role of cholate in relieving the "resistance" to
cleavage of GM1 by Clostridium perfringens neuraminidase
may be explained by an influence of its carbonyl function
on the "cage" structure of NeuAc and GalNAc, thus reducing
the strength of the interaction between these two sugars.
The enhancement of enzymic activity due to cholate with
GM1 and not with Des-GM1 as substrate , may be caused by
the hydrophobic ceramide moiety of the ganglioside resulting in the formation of mixed micelles or in alteration
of the configuration of the sialic acid moiety.

The difference in enzyme activity between GM1 below the critical micelle concentration and Des-GM1, may be the result of a difference in the configuration of the sialic acid residues in these two molecules, or to an influence of the enzyme itself on this configuration due to enzyme-ceramide interaction. Thus, it appears worthwhile to study the incorporation of neuraminidase molecules into ganglioside micelles with and without cholate.

REFERENCES

1. S.-S. Ng and J.A. Dain, The natural occurrence of sialic acids, in:"The Biological Roles of Sialic Acid", A. Rosenberg and C.-L. Schengrund, eds., Plenum Press, New York and London, 59-1o2 (1976)
2. R.W. Ledeen, Ganglioside structures and distribution: Are they localized at the nerve ending?, J. Supramol. Structure 8:1-17 (1978)
3. J. Haverkamp, R.W. Veh, M. Sander, R. Schauer, J.P. Kamerling, and J.F.G. Vliegenthart, Demonstration of 9-O-Acetyl-N-acetylneuraminic acid in brain gangliosides from various vertebrates including man, Hoppe-Seyler's Z. Physiol. Chem. 358:16o9-1612 (1977)
4. R.W. Veh, M. Sander, J. Haverkamp, and R. Schauer, Demonstration of O-acetyl groups in ganglioside-bound sialic acids and their effect on the action of bacterial and mammalian neuraminidases, in:"Proc. Fourth Int. Symp. Glycoconjugates", J.D. Gregory and R.W. Jeanloz, eds., Academic Press, New York, in press
5. G. Dawson, Glycolipid metabolism, in: "The Glycoconjugates", M.I. Horowitz and W. Pigman, eds., Academic Press, New York, Vol. II, 287-336 (1978)
6. K. Suzuki, Catabolism of sialyl compounds in nature, in: "Biological Roles of Sialic Acid", A. Rosenberg and C.-L. Schengrund, eds., Plenum Press, New York and London, 159-181 (1976)
7. A. Rosenberg and C.-L. Schengrund, Sialidases, in: "Biological Roles of Sialic Acid", A. Rosenberg and C.-L. Schengrund, eds., Plenum Press, New York and London, 295-359 (1976)
8. D.A. Wenger and S. Wardell, Action of neuraminidase from Clostridium perfringens on brain gangliosides in the presence of bile salts, J. Neurochem. 20:6o7-612 (1973)
9. H. Rauvala, Action of Clostridium perfringens neuraminidase on Gangliosides GM1 and GM2 above and below the critical micelle concentration of substrate, FEBS Letters 65:229-233 (1976)

10. K. Sugano, M. Saito, and Y. Nagai, Susceptibility of ganglioside GM1 to a new bacterial neuraminidase, FEBS Letters 89:321-325 (1978)

11. R. Schauer and H. Faillard, Das Verhalten isomerer N. O-Diacetyl-neuraminsäureglykoside im Submaxillarismucin von Pferd und Rind bei Einwirkung bakterieller Neuraminidase, Hoppe-Seyler's Z. Physiol. Chem. 349: 961-968 (1968)

12. S. Hakomori and T. Saito, Isolation and characterization of a glycosphingolipid having a new sialic acid, Biochemistry 8:5082-5088 (1969)

13. R. Schauer, Characterization of sialic acids, Methods Enzymol. 50:64-89 (1978)

14. J.P. Kamerling, J. Haverkamp, J.F.G. Vliegenthart, C. Versluis, and R. Schauer, Mass spectrometry of sialic acids, in: "Recent Developments in Mass Spectrometry in Biochemistry and Medicine", A. Frigerio, ed., Plenum Publ. Corp., Vol. 1, 503-520 (1978)

15. R.W. Veh, A.P. Corfield, M. Sander, and R. Schauer, Neuraminic acid-specific modification and tritium-labelling of gangliosides, Biochim. Biophys. Acta 486: 145-160 (1977)

16. R.W. Veh and R. Schauer, Interaction of human brain neuraminidase with tritium-labelled gangliosides, in: "Advances in Experimental Medicine and Biology, Vol. 101, Enzymes of Lipid Metabolism", S. Gatt, L. Freysz, and P. Mandel, eds., Plenum Press, New York and London, 447-462 (1978)

17. H. Wiegandt and H.W. Bücking, Carbohydrate components of extraneuronal gangliosides from bovine and human spleen and bovine kidney, Eur. J. Biochem. 15:287-292 (1970)

18. S. Nees, R.W. Veh, R. Schauer, and K. Ehrlich, Purification and characterization of neuraminidase from Clostridium perfringens, Hoppe-Seyler's Z. Physiol. Chem. 356:1027-1042 (1975)

19. T.L. Parker, R.W. Veh, and R. Schauer, Comparison of particulate neuraminidases from human heart and brain, in:"Proc. Fourth Int. Symp. Glycoconjugates", J.D. Gregory and R.W. Jeanloz, eds., Academic Press, New York, in press

20. M. Sander, R.W. Veh, and R. Schauer, Demonstration of glycoprotein- and glycolipid-specific neuraminidases in horse liver, in: Proc. Fourth Int. Symp. Glycoconjugates, Academic Press, New York, in press

21. M.F. Czarniecki and E.R. Thornton, Carbon-13 NMR Spin-Lattice Relaxation in the N-acylneuraminic acids, J. Am. Chem. Soc. 99:8273-8279 (1977)

22. P.L. Harris and E.R. Thornton, C-13 and Proton NMR Studies of Gangliosides, J. Am.Chem.Soc.100:6738-6745 (1978)

THE SPECIFICITY OF HUMAN N-ACETYL-ß-D-HEXOSAMINIDASES TOWARDS GLYCOSPHINGOLIPIDS IS DETERMINED BY AN ACTIVATOR PROTEIN

Ernst Conzelmann and Konrad Sandhoff

Max-Planck-Institut für Psychiatrie
D-8000 München 40, W.-Germany

ABSTRACT

It has been very difficult to correlate, on the basis of in vitro measurements of substrate specificities, the glycosphingolipid storage patterns observed in different variants of infantile G_{M2} gangliosidosis with the hexosaminidase (hex) isoenzyme deficiencies underlying these diseases.

However, the in vitro enzyme assays included detergents, which greatly enhanced the enzymic degradation of lipids by breaking down the large lipid micelles that cannot otherwise be attacked by the hydrolases. In vivo, the role of detergent is taken over by water-soluble, low molecular weight proteins, so-called activators, which bind the lipid monomers, thus solubilizing them. It can be shown that the activator protein for the enzymic degradation of ganglioside G_{M2} has a very strong preference for hex A over hex B; it also acts on glycolipid G_{A2} and, to a lesser extent, on kidney globoside. This isoenzyme specificity is much less prominent or even reversed when detergents are used to solubilize the substrates.

The substrate specificities of hex A and hex B measured in the presence of sufficient amounts of the activator protein most probably reflect the conditions occurring in vivo. They can explain the lipid storage patterns observed in different variants of infantile G_{M2}

gangliosidosis, especially the accumulation of ganglioside G_{M2} in variant B (where hex B is still present) and
the reduced storage of G_{A2} in the same variant as compared to variants O and AB. The physiological significance of the activator protein is demonstrated in variant
AB in which the activator is deficient, resulting in an
accumulation of glycolipids G_{M2} and G_{A2}.

INTRODUCTION

Infantile G_{M2}-gangliosidosis is a fatal inherited
storage disease which leads to death within the first
few years of life, due to excessive accumulation of
ganglioside G_{M2} and its asialo derivative, G_{A2}, in nervous tissue (Svennerholm, 1962; Sandhoff et al., 1971).
Three variants can be distinguished by the nature of the
biochemical defect underlying the disease: In variant B
(Tay-Sachs disease) N-acetyl-ß-D-hexosaminidase (hex) A
is deficient (Okada and O'Brien, 1969; Sandhoff, 1969),
whereas in variant O both major hex isoenzymes, A and B,
(E.C. 3.2.1.30) are missing (Sandhoff et al., 1968;
Sandhoff et al., 1971), presumably due to a defect of
the common subunit (Geiger and Arnon, 1976; Beutler et al.,
1976). The third variant, AB, in which the patients show
normal or even elevated levels of hexosaminidases
(Sandhoff et al., 1971; Conzelmann et al., 1978), has
recently been shown to be caused by the lack of a non-
enzymic protein cofactor necessary for the enzymic
degradation of glycolipids G_{M2} and G_{A2} (Conzelmann and
Sandhoff, 1978).

The differences in isoenzyme and activator protein
deficiencies found in these variants are paralleled by
variations in the lipid storage patterns observed in the
patients' tissues (Fig. 1).

Both hex isoenzymes have been purified from various
human tissues (Sandhoff and Waessle, 1971; Srivastava
et al., 1974; Geiger and Arnon, 1976; Sandhoff et al.,
1977) and have been tested in vitro for their ability to
degrade the accumulated substances (Sandhoff et al.,
1977). To achieve reasonable reaction rates detergents
had to be included in the assay mixtures, as the micelles
formed by the lipids in aqueous solution are practically
inaccessible to the enzymes. The values thus obtained
are listed in Table 1.

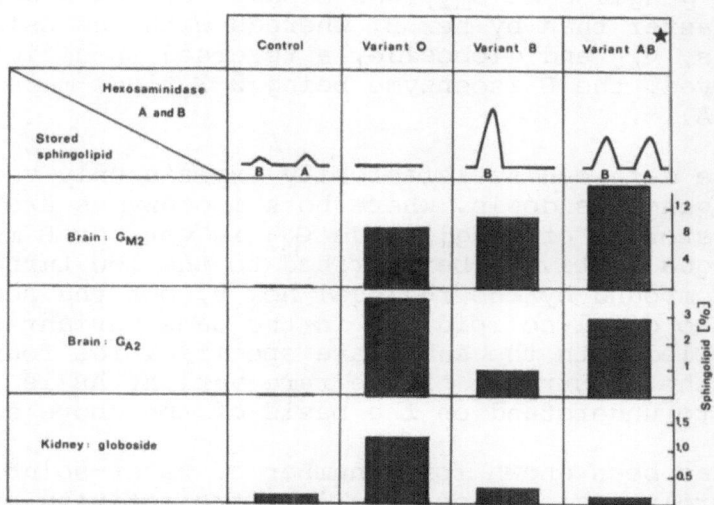

★ caused by deficiency of activator protein

Fig. 1. Hexosaminidase isoenzymes and lipid storage patterns in 3 variants of infantile G_{M2} gangliosidosis. Hexosaminidase distribution was measured with synthetic substrate after isoelectric focusing of brain extracts. The amounts of storage compounds are expressed as percent of dry weight of the respective organ (Sandhoff et al., 1971).

Table 1. Degradation of glycosphingolipids by human hexosaminidases in the presence of detergent (2 mM taurodeoxycholate). (nmoles split / hr x enzyme unit)

Substrate	Hex A	Hex B
Ganglioside G_{M2}	2.8	0.35
Glycolipid G_{A2}	51	170
Kidney globoside	142	242

Under the conditions employed the main storage compound, ganglioside G_{M2}, was cleaved by hex A about 8 times faster than by hex B, whereas with the asialo substrates, G_{A2} and globoside, a reversed specificity was observed, the B isoenzyme being 2-3 times more active than hex A.

These data can satisfactorily explain only variant O of G_{M2} gangliosidosis, where both isoenzymes are missing. The storage of ganglioside G_{M2} in variant B might at least qualitatively be ascribed to the low turnover of this compound by the residual hex B, but the accumulation also of glycolipid G_{A2} in the same variant cannot be reconciled with the substrate specificities found. Finally, the occurrence of the rare variant AB is impossible to understand on the basis of the above data.

It has been shown for a number of water-soluble, lipid-hydrolyzing enzymes, the in vitro activity of which strongly depends on the presence of suitable detergents, that in vivo they are stimulated by low molecular weight protein cofactors, the so-called activators (Fischer and Jatzkewitz, 1975; Ho, 1975; Mraz et al., 1976; Li and Li, 1976). Such an activator protein could also be demonstrated to exist for the degradation of ganglioside G_{M2} and glycolipid G_{A2} (Conzelmann and Sandhoff, 1978). The substrate specificities of hex isoenzymes measured with this activator instead of detergent are much more likely to represent the conditions occurring in vivo. We have therefore examined the substrate specificities of hex A and hex B in the presence of activator protein and related these to the accumulation of lipids observed in different variants of G_{M2}-gangliosidosis.

MATERIALS AND METHODS

Enzymes

Hexosaminidases were isolated from normal human liver as previously described (Sandhoff et al., 1977).

Substrates

Ganglioside G_{M2} was isolated from the brain of a Tay-Sachs patient as previously described (Sandhoff and Waessle, 1971) and tritiated in its N-acetyl galactosamine moiety by the method of O'Brien and coworkers (1977) to a specific activity of 18 μCi/μmole. Glycolipid G_{A2} and kidney globoside were purified from Tay-

Sachs brain and kidney, respectively, and labelled by catalytic reduction of the sphingosine double bond with tritium gas (Sandhoff et al., 1971). Specific activities were 190 and 200 μCi/μmole respectively.

Activator Protein

Human kidneys were homogenized, extracted with 10 vol. of distilled water and centrifuged (13000 x g, 30 min). The supernatant was heated to 60°C for 1 hr, precipitated protein was spun off and discarded. The supernatant was dialyzed against 10 mM phosphate buffer, pH 6.0, and loaded onto a DEAE-cellulose column equilibrated with the same buffer (4 mg of protein/ml gel volume). The column was eluted with a linear gradient of NaCl, 0-0,5 M, in the same buffer. Fractions were assayed for activator as described below.

Enzyme Assays

Hexosaminidase activity was tested as described, using 4-methylumbelliferyl N-acetyl-ß-D-glucosaminide as substrate (Sandhoff et al., 1977). One unit of hex activity releases 1 μmole of 4-methylumbelliferone per min at 37°C.

Ganglioside G_{M2} N-acetyl-ß-D-galactosaminidase activity was essentially determined according to O'Brien et al. (1977) as previously described (Conzelmann and Sandhoff, 1978): Ganglioside G_{M2} (10 nmoles), labelled with ^{3}H in its galactosamine moiety, was incubated with 0.1 U of pure hex A or 1 U of hex B, 20 μg of bovine serum albumin and the indicated amount of activator in a total volume of 40 μl of 0.1 M citrate buffer, pH 4.2, for 1 hr at 37°C. N-acetyl galactosamine liberated was separated from unreacted substrate on a small (1 ml) column of DEAE-cellulose and quantified by liquid scintillation counting. One activator unit (AU) was defined as the amount of activator that stimulated ganglioside G_{M2} degradation by 1 nmole/hr x unit hex A under the conditions mentioned above.

Degradation of glycolipid G_{A2} was measured as given before (Conzelmann and Sandhoff, 1978), employing 0.1 U of pure hex A or 3 U of pure hex B and 12.5 nmoles of tritiated substrate in a total volume of 100 μl. For kidney globoside the same assay system was used, with, however, 0.65 U hex A or 3.3 U hex B.

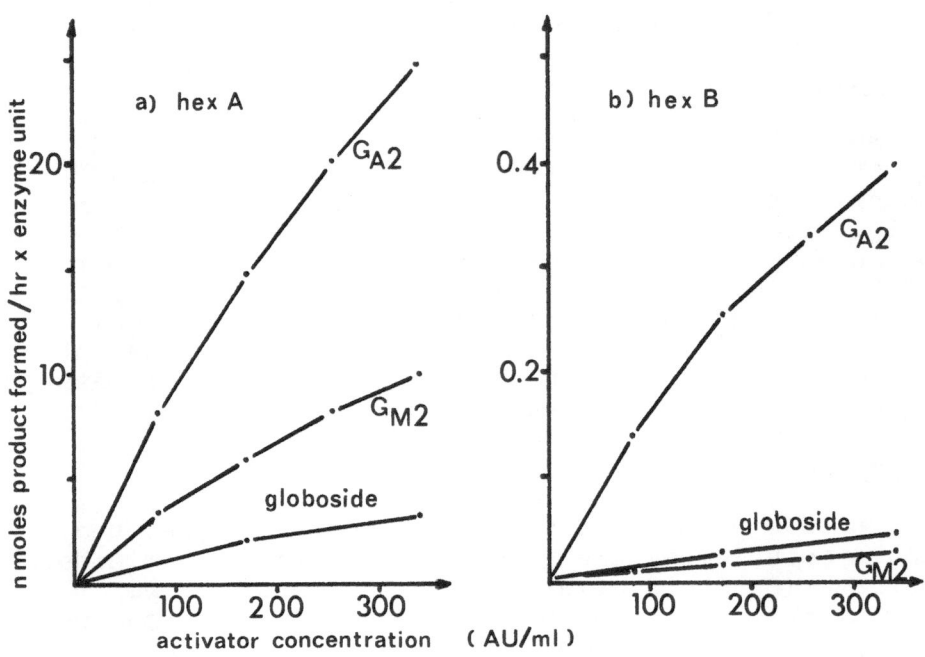

Fig. 2.   Degradation of glycosphingolipids by hex A and
          B in the presence of the activator protein.
          Ganglioside G_{M2} (10 nmoles), labelled with [3]H
          in its galactosamine part, was incubated for
          1 hr at 37°C with the indicated amount of
          activator protein and 0.1 U pure hex A or 1 U
          of pure hex B, in a total volume of 40 μl. The
          N-acetyl-D-galactosamine liberated was deter-
          mined as described in the materials and methods
          section. Glycolipid G_{A2} (12,5 nmoles), tritiated
          in the sphingosine moiety was incubated with
          activator at the concentrations indicated, and
          0.1 U hex A or 3 U hex B for 1 hr at 37°C in a
          total volume of 100 μl. Similar assays were
          performed with kidney globoside as substrate,
          but 0.5 U of hex A were employed. In both cases
          the product formed was separated from substrate
          by TLC and quantified as given before (Sandhoff
          et al., 1977). (a) degradation by hex A;
          (b) degradation by hex B.

RESULTS AND DISCUSSION

The degradation rates for ganglioside G_{M2}, glycosphingolipid G_{A2} and kidney globoside by hex A and B as
obtained with increasing concentrations of the partially
purified activator protein are shown in Fig. 2. It is
evident that the activator interacts almost exclusively
with hex A. The rates of hydrolysis by hex B of the asialo
substrates and ganglioside G_{M2} were, respectively, 60-
and 350-times slower than the rates obtained with hex A.
These results are entirely different from those measured
in the presence of detergent (cf. Table 1). If the activator protein is used to stimulate glycolipid breakdown,
the substrate specificities of the isoenzymes are obviously determined by the properties of the activator which
specifically recognizes only hex A.

The curve obtained by plotting the reaction rates
vs. the activator concentration resembles the substrate
saturation kinetics described by the Michaelis-Menten
equation. The double reciprocal plot according to
Lineweaver and Burk also yields straight lines (Fig. 3).
Together with the fact that the turnover of water-soluble
substrates such as 4-methylumbelliferyl glycosides is
not at all affected by the activator protein (not shown),
these results suggest that the activator does not act
directly on the enzyme, but rather binds to the lipid to
give a complex which is then recognized by the enzyme.
The binding of the substrate to the activator protein can
also be directly demonstrated: In the polyacrylamide gel
electrophoresis system of Williams and Reisfeld (1964),
glycolipid G_{A2} does not migrate since it is uncharged.
If, however, the activator protein is added to the sample,
part of the substrate, which can be traced by its radioactivity, moves to a position from which the activator
protein can also be recovered (Conzelmann and Sandhoff,
in preparation).

The mechanism by which the activator stimulates
glycolipid breakdown thus seems to be analogous to the
model described by Fischer and Jatzkewitz (1978) for the
activation of sulfatide hydrolysis by the activator
protein for cerebroside sulfatase: In an aqueous environment, due to their hydrophobic nature, the glycolipids
form tightly packed micelles or are incorporated into
membranes. Neither structure can be significantly attacked
by water-soluble hydrolases. The activator protein binds
lipid monomers, thus solubilizing them, and the resulting
complex is the true substrate for the enzyme (Fig. 4).

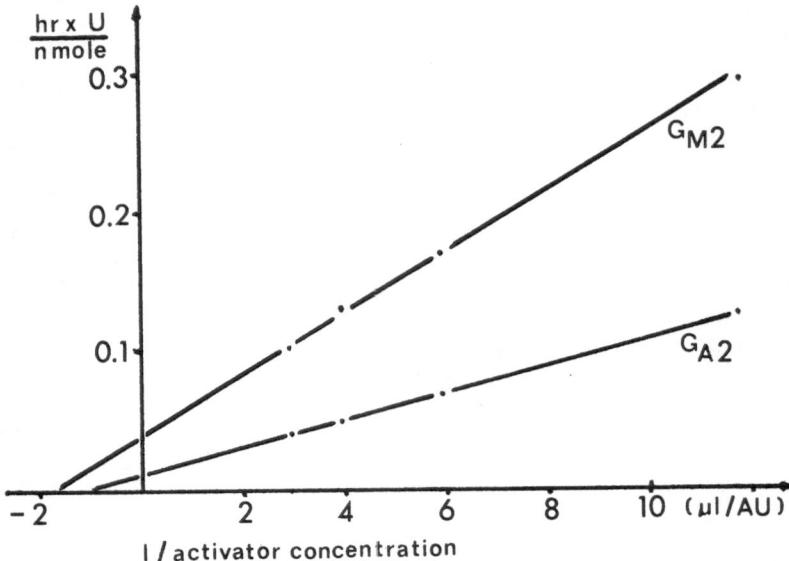

Fig. 3.   Double reciprocal plots of glycolipid degradation
          by hex A versus activator concentration. Data
          for glycolipids G_{M2} and G_{A2} of Fig. 2a were
          replotted in a double reciprocal plot according
          to Lineweaver and Burk.

        In contrast to the reaction rates measured with
detergents, the velocities observed in the presence of
the activator protein fully account for the lipid storage
patterns of the 3 variants of G_{M2}-gangliosidosis: The
absence of both isoenzymes in variant O leads to the
accumulation of all their substrates, including glycolipids. In variant B, where only hex A is deficient, the
residual hex B is unable to interact with the activator-
G_{M2} complex. The hydrolysis of glycolipid G_{A2} by this
isoenzyme is also slow as compared to the hex A catalyzed
reactions, but appreciably faster (approx. 10 times) than
that of the other substrates (Fig. 2b), so the accumulation of glycolipid G_{A2} in this variant is lower than in
variants O and AB (Fig. 1). The rare variant AB is caused
by the deficiency of the activator protein (Conzelmann
and Sandhoff, 1978); therefore those components which
can only be degraded in the presence of activator cannot
be catabolized any further and are deposited in the cells
where they are synthesised. Conversely, the nature of
the glycolipids found to accumulate in patients of
variant AB provides additional information about the

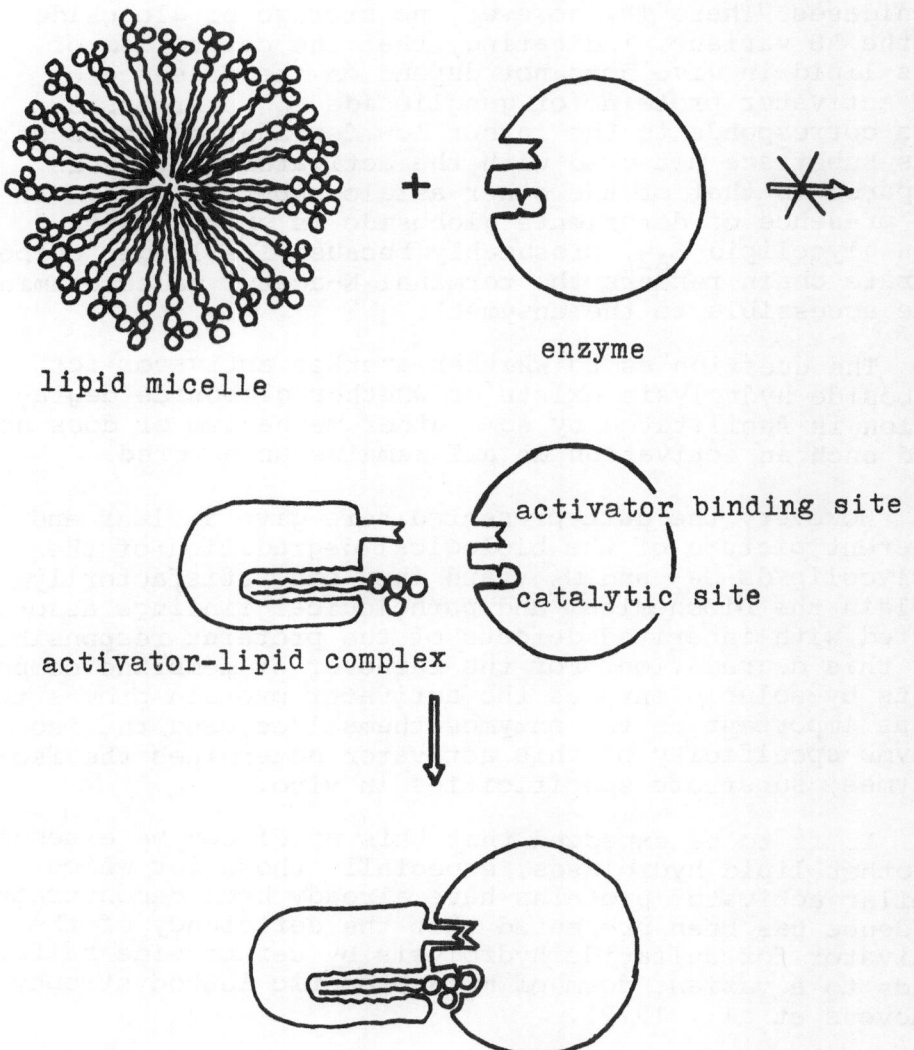

lipid micelle

enzyme

activator binding site

catalytic site

activator-lipid complex

Fig. 4. Model for the activation of enzymatic glycolipid hydrolysis by an activator protein. For
the system considered in this paper it is
assumed, that hex A possesses a binding site
for the activator-lipid complex, whereas hex B
does not (lipid = ganglioside G_{M2}:).

specifity of the activator. The fact, that kidney globoside accumulates in the visceral tissues in variant O
indicates that this glycolipid is degraded by hexosaminidases. There is, however, no storage of globoside
in the AB variant, indicating, that the catabolism of
this lipid in vivo does not depend on the presence of
the activator protein for ganglioside G_{M2} hydrolysis.
This corresponds to the rather low degradation rates of
this substrate measured with the activator protein as
compared to that of the other asialo substrate, G_{A2}. (In
the presence of detergents globoside is split faster
than glycolipid G_{A2}, presumably because its longer carbohydrate chain renders the terminal N-acetyl galactosamine
more accessible to the enzymes).

The question as to whether another activator for
globoside hydrolysis exists or whether globoside degradation is facilitated by some other mechanism or does not
need such an activation at all remains unanswered.

However, the data presented here give a clear and
coherent picture of the biological degradation of the
2 glycolipids G_{M2} and G_{A2}, and they can satisfactorily
explain the biochemical and pathological findings associated with inherited defects of the proteins responsible
for this degradation. For the turnover of membrane components by soluble enzymes the activator protein proves to
be as important as the enzymes themselves, and the isoenzyme specificity of this activator determines the isoenzymes' substrate specificities in vivo.

It is to be expected that this model can be extended
to other lipid hydrolases, especially those for which
similar activator proteins have already been demonstrated.
Evidence has been presented that the deficiency of the
activator for sulfatide hydrolysis by cerebroside sulfatase
leads to a variant form of metachromatic leukodystrophy
(Stevens et al., 1979).

REFERENCES

Beutler, E., Yoshida, A., Kuhl, W., and Lee, J.E.S., 1976,
        The subunits of human hexosaminidase A,
        Biochem. J., 159:541.
Conzelmann, E., and Sandhoff, K., 1978, AB variant of
        infantile G_{M2} gangliosidosis: Deficiency of a
        factor necessary for stimulation of hexosamini-
        dase A-catalyzed degradation of ganglioside G_{M2}
        and glycolipid G_{A2}, Proc. Natl. Acad. Sci. USA,
        75:3979.

Conzelmann, E., Sandhoff, K., Nehrkorn, H., Geiger, B., and Arnon, R., 1978, Purification, biochemical and immunological characterisation of hexosamini-dase A from variant AB of infantile G_{M2} ganglio-sidosis, Eur. J. Biochem., 84:27.

Fischer, G., and Jatzkewitz, H., 1975, The activator of cerebroside sulfatase. Purification from human liver and identification as a protein, Hoppe-Seyler's Z. Physiol. Chem., 356:605.

Fischer, G., and Jatzkewitz, H., 1978, The activator of cerebroside sulfatase. A model of the activation, Biochim. Biophys. Acta, 528:69.

Geiger, B., and Arnon, R., 1976, Chemical Characteriza-tion and subunit structure of human N-acetyl-hexosaminidases A and B, Biochemistry, 15:3484.

Ho, M.W., 1975, Specificity of low molecular weight glycoprotein effector of lipid glycosidase, FEBS Lett., 53:243.

Li, S.C., and Li, Y.T., 1976, An activator stimulating the enzymic hydrolysis of sphingoglycolipids, J. Biol. Chem., 251:1159.

Mraz, W., Fischer, G., and Jatzkewitz, H., 1976, Low molecular weight proteins in secondary lysosomes as activators of different sphingolipid hydro-lases, FEBS Lett., 67:104.

O'Brien, J.S., Norden, A.G.W., Miller, A.L., Frost, R.G., and Kelly, T.E., 1977, Ganglioside G_{M2}-ß-acetyl-galactosaminidase and asialo G_{M2} (G_{A2})-ß-acetyl-galactosaminidase; studies in human skin fibro-blasts, Clin. Genet., 11:171.

Okada, S., and O'Brien, J.S., 1969, Tay-Sachs disease: generalized absence of a ß-D-N-acetylhexosamini-dase component, Science, 165:698.

Sandhoff, K., Andreae, U., and Jatzkewitz, H., 1968, Deficient hexosaminidase activity in an excep-tional case of Tay-Sachs disease with additional storage of kidney globoside in visceral organs, Life Sci., 7:283.

Sandhoff, K., 1969, Variation of ß-N-acetylhexosamini-dase pattern in Tay-Sachs disease, FEBS Lett., 4:351.

Sandhoff, K., Harzer, K. Waessle, W., and Jatzkewitz, H., 1971, Enzyme alterations and lipid storage in three variants of Tay-Sachs disease, J. Neurochem., 18:2469.

Sandhoff, K., and Waessle, W., 1971, Anreicherung und Charakterisierung zweier Formen der menschlichen N-Acetyl-ß-D-hexosaminidase, Hoppe-Seyler's Z. Physiol. Chem., 352:1119.

Sandhoff, K., Conzelmann, E., and Nehrkorn, H., 1977,
        Specificity of human liver hexosaminidases A and
        B against glycosphingolipids G_{M2} and G_{A2}. Purifi-
        cation of the enzymes by affinity chromatography
        employing specific elution, Hoppe-Seyler's Z.
        Physiol. Chem., 358:779.
Srivastava, S.K., Awasthi, Y.C., Yoshida, A., and
        Beutler, E., 1974, Studies on human ß-D-N-acetyl-
        hexosaminidases. I. Purification and properties,
        J. Biol. Chem., 249:2043.
Stevens, R.L., Fluharty, A.L., Kihara, H., Kaback, M.M.,
        Shapiro, L.J., Sandhoff, K., and Fischer, G.,
        1979, Metachromatic leukodystrophy - variant with
        apparent cerebroside sulfatase activator deficiency,
        Clin. Res., 27:A104.
Svennerholm, L., 1962, The chemical structure of normal
        human brain and Tay-Sachs gangliosides, Biochem.
        Biophys. Res. Commun., 9:436.
Williams, D.E., and Reisfeld, R.A., 1964, Disc electro-
        phoresis in polyacrylamide gels: extension to
        new conditions of pH and buffer, Ann. N.Y. Acad.
        Sci., 121:373.

THE SPECIFICITY OF β-GALACTOSIDASE

IN THE DEGRADATION OF GANGLIOSIDES

Kunihiko Suzuki, Harumi Tanaka*, Tatsuhiro Yamanaka,
and Olga Van Damme

The Saul R. Korey Department of Neurology, Department of
Neuroscience, and the Rose F. Kennedy Center for Research
in Mental Retardation and Human Development, Albert
Einstein College of Medicine, Bronx, N.Y. 10461, U.S.A.

INTRODUCTION

Degradation of gangliosides both in vivo and in vitro appears
to proceed by sequential hydrolysis of the terminal carbohydrate
residues, rather than by removal of oligosaccharides of two or
more sugars or by initial de-acylation. These catabolic steps are
catalyzed by lysosomal acidic glycosidases. Within the degradative
pathways of known gangliosides, there are four compounds which
possess a β-galactosyl residue at the non-reducing terminal.

gal-galNAc-gal[NeuNAc]-glc-ceramide   (G_{M1}-ganglioside)

gal-galNAc-gal-glc-ceramide   (asialo G_{M1}-ganglioside)

gal-glc-ceramide   (lactosylceramide)

gal-ceramide   (galactosylceramide)

A priori, three possibilities can be considered regarding the
substrate specificities of the β-galactosidase(s) responsible for
the degradation of these compounds: a single β-galactosidase with
a broad substrate specificity that can hydrolyze all of these
compounds; four distinct β-galactosidases, each specific for hydrolysis of each of the four compounds; and more than one but less
than four β-galactosidases with relatively broad and possibly

---

* The present address of Dr. H. Tanaka is Department of Child
Neurology, National Musashi Research Institute for Mental and
Nervous Diseases, Kodaira, Tokyo 187, Japan.

overlapping specificities. Several lines of evidence are now available to at least partially answer this question. The answer, however, is not yet complete and several important points of uncertainty remain. We will attempt an objective assessment of the available evidence regarding the substrate specificities of the β-galactosidases with respect to these glycosphingolipid substrates.

GENETIC β-GALACTOSIDASE DEFICIENCIES

There are at least two disorders which are caused by genetically determined deficiencies of β-galactosidases. The knowledge on what occurs in organs of patients with the respective disorders provides important, if circumstantial, evidence that hydrolysis of the terminal β-galactosyl residues of the glycosphingolipids are not catalyzed by a single β-galactosidase.

A genetic deficiency of galactosylceramide β-galactosidase causes globoid cell leukodystrophy (Krabbe's disease) (SUZUKI & SUZUKI, 1978). The disease occurs almost exclusively in infancy, characterized by rapidly progressive signs of the white matter and of the peripheral nerves, elevated protein in the cerebrospinal fluid, and the relatively late occurrence of neuronal dysfunction. There are no clinical signs and symptoms of systemic organ involvement. Pathologically, the lesions are again limited to the nervous system, particularly to the white matter and the peripheral nervous system. At the terminal stage, the central white matter is almost completely devoid of the myelin sheath and the oligodendroglia, which are replaced by reactive astrocytes and the characteristic globoid cells of the mesodermal origin. Compared to the white matter, the gray matter is well preserved morphologically. The main pathological lesions in peripheral nerves are segmental demyelinations. Galactosylceramide is severely reduced in the white matter due to the lack of myelin (VANIER & SVENNERHOLM, 1975, 1976; SUZUKI & SUZUKI, 1978; SVENNERHOLM, et al., personal communication). Galactosylsphingosine which is essentially undetectable in normal brain is increased to at least 100 times the normal level (VANIER & SVENNERHOLM, 1975, 1976; SVENNERHOLM, et al., personal communication). Gangliosides are not increased to any significant degree. Lactosylceramide was reported to be slightly increased in white matter of one case but not in the gray matter (ETO & SUZUKI, 1971). A recent, more comprehensive study with 18 cases indicated that no abnormal lactosylceramide accumulation occurs in this disorder (SVENNERHOLM, et al., personal communication). The genetic deficiency of the β-galactosidase can be demonstrated using galactosylceramide (SUZUKI & SUZUKI, 1970, 1971), galactosylsphingosine (MIYATAKE & SUZUKI, 1972), monogalactosyl-diglyceride (WENGER, et al., 1973), and , under certain specific assay conditions, lactosylceramide (WENGER, et al., 1974; TANAKA & SUZUKI, 1975; SVENNERHOLM, et al., 1975). If, on the other hand, one uses G_{M1}-ganglioside,

asialo G_{M1}-ganglioside, or the conventional artificial substrates
for β-galactosidase, such as p-nitrophenyl- or 4-methylumbelliferyl
β-galactoside, this deficiency cannot be demonstrated (SUZUKI &
SUZUKI, 1970, 1971, 1974a).

G_{M1}-gangliosidosis is also primarily a neurological disorder
of infants. However, there are varying degrees of systemic organ
involvement, and there are phenotypically distinct forms of later
onset. Unlike in globoid cell leukodystrophy, the neurological
manifestations are primarily those of the gray matter with little
evidence of the white matter or peripheral nerve involvement.
Patients with G_{M1}-gangliosidosis, particularly of the infantile
form, show clinical signs of systemic organ involvement, such as
facial deformity, macroglossia, radiologic bone abnormalities, and
visceromegaly. The pathological lesions within the nervous system
are largely limited to the gray matter in which almost all neurons
are greatly distended. The white matter commonly shows mild to
moderate myelin loss with sudanophilia, generally considered to be
secondary to the extensive neuronal loss. In systemic organs,
swollen foamy histiocytes are found in the liver, spleen, lymph-
nodes, bone marrow, intestine and lungs. In contrast to globoid
cell leukodystrophy, the main glycosphingolipid that accumulates in
the brain is G_{M1}-ganglioside with associated moderate increases of
its asialo-derivative (SUZUKI, et al., 1969; SUZUKI & CHEN, 1967).
Although there are abnormally increased amounts of G_{M1}-ganglioside
in the systemic organs, the main storage materials in these organs
are fragments of glycoproteins of wide-ranging molecular weights
(SUZUKI, 1968; WOLFE, et al., 1976; TSAY & DAWSON, 1976). The
genetic deficiency of the β-galactosidase in G_{M1}-gangliosidosis can
be readily demonstrated using p-nitrophenyl- or 4-methylumbelliferyl
β-galactoside, G_{M1}-ganglioside, asialo G_{M1}-ganglioside, or appropriate glycoprotein fragments with exposed terminal β-galactose
(OKADA & O'BRIEN, 1968; MACBRINN, et al., 1969; SUZUKI & SUZUKI,
1974b). With lactosylceramide as the substrate, the deficiency is
undetectable under the assay conditions suitable for diagnosis of
globoid cell leukodystrophy. It is, however, possible to demonstrate
the deficient lactosylceramide-cleaving activity in G_{M1}-ganglio-
sidosis by employing assay conditions appropriate for the purpose
(SUZUKI & SUZUKI, 1974b; TANAKA & SUZUKI, 1974, 1977a). The β-
galactosidase deficiency in G_{M1}-gangliosidosis cannot be detected
with galactosylceramide as the substrate.

The existence of the two entirely dissimilar disorders each of
which is caused by a genetic mutation of a β-galactosidase provides
a priori evidence for at least two genetically distinct β-galactosidases with distinct substrate specificities. From the known
findings about these disorders, particularly from the nature of the
stored materials and the usefulness of the respective natural substrates for the enzymatic diagnosis, the following conclusions can
be drawn. (1) There are at least two genetically distinct

β-galactosidases in human tissues. (2) one of them catalyzes hydrolysis of galactosylceramide, galactosylsphingosine, and monogalactosyl-diglyceride. (3) The other enzyme is active against asialo
G_{M1}-ganglioside, G_{M1}-ganglioside, and glycoproteins. (4) Lactosylceramide is hydrolyzed by either of the two β-galactosidases.
Available and confirmed evidence from other experimental approaches
to be examined below is generally consistent with these conclusions.

SPECIFICITIES OF PURIFIED ENZYMES

The conventional and seemingly the most definitive approach to
the question of substrate specificity of an enzyme would be to
purify it and then examine its specificity against varieties of
potential substrates. As we shall see, the information that can be
obtained from this approach introduces its own complications and is
not as definitive as one wishes. Nevertheless, a great deal of
important information can be and has been gathered through this
route.

Of the two acidic β-galactosidases, the one which is deficient
in G_{M1}-gangliosidosis -- G_{M1}-ganglioside β-galactosidase -- is the
easier to purify. NORDEN et al. (1974) purified the enzyme from
human liver to electrophoretic homogeneity and reported its substrate
specificities. It was active against G_{M1}-ganglioside, asialo-fetuin,
and a host of artificial chromogenic or fluorogenic substrates but
was inactive against galactosylceramide and monogalactosyl-diglyceride. Although their initial report indicated no activity toward
lactosylceramide, this was due to the inappropriate assay conditions.
Later re-examination showed high activity of the purified G_{M1}-
ganglioside β-galactosidase toward lactosylceramide when proper
assay conditions were employed (MILLER, et al., 1977). These
findings are in agreement with those of TANAKA, et al. (1975), who
in addition showed good activity of their purified human hepatic
enzyme toward asialo G_{M1}-ganglioside.

Purification of the other β-galactosidase -- galactosylceramide
β-galactosidase -- is much more difficult. Purification comparable
to that for G_{M1}-ganglioside β-galactosidase has not been achieved
except for one report which claimed a 3900-fold purification of the
enzyme from human placenta to electrophoretic and immunological
homogeneity (BEN-YOSEPH, et al., 1978). However, the specific
activity of their preparation was substantially lower than that of
the preparations in our laboratory from human brain, which still
showed multiple protein bands in polyacrylamide gel electrophoresis
(HANADA & SUZUKI, unpublished). Therefore, the preparation of
BEN-YOSEPH et al. may not be as pure as claimed, or substantial
inactivation of the enzyme occurred during purification. Since no
details of their purification procedures have been published, exact
duplication of their results is difficult, and the best efforts of

at least one laboratory were unsuccessful (WENGER, personal communication). Despite these uncertainties, their human placental galactosylceramidase was active toward galactosylceramide, and under the conditions for lactosylceramidase I, was also active toward lactosylceramide. It was inactive toward lactosylceramide when the assay conditions for lactosylceramidase II were used.

In a series of studies, TANAKA & SUZUKI (1976, 1977b) took a slightly different approach. Rather than attempting at high purification of either of the two β-galactosidases, they attempted to obtain the two enzyme preparations, free from each other, with minimum of loss. This could be done by Sephadex G-200 gel filtration of the high-speed supernatant obtained from defatted human brain tissues. The galactosylceramidase fraction (Fraction I) was eluted at or near the void volume, while the G_{M1}-ganglioside β-galactosidase fraction (Fraction II) was eluted later. The substrate specificities of these two fractions were then examined under varieties of assay conditions against several natural and artificial substrates (Table 1).

First, the assay conditions were optimized for galactosylceramide hydrolysis by the Fraction I enzyme, and for G_{M1}- and asialo G_{M1}-ganglioside hydrolysis by the Fraction II enzyme. When the Fraction II enzyme was assayed for galactosylceramide-cleaving activity with the assay system optimized for the Fraction I enzyme, no activity could be detected. On the other hand, the Fraction I enzyme showed no activity toward G_{M1}-ganglioside or asialo G_{M1}-ganglioside when assayed with the system optimized for the Fraction II enzyme. These findings assured that the Fraction I contained galactosylceramidase and the Fraction II contained G_{M1}-ganglioside β-galactosidase and that each of these fractions were pure with respect to the other β-galactosidase.

Either of the two β-galactosidases was able to hydrolyze, under appropriate conditions, all of the substrates tested. Generally, galactosylceramidase was surprisingly active toward other substrates. Under the optimized conditions, specific activities toward G_{M1}- and asialo G_{M1}-ganglioside were 40% of that toward galactosylceramide. On the other hand, galactosylceramide was a poor substrate for G_{M1}-ganglioside β-galactosidase. Although detectable, the maximum activity was less than 1% of that toward G_{M1}-ganglioside. Generally, the optimal conditions were similar for a particular enzyme regardless of the nature of the substrates, suggesting that it is the nature of the enzyme, rather than the substrates, that determines the optimal conditions for the hydrolytic activity. These in vitro studies indicated that the substrate specificities of the two β-galactosidases are not absolute and that the relative activities toward different substrates depend on the assay conditions.

Table 1. Substrate Specificities of the Two Human Brain β-Galactosidases

Pooled fractions of the two β-galactosidases from the Sephadex G-200 column were assayed for their hydrolytic activities with both of the assay systems optimized for the respective enzymes.

| Substrate | Galactosylceramidase (Fraction I) | | | G_{M1}-ganglioside β-galactosidase (Fraction II) | | |
| | Fr. I assay | | Fr. II assay | Fr. I assay | Fr. II assay | |
	nmol/hr/ mg protein	R.S.A.*	nmol/hr/ mg protein	nmol/hr/ mg protein	nmol/hr/ mg protein	R.S.A.**
Gal-cer	7.5	100	1.8	0	2.3	0.3
Lact-cer ($C_{18:0}$)	38	510	5.4	36	580	78
Lact-cer ($C_{24:0}$)	19	250	1.2	13	140	19
G_{M1}-ganglioside	2.9	39	0	180	740	100
Asialo G_{M1}	3.1	41	0	200	1100	150
4-MU β-galactoside	8.7	120	---	---	1800	240

* Percent of activity relative to that toward galactosylceramide.
** Percent of activity relative to that toward G_{M1}-ganglioside.
Reproduced by permission from TANAKA & SUZUKI (1977b).

IN VITRO AND IN VIVO SPECIFICITIES

Some aspects of the findings of the above in vitro studies of
our laboratory are difficult to reconcile with what we expect from
the known analytical findings in the two genetic β-galactosidase
deficiencies. The analytical data suggest that each of the two β-
galactosidases functions in vivo within the environment of the
biological membrane with relatively clear-cut substrate specificities. The in vitro studies indicated, however, that under certain
conditions there are substantial overlappings of substrate specificities between the two enzymes. The precise in vivo reaction
mechanisms of these lysosomal acidic hydrolases are difficult to
assess. A part of the complications arises from the fact that all
of the assays above were done in the presence of large amounts of
bile salts of one kind or another, surely a highly artificial
environment. One possible approach to this problem might be to
examine the substrate specificities of these β-galactosidases in
assay systems which include the natural activators of these enzymes.
It is conceivable that the natural activators, some of which are
under active investigation in several laboratories, participate
actively in determining the substrate specificities of lysosomal
acid hydrolases in vivo.

In vivo degradation of lactosylceramide poses a particularly
interesting question. Under appropriate assay conditions, either
of the two β-galactosidases can readily hydrolyze lactosylceramide.
Lactosylceramide accumulation does not occur to any significant
extent in either globoid cell leukodystrophy or G_{M1}-gangliosidosis.
This suggests that both β-galactosidases participate in the degradation of lactosylceramide in vivo. However, it is difficult to
design experiments to assess the relative extent of contribution
by either of the enzyme to in vivo degradation of lactosylceramide.
As a compromise attempt, TANAKA & SUZUKI (1978) recently examined
hydrolysis of [^{3}H]lactosylceramide by intact cultured fibroblasts
from normal controls and from patients with globoid cell leukodystrophy. The rationale for this experiment was that the metabolic
activity of intact cultured cells should be close to that of organs
in situ and that fibroblasts of globoid cell leukodystrophy lack
activity of galactosylceramidase while control cells contain both
β-galactosidases. The cells were allowed to take up tritium-
labelled galactosylceramide or lactosylceramide added to the culture
medium. After the excess substrates in the medium were removed,
release of free tritium-labelled galactose was monitored for three
consecutive 2-day period. The total water-soluble radioactivity
was taken as the indicator of sphingolipid hydrolysis and was
expressed relative to the total intracellular radioactivity. When
galactosylceramide was the substrate, the results were 12.7±7.4
(SD)% for the control cells and 3.1±2.1% for globoid cell leukodystrophy. These results were expected in view of the known
galactosylceramidase deficiency. In contrast, the results with

lactosylceramide were 24.8±8.3% for controls and 17.5±5.1% for
globoid cell leukodystrophy. Unlike for galactosylceramide, the
difference between the control and globoid cell leukodystrophy was
at the borderline of statistical significance. Since the in vivo
system is highly complex, it is difficult to conclude quantitatively
that G_{M1}-ganglioside β-galactosidase is responsible for 70% of the
lactosylceramide hydrolysis in intact fibroblasts. It is possible
to conclude, however, that G_{M1}-ganglioside β-galactosidase is
active in hydrolyzing lactosylceramide in vivo.

When assayed in the optimized in vitro systems, only 10% of
the total lactosylceramide-cleaving activity of normal fibroblasts
can be attributed to galactosylceramidase, the remainder being due
to G_{M1}-ganglioside β-galactosidase (TANAKA & SUZUKI, 1977a). Therefore, fibroblasts from patients with G_{M1}-gangliosidosis would lack
90% of the lactosylceramide-hydrolyzing activity of normal cells.
With this point in mind, we have recently carried out a similar
experiment with G_{M1}-gangliosidosis and control fibroblasts (SUZUKI,
YAMANAKA & VAN DAMME, unpublished). G_{M1}-gangliosidosis cells
released on the average 60% of the galactose that was released by
the control cells. Therefore, it appears that in the absence of
G_{M1}-ganglioside β-galactosidase, the remaining galactosylceramidase
can maintain nearly normal degradative activity toward lactosylceramide.

QUESTION OF A SPECIFIC LACTOSYLCERAMIDASE

The lack of lactosylceramide accumulation in the two β-
galactosidase deficiency states and the findings of the above fibroblast experiments are certainly consistent with the working hypothesis that both of the two acidic β-galactosidases actively
participate in lactosylceramide degradation in vivo and that the
activity of either one is adequate to maintain normal turnover of
lactosylceramide. However, these findings can be equally well
explained by postulating a third β-galactosidase specific for lactosylceramide. Two groups of investigators have proposed such a
specific lactosylceramidase. NISHIMURA & AMANO (1976) purified
porcine thymus β-galactosidase to 3600-4000 fold. The enzyme gave
two peaks on electrofocusing but one band on electrophoresis. It
was active against lactosylceramide with a pH optimum of 4.6 in
the presence of sodium taurocholate. It was inactive toward
galactosylceramide and G_{M1}-ganglioside. We are not aware of
confirmation of their results in other laboratories.

The other report of a specific lactosylceramidase came from
BEN-YOSEPH, et al. (1977). Their enzyme belongs to an entirely
different category from the lysosomal acidic hydrolase we have been
discussing so far. They reported that purified human hepatic
neutral β-galactosidase, which is a non-lysosomal, non-glycoprotein

enzyme, can catalyze hydrolysis of lactosylceramide but is inactive toward galactosylceramide or G_{M1}-ganglioside. The assay condition was that for lactosylceramidase II of TANAKA & SUZUKI (1975). When lactosylceramidase I assay system was used, their preparation was inactive toward lactosylceramide. While this finding is intriguing, it is also controversial. At least two laboratories have been unable to confirm their results. One of them utilized an assay system different from that of BEN-YOSEPH et al. and this could explain the discrepancy (O'BRIEN, personal communication). However, the other laboratory attempted an exact duplication of the report without success (WENGER, personal communication). The lactosylceramidase II assay mixture contains a high concentration of sodium chloride which is known to be inhibitory to the neutral β-galactosidase (O'BRIEN, 1975). Because of the potential importance of their finding, independent confirmation is urgently needed. More recently the same group of investigators suggested that lack of the neutral β-galactosidase possessing the specific lactosylceramidase activity might explain the case of so-called lactosylceramidosis (BURTON, et al., 1978), in which both of the acid β-galactosidases-- lactosyl-ceramidase I and II -- are known to be normal (WENGER, et al., 1975). This is a bold proposal bound to remain controversial. Aside from the still controversial lactosylceramidase activity of the neutral β-galactosidase above, an inevitable consequence of this proposal is the break-down of a generally accepted minor dogma in cell biology -- lysosomal acidic hydrolases are responsible for catabolism of cellular constituents. Since the proportion of the total lactosylceramidase activity in tissues attributable to the neutral β-galactosidase is very minor, it follows that the two highly active lysosomal acidic β-galactosidases must be inactive in vivo for degradation of lactosylceramide in order for the proposed neutral β-galactosidase deficiency to cause an abnormal accumulation of lactosylceramide. The morphology of the patient's organs showed storage of abnormal material(s) in what appeared to be residual lysosomes.

CONCLUDING REMARKS

Available evidence indicates that at least two genetically distinct acidic lysosomal β-galactosidases are present in mammalian tissues. One of them, galactosylceramidase, is primarily responsi-ble for degradation of galactosylceramide, galactosylsphingosine, and monogalactosyl-diglyceride, while the other, G_{M1}-ganglioside β-galactosidase, degrades G_{M1}-ganglioside and asialo G_{M1}-ganglioside. Lactosylceramide can be hydrolyzed by either of the two enzymes. These substrate specificities of the two β-galactosidases can adequately explain the known findings in the two genetic β-galacto-sidase deficiency diseases. The possibilities of the specific lactosylceramidase have not yet received the necessary independent confirmation.

ACKNOWLEDGEMENT

The work described in this chapter that was carried out in the authors' laboratory was supported by research grants, NS-10885, NS-03356, and HD-01799 from the United States Public Health Service.

REFERENCES

BEN-YOSEPH, Y., HUNGERFORD, M. & NADLER, H. L. (1978): The nature of mutation in Krabee's disease. Am. J. Human Genet. 30, 644-652.

BEN-YOSEPH, Y., SHAPIRA, E., EDELMAN, D., BURTON, B. K. & NADLER, H. (1977): Purification and properties of neutral β-galactosidases from human liver. Arch. Biochem. Biophys. 184, 373-380.

BURTON, B. K., BEN-YOSEPH, Y. & NADLER, H. L. (1978): Lactosylceramidosis: Deficient activity of neutral β-galactosidase in liver and cultured fibroblasts? Clin. Chim. Acta 88, 483-493.

ETO, Y. & SUZUKI, K. (1971): Brain sphingoglycolipids in Krabbe's globoid cell leukodystrophy. J. Neurochem. 18, 503-511.

MACBRINN, M. C., OKADA, S., HO, M. W., HU, C. C. & O'BRIEN, J. S. (1969): Generalized gangliosidosis: Impaired cleavage of galactose from a mucopolysaccharide and a glycoprotein. Science 163, 946-947.

MILLER, A. L., FROST, R. G. & O'BRIEN, J. S. (1977): Purified human liver acid β-D-galactosidase possessing activity towards G_{M1}-ganglioside and lactosylceramide. Biochem. J. 165, 591-594.

MIYATAKE, T. & SUZUKI, K. (1972): Globoid cell leukodystrophy: Additional deficiency of psychosine galactosidase. Biochem. Biophys. Res. Commun. 48, 538-543.

NISHIMURA, K. & AMANO, R. (1976): Partial purification and properties of porcine thymus lactosylceramide β-galactosidase. J. Biochem. 80, 209-215.

NORDEN, A. G. W., TENNANT, L. L. & O'BRIEN, J. S. (1974): G_{M1}-ganglioside β-galactosidase A: Purification and studies of the enzyme from human liver. J. Biol. Chem. 249, 7969-7976.

O'BRIEN, J. S. (1975): Molecular genetics of G_{M1}-β-galactosidase. Clin. Genet. 8, 303-313.

OKADA, S. & O'BRIEN, J. S. (1968): Generalized gangliosidosis: β-galactosidase deficiency. Science 160, 1002-1004.

SUZUKI, K. (1968): Cerebral G_{M1}-gangliosidosis: Chemical pathology of visceral organs. Science 159, 1471-1472.

SUZUKI, K. & CHEN, G. C. (1967): Brain ceramide hexosides in Tay-Sachs disease and generalized gangliosidosis (G_{M1}-gangliosidosis) J. Lipid Res. 8, 105-113.

SUZUKI, K. & SUZUKI, Y. (1970): Globoid cell leukodystrophy (Krabbe's disease): Deficency of galactocerebroside β-galacto-sidase. Proc. Nat. Acad. Sci., U.S.A. 66, 302-309.

SUZUKI, Y. & SUZUKI, K. (1971): Krabbe's globoid cell leukodystro-phy: Deficiency of galactocerebrosidase in serum, leukocytes and fibroblasts. Science 171, 73-75.

SUZUKI, Y. & SUZUKI, K. (1974a): Glycosphingolipid β-galactosidases II. Electrofocusing characterization of the enzymes in human globoid cell leukodystrophy (Krabbe's disease). J. Biol. Chem. 249, 2105-2108.

SUZUKI, Y. & SUZUKI, K. (1974b): Glycosphingolipid β-galactosidases IV. Electrofocusing characterization in G_{M1}-gangliosidosis. J. Biol. Chem. 249, 2113-2117.

SUZUKI, K. & SUZUKI, Y. (1978): Galactosylceramide lipidosis: Globoid cell leukodystrophy (Krabbe's disease), in "The Metabolic Basis of Inherited Disease", STANBURY, J. B., WYNGAARDEN, J. B. & FREDRICKSON, D. S., Eds., McGraw-Hill, (New York), pp. 747-769.

SUZUKI, K., SUZUKI, K. & KAMOSHITA, S. (1969): Chemical pathology of G_{M1}-gangliosidosis (generalized gangliosidosis). J. Neuropath. Exp. Neurol. 28, 25-73.

SVENNERHOLM, L., HAKANSSON, G. & VANIER, M.-T. (1975): Chemical pathology of Krabbe's disease IV. Studies of galactosyl ceramide and lactosylceramide β-galactosidases in brain, white blood cells and amniotic fluid cells. Acta Paediat. Scand. 64, 649-656.

TANAKA, H. & SUZUKI, K. (1975): Lactosylceramide β-galactosidase in human sphingolipidoses: Evidence for two genetically distinct enzymes. J. Biol. Chem. 250, 2324-2332.

TANAKA H. & SUZUKI, K. (1976): Specificities of the two genetically distinct β-galactosidases in human sphingolipidoses. Arch. Biochem. Biophys. 175, 332-340.

TANAKA, H. & SUZUKI, K. (1977a): Lactosylceramidase assays for diagnosis of globoid cell leukodystrophy and G_{M1}-gangliosidosis. Clin. Chim. Acta 75, 267-274.

TANAKA, H. & SUZUKI, K. (1977b): Substrate specificities of the two genetically distinct human brain β-galactosidases. Brain Res. 122, 325-335.

TANAKA, H. & SUZUKI, K. (1978): Globoid cell leukodystrophy (Krabbe's disease): Metabolic studies with cultured fibroblasts. J. Neurol. Sci. 38, 409-419.

TANAKA, H., MEISLER, M. & SUZUKI, K. (1975): Activity of human hepatic β-galactosidase toward natural glycosphingolipid substrates. Biochim. Biophys. Acta 398, 452-463.

VANIER, M.-T. & SVENNERHOLM, L. (1975): Chemical pathology of Krabbe's disease III. Ceramide hexosides and gangliosides of brain. Acta Paediat. Scand. 64, 641-648.

VANIER, M.-T. & SVENNERHOLM, L. (1976): Chemical pathology of Krabbe disease: The occurrence of psychosine and other neutral sphingoglycolipids. in "Current Trends in Sphingolipidoses and Allied Disorders", VOLK, B. W. & SCHNECK, L., Eds., Plenum (new York), pp. 115-126.

WENGER, D. A., SATTLER, M. & MARKEY, S. P. (1973): Deficiency of monogalactosyl diglyceride β-galactosidase activity in Krabbe's disease. Biochem. Biophys. Res. Commun. 53, 680-685.

WENGER, D. A., SATTLER, M. & HIATT, W. (1974): Globoid cell leukodystrophy: Deficiency of lactosyl ceramide beta-galactosidase. Proc. Nat. Acad. Sci., U.S.A. 71, 854-857.

WENGER, D. A., SATTLER, M., CLARK, C., TANAKA, H., SUZUKI, K. & DAWSON, G. (1975): Lactosyl ceramidosis: Normal activity for two lactosyl ceramide β-galactosidases. Science 188, 1310-1312.

IMMUNOLOGICAL METHODS
FOR THE IDENTIFICATION OF GANGLIOSIDES

PREPARATION AND PROPERTIES OF ANTIBODIES TO GANGLIOSIDES

Donald M. Marcus and Samar K. Kundu

Departments of Medicine, Microbiology and Immunology
Albert Einstein College of Medicine
Bronx, New York 10461

In the limited space available I would like to consider some of the major problems encountered in immunological studies of gangliosides and neutral glycosphingolipids (GSLs), and some new approaches to these problems. Several comprehensive reviews of glycolipid immunology have been published recently[1-5]. The problems arise from the weak immunogenicity of many GSLs and from the amphipathic nature of these compounds.

Immunogenicity

Pure GSLs in Freund's adjuvant are very poor immunogens and most investigators administer the GSLs in the form of a noncovalent complex with albumin or methylated albumin. We have recently produced antibodies that react well with native gangliosides by immunization with G_{D3} or G_{M1} (nomenclature of Svennerholm[6]) coupled through their carboxyl groups to protein carriers[7]. Fifty percent fixation of complement was obtained with 20 ng of G_{D3} and 17 ng of purified IgG anti-G_{D3} antibodies, and no complement fixation was observed with 500 ng of the same antibody and 20 ng of G_{M3}, G_{M1}, G_{D1b} or CDH. An alternative method for coupling neutral GSLs or gangliosides to protein carriers or insoluble supports is the technique of Hakomori and collaborators[8] in which a carboxyl group is generated by oxidation of a double bond in the ceramide portion of the glycolipid.

Many protocols are used for administration of antigen; our present procedure consists of initial administration of approximately 0.5 mg of antigen in complete Freund's adjuvant in two foot pads and multiple intradermal sites, followed by an intravenous booster of 0.5 mg approximately 3-4 weeks later.

Detection

The simplest technique for detecting antibodies to GSLs is double diffusion in agar gel, but the insensitivity of this method limits its use to high titered antisera. We have recently used a counter-electrophoresis technique in which liposomes containing gangliosides are forced to migrate towards antibodies in an electrophoretic field. This technique is useful but non-specific precipitates form between the liposomes and basic serum proteins, predominantly immunoglobulins, unless the ganglioside concentration in the liposomes is maintained below 50 μg/ml and phosphatidyl choline, 2.5 x the weight of ganglioside, is incorporated in the liposome. We have also used neutral glycolipids in liposomes to which dicetylphosphate is added to provide a negative charge. The composition of these liposomes is dicetyl-phosphate 0.03 mg/ml, phosphatidyl choline 0.1-0.2 mg/ml, glyco-lipid 0.4-0.8 mg/ml[9]. A semiquantitative complement fixation technique performed in a microtiter apparatus[10] is another simple means of detecting anti-GSL antibodies, but it is not useful with weak antisera, e.g. titers less than 1:64, because of the anti-complementary activity of most rabbit sera.

We feel that radioimmunoassays (RIAs) provide the best approach to detection and quantitation of antibodies to GSLs. The solid phase procedure of Holmgren[11], in which the GSL is adsorbed to a plastic tube, is an extremely useful technique because of its simplicity. Hakomori and coworkers have devised another type of RIA in which the oxidized GSL derivative mentioned above is coupled to polyacrylic hydrazide[12]. The complex can be labeled either in the GSL moiety by galactose oxidase-borohydride or by acetylation of the polyacrylic hydrazide with ^{14}C-acetic anhydride. The antigen-antibody complex is precipitated by addition of heat killed Staphylococcus aureus organisms. A representative curve of one of their radioassays is presented in Figure 1. The major advantage of the RIA technique is its sensitivity and the fact that it measures the primary event, formation of an antigen-antibody complex, rather than a secondary manifestation such as precipitation or complement fixation. Another technique that has been widely used, but with which we have no personal experience, is the release of markers from antigen-containing liposomes[13,14].

Specificity

Antibodies to GSLs are finding increasing applications in the study of cell membranes, for ultrastructural localization of these compounds, and in neurophysiology. In these studies it is important to bear in mind certain limits to the interpretation of immunological data, and the complex specificity of conventional polyclonal antibodies. Antibodies to carbohydrate determinants

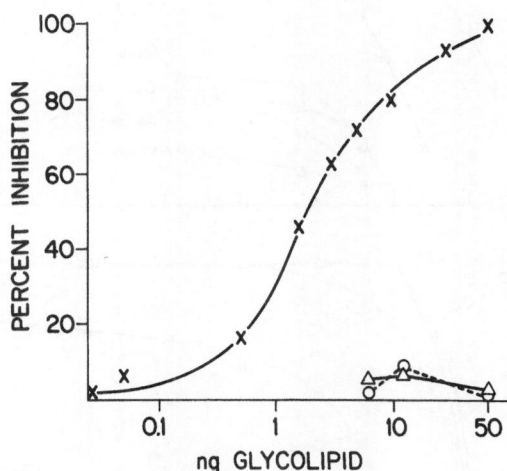

Fig. 1. Inhibition of precipitation of (^{14}C/^{3}H)-labeled Forssman
polyacrylic hydrazide by unlabeled glycolipid antigens.
x-x, Forssman; 0-0 globoside; Δ-Δ, blood group A glycolipid; mole ratios were glycolipid-sphingomyelin-
cholesterol, 1:250:200[12].

"recognize" a specific sugar sequence and will react with this
sequence, or a similar one, regardless of whether the sugars are
attached to a protein, lipid or even nucleic acid. By comparing
the reaction of anti-carbohydrate antibodies with oligosaccharides in which the configuration of hydroxyl groups, anomeric
configuration and sites of substitution are varied, it is possible
to distinguish two general types of specificities: antibodies
directed against the terminal non-reducing end of the chain, and
antibodies that can combine with an internal sequence[15]. An
example of the former type would be antibodies directed against
the blood group ABH determinants, and the latter type would be
exemplified by antibodies against polysaccharide chains with
repeating structures, such as the lipopolysaccharides of Gram-
negative bacteria. Recent studies have demonstrated that even
antibodies against relatively short carbohydrate chains may
include subpopulations that react with internal sequences, and
the proportion of these subpopulations may vary from animal to
animal and even between bleedings obtained from the same animal

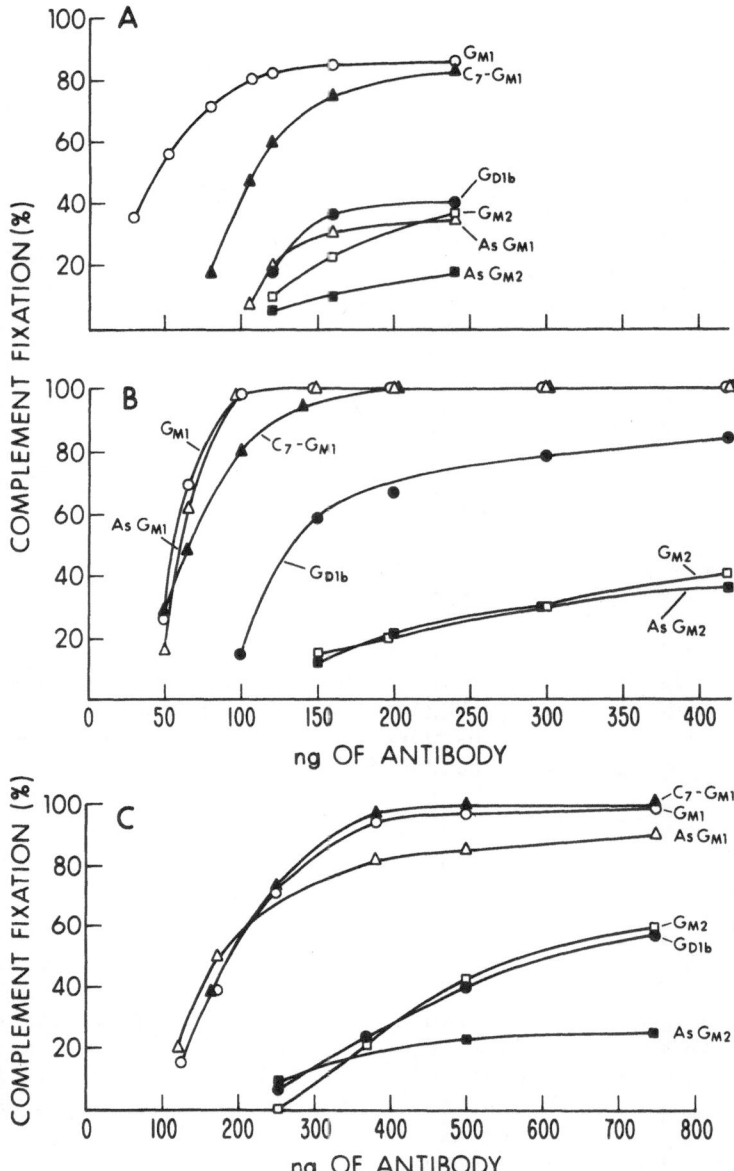

Fig. 2. Quantitative microcomplement fixation curves obtained with 20 ng of glycolipid antigens and purified IgG rabbit anti-glycolipid antibodies. The antigen preparations contained glycolipid, egg lecithin and cholesterol in a ratio of 1:2:10 (w/w). Panel A – antibodies from rabbit 525; panels B and C, antibodies purified from two different bleedings of rabbit 596.

over a period of months. For example, consider the reaction of
anti-G_{M1} antibodies with a series of gangliosides. These antibodies were raised against purified G_{M1} and they were purified by
affinity chromatography on a G_{M1} column[16]. Antibodies obtained
from serum 525 (Figure 2A) show a relatively weak cross-reaction
with asialo G_{M1} and G_{D1b} compared to antibodies obtained from
serum 596 (Figures 2B and 2C). The antibodies obtained from two
different bleedings of serum 596 also differ in their relative
cross-reactions with asialo G_{M1}, G_{D1b} and G_{M2}. In addition, the
data indicate that there are subpopulations of antibodies that
can react with internal sequences of G_{M1} that are represented by
G_{M2} and asialo G_{M2}. A similar phenomenon has been observed with
rabbit and human antibodies to globoside that can react with the
globoside portion of Forssman glycolipid[17]. These data indicate
the necessity for extensive characterization of each preparation
of purified antibodies used in the study of complex biological
materials, and emphasize the desirability of obtaining monoclonal
antibodies, as discussed below.

Monoclonal Antibodies

     Antibodies obtained by conventional procedures of immuniza-
tion are heterogeneous with regard to their specificity and
affinity for antigen. The recent development of techniques for
production of homogeneous monospecific antibodies by "hybridoma"
cells is revolutionizing immunology. Milstein and his collaborators[18] demonstrated that it was possible to obtain continuous
cell lines making homogeneous antibodies by fusing mouse myeloma
cells to spleen cells from immunized animals. The hybrid cells
can be grown in vitro or they can be injected into the peritoneal
cavities of mice, and large amounts of ascitic fluid containing
high concentrations of antibody can be obtained . The abundance
of these antibodies and their uniform serological properties make
them ideal reagents for immunological assays, for ultrastructural
studies, and for exploring the biological properties of glycolipids.

REFERENCES

1. M. M. Rapport, and L. Graf, Immunochemical reactions of lipids,
       Progr. Allergy, 13:273 (1969).
2. S. I. Hakomori, Structures and organization of cell surface
       glycolipids dependency on cell growth and malignant
       transformation, Biochim. Biophys. Acta, 417:55 (1975).
3. D. M. Marcus, Applications of immunological techniques to the
       study of glycosphingolipids, in: "Glycolipid Methodology",
       L.A. Witting, ed., American Oil Chemists Society, Champaign,
       Ill. (1976).

4. D. M. Marcus, and G.A. Schwarting, Immunochemical properties
    of glycolipids and phospholipids, Adv. Immunol. 23:203 (1976).

5. C. R. Alving, Immune reactions of lipids and lipid model mem-
    branes, in: "The Antigens", Vol.IV, M. Sela, ed , Academic
    Press, New York (1977).

6. L. Svennerholm, Chromatographic separation of human brain
    gangliosides, J. Neurochem. 10:613 (1963).

7. S. K. Kundu, and D. M. Marcus, Production of antibodies to G_{D3},
    Fed. Proc. 37:1766 (1978).

8. W. W. Young, R. A. Laine, and S. I. Hakomori, Covalent attach-
    ment of glycolipids to solid supports and macromolecules,
    in: "Methods in Enzymology", Vol. L, V. Ginsburg, ed.,
    Academic Press, New York (1978).

9. J. Eng, and D. M. Marcus, unpublished observations.

10. M. Naiki, and T. Taketomi, Chemical and immunochemical proper-
    ties of glycolipids from pig spleen, Jap. J. Exp. Med.
    39:549 (1969).

11. J. Holmgren, Comparison of the tissue receptors for Vibrio
    cholerae and Escherichia coli enterotoxins by means of
    gangliosides and natural cholera toxoid, Infect. Immun.
    6:851 (1973).

12. W. W. Young, Jr., J. W. Regimbal, and S. Hakomori, Radioimmuno-
    assay of glycosphingolipids: application for the detection
    of Forssman glycolipid in tissue extracts and cell membranes,
    J. Immunol. Meth. (in press).

13. S. C. Kinsky, Antibody-complement interaction with lipid model
    membranes, Biochim. Biophys. Acta 265:1 (1972).

14. H. R. Six, W. W. Young, K. Uemura, and S. C. Kinsky, Effect
    of antibody-complement on multiple vs. single compartment
    liposomes. Application of a fluorometric assay for follow-
    ing changes in liposomal permeability. Biochemistry
    13:4050 (1974).

15. E. A. Kabat, Structure of antibody combining sites, Ann.
    Immunol. (Inst. Pasteur) 127:239 (1976).

16. S. K. Kundu, and D. M. Marcus, Properties of antibodies to
    gangliosides G_{D3} and G_{M1}, in preparation.

17. G. A. Schwarting, S.K. Kundu, and D. M. Marcus, Reactions of
    antibodies that cause paroxysmal cold hemoglobinuria (PCH)
    with globoside and Forssman glycosphingolipids, Blood
    53:186 (1979).

18. G. Kohler, and C. Milstein, Continuous cultures of fused cells
    secreting antibody of predefined specificity, Nature
    256:495 (1975)

19. F. Melcher, M. Potter, N.L. Warner, "Lymphocyte Hybridomas,
    Current Topics in Microbiology and Immunology", Vol. 81,
    Springer-Verlag, New York (1978).

ANTIBODIES TO TOTAL BRAIN GANGLIOSIDES: TITER AND SPECIFICITY

OF ANTISERA

Maurice M. Rapport, Liselotte Graf, Yung-yu L. Huang,
William Brunner, and Robert K. Yu

Division of Neuroscience, New York State Psychiatric
Institute, Departments of Biochemistry and Pathology,
Columbia University College of Physicians and Surgeons,
New York, N.Y. 10032 and Department of Neurology, Yale
University School of Medicine, New Haven, Conn.

INTRODUCTION

Our sustained studies of the immunological activity of lipids[1,2]
led to the important generalization that such activity was intimately
related to glycosphingolipid structures. We therefore had an early
interest in the immunological activity of gangliosides and an early
appreciation that they exhibited only weak immunogenic properties,
much weaker than those of neutral glycosphingolipids.

Studies of antibody specificity for glycolipid structures require antisera with a content of antibody high enough to permit
evaluation of specificity in terms of reactive quantities of different antigens. The first reported preparation of antisera against
gangliosides that appeared to satisfy this requirement involved
repeated injections of large amounts of the antigen over long periods
of time[3]. We prepared a number of antisera against total brain
gangliosides in this way and then tested their specificity with pure
ganglioside species supplied by Dr. Robert Ledeen. The results
showed that, of the 4 major gangliosides in brain, the antisera reacted with GM_1 and GD_{1b} but not with GD_{1a} and GT_1. We also established that these antiganglioside sera reacted with the asialo-GM_1
molecule, the neutral glycosphingolipid parent structure. We published these results as abstracts[4,5] but never in extenso. A number
of puzzling observations may have contributed to this desultory
behavior, one of which was that the complement-fixation reactions
with the total ganglioside preparation, the immunizing antigen,
were weak, much weaker and less reliable than those with the pure

molecules. Another was that one or another of these antisera gave
quixotic reactions with other glycosphingolipids, such as cytolipin R
or cytolipin F (Forssman hapten). Since the supplies of pure gangliosides were at that time too limited to resolve these problems,
we diverted our attention until they would be more readily approach-
able or the need for resolution would be more compelling. This need
became apparent about 4 years ago when we discovered that antiganglioside antibodies, passively administered by intracerebral injection, were able to perturb CNS functions in a manner similar to that
of antisynaptic membrane antibodies[6,7]. One of the major demands
of this intriguing area of immunoneurological research is for a
readily available, reproducible reagent[8], and we believe that the
antiganglioside-ganglioside system is the most promising candidate
that has yet appeared. The antigen is obtainable in large quantities, its chemistry is well known, and the immune response in
rabbits occurs with sufficient regularity to obtain as many antisera as may be required to evaluate their biological variability.

    Faced with the problem of reproducibility of the biological
properties of antibodies, one of the most important questions that
should be considered is the administered dose. It is clear that
the quantities of a foreign agent that can be injected into the
brain are especially limited, and it cannot be assumed that all
antisera will have sufficient content of the requisite antibodies
to be biologically active. An early step in achieving reproducibility is to evaluate the variability of the reagent, i.e. to prepare an appreciable number of antisera and determine their relative
content (titer) of antiganglioside antibodies. Another important
question is that of antibody specificity, particularly since the
immunizing antigen contains several ganglioside species. Data providing answers to both of these questions will be presented.

MATERIALS AND METHODS

Gangliosides. The total ganglioside fraction from bovine brain
gray matter, prepared as described by KANFER[9], contained 27% sialic
acid. GM_1 ganglioside was prepared from the total ganglioside fraction either by enzymic hydrolysis with C. perfringens neuraminidase[10] or by mild acid hydrolysis[11,12]. GD_{1b} ganglioside was isolated from human brain[13]. Asialo-GM_1 was prepared from GM_1 ganglioside by either enzymic hydrolysis[14] or acid hydrolysis[12]. These
individual lipids contained less than 3% contamination based on
thin layer chromatography.

Immunization of rabbits. Over several years we have tried a number
of different methods, most giving very unimpressive results. In
this report we have compared different procedures (Table I). One
is that of PASCAL et al.[3] at somewhat reduced dosage: each rabbit
received an injection (i.m.) of 10 mg of ganglioside combined with

Freund's complete adjuvant (FCA) 3 times weekly for 3 weeks, followed after 3 weeks by injections of 1 mg (i.v.) 3 times weekly for
2 weeks and weekly until the response was satisfactory. About 100
mg of ganglioside was required per rabbit. In the second method
we followed suggestions derived from the literature[15,16] and emulsified the gangliosides with bovine serum albumin (BSA) and
Freund's incomplete adjuvant (FIA) fortified with a much larger
than usual amount of mycobacteria. The immunizing antigen contained 1.5 mg of ganglioside and 6 mg of BSA in 1.0 ml of saline,
combined with an equal volume of FIA containing 6 mg of mycobacteria; 0.5 ml was injected into each footpad. The protocol is
shown in Table I.

TABLE I.   IMMUNIZATION PROTOCOLS

	Antigen	Dose per injection	Injection route	Comment
Method I	Brain ganglioside + FCA in phosphate buffer	10 mg	i.m. 3x weekly for 3 weeks	rest 3 weeks
	Brain ganglioside in buffer	1 mg	i.v. 3x weekly for 2 weeks, then 1x weekly	until response is satisfactory
Method II	Brain ganglioside + BSA + FIA + extra myco-bacteria	4x0.5ml (1.5 mg)	in each footpad	then wait 10 days
	Same but with FCA (instead of FIA + mycobacteria)	2.0 ml (1.5 mg)	i.m.	wait 14 days; and repeat

Antiserum titration. The titer was established conveniently and
sensitively by complement-fixation provided two conditions were met.
The first was that the test antigen had to be a homogeneous ganglioside species rather than the total ganglioside preparation used
for immunization, and the second was that the test ganglioside had
to be combined with auxiliary lipids. Proportions of auxiliary
lipids were adjusted so that maximal effectiveness was observed in
the complement-fixation test. Antibody titrations were carried out
with 100 nanograms of pure GM_1 ganglioside combined with 100 parts
of lecithin-cholesterol (1:2 w/w) using four 50% units of complement
as described[17]. Both GD_{1b} and asialo-GM_1 were also combined with
the auxiliary lipid in the same proportions.

Antibody absorption with pure GM_1 ganglioside. A mixture of 25 μg
of pure GM_1 ganglioside was combined with 50 μg of pure egg lecithin

(Sylvana Chemical Co.) and evaporated to dryness under nitrogen. The residue was dissolved in 0.05 ml ethanol followed by 0.45 ml of saline and 0.50 ml of antiserum. After stirring at room temperature (25°) for 2 hours and refrigerating overnight, the mixture was centrifuged for 1 hr. at 24,000 xg at 4°. The supernatant (0.95 ml) was removed and used for testing.

RESULTS

Antibody titers obtained in 41 rabbits are presented, 27 immunized by method I and 14 immunized by method II. For discussion, these antisera are separated into 4 groups (with progressively increasing titers bearing a relation to their biological activities), namely, those having titers (against pure G_{M1} ganglioside) of less than 50, 50-200, 200-350, and over 350. The groups with titers above 200 are, we believe, almost certain to show biological activity when injected intracerebrally. The lowest group appears to be inactive, whereas the group in the range of 50 to 200 is still associated with uncertain biological activity. It is perhaps worth noting that a number of antiganglioside sera that were prepared elsewhere and given to us for determination of titer all fell into the two lowest groups.

The two methods of immunization did not differ much in inducing an antibody response to G_{M1} ganglioside (Fig. 1). With both, antisera from 35 to 40% of the rabbits had titers above 200 and from 20 to 30% had titers below 50. The trend favored the longer, more demanding method I and clearly more of the antisera with titers above 350 were obtained by this method. The more important advantage of this method, however, is that the antisera obtained with it did not contain antibodies to serum albumin, since the immunizing antigen did not contain this protein. Antibodies to albumin have presented obstacles in subsequent usage of the antisera, whether for immunohistochemical localization of ganglioside[18], perturbation of CNS functions[19], or even preparation of purified antibodies by affinity chromatography[20].

Four rabbits that were good antibody producers against the total ganglioside preparation by method II were studied more extensively (Fig. 2). The results indicate that five or more weeks may be required to obtain high titers according to this method, even though it is more efficient than method I in eliciting the antibody response. With 3 of the 4 rabbits more than 1 injection was necessary. Although each of these rabbits received 6 injections over a 28 week period, the data do not show a decisive relationship between reinjections and the boosting of antibody titers. However, the responses of rabbits 2273 and 2275 may be interpreted differently but not with conviction. Antibody responses by method I are slower than by method II.

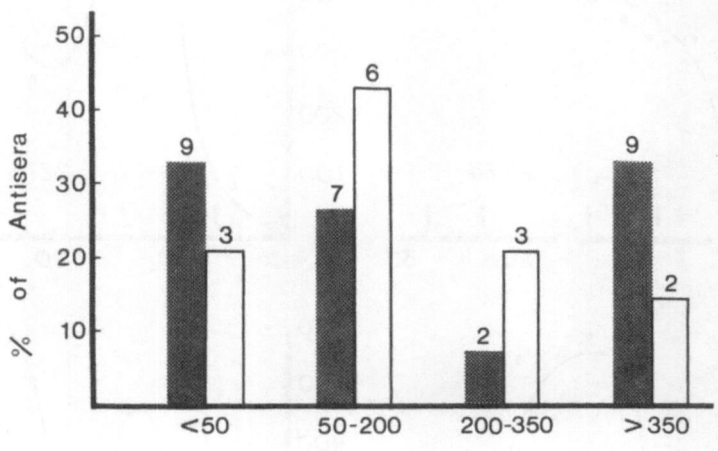

Fig. 1.   Percent distribution of responses in rabbits to immunization with total brain gangliosides by two methods. Solid
bars, 27 rabbits immunized by method I. Clear bars, 14
rabbits immunized by method II. Number above bar indicates
rabbits in each group.

Antibody specificity

The specificity of 7 of the highest titered antisera was examined by determining the loss of antibody titer against three molecules (GM_1 ganglioside, G_{D1b} ganglioside and asialo-GM_1) after sub-
total absorption of the antisera with pure GM_1 ganglioside. The
results (Table II) clearly indicate that removal of antibodies reacting with GM_1 ganglioside eliminates most of the reactivity with
G_{D1b} ganglioside and asialo-GM_1. With antisera 1849, 1825 and 1634,
the loss in reactivity is similar for all three molecules. With
the exception of antiserum 1856, the loss of reactivity against
asialo-GM_1 is equal or greater than that against GM_1. Antisera
1856, 1850, 1824 and 1853 show a somewhat smaller loss in reactivity
with G_{D1b} than with GM_1 indicating that they contain a small quantity
of antibodies reacting more specifically with G_{D1b} ganglioside than
with GM_1 ganglioside. Such antisera represent a potential source
of antiganglioside antibodies with distinctive specificity.

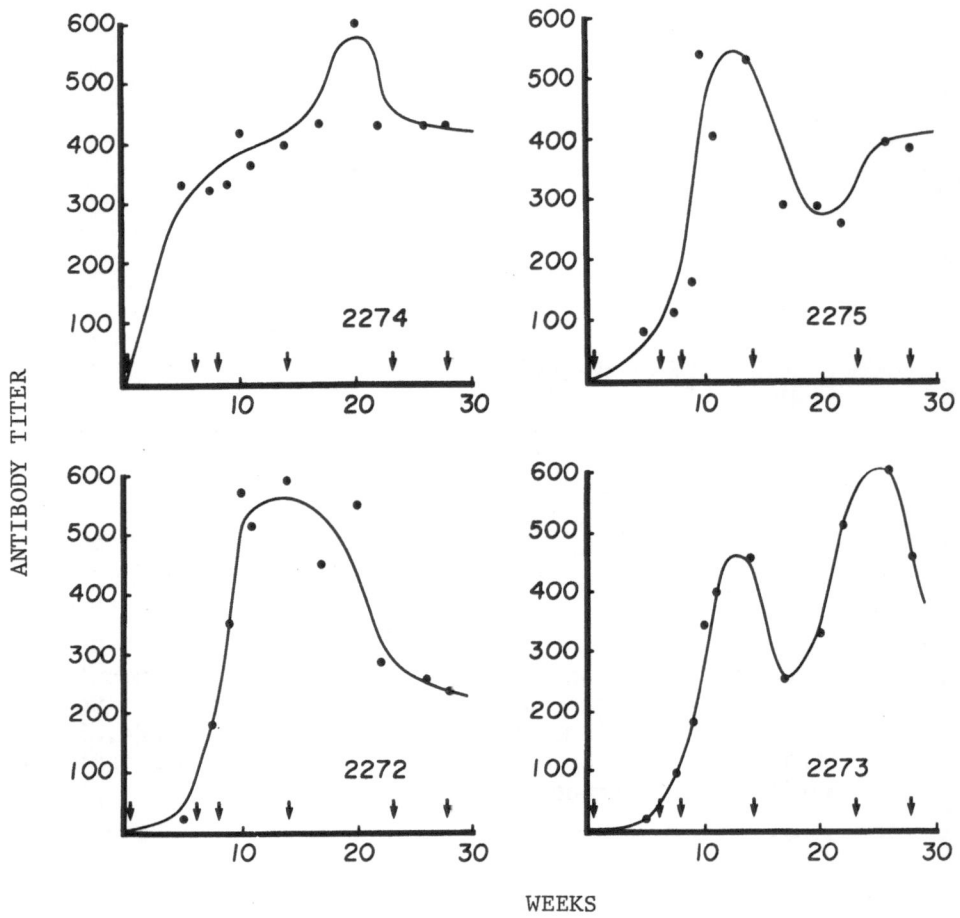

Fig. 2. Individual responses over time of 4 rabbits immunized with
total brain ganglioside by method II. Arrows indicate injections with immunizing antigen. Antibody titers were
determined by complement-fixation with G_{M1} ganglioside.

DISCUSSION

It is generally appreciated that different antisera against the
same antigen may exhibit different characteristics, expressed in
both reactivity and specificity, since the antisera contain many different types of antibodies. When antisera are used as chemical or
biological reagents, their standardization becomes important if
they are to be relied upon to give reproducible results. It is one
of the great promises of the new techniques for preparing monoclonal
antibodies in the mouse that we may be able to avoid the need for
empirical efforts to achieve such standardization. However, for

TABLE II. ANTIBODY TITERS OF ANTI-GANGLIOSIDE SERA BEFORE AND AFTER ABSORPTION WITH G_{M1} GANGLIOSIDE

Anti-Serum	G_{M1}			G_{D1b}			Asialo-G_{M1}		
	Before	After	% Loss	Before	After	% Loss	Before	After	% Loss
1856	740	12	98	650	180	73	635	73	88
1850	730	86	89	650	185	72	635	<20	>97
1849	535	83	84	430	54	87	470	62	87
1825	440	51	88	328	46	86	235	<20	>91
1824	425	95	78	135	45	67	85	<20	>76
1634	420	94	78	325	91	76	405	86	79
1853	370	40	89	225	53	76	70	<20	>72

the present we must still cope with a problem which is particularly onerous when dealing with antigens that are only weakly immunogenic such as the gangliosides.

The importance of antiserum selection to obtain reproducibility can be readily appreciated from the weak antibody responses of some rabbits and the lack of response of others. The incidence of good responses is, however, frequent enough to regard antiganglioside sera with high titers as readily obtainable reagents. This is especially true since the immunogen, total brain ganglioside, can be conveniently prepared in large quantity by a number of methods.

The specificities of these antisera are very similar. All high-titered antisera to total brain ganglioside react well with G_{M1} and G_{D1b} gangliosides but not with G_{D1a} and G_{T1} gangliosides. They also react to a considerable degree with the neutral glycosphingolipid asialo-G_{M1}. By and large, the antibodies reacting with these 3 molecules are the same since absorption with pure G_{M1} ganglioside removes most of the antibodies reacting with G_{D1b} ganglioside and asialo-G_{M1}. This specificity is not very different from that previously reported for antibodies prepared against G_{M1} ganglioside[15].

When antiganglioside sera are tested for their reactivity against a number of different neutral glycosphingolipids, one or another may show unusual cross reactions. We have found such reactions with cytolipin R[2], asialo-G_{M2}, and cytolipin F (Forssman hapten). These reactions are idiosyncratic and unexplainable. Another interesting and puzzling observation we have made is that antibodies prepared against asialo-G_{M1} do not react with G_{M1} ganglioside.

Evaluations of specificity based on differences in reactivity

are critically dependent on measuring reactivity under optimal conditions. For reactions based on complement-fixation this requires
that auxiliary lipids be adjusted both for the antigen and to some
degree for a particular antiserum. To what extent this applies to
measurements based on techniques other than complement-fixation is
not known. It is however worth emphasizing that some glycosphingolipid systems may be quite sensitive to inhibition by the arbitrary
addition or omission of auxiliary lipids, and the weak serological
reactions obtained with the mixture of gangliosides used for immunization compared with those obtained with the separate molecular
species could be an indication of such inhibition.

The results described in this report provide a reasonable basis
for predicting the results to be expected in the preparation of
antiganglioside sera in rabbits and the application of such sera
to various biological problems.

1. RAPPORT, M.M. (1961): J. Lipid Res. 2, 25-36.
2. RAPPORT, M.M. and GRAF, L.(1969): in "Progress in Allergy",
   Kallos, P. and Waksman, B.H., eds.Vol.13, Karger,Basel pp.273-331
3. PASCAL, T.A., SAIFER, A. and GITLIN, J.(1966): Proc. Soc. Exp.
   Biol. & Med. 121, 739-743.
4. RAPPORT, M.M., GRAF, L. and LEDEEN, R. (1968): Fed. Proc. 27, 463
5. RAPPORT, M.M., GRAF, L. and LEDEEN, R. (1970): Fed. Proc. 29, 573
6. KARPIAK, S.E., GRAF, L. and RAPPORT, M.M.(1976): Sci. 194,735-737
7. KARPIAK, S.E., GRAF, L. and RAPPORT, M.M.(1978): Brain Res.151,
   637-640.
8. RAPPORT, M.M.(1973): in "Psychopathology", Hammer, M.,Salzinger,
   K.,Sutton, S. eds. John Wiley & Sons, New York, pp. 309-316.
9. KANFER, J.N.(1969): in "Methods in Enzymology", Lowenstein,J.M.,
   ed. Vol. XIV, Academic Press, N.Y. pp. 660-664.
10. BURTON, R.M. (1963): J. Neurochem. 10, 503-522..
11. SVENNERHOLM, L. and RAAL, A.(1961):Biochim.Biophys.Acta,53,422-424
12. EPSHTEIN, Y. and MARKAEV, T.(1969): Biochemistry (USSR) 34,735-738
13. ANDO, S. and YU, R.K.(1977): J. Biol. Chem. 252, 6247-6250.
14. WENGER, D.A. and WARDELL, S.(1973): J. Neurochem. 20, 607-612.
15. NAIKI, M., MARCUS, D.M. and LEDEEN, R.(1974): J. Immunol.113,84-93
16. KOSCIELAK, J., HAKOMORI, S. and JEANLOZ, R.W.(1968):Immunochem.
    5, 441-455.
17. RAPPORT, M.M. and GRAF, L.(1967):in "Methods in Immunology and
    Immunochemistry",Vol.1, Academic Press, New York, pp. 187-196
18. LAEV, H., RAPPORT, M.M., MAHADIK, S.P. and SILVERMAN, A.J.(1978):
    Brain Res. 157, 136-141
19. RAPPORT, M.M. and KARPIAK, S.E.(1978): in "Senile Dementia",
    Nandy, K. ed., Elsevier, New York, pp. 73-88
20. MARCUS, D.M.(1977): in "Glycolipid Methodology", Witting, L.A.
    ed., Am. Oil. Chem. Soc., Champaign, Illinois, pp. 242-243

Supported in part by NIH grant NS 13762

PERTURBATION OF CNS FUNCTIONS BY ANTIBODIES TO GANGLIOSIDES.

SPECULATIONS ON BIOLOGICAL ROLES OF GANGLIOSIDE RECEPTORS

Maurice M. Rapport, Stephen E. Karpiak, and Sahebarao P. Mahadik

Division of Neuroscience, N.Y. State Psychiatric Institute and
Departments of Biochemistry and Psychiatry, Columbia University
College of Physicians and Surgeons, New York, N.Y. 10032

Since gangliosides are found in the plasma membranes of numerous cell types and interact with apparent specificity with a number of different agents such as toxins, antibodies, and hormones, it is reasonable to regard them as receptor molecules. Our interest in ganglioside receptors stems from their reactions with antibodies. Gangliosides represent only one class of molecules in vertebrate synapses that can be identified as functional components through perturbation of CNS functions by intracerebral administration of antibodies. Since gangliosides are relatively small, stable molecules whose chemistry has been well established, they provide a better focus than other antigenic components of the synapse for developing methods to detect and identify functional molecules. GM_1 ganglioside may be especially useful for this purpose since, of the various ganglioside species present in the synaptic membrane, more of the GM_1 is exposed on the outer surface and is therefore accessible to antibodies in the extracellular space.[1]

In searching for functions, one looks for evidence of specificity. Using antisera directed against defined antigens, we have been able to induce functional alterations with antisera to gangliosides and to S-100 protein but not with antisera to galactocerebroside or 14-3-2 protein.[2] Furthermore, antibodies to GM_1 ganglioside induce EEG spiking following intracortical injection whereas antibodies to S-100 protein do not. These results indicate that the technique of passive administration of antibodies shows specificity for function as well as for different molecular constituents of the CNS. However, available knowledge suggests that gangliosides are present in all synaptic connections, and if the hypothesis is correct that disturbances of CNS functions by antibodies result from perturbation of synaptic connections, we should then expect that antiganglioside antibodies would cause disturbances in every function examined. This is not the case. In our studies we have been restricting our attention to antibodies against GM_1 ganglioside since the control for the antiganglioside serum is the very same antiserum absorbed with pure GM_1 ganglioside to remove specifically only the anti-GM_1 ganglioside antibodies. Using this control we have shown that antibodies to GM_1 ganglioside induce seizure activity[3], inhibit passive avoidance learning by blocking consolidation[4], retard development in very young animals[5,6], and block morphine analgesia[7]. In addition, antibodies to GM_1 ganglioside were found to block reserpine sedation. In contrast, in a number of other behavioral tests no interference by antiganglioside serum was seen, and these tests include pattern discrimination, (diurnal) activity levels, fixed ratio conditioning, pain thresholds, and self-stimulation. In independent studies using our reagents Drs. Williams and Schupf found that antiganglioside antibodies injected into the hypothalamus did not affect eating and drinking but did block cholinergic stimulation of drinking (personal communication).

The results, summarized in Table 1, indicate that although there is a considerable range of behavioral activities that are affected by antibodies to GM_1 ganglioside, there are also a substantial number of behaviors that are unaffected. The implication is that ganglioside receptors and particularly GM_1 ganglioside receptors may provide a chemical basis for discriminating among different behaviors. How could this occur? One possible explanation is that there are differences in topographical distribution of the receptors in different synaptic connections, differences both in number and distribution (clustering). Another is that the processes in membranes of different synapses that are triggered by the binding of antibody ligands to ganglioside receptors may be quite different. Models exist for at least a number of possible mechanisms including 1) a change in conformation that results in altered gating mechanisms 2) an increased metabolic turnover of receptors 3) an induction of membrane destruction by complement components 4) internalization via endocytosis 5) an alteration of membrane fluidity 6) a redistribution of membrane molecules 7) simple disruption of the interaction between gangliosides and other macromolecules, etc. One way in which differences in mechanism might be looked for is through differential effects of the antibodies on release or reuptake of the different neurotransmitters present in the CNS.

Aside from their experimental and possible disease-inducing roles as receptors for antibodies, gangliosides may provide some of the signaling mechanisms that regulate sequential processes in the CNS during development. Some indication of such signaling emerged from a recent behavioral, morphological, and chemical study of the effects of intracisternal injection of antibodies to GM_1 ganglioside into neonatal rats[5,6]. When these animals reached adulthood and were tested on a complex learning task (DRL: differential reinforcement at low rates) a deficit in learning ability was detectable if the task was made sufficiently difficult. Golgi preparations showed a loss of thin spines on pyramidal cell dendrites. Chemical analysis of somatosensory cortex showed no significant change in total solids, total protein, or DNA but a 30% loss (p <.01) in ganglioside sialic acid, galactocerebroside, and RNA. The loss of ganglioside sialic acid is readily understandable, but the losses of galactocerebroside and RNA are puzzling. The decrease in galactocerebroside indicates a loss in myelin[8]. Since dendritic arborization precedes myelinogenesis, Drs. Kasarskis and Bass have proposed that gangliosides on the dendritic surface generate the signal for local myelinogenesis. Indirect support for this hypothesis might be obtained if antibodies against a non-ganglioside component of dendrites were to alter dendrogenesis without causing a concomitant decrease in galactocerebroside. The decrease in RNA must indicate

TABLE 1. BIOLOGICAL EFFECTS OF ANTIGANGLIOSIDE SERUM

EFFECTIVE		INEFFECTIVE	
Response	Target	Test Procedure	Target
1. EEG seizures	cortex: sensorimotor, frontal, visual; hippocampus; amygdala	1. EEG seizures	hypothalamus
2. Inhibition of learning (passive avoidance)	? i.vc.	2. Pattern discrimination	? i.vc.
3. Inhibition of morphine analgesia	PAG	3. Activity levels and diurnal cycle	? i.vc.
4. Blockade of reserpine sedation	? i.vc.	4. Fixed ratio conditioning	? i.vc.
5. Developmental interference (DRL behavior; dendrogenesis of pyramidal cells in cortex)	? i.cist.	5. Self-stimulation	lateral hypothalamus
		6. Pain threshold	a) PAG b) ? i.vc.
		7. Eating and drinking	lateral hypothalamus
6. Blockade of cholinergic stimulation of drinking	lateral hypothalamus		

a loss of rough endoplasmic reticulum, and it will be of interest
to determine whether this is correlated with a decrease in capacity
for protein synthesis. Certainly one of the more interesting questions concerns differential losses of one or another of the ganglioside species, since this may offer an important clue to signaling mechanisms.

In conclusion, the availability of antiganglioside antibodies
with a considerable degree of specificity for individual ganglioside molecules and the biological effects of such antibodies in
perturbing CNS functions offer a unique tool for further investigation of the functional roles of gangliosides.

1. RAPPORT, M.M. and MAHADIK, S.P.(1977): Topographic studies of
   glycoproteins and gangliosides in synaptosomes in "Mechanisms
   of Regulation and Special Functions of Protein Synthesis in the
   Brain" Roberts et al. eds. Elsevier, New York, pp. 221-230.
2. RAPPORT, M.M. and KARPIAK, S.E.(1978): Immunological perturbation of neurological functions in "Senile Dementia" Nandy, K.
   ed., Elsevier, North Holland, New York, pp. 73-88.
3. KARPIAK, S.E., GRAF, L., AND RAPPORT, M.M.(1976): Antiserum to
   brain gangliosides produces recurrent epileptiform activity.
   Science 194, 735-737.
4. KARPIAK, S., SOWIN, T., and RAPPORT, M.M.(1977): Antiserum to
   brain gangliosides inhibits consolidation phases of learning.
   Neurosci. Abstr. 3, 236.
5. RAPPORT, M.M., KARPIAK, S.E., KASARSKIS, E.J. and BASS, N.H.
   (1979): Behavioral changes after perinatal exposure to anti-G_{M1}
   ganglioside. Trans. Am. Soc. Neurochem. 10, 234.
6. KASARSKIS, E.J., KARPIAK, S.E., RAPPORT, M.M., and BASS, N.H.
   (1979): Abnormal cortical maturation after neonatal anti-G_{M1}
   ganglioside exposure. Trans. Am. Soc. Neurochem. 10, 233.
7. KARPIAK, S.E., BODNAR, R.J., HANSON, S., GLUSMAN, M., and
   RAPPORT, M.M.(1978): Antibodies to G_{M1} ganglioside block morphine analgesia. Neurosci. Abstr. 4, 460.
8. HESS, H.H. and POPE, A.(1972): Quantitative neurochemical histol-
   ogy in "Handbook of Neurochemistry" Lajtha, A. ed., Plenum,
   New York Vol. VII, pp. 289-329.

Supported in part by USPHS grant NS 13762.

IMMUNOASSAYS BASED ON PLASTIC-ADSORBED GANGLIOSIDES

J. Holmgren[1], H. Elwing[1], P. Fredman[2] and
L. Svennerholm[2]

Institute of Medical Microbiology[1] and
Department of Neurochemistry[2], University
of Göteborg, S-413 46 Göteborg, Sweden

INTRODUCTION

Important qualities of good immunoassays include high specificity, sensitivity, accuracy and simplicity in performance. These
criteria are only met to a limited extent by the traditional
immunological methods registering e.g. precipitation, agglutination or complement fixation. Therefore, in recent years these
methods have gradually been replaced by various "primary binding"
assays. These methods often use insolubilized antigen (or antibody) as a means to enable effective separation of bound and
unbound reagents and a radioisotope- or enzyme-label for the
detection step.

We have worked out some versatile, sensitive methods allowing
simple, quantitative determinations of the specific binding of
various ligands to gangliosides. The methods are based on the use
of plastic-attached gangliosides. The amphipathic ganglioside
molecules adsorb spontaneously with strong hydrophobic bonds to
plastic surfaces via their ceramide moiety, which leaves the
oligosaccharide portion free to react with specific ligands.
Separation of bound from free ligand and from unrelated material
can then be done by simple washing. This article briefly describes
these methods and discusses their usefulness in receptor studies
and for assaying gangliosides and antibodies to gangliosides.

ATTACHMENT AND BINDING PROPERTIES OF GANGLIOSIDES

The attachment of gangliosides to polystyrene test tubes

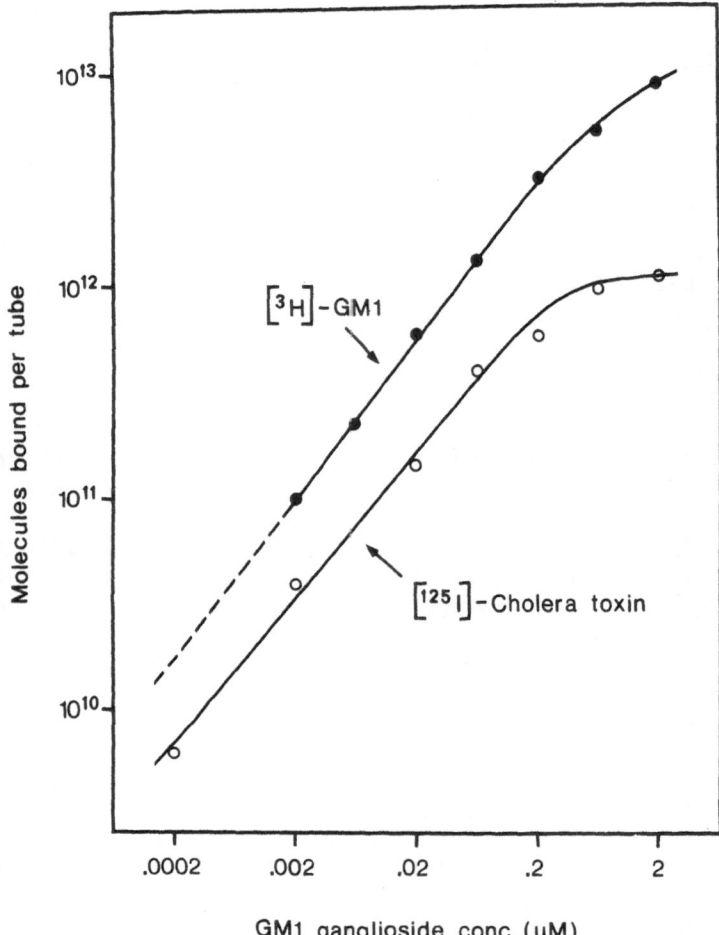

Fig 1.   Attachment of GM1 ganglioside to polystyrene test tubes
          as a function of the ganglioside concentration used for
          coating (0.2 ml incubation volume), and relationship
          between bound ganglioside and the amount of cholera
          toxin the coated tubes can bind specifically.

(inner diameter 9 mm; Heger Plastics AB, Stallarholmen, Sweden) was analyzed by means of H^3-labelled GM1 and GM2. A linear relationship was seen between the amount of ganglioside bound and a logarithmic time scale. Adsorption was similar at temperatures within the range $4-37°C$. Over the whole range of ganglioside concentrations within which accurate measurements were possible (2 nM - 2 μM) the ganglioside attachment increased almost linearly with increasing concentration, so that for every 10-fold rise in concentration there was approximately a 3-fold increase in bound ganglioside (Fig 1). Polystyrene from different manufacturers differed markedly in the amount of ganglioside bound.

The plastic-attached ganglioside retains its specific ligand-binding properties as indicated by the results obtained with ^{125}I-labelled cholera toxin. This toxin bound specifically to the GM1-coated tubes in amounts directly proportional to the attached amounts of GM1 up to a ganglioside coating concentration of about 0.2 μM; at higher coating concentrations probably steric hindrance put a limit for further binding of toxin. The molar ratio of bound toxin to GM1 was about 1:3 at 0.2 μM GM1 approaching 1:2 at 0.2 nM GM1 (Fig 1). The K_A of binding determined (1.2 nM^{-1}) was very similar to that found for binding to viable cells.

Gangliosides GM1 and GM2 had indistinguishable adsorption properties. The attachment of other gangliosides to the polystyrene tubes was studied in an indirect way employing *Vibrio cholerae* sialidase and ^{125}I-labelled cholera toxin. Cholera toxin binds specifically to GM1 ganglioside and only minimal binding of ^{125}I-cholera toxin thus took place to tubes coated with other gangliosides than GM1. However, after sialidase treatment of tubes coated with di- or trisialogangliosides cholera toxin bound in a similar amount as in GM1-coated tubes. This indicates that the various gangliosides were adsorbed similarly to the plastic.

GANGLIOSIDE-ELISA AND WATER CONDENSATION-ON-SURFACE TECHNIQUES

The principles of the two main methods elaborated - the ganglioside enzyme-linked immunosorbent assay (ganglioside-ELISA) and the water vapour condensation-on-surface (VCS) assay - are outlined in Fig 2.

In the ganglioside-ELISA method the glycolipid (0.1 or 0.2 ml) is attached to the polystyrene test tubes (or to the wells of microtiter trays) by incubation at room temperature overnight. Unoccupied binding sites on the plastic surface are blocked by incubation with 1% serum albumin. The tubes are then incubated for 1-5 h at room temperature with dilutions of the test samples (usually 2 h for nonantibody ligands and 5 h for antibodies) in

GM1-ELISA method

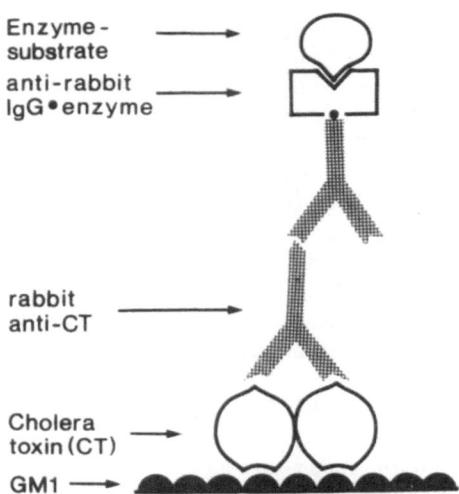

Enzyme-
substrate

anti-rabbit
IgG•enzyme

rabbit
anti-CT

Cholera
toxin (CT)

GM1

Vapour condensation-on-surface method

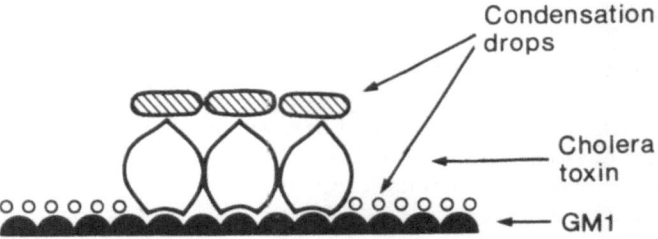

Condensation
drops

Cholera
toxin

GM1

<u>Fig 2.</u>   Principles of the detection methods as illustrated for
the cholera toxin GM1-system.

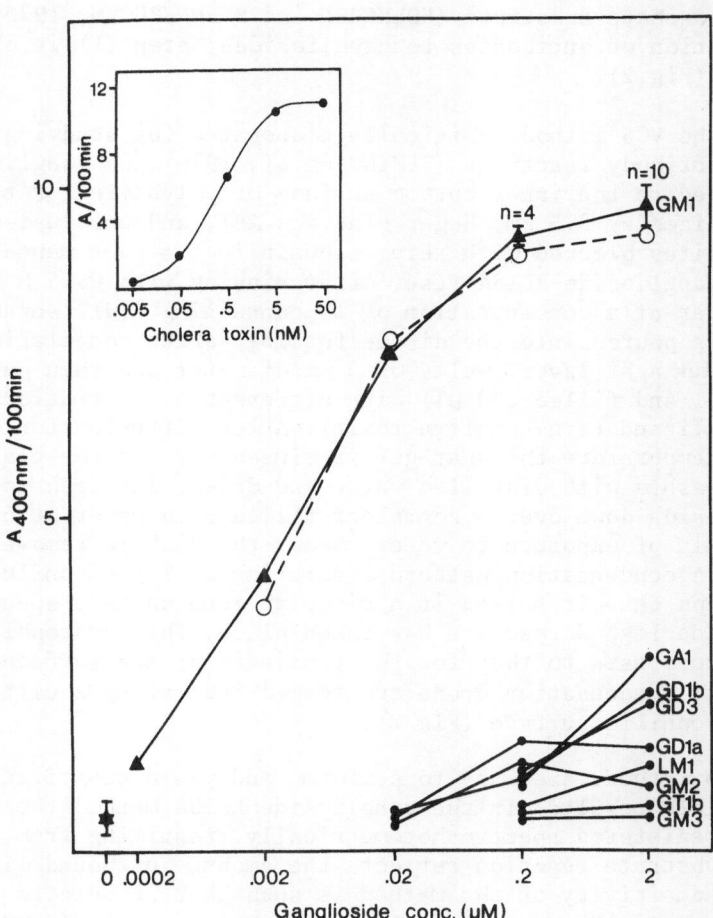

<u>Fig 3</u>.   Specific binding of purified (▲) and crude (o) cholera
toxin to GM1-coated plastic tubes as measured with the
ganglioside-ELISA method. Insert shows the binding
activity to GM1-tubes for various cholera toxin concentrations.

buffer pH 7.0-7.2 supplemented with 1% serum albumin or in the case
of antibodies with 0.05% Tween 20. Unbound material is removed by
rinsing the tubes with buffer containing 0.05% Tween 20. Bound
non-antibody ligands are thereafter demonstrated by sequential
incubations of the tubes with: (1) specific antibody, (2) anti-
-immunoglobulin coupled to alcaline phosphatase, and (3) nitro-
phenyl-phosphate substrate (HOLMGREN and SVENNERHOLM, 1973). For
demonstration of antibodies to gangliosides, step (1) is obviously
excluded (Fig 2).

     In the VCS method, originally elaborated for studying
antigen-antibody reactions (ELWING et al. 1976) the ganglioside
is attached to the inner bottom surface of polystyrene Petri
dishes (diameter 3.5 cm; Heger Plastics AB), and unoccupied
binding sites blocked with serum albumin in the same manner as
for the ganglioside-ELISA test. After rinsing with 0.15 M NaCl,
melted agar at a concentration of 1% containing 0.01% serum
albumin is poured into the dishes forming, after congelation, a
2.5-mm thick agar layer. Wells of 5 mm diameter are then punched
in the gel and filled (50 μl) with different concentrations of
the test ligand (e.g. cholera toxin). After diffusion for 20 h
at room temperature the agar gel is rinsed off and the plastic
surface washed with distilled water and dried. The dish is then
placed upside down over a container filled with water at 60°C.
After 1 min of exposure to water vapour the dish is removed and
covered. A condensation pattern consisting of large confluent
water drops then is formed in a circular zone where a specific
ganglioside-ligand reaction has taken place. This hydrophilic
pattern contrasts to that for the remainder of the surface where
only small condensation drops are formed indicating a distinctly
less hydrophilic surface (Fig 2).

     Both methods are easy to perform, and yield quantitative,
reproducible results. In the ganglioside-ELISA method the colour
change, registered spectrophotometrically, resulting from the
enzyme-substrate reaction reflects the amount of bound ligand,
and the sensitivity of the method is such that it detects as
little as e.g. 10 pM concentrations of cholera toxin (<100 pg)
(Fig 3). The ganglioside-ELISA seems to be a useful versatile
method for all ligands against which antisera can be prepared.
This assay is applicable to purified as well as highly impure
materials as illustrated by the indistinguishable results
obtained with pure cholera toxin and a crude preparation containing 99.9% unrelated material (Fig 3). The VCS technique is
comparatively insensitive but allows direct visualization of
the specific ganglioside-ligand binding without any need for
purification of the ligand or preparation of the immunoreagents.
In the described gel-diffusion variant the "wet" area is directly
related to the amount of binding material applied (Fig 4). In

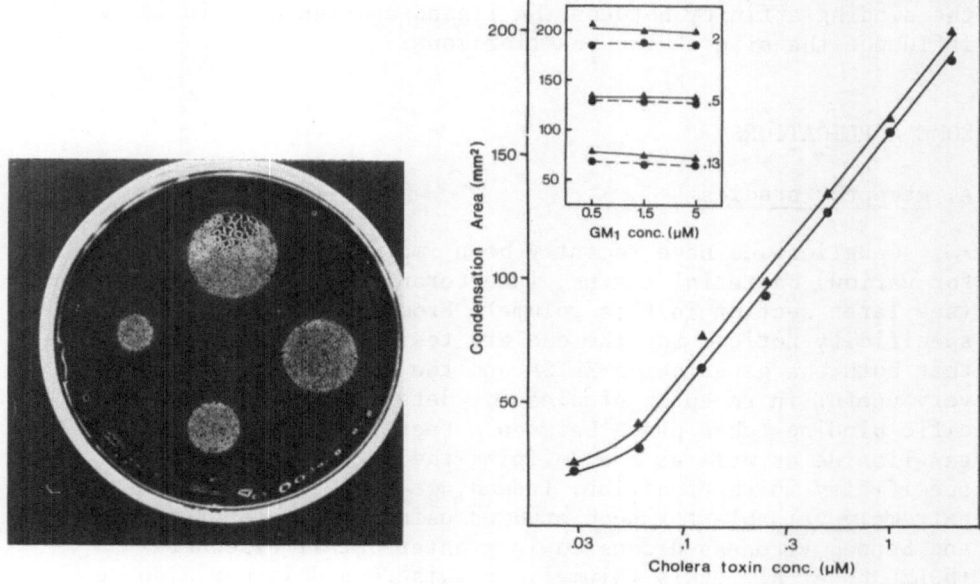

Fig 4. Water condensation patterns (photograph) and area-
-concentration relationships for cholera toxin binding
to GM1 ganglioside as tested with the VCS method.
● purified toxin, ▲ crude toxin containing a 1000-fold
excess concentration of unrelated material.

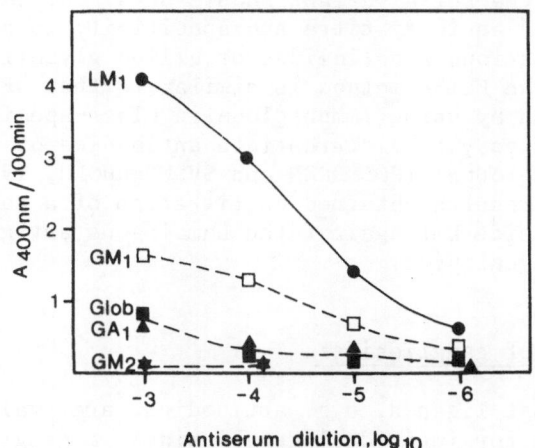

Fig 5. Reactivity of rabbit antiserum to LM1 ganglioside with
homologous and heterologous glycolipid antigens as
tested with the ELISA method.

addition, of course, the diffusion properties of the ligand and
the binding affinity between the ligand and the ganglioside will
influence the size of the reaction zone.

SOME APPLICATIONS

A. Receptor studies

     Gangliosides have recently been implicated as receptors
for various bacterial toxins, interferon, viruses and hormones
(see later section in this volume). From the excellent binding
specificity noticed for the cholera toxin system it is obvious
that both the ganglioside-ELISA and the VCS methods should be
very useful in receptor studies for determining whether a specific binding takes place between a test ligand and a certain
ganglioside as well as for defining the degree of structure
specificity in the reaction. Indeed, we have found these methods
extremely valuable for such studies using e.g. tetanus toxin
and Sendai virus as discussed in greater detail elsewhere
(HOLMGREN et al., this volume). In this regard it can also be
mentioned that a GM1-ELISA method has become a valuable diagnostic
tool for quantitating not only cholera toxin but also the
structurally and immunologically related *Escherichia coli* heat-
-labile enterotoxin in bacterial culture filtrates or diarrhoeal
fluids (SVENNERHOLM and HOLMGREN, 1978).

B. Antibody assays

     Especially the ELISA variant should be almost ideal for
determinations of antibody titre and specificity in antisera
raised against various gangliosides or allied glycolipids. The
sensitivity of the ELISA method is similar to that of radioimmunoassays, and by using immunoglobulin class-specific enzyme
conjugates it is easy to differentiate antibodies of various
classes (and subgroups) (HOLMGREN and SVENNERHOLM, 1973). Fig 5
illustrates the results obtained on titration of a rabbit antiserum to ganglioside LM1 against the homologous antigen and
various other glycolipids.

C. Quantitation of gangliosides

     Provided that ligands, e.g. antibodies, are available that
are monospecific for individual gangliosides, a ganglioside-
ELISA inhibition test should be a very sensitive method to
quantitate the ganglioside in question in biological fluids,
tissue extracts or cell suspensions. Fig 6 illustrates the

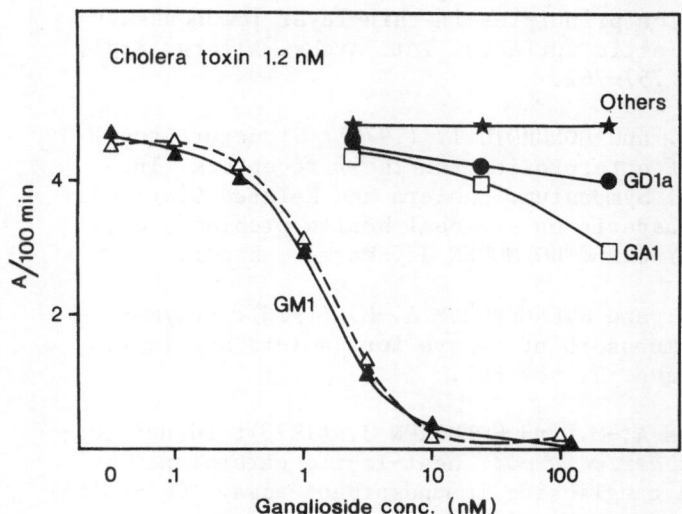

<u>Fig 6</u>. Specific inhibition of cholera toxin by GM1 ganglioside
in a ganglioside-ELISA inhibition test. Cholera toxin
was inhibited equally well by GM1 diluted in buffer
supplemented with 0.5 μM GM2 (---) as by GM1
diluted in buffer only.

outcome of such inhibition tests in a monospecific system: it
can be seen that cholera toxin was effectively inhibited by
GM1 ganglioside but not by any of the other tested substances.
Quantitation of GM1 was unaffected by the presence of other
gangliosides as indicated by e.g. the unchanged inhibition
curve obtained when dilutions of GM1 were tested in a large
excess concentration of GM2 (Fig 6). For total inhibition of
cholera toxin an approximately 5-fold molar excess of GM1 was
required, consistent with the presence of 5 identical binding
subunits in the cholera toxin molecule (HOLMGREN and LÖNNROTH,
1979). In the performance illustrated in Fig 6, GM1 concentrations down to 0.5 nM could be quantitated and if needed it is
possible to increase the sensitivity at least 10-fold by
reducing the amount of toxin in the assay.

ACKNOWLEDGEMENTS

The financial support of the Swedish Medical Research Council
and the skilful technical assistance of Ms Gudrun Andersson
are gratefully acknowledged.

REFERENCES

ELWING H., NILSSON L.-Å. and OUCHTERLONY Ö. (1976):
Visualization principles in thin-layer immunoassays
(TIA) on plastic surfaces. Int. Archs. Allergy appl.
Immun. 51, 757-762.

HOLMGREN J. and LÖNNROTH I. (1979): Structure and
function of enterotoxins and their receptors. In
"43rd Nobel Symposium. Cholera and Related Diarrheas.
Molecular aspects on a global health problem", eds.
OUCHTERLONY Ö and HOLMGREN J., Karger, Basel.

HOLMGREN J. and SVENNERHOLM A.-M. (1973): Enzyme-
-linked immunosorbent assays for cholera serology.
Infect. Immun. 5, 662-667.

SVENNERHOLM A.-M. and HOLMGREN J. (1978): Identification of *Escherichia coli* heat-labile enterotoxin by
means of a ganglioside immunosorbent assay (GM1-ELISA)
procedure. Curr. Microbiol. 1, 19-27.

IMMUNOLOGICAL PROPERTIES OF GLYCOLIPIDS INCLUDING SOME

GANGLIOSIDES IN MAMMALIAN ERYTHROCYTE MEMBRANES

Tamotsu Taketomi and Kei-ichi Uemura

Department of Biochemistry, Institute of Adaptation
Medicine, Shinshu University, Matsumoto 390, Japan

It is well known that various mammalian erythrocyte membranes contain different types of major glycosphingolipid which are considerably species-specific (Table I). These glycosphingolipids are not only structural components of membranes but also important antigenic sites on cell surfaces. Specific antibodies against individual glycosphingolipids are useful for various purposes in immunochemical and membrane research.

In this paper we report the preparation, detection and immunological properties of antibodies against neutral glycosphingolipids and gangliosides, which have been isolated from various mammalian erythrocyte membranes and characterized in our laboratory. Immunochemical methods are also presented in which glycolipid haptens are incorporated into liposomal bilayer membranes.

MATERIALS AND METHODS

A. The following immunization procedures were employed to obtain antibodies against each glycosphingolipid shown in parentheses.

1. Each rabbit received a total of 470 mg wet weight of erythrocyte stroma over a period of 4 weeks by twelve intravenous or intraperitoneal injections of increasing amounts from 5 to 100 mg. Sera were obtained at week 5. Erythrocyte stroma of pig (anti-globoside), sheep (anti-Forssman) or guinea pig (anti-asialo G_{M2}) were used.

2. The suspension of guinea pig erythrocyte stroma (20 mg in 1 ml of saline) was emulsified with 1 ml of complete

Table I. Major glycosphingolipids in mammalian erythrocyte
membranes.

I. Neutral glycosphingolipids

GalNAcα1-3GalNAcβ1-3Galα1-4Galβ1-4Glc-Cer   Forssman glycolipid
                                              (goat*, sheep)

GalNAcβ1-3Galα1-4Galβ1-4Glc-Cer   globoside
                                    (human*, pig*)

GalNAcβ1-4Galβ1-4Glc-Cer   asialo G_{M2}
                            (guinea pig*)

Galα1-3Galβ1-4GlcNAcβ1-3Galβ1-4Glc-Cer   α-Gal-paragloboside
                                           (ox*, rabbit)

II. Gangliosides

NeuGcα2-3Galβ1-4Glc-Cer   NeuGc-hematoside
                           (horse*, Japanese dog)

NeuAcα2-3Galβ1-4Glc-Cer   NeuAc-hematoside
                           (dog)

NeuGcα2-3Galβ1-4GlcNAcβ1-3Galβ1-4Glc-Cer   sialosylparagloboside
                                             (ox*)

---

* Individual glycosphingolipids were isolated from these erythrocyte
membranes (1-5). NeuAc-hematoside was obtained from bovine erythrocytes (4).

Freund's adjuvant (CFA) containing 5 mg of <u>Mycobacterium</u>
<u>bovis</u>. The emulsion was injected into foodpads and intra-
cutaneous sites on the flanks of each rabbit. Sera were
obtained 4 weeks after the single injection (anti-asialo
G_{M2}).

3. Purified glycolipid (1 to 3 mg) was suspended in 0.5 ml of
water and was mixed with 5 mg of bovine serum albumin dissolved in 0.5 ml of saline. The mixture was emulsified with
1 ml of CFA and was injected into each rabbit as procedure
2. Sera were obtained after 4 weeks (anti-Forssman, anti-
asialo G_{M2}), or after 7 weeks with two booster injections
using incomplete Freund's adjuvant at week 3 and 5 (anti-
globoside, anti-NeuGc-hematoside).

4. This procedure is the same as procedure 3 except that purified NeuGc-hematoside (1 mg) was incorporated in liposomal
membranes with 5.9 mg of egg lecithin and 2.2 mg of cholesterol. The dried film of lipid mixture was dispersed in
0.5 ml of saline using a Vortex mixer. Sera were obtained
at week 7 after two booster injections (anti-NeuGc-hematoside).

5. This procedure is the same as procedure 3 except that bovine serum albumin was omitted. Sera were obtained 5 weeks after a single injection of Forssman glycolipid (1 mg) with CFA (anti-Forssman).

B. Fractions containing IgM or IgG antibodies were separated by Sephadex G-200 gel filtration of antisera. Purification of antibodies by adsorption to glycolipid-containing liposomes and elution with 1 M NaI was done according to the method of Alving and Richards (6).

C. The following immunological assay procedures were employed.

1. Immunodiffusion analysis in 0.8% agarose gel (5).

2. Hemagglutination, hemolysis, hemagglutination inhibition and complement fixation assays on microtiter plates (5,7). For antigen preparations each glycolipid was incorporated into liposomal membranes according to the method of Kinsky et al. (8) in a molar ratio of egg lecithin, cholesterol, and glycolipid, of 1:0.75:0.1.

3. Liposome agglutination assay : Liposomes were generated by dispersing dried lipid film (composition : 1 µmol of egg lecithin, 0.75 µmol of cholesterol, 0.1 µmol of dicetyl phosphate and 0.1 µmol of glycolipid) in 1 ml of saline using a Vortex mixer. Liposomes (50 µl) were added to a cuvette (5 mm light path) containing sufficient saline to give a final volume of 1.5 ml, and the reaction was started by the addition of IgG or IgM antibody preparation. Increase of turbidity due to liposome agglutination was fol- lowed spectrophotometrically at 340 nm in an absorbance range of 0-0.1 during an incubation period of 30 min at room temperature (18-20°C). The readings were corrected for di- lution and absorbance of antibody preparation by running in parallel control cuvettes containing either liposome or antibody. Liposome agglutination was also examined on mi- crotiter plates by mixing serial dilutions of antibody (25 µl) and 25 µl of liposomes (2 mM lecithin) (5).

4. Immune damage to liposomes (9) with a slight modification : Multi-compartment liposomes were prepared from sphingo- myelin, cholesterol, dicetyl phosphate and glycolipid hapten (1, 0.75, 0.1 and 0.05 µmol, respectively) by dispersing dried lipids in 250 µl of 5 mM 4-methylumbelliferyl phosphate-285 mM glucose. Release of fluorogenic markers from 5 µl of liposomes was determined after 30 min incuba- tion with antibody preparation, guinea pig serum (20 µl) as complement and alkaline phosphatase (0.4 U) in a final volume of 1 ml.

RESULTS

Each antibody against three neutral glycosphingolipids bearing N-acetylgalactosamine as the same non-reducing terminal sugar residue could be obtained and showed no cross-reaction with each other compounds. Antisera against asialo G_{M2} were obtained by either one of the three procedures (A-1,2,3), with agglutination titers against guinea pig erythrocytes ranging from 256 to 512. Anti-globoside sera were also obtained by two procedures (A-1,3), their titers in agglutination of trypsinized human 0 erythrocytes reaching up to 256. Anti-Forssman antibodies were elicited without the use of bovine serum albumin as a carrier (procedure A-5). Hemagglutination and hemolytic titers against sheep erythrocytes of around 64 and 5,000, respectively, were easily obtained. Hemagglutinating activity of each antiserum was completely inhibited by the corresponding glycosphingolipid but not by others (Table II).

The specificity of each antiserum was also demonstrated by agglutination of liposomes containing purified glycolipids and by immunodiffusion analyses. Liposome agglutination assays on microtiter plates gave about the same titers as hemagglutination tests. In immunodiffusion, anti-Forssman, anti-globoside and anti-asialo G_{M2} antisera produced a precipitation line only with the corresponding glycolipid. Asialo G_{M2} should be mixed with 0.1% sodium taurocholate to give a clear micellar solution.

Antibodies to NeuGc-hematoside (G_{M3}) were prepared. Incorporation of glycolipid into liposomes (immunization procedure A-4) had

Table II.   Inhibition of hemagglutination by purified glyco-
            lipids.

Rabbit immune sera vs. red blood cells (RBC)	Concentration of inhibitors (µg/ml)		
	asialo G_{M2}*	globoside	Forssman
Anti-asialo G_{M2} vs. guinea pig RBC	2.4	>600	>600
Anti-globoside vs. human 0 RBC	>300	2.4	>600
Anti-Forssman vs. sheep RBC	>300	>600	0.15

* incorporated in liposomes.

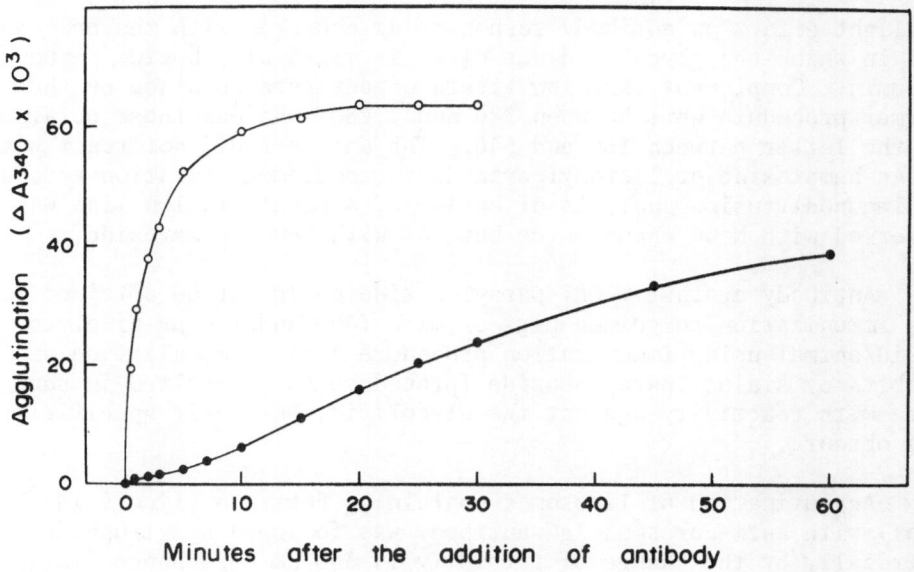

Fig. 1. Agglutination of liposomes (o——o) or micelles (●——●) containing 5 nmol of Forssman glycolipid followed by turbidity change after the addition of anti-Forssman antibodies (50 μl of IgG fraction).

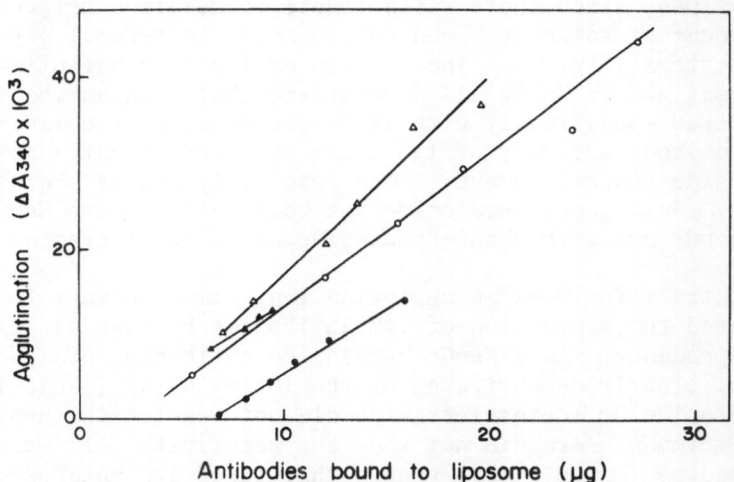

Fig. 2. Plot of liposome agglutination (ΔA_{340} values) against the amount of antibodies bound to liposomes. Liposomes containing 5 nmol of glycolipid were incubated with various amounts of antibody preparation. Asialo G_{M2} vs. IgG anti-asialo G_{M2} (O), asialo G_{M2} vs. IgM anti-asialo G_{M2} (●), Forssman glycolipid vs. IgG anti-Forssman (Δ), NeuGc-hematoside vs. IgG anti-NeuGc-hematoside (▲).

a slight effect on antibody responses as compared with the procedure
A-3 in which the glycolipid was directly mixed with bovine serum
albumin. Complement fixation titers of antisera obtained by the
former procedure were between 320 and 1,280, whereas those obtained
by the latter between 160 and 640. The antisera did not react with
NeuAc-hematoside or lactosylceramide in complement fixation assays.
In immunodiffusion analysis of antisera, a precipitation line was
observed with NeuGc-hematoside but not with NeuAc-hematoside.

Antibody against α-Gal-paragloboside could not be obtained by
its immunization to guinea pigs or mice (400 and 120 μg of glyco-
lipid/animal using immunization procedure A-3). Immunization of
rabbits by sialosylparagloboside (procedure A-3) resulted in some
sera with reactivity against the glycolipid, but their specificity
was obscure.

Agglutination of liposomes containing Forssman glycolipid (5
nmol) with anti-Forssman IgG antibody was followed spectrophotometrically by the change of turbidity in 340 nm absorbance (ΔA_{340}).
The ΔA_{340} reached a plateau after about 20 min, whereas the same
amount of glycolipid in micellar form showed an increasing curve
even 60 min after antibody was added (Fig. 1). Liposomes alone or
control liposomes (without glycolipid) plus antibody showed no appreciable change. Thus, the determination of ΔA_{340} in 30 min might
be a convenient and reliable method to detect glycolipid-antibody
interaction. Liposome agglutination can be expressed quantitatively
by ΔA_{340} values, since there was a linear correlation between ΔA_{340}
and the amount of antibody bound to liposomes in several glycolipid-
antibody systems (Fig. 2). The results of liposome agglutination
assays summarized in Table III demonstrate that each antibody preparation reacted specifically with its corresponding glycolipid and
that the antibody activity of IgG class was predominantly detected.
Anti-globoside showed, however, some reactivity against Forssman
glycolipid. Anti-NeuGc-hematoside antibody reacted with NeuGc-
hematoside but not with NeuAc-hematoside or lactosylceramide (CDH).

In contrast to liposome agglutination, immune damage to liposomes favored the expression of IgM antibodies because of its com-
plement-dependency. Anti-NeuGc-hematoside antibodies of IgM class
were, thus, clearly demonstrated by the latter assay (Table IV).
NeuAc-hematoside or lactosylceramide did not react with these antibodies. Preimmune sera did not show any reactivity. It was also
demonstrated by immune damage assays that three IgG antibody preparations against neutral glycosphingolipids (all having a terminal
N-acetylgalactosamine), anti-Forssman, anti-globoside, or anti-
asialo G_{M2}, caused specific marker release only from liposomes containing the corresponding glycolipid. However, IgM fractions of
these antisera showed various degrees of cross-reactivity. This is
probably ascribed to the presence of natural antibodies in IgM fractions rather than to the cross-reactivity of antibodies elicited by

Table III. Agglutination of liposomes.

Antibody		$\Delta A_{340} \times 10^3$ in 30 min of liposomes containing			
		NeuGc-G_{M3}	NeuAc-G_{M3}	CDH	No glycolipid
Anti-NeuGc-G_{M3}	IgM	0	0		
Anti-NeuGc-G_{M3}	IgG	10.6	0	0.5	0.6
Preimmune	IgG	0.5	0.5		
None		0	0.1	0	0.3

Antibody		$\Delta A_{340} \times 10^3$ in 30 min of liposomes containing			
		asialo G_{M2}	globoside	Forssman	No glycolipid
Anti-asialo G_{M2}	IgM	16.8	0	0	0.5
Anti-asialo G_{M2}	IgG	40.9	0	0	0.5
Anti-globoside		0	31	7	0
Anti-Forssman	IgM			2.3	0.5
Anti-Forssman	IgG	0.3	0.2	29.1	0
None		0.5	0	0.4	0.9

Table IV. Immune damage to liposomes.

Antibody		% Marker released from liposomes containing		
		NeuGc-G_{M3}	NeuAc-G_{M3}	CDH
Anti-NeuGc-G_{M3}	whole	55.2	5.5	2.0
Anti-NeuGc-G_{M3}	IgM	68.3	1.7	
Anti-NeuGc-G_{M3}	IgG	9.0	1.3	
Preimmune	IgM	1.3	0.6	
None		6.5	0.9	0.2

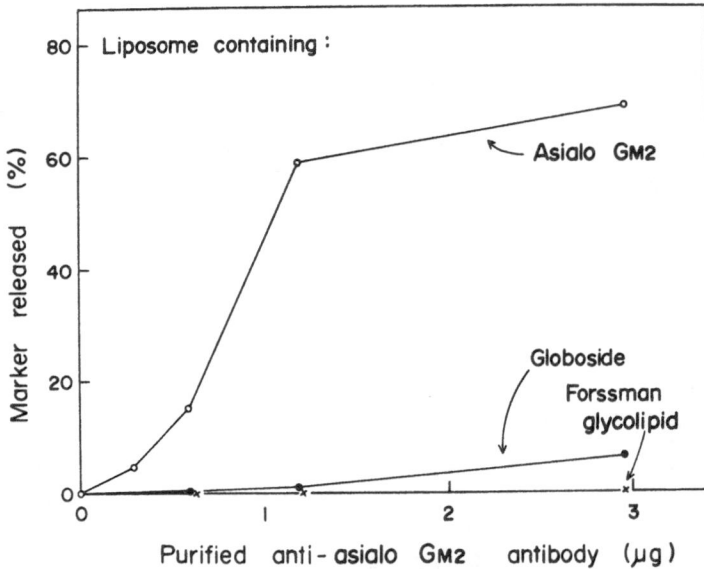

Fig. 3. Immune damage to liposomes by purified anti-asialo G_{M2} antibody.

immunization. Normal rabbit sera actually contained natural anti-bodies to various neutral glycosphingolipids when examined by sensitive immune damage assays. Activities against Forssman glycolipid, globoside and asialo G_{M2} were found in most normal rabbit sera ; antibodies to trihexosylceramide (CTH) and lactosylceramide were less frequent ; anti-galactosylceramide in one serum out of 13 rabbits tested. When specific antibodies against asialo G_{M2} were purified by adsorption to and elution from liposomes containing asialo G_{M2}, the purified antibodies showed no cross-reactivity with globoside or Forssman glycolipid (Fig. 3).

DISCUSSION

Immunological properties of glycolipids have been found to some extent by accumulation of various studies (reviewed in 10-12). Immunochemical studies on a variety of glycolipids, however, should be contributed to the elucidation of the role and function of glyco-lipids in cell membranes. Antibodies against three neutral glyco-sphingolipids, asialo G_{M2}, globoside and Forssman glycolipid, can be easily obtained and they discriminate these compounds having the same non-reducing terminal N-acetylgalactosamine due to differences in the configuration, sequence or sites of substitution of internal residues. Specific purification of antibodies or the use of IgG fraction may be, however, necessiated because natural antibodies are present in most rabbit sera. Natural antibodies against CDH and CTH

have been also reported by Alving and Richards (13). Purified antibodies against asialo G_{M2} showed no cross-reactivity with globoside
or Forssman glycolipid. An attempt has been made by Rosenfelder et
al. (14) to prepare anti-asialo G_{M2} antibodies and use them for detection of cell surface glycolipids by fluorescent antibody technique. We have reported the immunological properties of Forssman
glycolipid and globoside elsewhere (3,7,15,16).

Antibodies against erythrocyte membrane gangliosides were more
difficult to be obtained, probably because sialic acid-containing
heterosaccharides occur in rabbit blood. However, a fairly high
titer of antibody against equine NeuGc-hematoside could be obtained.
Anti-NeuGc-hematoside antibodies have previously been reported (17,
18) and shown to be directed toward the N-glycolylneuraminosyl residue.

Glycolipids incorporated in liposomal bilayer membranes are
immunologically responsive as those on natural cell surfaces. The
liposome system, rather than micellar aggregates, gives glycolipid
haptens an environment similar to natural membranes, and antigen
concentration as well as constituent lipids can be easily controlled. Of the two assay procedures presented here, "immune damage
to liposomes" favors the expression of IgM antibodies and "liposome
agglutination" detects IgG antibodies more effectively. Each assay
requires only 1 nmol and 5 nmol of glycolipid hapten, respectively ;
as little as 0.1-0.01 nmol of glycolipid is detectable by immune
damage assay.

ACKNOWLEDGEMENTS

We gratefully acknowledge the excellent technical assistance of
Mrs. Matsuko Yuzawa-Watanabe.

REFERENCES

1. Taketomi, T. and Kawamura, N. (1970) J. Biochem. 68: 475
2. Taketomi, T. and Kawamura, N. (1972) J. Biochem. 72: 791
3. Taketomi, T. and Kawamura, N. (1972) J. Biochem. 72: 799
4. Uemura, K., Yuzawa, M. and Taketomi, T. (1978) J. Biochem. 83:
    463
5. Uemura, K., Yuzawa, M. and Taketomi, T. (1978) J. Biochem. 83:
    1199
6. Alving, C.R. and Richards, R.L. (1977) Immunochemistry 14: 373
7. Uemura, K., Yuzawa, M. and Taketomi, T. (1979) Jap. J. Exp. Med.
    49: (in press)
8. Kinsky, S.C., Haxby, J.A., Zopf, D.A., Alving, C.R. and Kinsky,
    C.B. (1969) Biochemistry 8: 4149
9. Six, H.R., Young, W.W., Jr., Uemura, K. and Kinsky, S.C. (1974)

            Biochemistry 13: 4050
10.   Rapport, M.M. and Graf, L. (1969) Prog. Allergy 13: 273
11.   Marcus, D.M. and Schwarting, G.A. (1976) Adv. Immunol. 23: 203
12.   Alving, C.R. (1977) The Antigens Vol. IV, pp. 1-72, Academic
       Press, New York
13.   Alving, C.R. and Richards, R.L. (1977) Immunochemistry 14: 383
14.   Rosenfelder, G., Young, W.W., Jr. and Hakomori, S. (1977)
       Cancer Res. 37: 1333
15.   Taketomi, T., Hara, A. and Uemura, K. (1975) Jap. J. Exp. Med.
       45: 293
16.   Uemura, K., Hara, A. and Taketomi, T. (1976) J. Biochem. 79:
       1253
17.   Hakomori, S., Teather, C. and Andrews, S. (1968) Biochem. Bio-
       phys. Res. Commun. 33: 563
18.   Laine, R.A., Yogeeswaran, G. and Hakomori, S. (1974) J. Biol.
       Chem. 249: 4460

DETECTION OF ANTIBODIES TO GANGLIOSIDES IN PATHOLOGIC HUMAN SERA.

SERUM-SICKNESS TYPE HETEROPHILE ANTIBODIES

Masaharu Naiki and Hideyoshi Higashi

Department of Biochemistry, Faculty of Veterinary
Medicine, Hokkaido University
Sapporo, 060, Japan

INTRODUCTION

The production of heterophile antibodies in sera of patients
receiving therapeutic injections of foreign serum was first described
by HANGANUTIZIU (1924) and DEICHER (1926). Thereafter, these antibodies were called serum-sickness or H-D antibodies. The antibodies
differ in specificity from other kinds of heterophile antibodies,
such as FORSSMAN (1911) antibodies and PAUL-BUNNELL (1932) antibodies
(P-B antibodies) of infectious mononucleosis. H-D antibodies react
with erythrocytes and sera of various animal species (horse, sheep,
ox and rabbit) and are absorbed by sediment of guinea-pig kidney
homogenate (DAVIDSON and WALKER, 1935). Recently, "H-D antibodies"
were detected in sera from almost of all patients who received γ-
globulin fraction of goat anti-human thymocyte serum (PIROFSKY,
RAMIREZ-MATEOS and AUGUST (1973) and also in sera from some patients,
suffering from various diseases, who had never received a therapeutic
injection of foreign serum (KASUKAWA et al., 1976). The nature of
H-D antigen was unknown, except for some properties: heat-stable,
extractable with hot ethanol (SCHIFF, 1937) and precipitable by 75 %
ethanol solution (KASUKAWA et al., 1976). Our previous studies
(HIGASHI et al., 1977) succeeded in the isolation of H-D antigen
from bovine and equine erythrocytes and demonstrated that this
antigen is a ganglioside with N-glycolylneuraminic acid (GcNeu). The
antigen of equine erythrocytes was identified as GcNeu-hematoside
(YAMAKAWA and SUZUKI, 1951), and the antigen of bovine erythrocytes
was identified as GcNeu-sialosylparagloboside (WIEGANDT and SCHULZE,
1969). The structures of both the gangliosides are shown in Table 1.

Table 1.   Structures of Gangliosides

Compound	Chemical structure
GcNeu-hematoside	GcNeu(α2–3)Gal(β1–4)Glc–Cer
GcNeu-sialosylparagloboside	GcNeu(α2–3)Gal(β1–4)GlcNAc(β1–3)Gal(β1–4)Glc–Cer
GcNeu-disialoganglioside	GcNeu(α2–8)GcNeu(α2–3)Gal(β1–4)Glc–Cer
GcNeu-fucoganglioside	Fuc GcNeu | (α1–2) | (α2–3) Gal(β1–3)GalNAc(β1–4)Gal(β1–4)Glc–Cer
AcNeu-hematoside	AcNeu(α2–3)Gal(β1–4)Glc–Cer
AcNeu-sialosylparagloboside	AcNeu(α2–3)Gal(β1–4)GlcNAc(β1–3)Gal(β1–4)Glc–Cer

Abbreviations: Gal, D-galactose; Glc, D-glucose; Fuc, L-fucose;
GlcNAc, N-acetyl-D-glucosamine; GalNAc, N-acetyl-D-galactosamine;
GcNeu, N-glycosylneuraminic acid; AcNeu, N-acetylneuraminic acid;
Cer, ceramide (N-acylsphingosine).

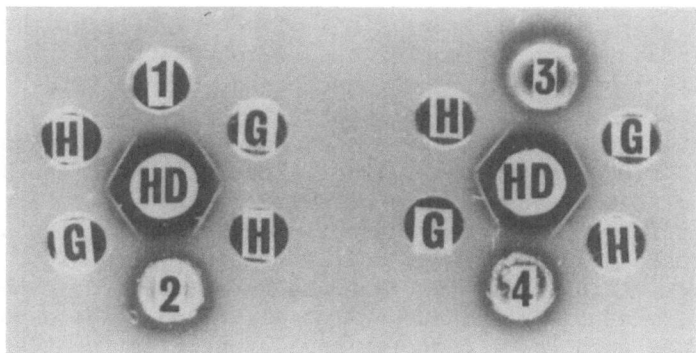

Fig. 1. Double diffusion test of H-D serum and P-B sera with GcNeu-
hematoside and GcNeu-sialosylparagloboside. HD, serum from a patient
"Z.M." containing H-D antibodies; G, GcNeu-sialosylparagloboside
(250 μg/ml); H, GcNeu-hematoside (250 μg/ml); 1 to 4, sera from 4
patients with infectious mononucleosis containing P-B antibodies.

DETECTION OF THE GANGLIOSIDE WITH H-D ANTIGEN-ACTIVITY

The ganglioside fraction from equine and bovine erythrocytes was
purified by a standard procedure (HIGASHI et al., 1977). H-D antigen-
activity from equine erythrocytes was purified and determined to be

Table 2. Immunological Reactions of GcNeu-Hematoside with H-D Sera from Patients with Different Diseases

Patient	Age	Sex	Disease	Gel-precipitation[a]	50 % C'F units per ml of serum[b]
Z.M.	50	M	Liver disorder	++++	N[c]
T.M.	46	M	Liver cirrhosis	+	1220
C.M.	30	F	Congenital heart disease	+	1852
Y.K.	58	F	Sub-pharyngeal tumor	+	1389
C.S.	42	F	Chronic hepatitis	++	2222
M.T.	69	M	Throat tumor	+±	455
T.M.	11	M	Implantation of large blood vessel	++	N
T.T.	56	M	Tabes dorsalis, progressive paralysis	+±	294
T.H.	68	M	Brain tumor	++	N
I.N.	57	M	Tumor of oral cavity	+	526
Ten healthy persons				-	<100

[a]Double diffusion tests were carried out as described in Fig. 1. [b]Micro-complement fixation tests were performed by the method of WASSERMAN and LEVINE (1961). An antigen preparation consisting of GcNeu-hematoside, lecithin and cholesterol (1:2:35, w/w), 125 ng of that in the hematoside amount was used for an assay. [c]Not examined.

GcNeu-hematoside from its inhibitory potency in the hemagglutination
of equine erythrocytes by H-D antibodies and from chemical analyses.
GcNeu-hematoside was also isolated from bovine erythrocytes, but the
major antigenic component was GcNeu-sialosylparagloboside. Both
gangliosides gave a specific precipitin line in a double diffusion
test (Fig. 1) and fixed complement with all H-D antibody-positive
sera examined (Table 2). Neither of the compounds, however, cross-
reacted with sera from 4 patients with infectious mononucleosis,
which contain high levels of P-B antibodies (a gift from Dr. KANO,
School of Medicine, State University of New York). For complement
fixation, it was necessary to add lecithin and cholesterol as auxiliary lipids to GcNeu-sialosylparagloboside or GcNeu-hematoside. The
most active composition of each antigen preparation consisted of the
ganglioside or the hematoside, lecithin and cholesterol at the ratio
of 1:2:10, w/w or 1:2:35, w/w, respectively. MERRICK *et al.* (1977)
also isolated a ganglioside fraction with H-D antigen-activity from
bovine erythrocytes, but they could not elucidate the structure since
their fraction was still a mixture of gangliosides. A few months
later, our identification of the ganglioside structure was confirmed
by them (MERRICK, ZADARLIK and MILGROM, 1978).

Specificity of both gangliosides was examined by a hemagglutination inhibition test using other gangliosides as inhibitors. GcNeu-
hematoside and GcNeu-sialosylparagloboside were the most potent inhibitors. GcNeu-disialohematoside (HANDA and HANDA, 1965) and GcNeu-
fucoganglioside (OHASHI and YAMAKAWA, 1977), both of which have the
identical trisaccharide structure as GcNeu-hematoside as regards the
internal sequence of the carbohydrate chain, also inhibited hemagglutination but these compounds were about 8 times and 128 times less
active, respectively (Table 3). AcNeu-hematoside and AcNeu-sialosyl-
paragloboside from human erythrocytes (NAIKI and MARCUS, 1974) and
other gangliosides, GM1, GM2 and GD1a from bovine brain gave no inhibition at 250 times higher concentration. The findings suggest
that the terminal N-glycolyl-neuraminyl residue is essential for H-D
immunological activity and that substitution of its 8-carbon does not
influence this activity, but substitution of the position 4-carbon
of the internal galactosyl residue dramatically diminishes the antigenic activity.

GLYCOPROTEINS WITH H-D ANTIGEN-ACTIVITY

Glycoproteins from submaxillary gland and fetuin from bovine
fetal serum had H-D antigenic activity. Porcine submaxillary gland
mucin has N-glycolylneuraminic acid at a terminal of carbohydrate
chains (CARLSON, 1968) and fetuin has the same oligosaccharide chain
as that of GcNeu-hematoside (SPIRO, 1960, 1964). We recently had an
opportunity to examine H-D antigen activity of bovine erythrocyte
glycoproteins and found that these are also as active as GcNeu-
hematoside and GcNeu-sialosylparagloboside. A glycopeptide was

Table 3. Hemagglutination inhibition[a] by various H-D
antigen-active substances

Substance	H-D antibodies	
	N.C. Serum	S.K. Serum
GcNeu-hematoside	1 (240 ng/ml)	1 (480 ng/ml)
GcNeu-sialosylparagloboside	1	0.5
GcNeu-disialohematoside	8	8
GcNeu-fucoganglioside	N[b]	128
Bovine RBC glycoproteins	8	0.016
Bovine RBC glycopeptide	512	1

[a]The lowest concentration (ng/ml) giving complete inhibition in
hemagglutination of equine erythrocytes with 4 units of two different H-D sera was expressed as a relative ratio to that of GcNeu-
hematoside.   [b]Not examined.

obtained from the glycoproteins by trypsin digestion. Both the compounds were kindly donated by Dr. UHLENBRUCK, Medizinischen Univers-
itätsklinik, Köln (NEWMAN and UHLENBRUCK, 1977). As shown in Table 2,
the glycoproteins inhibited hemagglutination with one H-D serum less
effectively than GcNeu-hematoside, but inhibited hemagglutination
with another H-D serum more effectively. The glycopeptide was approximately 60 times less active than the glycoproteins.

ANTIGENIC DETERMINANT STRUCTURE

In order to elucidate the antigenic determinant of H-D gangliosides, the trisaccharide and the pentasaccharide were obtained by
removing the ceramide by ozonolysis of GcNeu-hematoside and GcNeu-
sialosylparagloboside, respectively. A portion of the trisaccharide
was reduced with sodium borohydride and used as an inhibitor. The
structures of sialosugars employed are shown (Table 4). The trisaccharide and the pentasaccharide were the most active inhibitors in
both hemagglutination and complement fixation. The reduced trisaccharide and N-glycolylneuraminic acid were approximately 5 times and
12.5 times less active, respectively. Therefore, a common portion
between glucose and the N-acetylglucosamine residues in GcNeu-
hematoside and GcNeu-sialosylparagloboside, may also contribute to
the antigenic determinant. In future studies, however, it will be
necessary to examine how the antigenic determinant differs with
other H-D antibody-positive sera.

Table 4.   Inhibition of Hemagglutination and Complement Fixation
Sialosugars

Mono- and oligosaccharide	Hemagglutination[a] inhibition (mM)	50 % C'F[b] inhibition (μM)
GcNeu(α2-3)Gal(β1-4)Glc	0.5	110
GcNeu(α2-3)Gal(β1-4)GlcNAc(β1-3)Gal(β1-4)Glc	0.5	70
GcNeu(α2-3)Gal(β1-4)glucitol	2.5	370
N-glycolylneuraminic acid (GcNeu)	6.2	N[c]
AcNeu(α2-3)Gal(β1-4)Glc	-(>5)[d]	-(>2500)
N-acetylneuraminic acid (AcNeu)	-(>100)	N

[a]The lowest concentration giving complete inhibition under the same
condition that S.K. serum was employed as H-D hemagglutinin in
Table 3. [b]A concentration giving 50 % inhibition of complement
fixation under the same condition as described in Table 2. [c]Not
examined. [d]Negative at a concentration of less than 5 mM.

The oligosaccharides inhibited hemagglutination, but the inhibitory potency was dramatically lower than that of GcNeu-hematoside
and GcNeu-paragloboside (Tables 3 and 4). The glycopeptide obtained
by trypsin digestion was poorer in inhibitory activity than the intact
glycoproteins from bovine erythrocytes. This finding suggests that a
non-specific interaction of a hydrophobic part of the hematoside,
the ganglioside or the glycoproteins is important for enhancement of
the reaction with antibody.

DISCUSSION

The chemical structures of the two H-D antigenically-active
gangliosides were first established by us (HIGASHI et al., 1977).
Both gangliosides have a common disaccharide moiety, GcNeu(α,2-3)Gal
at a terminal. But in inhibition study, the trisaccharide GcNeu-
(α,2-3)Gal(β,1-4)Glc was more potent than the reduced compound,
GcNeu(α,2-3)Gal(β,1-4)glucitol. Therefore, it can be concluded that
GcNeu (α,2-3)Gal(β,1-4)Glc or GlcNAc is the specific determinant
group of H-D antigens. Participation of the hydrophobic part of both
the gangliosides and bovine erythrocyte glycoproteins in the antibody
reaction was also important. Such hydrophobic interaction had been
pointed out for other glycolipid antigens, such as Forssman antigen
(TAKETOMI, HARA and UEMURA, 1975) or blood group P[k] antigen (NAIKI
and KATO, in press), and for glycoprotein antigens, such as blood
group M N antigens (SPRINGER, 1970).

Various species of animals including horse, calf, sheep, rabbit, pig, elk, dog, cat, mouse, hamster and primate monkey (but not chicken) have N-glycolylneuraminyl residues in complex carbohydrates such as mucin, immunoglobulins, glycoproteins and gangliosides (GOTTSCHALK, 1972). But in humans, N-glycolylneuraminic acid has never been detected in any substances from body fluids and organs (NG and DAIN, 1976). Therefore, if some foreign substance containing such an antigenic determinant invades the human either by injection of serum from a foreign species or infection by microorganisms, or absorption of milk and meats from abnormal intestinal sites or by inhalation of animal dandruff, H-D antibody production may be stimulated. However, we must eventually establish why such antibodies are elevated only in pathologic human sera.

ACKNOWLEDGEMENTS

We should like to thank Drs. S. MATUO and K. OKOUCHI, Hospital of Kyushu University, for their enthusiastic support and supply of H-D sera from patients. This investigation was supported in part by a Research Grant N° 310615 from the Ministry of Education, Japan.

REFERENCES

CALAHAN H.J. (1976): Preparation of an infectious mononucleosis receptor from sheep erythrocyte stroma. Int. Aechs. Allergy appl. Immun. 51, 696-708.

CARLSON D.M. (1968): Structures and immunochemical properties of oligosaccharides isolated from pig submaxillary mucins. J. biol. Chem. 243, 616-626.

DAVIDSON I. and WALKER P.H. (1935): The nature of heterophile antibodies in infectious mononucleosis. Amer. J. clin. Path. 5, 455-465.

DEICHER H. (1926): Über die Erzeugung heterospezifischer Hämagglutinine durch Injektion artfremden Serums. Z. Hyg. Infektionskr. 106, 561-579.

FORSSMAN J. (1911): Die Herstellung hochwertiger spezifischer Schafhämolysine ohne Verwendung von Schafblut. Biochem. Z. 37, 78-115.

GOTTSCHALK A. (1972): Glycoproteins, their composition, structure and function. Elsevier Publishing Co. (Amsterdam).

HANDA N. and HANDA S. (1965): The chemistry of lipids of posthemolytic residue or stroma of erythrocytes. XIV. Chemical structure of glycolipid of cat erythrocyte stroma. Jap. J. exp. Med. 35, 331-341.

HANGANUTZIU M. (1924): Hémagglutinines hétérogénétiques après
    injection de sérum de cheval. C. R. Soc. Biol. 91, 1457-1459.

HIGASHI H., NAIKI M., MATUO S. and ŌKOUCHI K. (1977): Antigen of
    "serum-sickness" type of heterophile antibodies in human sera:
    identification as gangliosides with N-glycolylneuraminic acid.
    Biochem. Biophys. Res. Commun. 79, 388-395.

KASUKAWA R., KANO K., BLOOM M.L. and MILGROM F. (1976): Heterophile
    antibodies in pathologic human sera resembling antibodies stimul-
    ated by foreign species sera. Clin. exp. Immunol. 25, 122-132.

MERRICK J.M., SCHIFFERLE R., ZADARLIK K., KANO K. and MILGROM F.
    (1977): Isolation and partial characterization of the heterophile
    antigen of infectious mononucleosis from bovine erythrocytes.
    J. supramol. Struct. 6, 257-290.

MERRICK J.M., ZADARLIK K. and MILGROM F. (1978): Characterization of
    the Hanganutziu-Deicher (serum-sickness) antigen as gangliosides
    containing N-glycolylneuraminic acid. Int. Archs. Allergy appl.
    Immun. 57, 477-480.

NAIKI M. and MARCUS D.M. (1974): Human erythrocyte P and P^k blood
    group antigens: identifications as glycosphingolipids. Biochem.
    Biophys. Res. Commun. 60, 1105-1111.

NEWMAN R.A. and UHLENBRUCK G.G. (1977): Investigation into the
    occurrence and structure of lectin receptors on human and bovine
    erythrocyte, milk-fat globule and lymphocyte plasma-membrane
    glycoproteins. Eur. J. Biochem. 76, 149-155.

NG S.S. and DAIN J.A. (1976): The natural occurrence of sialic acids,
    in "Biological Roles of Sialic Acid", ROSENBERG A. and SCHENGRUND
    C.L., Eds., Plenum Press (New York and London), pp. 59-102.

OHASHI M. and YAMAKAWA T. (1977): Isolation and characterization of
    glycosphingolipids in pig adipose tissue. J. Biochem. 81, 1675-1690.

PAUL J.R. and BUNNELL W.W. (1932): The presence of heterophile anti-
    bodies in infectious mononucleosis. Amer. J. Med. Sci. 183, 90-104.

PIROFSKY B., RAMIREZ-MATEOS L.C. and AUGUST A. (1973): "Foreign serum"
    heterophile antibodies in patients receiving antithymocyte antisera.
    Blood 42, 385-393.

SCHIFF F. (1937): Heterogenetic hemagglutinins in man following the
    therapeutic injections of immune sera produced in rabbits. J.
    Immun. 33, 305-313.

SPIRO R.G. (1960): Studies on fetuin, a glycoprotein of fetal serum.
    1. Isolation, chemical position, and physicochemical properties.
    J. biol. Chem. 235, 2860-2869.

SPIRO R.G. (1964): Periodate oxidation of the glycoprotein fetuin.
J. biol. Chem. 239, 567-573.

SPRINGER G.F. (1970): Role of human cell surface structures in interactions between man and microbes. Naturwissenschaften 57, 162-171.

TAKETOMI T., HARA A. and UEMURA K. (1975): Immunological studies of
lipids. IV. Chemical modification of Forssman globoside and immunological activity. Jap. J. Med. 45, 293-298.

WASSERMAN E. and LEVINE L. (1961): Quantitative microcomplement
fixation and its use in the study of antigenic structure by specific antigen-antibody inhibition. J. Immun. 87, 290-295.

WIEGANDT H. and SCHULZE B. (1969): Spleen gangliosides: the structure
of ganglioside GLNnT1 (NGNA). Z. Naturforschg. 24b, 945-946.

YAMAKAWA T. and SUZUKI S. (1951): The chemistry of the lipids of
posthemolytic residue or stroma of erythrocytes. I. Concerning the
ether-insoluble lipids of lyophilized horse blood stroma. J.
Biochem. 38, 199-212.

SIALIDOSES AND SIALIDASES

SIALIDOSES, NEW TYPES OF INBORN DISEASES

Gérard Strecker

Laboratoire de Chimie Biologique de l'Université des
Sciences et Techniques de Lille I and Laboratoire
Associé au CNRS n° 217, B.P. n° 36, 59650 - Villeneuve
d'Ascq (France)

INTRODUCTION

Glycoproteinoses are inherited metabolic diseases involving
the deficiency of tissue hydrolase activities which result in
an accumulation of incompletely degraded glycans in tissues and
urine. Six different types of glycoproteinoses are known at
present : mannosidosis, fucosidosis, aspartyl-glucosaminuria,
GM1-gangliosidosis, Sandhoff's disease and sialidoses.

This report describes the biochemical findings concerning
four different types of glycoproteinoses defined as sialidoses on
the basis of the specific α-neuraminidase deficiency and the
nature of the storage material.

A brief review of glycoprotein structures will be included
to provide a basis for understanding the significance of the
enzyme deficiency associated with this group of diseases.

ASPECTS OF THE PRIMARY STRUCTURE OF N-GLYCOPROTEIN GLYCANS

Two types of covalent linkages between glycans and proteins
lead to the definition of O- and N-glycoproteins. The N-glycoproteins can be classed in two families according the nature of
the carbohydrates which substitute the common pentasaccharidic
core of trimannosido-di-N-acetylchitobiose (Montreuil, 1975).

In the first family, the pentasaccharide is only substituted
by mannose residues : these glycans are called the oligomannosidic type. In the second family or N-acetyllactosaminic type, the
pentasaccharide is substituted by a variable number of N-acetyl-

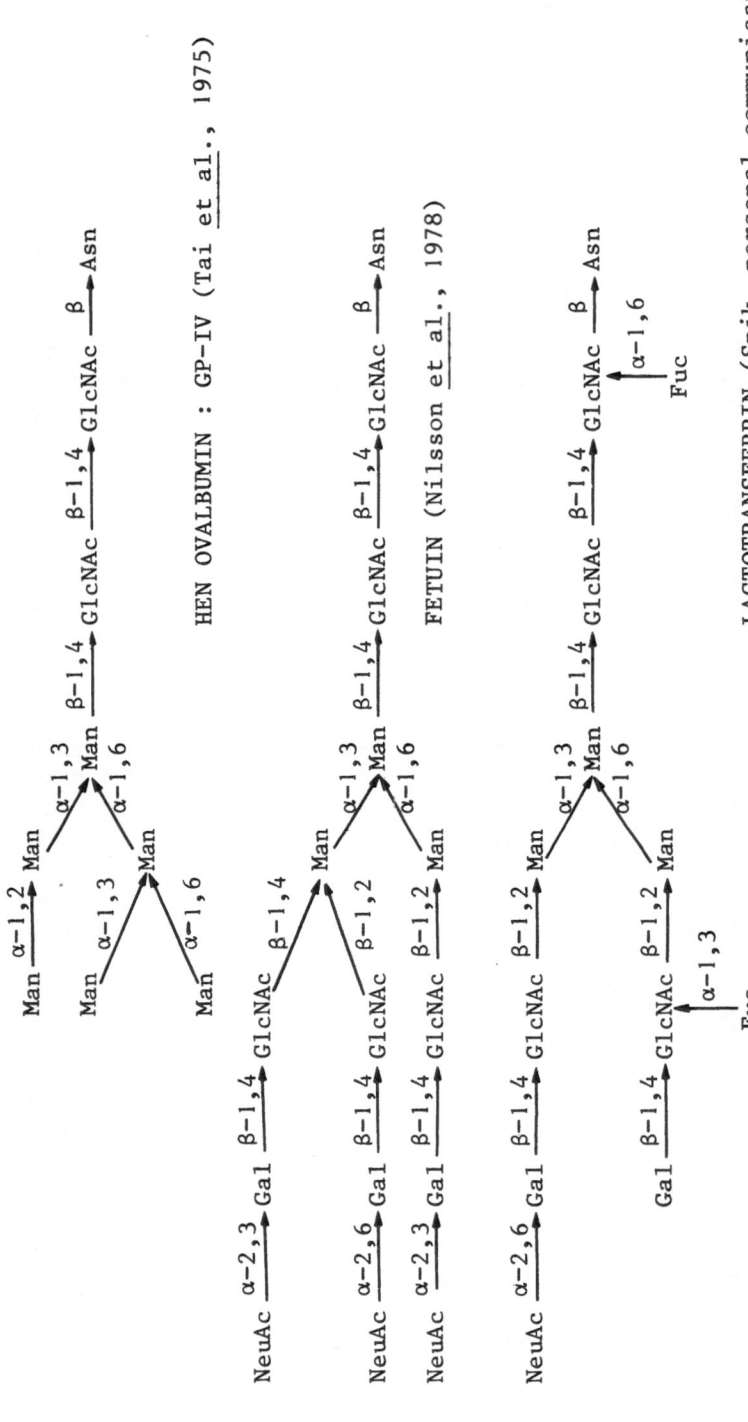

Figure 1 : Structures of oligomannosidic and N-acetyllactosamiric type glycans

lactosamine residues, themselves substituted by sialic acid and/or fucose. These definitions are illustrated by the examples of oval-bumine, fetuine and lactotransferrin given in Fig. 1.

SIALIDOSES

Mucolipidosis II, or I-Cell disease

In 1974, our attention was drawn by the fact that the urine of a patient with mucolipidosis II contained a large quantity of sialyl-oligosaccharides (Hondi-Assah et al., 1975). This observa-tion was confirmed by studying the urine of two new cases of this affliction, and the structure of the three major sialyl-oligosac-charides was determined (Strecker et al., 1976a, 1976b). The structure of six other oligosaccharides was further elucidated (Strecker et al., 1977). This sialyl-oligosacchariduria has been confirmed in ten other cases of I-Cell disease (Strecker et al., 1978b ; Okada et al., 1978) and can be considered as an element of diagnosis. In addition, these sialyl oligosaccharides have been found to accumulate in liver and brain (Strecker et al., 1977), and bound sialic acid is also increased in cultured fibroblasts (Thomas et al., 1976 ; Cantz et al., 1977). The results are illustrated by Fig. 2 and Table I.

The determination of α-neuraminidase activity in leukocytes, using as a substrate the sialyl-oligosaccharide n° II (Fig. 2), allowed us to demonstrate a total deficiency of this hydrolase, while the other hydrolase activities were perfectly normal (Strecker et al., 1976c). On the basis of these observations, we defined mucolipidosis II as a sialidosis.

These results were not considered as characteristic of a primary defect in α-neuraminidase, though the same deficiency has been confirmed for cultured fibroblasts (Thomas et al., 1976, Cantz et al., 1977). Previously, many observations had been made concerning the deficiencies of several other fibroblast hydrola-ses, such as β-D-galactosidase, α-L-fucosidase, β-D-glucuronidase, α-D-mannosidase, β-D-hexosaminidase and arylsulfatase A.

Curiously, acid phosphatase, β-D-glucosidase and β-D-xylosidase activities are normal.

In contrast, high levels of these defective hydrolase activi-ties are observed in culture medium or in extracellular fluids. For these reasons, the α-neuraminidase defect was generally considered as being added to the long list of hydrolase deficien-cies characteristic of I-Cell disease.

The evidence for a primary defect of α-neuraminidase may be

TABLE I

Excretion of sialyl-oligosaccharides (mg/l) in Sialidoses

Compounds (a)	Normal urine	MLP II Case n°1	2	3	MLP III Case n°1	2	MLP I Case n°1	2	MLP:"W"	"De P"	"N"	"CRS" (b)
I	< 0.3	8	4	5	2	6	16	18	28	4	4	2
II	< 0.5	11	6	7	3	12	185	228	125	140	52	12
III	< 0.2	14	11	7	2	8	22	16	15	14	12	3
IV	< 0.1	15	9	9	4	11	8	11	5	2	3	< 1
V	< 0.5	12	10	11	3	13	32	29	70	20	32	4
VI	< 0.1	5	4	3	8	5	11	11	24	8	5	< 1
VII	< 0.1	7	9	8	7	8	11	14	25	1.5	8	2
VIII	< 0.3	16	15	11	7	14	6	2	48	5	11	2
IX	< 1	23	25	20	10	24	225	265	160	110	135	5
X	< 0.1	-	-	12	-	10	12	-	17	10	24	-
XI	< 0.1	-	-	4	-	5	17	12	8	-	11	-

(a) see Figure 2

(b) "CRS" : Cherry red spot-myoclonus syndrome.

brought by several observations. Despite the fact that α-L-fucosidase and α-D-mannosidase activities are also defective in I-Cell fibroblasts and liver, fucose or mannose-containing oligosaccharides do not accumulate in tissues and urine. In addition, the α-neuraminidase activity is not detectable in serum or culture medium. This last result was obtained by Potier et al. (1977), who used a fluorometric method (methylumbelliferyl-α-D-N-acetylneuraminic acid). Potier et al., found in the fibroblasts of one obligate heterozygote an α-neuraminidase activity of 50 % of normal. With sialoside n° II (Fig. 2) as a substrate, we found, in the leukocytes of two heterozygote activities of 30 and 70 % of normal, respectively.

An important biological observation was made by Hickman and Neufeld (1972) who showed that I-Cell fibroblast were able to pinocytose exogenous lysosomal hydrolases, while the enzymes excreted from I-Cell fibroblasts did not enter lysosomes of various hydrolase deficient fibroblast. These authors postulated that the excreted enzymes were deficient in a recognition marker necessary for normal cellular uptake. If the hypothesis of this physiological pathway has not been verified, the concept of a molecular lesion of lysosomal hydrolases has been accepted.

Several recognition markers have been identified : galactose (Ashwell and Morell, 1974), for hepatocytes, mannose or N-acetylglucosamine (Stahl et al., 1976) for Kupffer cells and phosphomannose for fibroblasts (Kaplan et al., 1977). In these three cases, a primary deficiency of α-neuraminidase may explain the non-uptake or the failure of hydrolase packaging by I-Cell, since the terminal sequence α-NeuAc → Gal → GlcNAc → Man → R is common to N-acetyllactosaminic type glycoproteins. Vladutiu and Rattazzi (1975) observed a higher electrophoretic mobility of lysosomal hydrolases excreted by I-Cell fibroblasts, due to additional sialic acid residues. It has also been shown that plasma hydrolases are normally sialylated (except for acid phosphatase, β-glucosidase and β-xylosidase !), in contrast to liver hydrolases (Willcox and Renwick, 1977). On the basis of these observations, we may conceive that the possible lack of phosphomannosyl recognition marker in I-Cell disease (Sly, this volume) may result from a primary defect of α-neuraminidase.

The glycanic structure of lysosomal hydrolases is not yet known. Recent results obtained by Alhadeff and Freeze (1977) show that human lysosomal α-fucosidase contains essentialy N-acetylglucosamine and mannose. Van Elsen and Leroy (personal communication) found that lysosomal β-N-acetylhexosaminidase isolated from normal cells or mannosidosis fibroblasts was bound by Con-A, but not by ricin ; in contrast the same hydrolase isolated from GM1 or I-Cell fibroblasts is bound by ricin.

As we know that this lectin has an affinity for the sequence Gal $\xrightarrow{\alpha1,4}$ GlcNAc → R and NeuAc $\xrightarrow{\alpha2,6}$ Gal $\xrightarrow{\beta1,4}$ GlcNAc → R (Debray et al., 1979) we can infer a modification of the glycanic structure of the lysosomal hydrolases in these two diseases, due to the primary deficiency, β-galactosidase and α-neuraminidase respectively.

Table II

α-Neuraminidase activity in leukocytes (in % of normal value)

| | | Substrate | |
		α-NeuAc-(2→3)-R	α-NeuAc-(2→6)-R
MLP II	case n° 1	n.d.	0
	case n° 2	0	0
	case n° 3	8	6
MLP III	case n° 1	12	14
	case n° 2	8	6
MLP I	case n° 1 [a]	100	0
	case n° 2 [b]	26	0
		126	0
		150	0
MLP "W"		80	0
MLP "De P"		60	6

(a) Maroteaux et al., (1978a) ; (b) Bakker et al., unpublished results.

The structures of sialyl-oligosaccharides accumulating in urine and tissues are given in Fig. 2. Table I shows that the compounds having respectively a sialic acid linkage in α-(2 → 3) or in α-(2 → 6) are approximatively in the ratio 1 to 1. The α-neuraminidase assay (Table II), using as natural substrates oligosaccharides I and II, confirmed a total lack of the two neuraminidase activities, α-(2 → 3) and α-(2 → 6), in leukocytes (Strecker and Michalski, 1978a).

Mucolipidosis III, or Hurler Pseudopolydystrophy

Mucolipidosis III is generally considered as a mild form of I-Cell disease, and the lysosomal hydrolase absormalities are very similar. As is shown in Table I and Fig. 2, the same oligosaccharides accumulate in the urine. The α-(2 → 3) and α-(2 → 6) neuraminidase activities are deficient in leukocytes, while the other hydrolase activities are normal.

Mucolipidosis I, or Lipomucopolysaccharidosis

Mucolipidosis I has been recently recognized as a sialidosis (Cantz et al., 1977 ; Spranger et al., 1977) on the basis of the primary defect in α-neuraminidase. Structures of the sialyl-oligosaccharides excreted in urine are identical to those of I-Cell disease (Michalski et al., 1977). In three other cases of mucolipidosis I, the same elevated sialyl-oligosacchariduria observed (Maroteaux et al., 1978a ; Bakker et al. and Chamoles et al., unpublished).

Other cases of sialyl-oligosaccharidosuria

In 1976, we described two "new cases" of sialidosis, named by the initials of the patients : mucolipidosis "W" and "De P" (Strecker et al., 1976a, 1976b, 1977 ; Durand et al., 1977). Mucolipidosis "W" is very similar to mucolipidosis I, with additional renal insufficiency, and Maroteaux et al. (1978b) proposed the term nephrosialidosis. The two patients "De P." are adolescent and adult, with a normal intelligence. They do not present significant symptoms, except a bilateral cherry-red spot, red-green colour blindness and vacuolated lymphocytes. The clinical picture of a fifth patient, "N" was described by Federico et al. (1977) as being very similar to that of the two siblings reported by Durand et al. (1977).

During these last two years, many other cases of sialyl-oligosacchariduria with α-neuraminidase deficiency were described by Kelly and Graetz (1977), O'Brien (1977, 1978), Thomas et al. (1978), Okada et al. (1978) and Koseki et al. (1978). The sialidoses described by O'Brien (1977) and Thomas et al. (1978) as "cherry red spot-myoclonus syndrome" are similar to the mucolipidosis "De P"., with additional progressive myoclonus.

From a biochemical point of view, all these new cases of mucolipidosis may be defined as variants of mucolipidosis I.

Biochemical findings in mucolipidosis I and variants

In all the cases of mucolipidosis I and its variants, an excessive sialyl-oligosacchariduria was observed (Strecker et al. 1976a,b, 1977 ; Michalski et al. 1977 ; O'Brien 1977) (see also Table I). The structures of the sialyl-oligosaccharides (Fig. 2) are identical to those excreted in I-Cell disease. Nevertheless, the ratio of $\alpha-(2 \to 6)$ to $\alpha-(2 \to 3)$ sialic acid linkages is in this case 5 to 30 (Table I). This observation suggested a single defect of $\alpha-(2 \to 6)$ neuraminidase activity. This specific deficiency was verified for mucolipidosis "W", "De P", "N" (Strecker and Michalski, 1979). In two cases of mucolipidosis I (Maroteaux et al. 1978a, Bakker et al. unpublished results), the $\alpha-(2 \to 3)$

neuraminidase activity was also found to be normal (Table II).
Nevertheless, in the first case of mucolipidosis I described by
Spranger et al. (1977), the two neuraminidase activities were
abolished (Cantz, this volume). Mucolipidosis I and related
sialidoses present too a variability in their clinical symptoms
for it to be possible to interpret this discordance. The patient
described by Bakker et al. (personnal communication) presented
all the symptoms described by Spranger et al. (1977). The α-(2 → 3)
neuraminidase activity was none for sonicated leukocytes, but was
found to be respectively 26, 126 and 150 % of normal when the
assays were carried out on freshly homogeneized leukocyte, using
a Dounce apparatus (Table II). In normal leukocytes, the α-(2 → 3)
neuraminidase activity was not so sensitive to physical treatment.

CATABOLIC PATHWAY OF N-ACETYLLACTOSAMINE TYPE GLYCOPROTEINS

The structures of accumulating sialyl-oligosaccharides are
related to the class of N-acetyllactosaminic type glycoproteins
(Fig. 1). The fact that all these oligosaccharides possess a N-
acetylglucosamine residue in the terminal reducing position leads
to the hypothesis that the catabolism of sialylated glycoproteins
or glycopeptides starts by the action of β-endo-N-acetylglucosa-
minidase which splits the residue of di-N-acetylchitobiose
(Montreuil 1975 ; Strecker et al. 1976a). Such a hydrolase activity
has never been described. The only known mammalian β-endo-N-acetyl-
glucosaminidases act on substrates with terminal non reducing
mannose (Nishigaki et al., 1974) or galactose (Pierce et al., personal communication). In rat liver these two activities have been
located in the cytosol but not in lysosomes (Pierce et al.,
personal communication).

$$\text{NeuAc} \xrightarrow{\alpha-2,3} \text{Gal} \xrightarrow{\beta-1,4} \text{GlcNAc} \xrightarrow{\beta-1,2} \text{Man}$$
$$\xrightarrow{\alpha-1,3} \text{Man} \xrightarrow{\beta-1,4} \text{GlcNAc} \qquad \text{I}$$

$$\text{NeuAc} \xrightarrow{\alpha-2,6} \text{Gal} \xrightarrow{\beta-1,4} \text{GlcNAc} \xrightarrow{\beta-1,2} \text{Man}$$
$$\xrightarrow{\alpha-1,3} \text{Man} \xrightarrow{\beta-1,4} \text{GlcNAc} \qquad \text{II}$$

$$\text{NeuAc} \xrightarrow{\alpha-2,6} \text{Gal} \xrightarrow{\beta-1,4} \text{GlcNAc} \xrightarrow{\beta-1,2} \text{Man}$$
$$\xrightarrow{\alpha-1,3} \text{Man} \xrightarrow{\beta-1,4} \text{GlcNAc} \qquad \text{III}$$
$$\xrightarrow{\alpha-1,6} \text{Man}$$

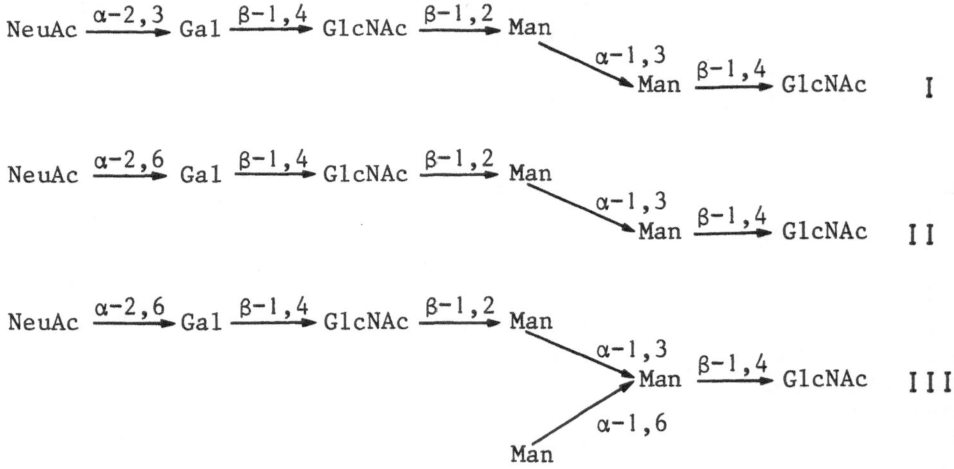

Fig. 2 [cont'd]

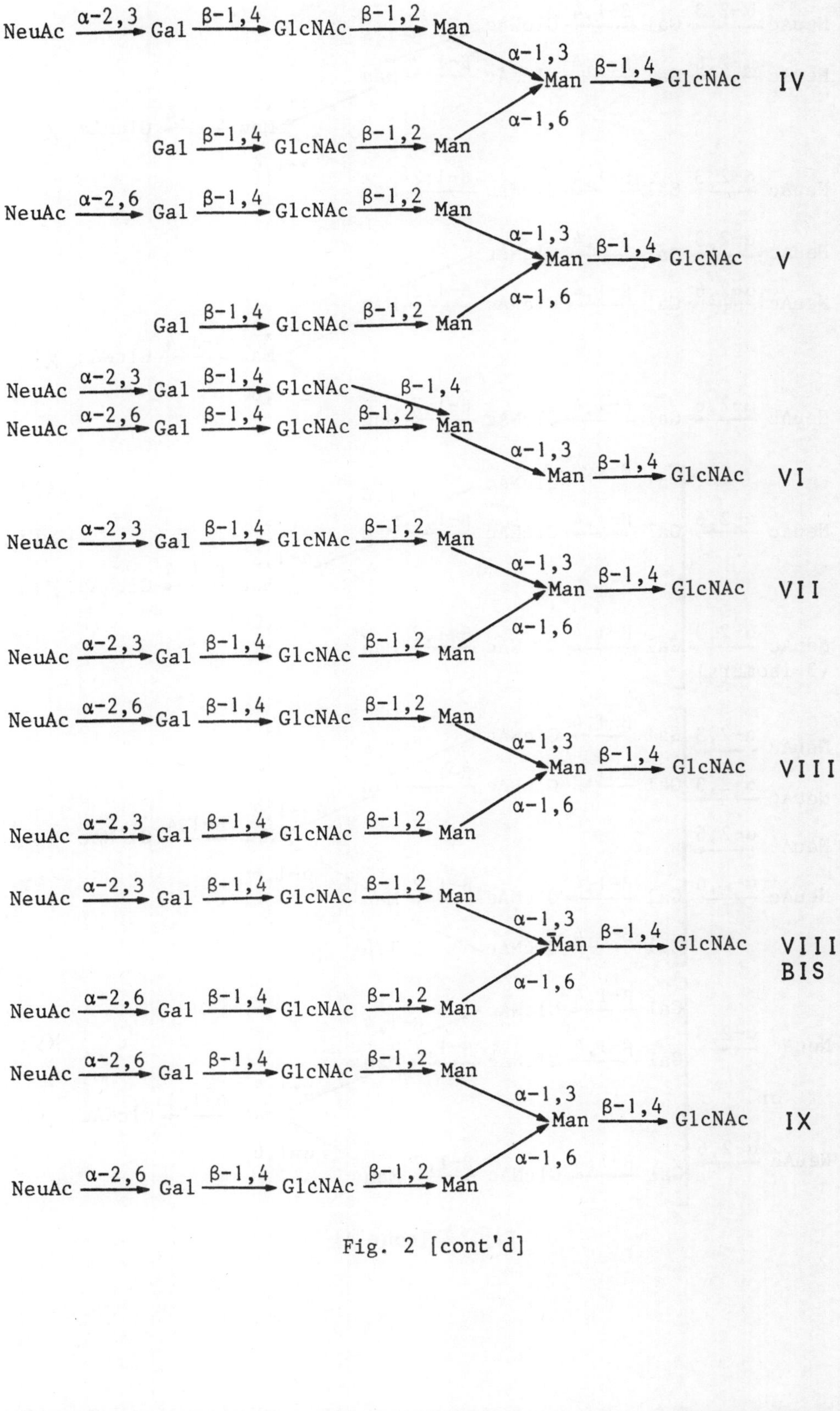

NeuAc $\xrightarrow{\alpha-2,3}$ Gal $\xrightarrow{\beta-1,4}$ GlcNAc $\xrightarrow{\beta-1,2}$ Man
$\xrightarrow{\alpha-1,3}$ Man $\xrightarrow{\beta-1,4}$ GlcNAc   IV
$\alpha-1,6$
Gal $\xrightarrow{\beta-1,4}$ GlcNAc $\xrightarrow{\beta-1,2}$ Man

NeuAc $\xrightarrow{\alpha-2,6}$ Gal $\xrightarrow{\beta-1,4}$ GlcNAc $\xrightarrow{\beta-1,2}$ Man
$\xrightarrow{\alpha-1,3}$ Man $\xrightarrow{\beta-1,4}$ GlcNAc   V
$\alpha-1,6$
Gal $\xrightarrow{\beta-1,4}$ GlcNAc $\xrightarrow{\beta-1,2}$ Man

NeuAc $\xrightarrow{\alpha-2,3}$ Gal $\xrightarrow{\beta-1,4}$ GlcNAc $\xrightarrow{\beta-1,4}$
NeuAc $\xrightarrow{\alpha-2,6}$ Gal $\xrightarrow{\beta-1,4}$ GlcNAc $\xrightarrow{\beta-1,2}$ Man
$\xrightarrow{\alpha-1,3}$ Man $\xrightarrow{\beta-1,4}$ GlcNAc   VI

NeuAc $\xrightarrow{\alpha-2,3}$ Gal $\xrightarrow{\beta-1,4}$ GlcNAc $\xrightarrow{\beta-1,2}$ Man
$\xrightarrow{\alpha-1,3}$ Man $\xrightarrow{\beta-1,4}$ GlcNAc   VII
$\alpha-1,6$
NeuAc $\xrightarrow{\alpha-2,3}$ Gal $\xrightarrow{\beta-1,4}$ GlcNAc $\xrightarrow{\beta-1,2}$ Man

NeuAc $\xrightarrow{\alpha-2,6}$ Gal $\xrightarrow{\beta-1,4}$ GlcNAc $\xrightarrow{\beta-1,2}$ Man
$\xrightarrow{\alpha-1,3}$ Man $\xrightarrow{\beta-1,4}$ GlcNAc   VIII
$\alpha-1,6$
NeuAc $\xrightarrow{\alpha-2,3}$ Gal $\xrightarrow{\beta-1,4}$ GlcNAc $\xrightarrow{\beta-1,2}$ Man

NeuAc $\xrightarrow{\alpha-2,3}$ Gal $\xrightarrow{\beta-1,4}$ GlcNAc $\xrightarrow{\beta-1,2}$ Man
$\xrightarrow{\alpha-1,3}$ Man $\xrightarrow{\beta-1,4}$ GlcNAc   VIII BIS
$\alpha-1,6$
NeuAc $\xrightarrow{\alpha-2,6}$ Gal $\xrightarrow{\beta-1,4}$ GlcNAc $\xrightarrow{\beta-1,2}$ Man

NeuAc $\xrightarrow{\alpha-2,6}$ Gal $\xrightarrow{\beta-1,4}$ GlcNAc $\xrightarrow{\beta-1,2}$ Man
$\xrightarrow{\alpha-1,3}$ Man $\xrightarrow{\beta-1,4}$ GlcNAc   IX
$\alpha-1,6$
NeuAc $\xrightarrow{\alpha-2,6}$ Gal $\xrightarrow{\beta-1,4}$ GlcNAc $\xrightarrow{\beta-1,2}$ Man

Fig. 2 [cont'd]

Fig. 2 [cont'd]

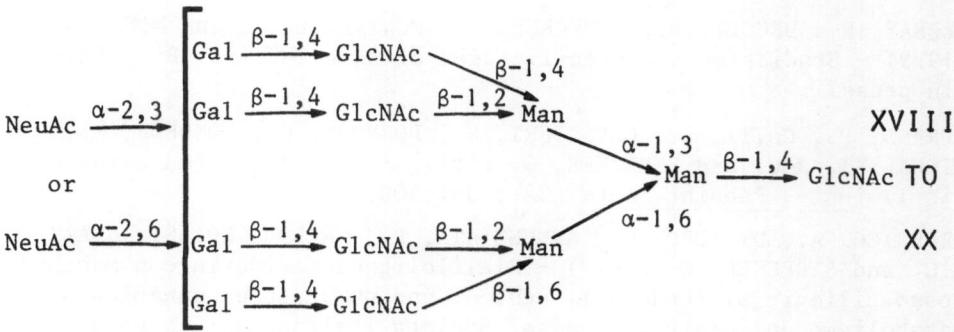

Figure 2 : Structures of sialyl-oligosaccharides accumulating in
the urine of patients with sialidosis.

CONCLUSION

    While mucolipidosis I and its variants are well defined as
sialidosis, the basic defect in mucolipidosis II and III is still
discussed. The neuraminidase defect in these disorders is generally
considered as one of many lysosomal hydrolase deficiencies, accompanied by a high extracellular activity of these enzymes. Our
hypothesis of a specific α-neuraminidase defect, in these cases,
is based on the fact that it is common to numerous tissues and
extracellular fluids, in opposite to other hydrolases activities.
The choice of the substrate for the α-neuraminidase determination
appears to be very important, in view of their peculiar specificities.

ACKNOWLEDGMENTS

    This work was supported by the Centre National de la Recherche
Scientifique (L.A. n° 217) and by the Institut National de la
Santé et de la Recherche Médicale (Contrat n° 78-1-052-3).

REFERENCES

ALHADEFF, J.A. and FREEZE, H. (1977) - Carbohydrate composition of
purified human liver α-L-fucosidase. Mol. Cell. Biochem., 18 : 33-
37

ASHWELL, G. and MORELL, A.G. (1974) - The role of surface carbohydrates in the hepatic recognition and transport of circulating
glycoproteins. Adv. Enz., 41 : 99-128

CANTZ, M., GEHLER, J. and SPRANGER, J. (1977) - Mucolipidosis I :
Increased sialic acid content and deficiency of an α-N-acetyl-
neuraminidase in cultured fibroblasts. Biochem. Biophys. Res.
Commun., 74 : 732-738

DEBRAY, H., DECOUT, D., STRECKER, G., MONTREUIL, J. and MONSIGNY, M.
(1979) - Studies on the specificities of some lectins. Biochimie
(in press)

DURAND, P., GATTI, S., CAVALIERI, S., BORRONE, C., TONDEUR, M.,
MICHALSKI, J-C. and STRECKER, G. (1977) - Sialidosis (mucolipidosis I). Helv. Paediat. Acta, 32 : 391-400

FEDERICO, A., DI TORO, G., ANNUNZIATA, P., CARLOMAGNO, S., GUAZZI,
G.C. and STRECKER, G. (1977) - Sialiloligosaccariduria con macchia
rosso ciliegia al fondo dell'ochio : una nuova turba genetica del
metabolismo delle glycoproteine. Societa Italiana per lo studio
degli errori congeniti del metabolismo, 2e Congresso Nazional San
Remo, 1-2 october 1977

HICKMAN, S. and NEUFELD, E.F. (1972) - A hypothesis for I-cell
disease : defective hydrolases that do not enter lysosomes.
Biochem. Biophys. Res. Commun., 49 : 992-999

HONDI-ASSAH, T., STRECKER, G. and FARRIAUX, J-P. (1975) - Carac-
teristics of sialyl-oligosaccharides isolated from the urine of a
patient with mucolipidosis II (I-cell disease). Abstr. Commun.
10th Meet. Eur. Biochem. Soc., Abs. n° 1560

KAPLAN, A., ACHORD, D.T. and SLY, W.S. (1977) - Phosphohexosyl
components of a lysosomal enzyme are recognized by pinocytosis
receptors on human fibroblasts. Proc. Natl. Acad. Sci. USA, 74 :
2026-2030

KELLY, T.E. and GRAETZ, G. (1977) - Isolated acid neuraminidase
deficiency : A distinct lysosomal storage disease. Am. J. Med. Gen.,
1 : 31-46

KOSEKI, M., WU, J.Y., TSURUMI, K. and NAGAI, Y. (1978) - Sialoglu-
ciduria in lysosomal diseases : quantitative and qualitative
analysis of urinary low molecular sialoglucides from patients with
mucopolysaccharidosis and with mucolipidosis. Tohoku J. Exp. Med.,
124 : 351-360

MAROTEAUX, P., POISSONNIER, M., TONDEUR, M., STRECKER, G. and
LEMONNIER, M. (1978a) - Sialidose par déficit en α-(2 → 6) neura-
minidase sans atteinte neurologique. Mucolipidose de type I ?.
Arch. Franç. Pediat., 35 : 280-291

MAROTEAUX, P., HUMBEL, R., STRECKER, G., MICHALSKI, J-C. and
MANDE, R. (1978b) - Un nouveau type de sialidose avec atteinte
rénale : la néphrosialidose. I. Etude clinique, radiologique et
nosologique. Arch. Franç. Pédiat., 35 : 819-829

MICHALSKI, J-C., STRECKER, G., FOURNET, B., CANTZ, M. and SPRANGER,
J. (1977) - Structure of sialyl-oligosaccharides excreted in the
urine of a patient with mucolipidosis I. FEBS-Lett., 79 : 101-104

MONTREUIL, J. (1975) - Recent data on the structure of the carbohydrate moiety of glycoproteins. Metabolic and biological implications. Pure and Appl. Chem., 42 : 431-477

NILSSON, B., NORDEN, N.E. and SVENSSON, S. (1978) - Structure of
the carbohydrate chains of fetuin. Proc. 4th Intern. Symp.
Glycoconjugates, Woods Hole, 1977, Acad. Press, New York (in press)

NISHIGAKI, M., MURAMATSU, T. and KOBATA, A. (1974) - Endoglycosi-
dases acting on carbohydrate moieties of glycoproteins : demonstration in mammalian tissue. Biochem. Biophys. Res. Commun., 59,
638-645

O'BRIEN, J.S. (1977) - Neuraminidase deficiency in the cherry red
spot-myoclonus syndrome. Biochem. Biophys. Res. Commun. 79 : 1136-
1141

O'BRIEN, J.S. (1978) - The cherry red spot-myoclonus syndrome.
Clin. Gen., 14 : 55-64

OKADA, S., KATO, T., MIURA, S., YABUUCHI, H., NISHIGAKI, M.,
KOBATA, A., CHIYO, H. and FURUKAMA, J.I. (1978) - Sialyloligosac-
chariduria in mucolipidosis : a method of diagnosis. Clin. Chim.
Acta, 86 : 159-168

POTIER, M., MAMELI, L., DALLAIRE, L. and MELANCON, B. (1977) -
Neuraminidase deficiency in mucolipidosis II (I-Cell disease)
established using 4-methylumbelliferyl-α-D-N-acetylneuraminic acid
substrate. Pediatr. Res., 11 : 462

SPRANGER, J., GEHLER, J. and CANTZ, M. (1977) - Mucolipidosis I :
A sialidosis. Am. J. Med. Gen., 1 : 21-29

SCHLESINGER, P.H., DOEBBER, T.W., MANDELL, B.F., WHITE, R.,
DESCHRYVER, C., RODMAN, J.S., MILLER, M.J. and STAHL, P. (1978) -
Plasma clearance of glycoproteins with terminal mannose and N-
acetylglucosamine by liver non-parenchymal cells. Biochem. J.,
176 : 103-109

STRECKER, G., HONDI-ASSAH, T., FOURNET, B., SPIK, G., MONTREUIL, J.,
MAROTEAUX, P., DURAND, P. and FARRIAUX, J-P. (1976a) - structure
des trois oligosaccharides majeurs accumulés dans l'urine de sujets
atteints de mucolipidose II et de deux nouveaux types de maladies
métaboliques. C.R. Acad. Sci., 282 : 671-673

STRECKER, G., HONDI-ASSAH, T., FOURNET, B., SPIK, G., MONTREUIL, J.,
MAROTEAUX, P., DURAND, P. and FARRIAUX, J-P. (1976b) - Structure
of the three major sialyl-oligosaccharides excreted in the urine of
five patients with three distinct inborn diseases : "I-Cell disease"
and two new types of mucolipidosis. Biochim. Biophys. Acta, 444 :
349-358

STRECKER, G., MICHALSKI, J-C., MONTREUIL, J. and FARRIAUX, J-P.
(1976c) - Deficit in neuraminidase associated with mucolipidosis II
(I-Cell disease). Biomed. Exp., 25 : 238-240

STRECKER, G., PEERS, M-C., MICHALSKI, J-C., HONDI-ASSAH, T., FOURNET, B., SPIK, G., MONTREUIL, J., FARRIAUX, J-P., MAROTEAUX, P. and DURAND, P. (1977) – Structure of nine sialyl-oligosaccharides excreted in urine of eleven patients with three different types of sialidosis. European J. Biochem., 75 : 391–403

STRECKER, G. and MICHALSKI, J-C. (1978a) – Biochemical basis of six different types of sialidosis. FEBS-Lett., 85 : 20–24

STRECKER, G., MICHALSKI, J-C., HERLANT-PEERS, M.C., FOURNET, B. and MONTREUIL, J. (1978b) – Structure of 40 oligosaccharides and glycopeptides accumulating in the urine from patients with catabolism defects of glycoconjugates (sialidosis, fucosidosis, mannosidosis and Sandhoff's disease). Proc. IVth Int. Symp. Glycoconjugates, 1977 (in press)

THOMAS, G.H., TIPTON, R.E., CH'IEN, L.T., REYNOLDS, L.W. and MILLER, C.S. (1978) – Sialidase (α-N-acetylneuraminidase) deficiency : the enzyme defect in an adult with macular cherry red spots and myoclonus without dementia. Clin. Gen., 13 : 369–379

THOMAS, G.H., TILLER, G.E., REYNOLDS, L.W., MILLER, C.S. and BACE, J.W. (1976) – Increased levels of sialic acid associated with a sialidase deficiency in I-cell disease (mucolipidosis II) fibroblasts. Biochem. Biophys. Res. Commun., 71 : 188–195

TAI, T., YAMASHITA, K., OGATA, A.M., KOIDE, N., MURAMATSU, T., IWASHITA, S., INOUE, Y. and KOBATA, A. (1975) – Structural studies of two ovalbumin glycopeptides in relation to the endo-β-N-acetylglucosaminidase specificity. J. Biol. Chem., 250 : 8569–8575

WILLCOX, P. and RENWICK, A.G.C. (1977) – Effect of neuraminidase on the chromatographic behaviour of eleven acid hydrolases from human liver and plasma. Eur. J. Biochem., 73 : 579–590

VLADUTIU, G.O. and RATTAZZI, M.C. (1975) – Abnormal lysosomal hydrolases excreted by cultured fibroblasts in I-cell disease (mucolipidosis II). Biochem. Biophys. Res. Commun., 67 : 956–964

SIALIDOSES (MUCOLIPIDOSES), CLINICAL PICTURES

Ulrich N. Wiesmann and Norbert Herschkowitz

Department of Pediatrics, University of Berne,
Freiburgstrasse 15
3010 Bern, Switzerland

INTRODUCTION

The mucolipidoses (I,II,III) have been recognized as a separate entity of diseases by Spranger and Wiedemann (1970) who separated them on the basis of biochemical, histological and clinical findings from the group of the mucopolysaccharidoses. Mucolipidosis I was first described in 1968 by Spranger et al. under the name lipomucopolysaccharidosis. The overall clinical picture was that of a Hurler-Hunter phenotype with normal muco-polysacchariduria, a cherry-red macular spot and signs of a neuro-degenerative process involving mainly the white matter (Spranger et al. 1968). Mucolipidosis II is identical with the so called inclusion cell disease (I-cell disease) according to the numerous cytoplasmic inclusions in the cultured fibroblasts of the patients as described by Leroy et al. (1969, 1971). It also is clinically resembling Hurler's syndrome with a remarkably stunted growth and severe retardation. Mucolipidosis III, formerly named pseudo-hurler polydystrophy by Maroteaux and Lamy (1966) presents clini-cally with short stature, mild Hurler-like dysmorphism and mental retardation. The term sialidosis was coined by Strecker et al. (1977) in analogy to mannosidosis and fucosidosis. He proposed to call the three types of mucolipidosis (I,II,II) sialidosis I, II, III respectively, in regard to the nature of the enzymatic defect and to the accumulation of sialic acid rich oligosaccharides.

SIALIDOSIS I

Two clearly distinct forms of Sialidosis I have been des-cribed (Spranger 1975, Durand et al. 1977; the first one described by Spranger is of the early onset and severe type, the other one

according to Durand et al. (1977) affects adolescent patients and
has a mild clinical course.

1. Early onset, severe type

At least four patients have been described in the last years
with this disorder (Spranger et al. 1975, Maroteaux et al. 1978).
In two of them a deficiency of α-neuraminidase was associated
with the syndrome. In another two cases described by Japanese
authors (Yamamoto et al. 1974, Orii et al. 1972) and in one case
published by Goldberg et al. (1971) the same disorder was suspected (Maroteaux et al. 1978).

Clinically this lysosomal storage disease may be difficult
to be delineated from other forms of oligosaccharidoses.
Specially at the onset of the symptoms other diseases like mano-
sidosis, fucosidosis as well as mucolipidosis II and III may have
to be considered. Table 1 summarizes the major clinical features
in four patients with Sialidosis I. The onset of the clinical
symptoms is in early childhood or late infancy, when psychomotor
retardation and Hurler like appearance are noticed. The degree of
mental retardation may be variable (Maroteaux et al. 1978) it may
aggravate because of impaired hearing. X-ray abnormalities of the
lumbar vertebrae and an enlarged or omega formed sella turtica
are common. Clinical symptoms that have been observed in all cases
are:

Table 1.   Major clinical features in 4 patients with Sialidosis I
           (Mucolipidosis I) (severe type)

	Spranger et al. (1975)		Bérard et al. (1968)	Maroteaux et al. (1978)
	R.H.	D.F.	?	?
Age(yrs)	12	12	15	8
Onset (yrs)	2	1	1	2
Hurler phenotype	+	+	+	+
Small stature	+	+	+	+
Hepatomegaly	(+)	(+)	+	+
Splenomegaly	−	−	+	+
Cherry red macular spot	+	+	+	+
Corneal opacities	−	+	+	−
Impaired hearing	(+)	+	+	+
Neurodegeneration	+	+	+	(+)
IQ	72	45	45	normal
Neuraminidase		↓↓		↓↓

1. The presence of a cherry-red macular spot indicating the storage of lipid material in the retina.
2. The enlargement of the liver also seems to be common feature. At the electronmicroscopical level the Kupffer cells accumulate large amounts of granular moderately osmiophilic material within vacuols. In contrast, the hepatocytes are only moderately affected (Maroteaux et al. 1978), they may contain some lipophilic material of a polymorphous character.
3. The bone marrow contains large cells that are ballooned by multiple vacuols that appear optically empty, similar to the ones found in Niemann-Pick's disease. The neurological abnormalities are shown in Table 2. Neurological symptoms, if present are hypotonia, cerebellar ataxia with nystagmus and tremor. Typical electroencephalographic alterations have been described by Doose et al. (1975). Our present knowledge of the inheritance of the disorder is compatible with an autosomal recessive transmission. All patients described excreted large amounts of sialic acid rich oligosaccharides in their urine. No elevation of acid-mucopolysaccharides was noticed. The final diagnosis is made by the determination of neuraminidase (α 2-6) neuraminidase which is found almost totally deficient (Cantz et al. 1977, Spranger et al. 1977, Maroteaux et al. 1978) in peripheral leukocytes and in cultured fibroblasts of the patients. All other lysosomal enzymes are normal or even elevated. There may be an exception in the activity of the β -galactosidase which was found high or normal in the liver of the two patients of Spranger (1975) but low in the japanese cases. There may be still further clinical heterogeneity among the patients with mucolipidosis I (Sialidosis I) since Maroteaux et al. (1978) described another patient with this disorder, lacking α(2-6) neuraminidase activity with a rapid mental deterioration and with renal insufficiency.

Table 2.   Neurological abnormalities in 4 patients with Sialidosis I (Mucolipidosis I) (severe type)

	Spranger et al. (1975)		Bérard et al. (1968)	Maroteaux et al. (1978)
	R.H.	D.F.		
Muscular hypotonia	+	+	+	+
Cerebellar ataxia	+	+	+	−
Tremor	+	+	+	−
"Specific EEG pattern"	+	+	+	−
Peripheral Neuropathy	+	+	?	?
Myelin degeneration	+	+	?	?

2. Late onset, mild type

Durand et al. (1977) described a 22 year old male and a 13 year old girl, children from consanguinous italian parents presenting with progressive visual impairment and with a cherry red macular spot as the leading symptoms (Table 3). They were both normally intelligent and of average stature with no skin, joint or bone abnormalities. Hyperreflexia was the only common neurological symptom. Laboratory examinations revealed cytoplasmic vacuoles in the peripheral lymphocytes and refringent inclusions in the cultured skin fibroblasts of the parents (Table 4).

Table 3.   Clinical manifestations of 2 patients with late onset mucolipidosis I (Sialidosis I) (Durand et al. 1977)

	De P.F.	De P.C.
Age (yrs) sex	22 male	13 female
Onset (yrs)	15	7
Progressive visual impairment	+	+
Green blindness	+	?
Cherry red macular spot	+	+
Hepatomegaly	-	+
Normal IQ	+	+
Normal hearing	+	+
Normal stature	+	+
Hyperreflexia	+	+

Table 4.   Laboratory findings in 2 patients with late onset mucolipidosis I (Sialidosis I) (Durand et al. 1977)

	De P.F.	De P.C.
Lymphocyte inclusions	+	+
Fibroblast inclusions	+	+
Sialo-oligosaccharide excretion(urine)	++	++
Leukocyte α-neuraminidase percent of normal	18%	20%

At the electronmicroscopical level Kupffer cells and hepatocytes in a liver biopsy were accumulating lysosomal material. In the urine very large amounts of sialic acid rich oligosaccharides were excreted. A profound deficiency of α-neuraminidase was found in the leukocytes and in the cultured fibroblasts. All other lysosomal enzyme activities in the leukocytes were either normal or slightly

elevated. In the leukocytes the residual activity of α-neuramini-
dase was about 20% of normal control values. Thus the mild clinical
course of the disorder in the two patients could be explained in
the same way as in many other lysosomal disorders where marked
differences in the onset and in the evolution of a disease is
found regardless of the fact that the diseases share the same
enzyme defect, but have different residual enzyme activities.

Another patient similar to the ones described by Durand et
al. (1977) was recently published by Endo et al. (1977) whose
clinical symptoms were beginning at 15 years of age with visual
troubles with a cherry red spot but who developed cerebellar
ataxia and died after a development of grand mal attacks at 22
years of age.

SIALIDOSIS II (MUCOLIPIDOSIS II)

Mucolipidosis II (Spranger et al. 1970) or I-cell disease
(Leroy et al. 1971) is an autosomal recessive disorder which is
expressed from birth by a facies of a mucopolysaccharidosis,
shallow orbits with protruding eyes and typical roentgenological
changes. Often orthopedic problems, such as clubfoot or bilateral
hip dislocations force the parents to seek professional help. In
boys, often inguinal hernias are operated which in one of our
cases led the pediatric surgeon suspect a mucopolysaccharidosis
(Wiesmann et al. 1973). In most patients the clinical picture is
fully developed at an age of 1 year (Table 5). The majority of
the patients described shows severe psychomotor retardation, some
may not even learn to sit upright without help and never walk.

Table 5.   Mucolipidosis II, leading clinical symptoms (Leroy
           et al. (1971)

- Facies reminiscent of a mucopolysaccharidosis
- shallow orbits
- clinical symptoms expressed from birth;
  slowly progressive, fatal outcome usually before
  4 years
- neonatal orthopedic problems and/or inguinal
  hernias in males
- psychomotor retardation
- severe growth failure
- early hypertrophy of gingival tissues
- progressive stiffness of all joints, first
  apparent in shoulders

The mental retardation usually is so severe that the patients may
not acquire vocalisation or speech. In an exceptional case however the patient was able to speak and to walk although presenting

all other somatic symptoms (Wiesmann et al. 1973). Linear growth
is stunted from birth and ceases about the age of one year so that
no patient with ML II has ever reached a length of more than 80 cm.
The growth may already be retarded in utero, since the birthweight
is often below 2500 g. In many patients the skin is thickened and
rigid especially in the cheeks. All patients are developing a
severe stiffness of all major joints beginning with the shoulder
joints. Severe corneal clouding and hepatomegaly are usually
absent (Leroy and Martin 1975). Some roentgenologic changes in
ML II are present at birth. During the progression of the disease
two successive stages can be differentiated. An early bone dysplasia typical for this disease is first expressed up to 6 months
of life as an extended periostal reaction of the long tubular
bones. A diffuse demineralisation of the bones also may be noted.
During the second phase (1 year) dysplasia may be centered in the
cranium, the spine and in the plevis, resulting in a retardation
of ossification, demineralisation and osteoporosis (Reither et al.
1977), (Patriquin et al. 1977).

     Recurrent pneumonia and upper respiratory infections are fre-
quent and become increasingly refractory to antibiotic treatment.
Death usually occurs in early childhood due to cardiorespiratory
failure. In a patient who died at 16 1/2 months of age a congestive heart failure due to coronary insufficiency accompanied
by mitral insufficiency was clinically noted (Patriquin et al.
1977). At autopsy the mitral valve, the left ventricule and the
coronary arteries were thickened by elastosis and deposition of
foam cells. Only the left side of the heart was involved. Another
patient who survived exceptionally long, died from congestive
heart failure at 11 years after several episodes of acute heart
failurees clinically presenting with a severe mitral insufficiency.

Organ manifestations

     The accumulation of material in ML II seems to be restricted
to primarily mesenchymal organs such as bone, skin, spleen and to
mesenchymal, mainly fibrocytic elements in all other organs. There
appears to be no important accumulation of substances in the brain
and in the peripheral nerves (Farriaux et al. 1976) although
Nagashima and coworkers (1977) found that the ganglia of the peripheral nervous system including the intestinal autonomic ganglia
were slightly swollen and contained intracellular lysosomal PAS
positive material. In the central nervous system neither morphological evidence of lipid storage nor the presence of degenerative
changes are severe enough to explain the grave mental retardation.
In the heart accumulation was observed in the fibrocytes of the
cardiac valves and also on the myocardial fibers in sarcolemmal inclusions containing a polymorphous material, often distending the
myofibril apart. Biochemically the main lipid corresponded to
ceramide trihexoside (CTH) (Nagashima et al. 1977). In the kidney

glomeruli great numbers of enlarged cells were found with a clear and foamy cytoplasm. They have been identified as visceral cells of the Bowman glomerular capsule with a hugely distended cytoplasma of the podocytes (Leroy and Martin 1975).

Biochemical changes

Because the affected cellular elements only represent a small fraction in most organs the biochemical abnormalities are expected to be detectable only in purely fibroblastic tissues. Lysosomal enzymes in organs have been measured only in a limited number of cases. In the skin a severe (Wiesmann et al. 1973, Leroy et al. 1972) in the liver (Farriaux et al. 1976) and in the brain (Martin et al. 1975) a 50% reduction of β-galactosidase activity was shown, whereas other lysosomal enzymes were essentially normal (Leroy et al. 1973, Suzuki et al. 1977) or slightly elevated. Studies on cultured fibroblasts from patients with ML II revealed the presence of large numbers of cytoplasmic inclusions (Leroy et al. 1969) who gave to this disorder the name inclusion cell disease or I-cell disease. A multiple lysosomal enzyme deficiency was later demonstrated in cultured fibroblasts by Lightbody et al. 1971 and by Leroy et al. 1972. Some lysosomal enzymes appear to be spared such acid phosphatase, α and β-glucosidase that are all three tightly membrane bound. The multiple lysosomal enzyme deficiency is due to the loss of lysosomal enzymes into the culture medium (Wiesmann et al. 1971a). We assume that this loss of enzyme is due to a structural modification of the lysosomal glycoprotein as the result of a post-translational error leading to the abnormal secretion of enzymes that in turn cannot be repinocytosed from cell (Wiesmann et al. 1974). Hickmann and Neufeld (1972) suggested as an alternative possibility that the lysosomal enymes are secreted and the re-

Table 6.   Changes of acid hydrolases in Mucolipidosis II
            according to Farriaux et al. 1976

	Serum	Leukocytes	Fibroblasts	Organs	Amniotic fluid
Hexoaminidase	↑↑	N	↓↓	N	↑↑
β-Glucuronidase	↑↑	N	↓↓	N	↑↑
β-Galactosidase	↑	N	↓↓	↓↓	↑↑
β-Glucosidase	N	N	N	N	N
α-Fucosidase	↑↑		↓↓	N	↑
Arylsulfatase A	↑↑↑		↓↓↓	N	↑↑
Mannosidase	↑↑	n.d.	↓↓	N	?
Neuraminidase	N	↓↓	↓↓↓	?	?

pinocytosed into the lysosomes as the normal route for incorporation into the lysosomes. In the case of mucolipidosis II
the lysosomal enzymes that are abnormal in their carbohydrate part
cannot be repinocytosed and therefore remain extracellular. This
observation of extracellularly increased lysosomal enzyme activities has not only be confirmed in vitro in cultured fibroblasts
but also in the extracellular fluids, urine, serum and cerebrospinal fluid of the patients (Wiesmann et al. 1971b, 1974) and is
presently used as a criterium in the diagnosis of ML II.

     Recently a profound deficiency of **α**-neuraminidase in cultured
fibroblasts together with a 3-4 fold increase of total neuraminic
acid over the controls have been reported by Thomas et al. (1976)
and by Strecker et al. (1976). Attempts to detect neuraminidase
activity in conditioned media from both I-cell and normal fibroblast cultures yielded negative results. Additionally no neuraminidase activity could be detected in serum from either I-cell
patients or normal individuals. Large amounts of neuraminic acid
rich oligosaccharides are excreted in the urines of the patients
(Strecker et al. 1977). Since **α**-neuraminidase is lacking in the
lysosomes there could be a higher proportion of neuraminic acid
(sialic acid) in the other lysosomal glycoproteins. This could
possibly explain the findings of Vladutiu and Rattazzi (1975) on
oversialidated lysosomal enzymes in serum of ML II patients,
demonstrating abnormal electrophoretic mobility,which could be
normalized by neuraminidase treatment. However it cannot explain
the observation of reduced absorptive pinocytosis of the lysosomal
enzymes in this disorder (Wiesmann et al. 1974, Hickman 1972) because neuraminidase treatment does restore normal electrophoretic
mobility but not the normal uptake into the cells. Thus the
primary defect in mucolipidosis II still remains to be elucidated.

Prenatal diagnosis

     Mucolipidosis II (Sialidosis II) can be diagnosed prenately
by the analyses of lysosomal hydrolases in amniotic fluid that
are manfold increased in their activity. In cultured amniotic
cells lysosomal enzyme activities are decreased and the accumulation and the degradation of $^{35}SO_4$ sulfate labelled mucopolysaccharides are greatly abnormal (J. Gehler et al. 1976, Matsuda
1975, Warren 1973). Lysosomal storage process starts in utero and
the placenta can be used for pathological studies (Erickson et al.
1975).

SIALIDOSIS III (MUCOLIPIDOSIS III)

     Mucolipidosis III is an autosomal recessive disorder that
resembles a mucopolysaccharidosis but the patients secrete normal
amounts of acid mucopolysaccharides in the urine. First described
by Maroteaux and Lamy (1966) as Pseudohurler polydystrophy the

syndrome was considered by Spranger and Wiedemann (1970) as type III of the clinical mucolipidoses presenting with short stature, restricted joint mobility, mild Hurler'like dysmorphism, mental retardation and peculiar skeletal abnormalities of the pelvis and spine. In a larger survey 1975 Kelly et al. (1975) described 12 cases of ML III. They state that the disorder becomes clinically manifest at 2-4 years of life by stiffness of the hands and shoulders which could be misdiagnosed for an attack of rheumatoid arthritis. Formation of claw hand deformities and progressively retarded growth becomes apparent before the age of 10 years. Unilateral or more often bilateral carpal tunnel syndrome seems to be a frequent complication, leading to secondary peripheral neuropathy. Table 7 shows the frequency by which the major clinical symptoms were present (Kelly et al. 1975).

Table 7. Clinical symptoms in 12 patients with Mucolipidosis III

Sex	7 males + 5 females		
Height below 3 Percentile	11	/	12
Corneal clouding	11	/	12
Carpal tunnel syndrome	8	/	12
Cardial murmor	12	/	12
IQ below 80	6	/	12
Claw hands	12	/	12
Severe restriction of mobility	8	/	12
Coarse facies (severe)'	8	/	12
Gibbus	4	/	12
Lordosis	5	/	12

(modified from Kelly et al. 1975)

The youngest patient at the time of the study was 6, the oldest patient 28 years old. The mental retardation, the restriction of joint mobility (claw hands) and coarseness of the facies are slowly progressing. Corneal clouding also can be missing (Leroy and Van Elsen 1975). The dysostosis multiplex seems to be more expressed in the males than in the females. The skeletal changes that can be similar to those in Marquio's syndrome begin with small irregular proximal epiphyses with subluxations, coxa valga and odontoid hypoplasia. Severe skeletal abnormalities in ML III before one year of life have been reported by Nolte and Spranger (1976).

Organ manifestations

Liver and spleen usually are not very much enlarged or normal in size. The organ manifestations in ML III appear to be similar as in ML II, but generally milder. The accumulation in lysosomes

also restricted to mesodermal organs and to mesenchymal cells in
other organs. Detailed histological studies on biopsies or
autopsy material are missing. Lymphocytic vacuoles and inclusions
in the cells of the bone marrow may be absent (Glaser et al. 1974).
Clinically the disorder can be mistaken for mucopolysaccharidosis
VI or for one of the adolescent forms of genetic β-galactosidase
deficiencies (Kelly et al. 1975).

Biochemical changes

Patients with ML III excrete in their urines markedly increased quantities of glycoproteins (Berman et al. 1974) containing an electrophoretically fast moving component which is absent in normal urine. Sialic acid rich oligosaccharidides were
also found to be excreted in excess of normal (Strecker et al.
1977).

In the patients described by Berman et al. (1974) elevated
amounts of protein, sialic acid, amino-sugars and neutral sugars
were present in the urine.

In the serum of the patients, lysosomal enzyme activities
are in great excess of normal. Specially that of arylsulfatase A
and of N-acetyl-β-glucosaminidase (Kelly et al. 1975, Thomas et
al. 1973, Leroy and Van Elsen 1975).

Cultured fibroblasts of ML III patients, similarly to fibroblasts of ML II (I-cell disease) patients also show the I-cell phenomenon which consists of numerous cytoplasmic inclusions representing enlarged lysosomes (Leroy et al. 1968). The fibroblasts
also accumulate $^{35}SO_4$ labelled mucopolysaccharides due to faulty
degradation (Glaser et al. 1974, Wiesmann et al., unpublished results).

In the cultured fibroblasts lysosomal enzyme activities are
markedly reduced (Berman et al. 1974, Kelly et al. 1975, Leroy
and Van Elsen 1975) with the exception of acid phosphatase and
β-glucosidase (Wiesmann unpublished). Concomitantly with the reduction of cellular acid hydrolase activities the conditioned
culture media contain acid hydrolases in great excess (Bermann
et al. 1974, Kelly et al. 1975). Although it appears the residual
activities of lysosomal enzymes in fibroblasts from ML III
patients are somewhat higher than in ML II, this as a whole seems
to be a soft diagnostic criterium. Leroy and O'Brien however (1976)
have suggested that a real difference in residual enzyme activity
may present for β-galactosidase being 3 times higher in ML III
fibroblasts than in ML II cells. At present the biochemical
diagnosis of ML III is based on the finding of elevated activities
of acid hydrolases in the sera of the patients and on the biochemical changes in the cultured fibroblasts. According to those

criteria a prenatal diagnosis seems possible.

Heterozygotes for ML III according to Leroy and Van Elsen (1975) can be diagnosed by difference in the heat inactivation of N-acetly- β -glucosaminidase activity in the serum. The percent of heat labile enzyme in the sera of ML III patient is of the order of 25%. In normal sera this portion of enzyme activity is around 50-60%. In the heterozygote intermediate values were found.

Pathogenetic considerations

As the multiple acid hydrolase deficiencies in cultured fibroblasts as well as the concomitantly increased activities in the culture media and in the sera of the patients are only secondary manifestations the primary genetic defect remains to be discovered. Similarly as in ML II, lysosomal enzymes found increased in the extracellular fluids in vitro and in vivo in ML III are not normally taken up by cultured fibroblasts (Glaser et al. 1974). This is likely due to an abnormal composition of the carbohydrate portion of the enzyme glycoproteins which serves as a recognition marker for these enzymes (Hickman et al. 1972, Wiesmann et al. 1974). There is an additional interesting observation made by Glaser et al. (1974) using white blood cell extract from controls and from ML III patients as a source of β -glucuronidase enzyme to be added to culture medium and to be taken up by β -glucuronidase deficient fibroblasts. Although the activity of β -glucuronidase in the patient white cell was normal the enzyme was taken up very poorly compared to normal white cell β -glucuronidase. This might suggest that the enzyme abnormalities also extends to non fibroblast cells, although the mechanism leading to enzyme loss is not expressed. The posttranslational defect in lysosomal enzymes on the level of the carbohydrate portion responsible for the impaired cellular uptake also could result in the instability of some lysosomal enzymes e.g. α -neuraminidase and thus explain the total absence of that enzyme both in the leukocytes and in the fibroblasts without concomitant increase of its activity in the extracellular fluids.

Final considerations of the clinical and the biochemical criteria for the diagnosis of sialidosis

In order to summarize and to compare the clinical expressions of the three mucolipidoses the clinical diagnostic criteria are compiled in Table 8. Although all three disorders seems to share amoung variable other biochemical abnormalities (Table 9), the deficiency in α -neuraminidase and the excretion of large amounts of sialic acid rich oligosaccharides it appears at this time to be very difficult to explain the clinical expression of the disease by a common pathogenetic mechanism in respect to α - neuraminidase. Therefore further research is necessary in the

Table 8.   Clinical diagnostic criteria for ML I, II, III

	ML I		ML II	ML III
	S	M		
Onset (yrs)	1-2	7-5	1	2-4
Short stature	++	-	++++	++
Progressive joint problems	-	-	+++	++
Mental retardation	severe	none	severe	mild
Corneal opacities	(+)	-	(+)	(+)
Dysostosis multiplex	+	-	+++	++
Heart involvement	-	-	+++	+
Neurological symptoms	Ataxia	-	-	-
Cherry red macular spot	++	++	-	-
Hearing problems	++	-	+	-

S = severe form, M = mild form

Table 9.   Biochemical, diagnostic criteria for ML I, II, III

	ML I		ML II	ML III
	S	M		
Normal MPS-uria	+	+	(+)	+
Increased urinary sialic acid rich oligosaccharides	++	++	++	++
Elevation of serum acid hydrolases	-	-	++	+
Cultured fibroblasts				
- cytoplasmic inclusions	+	+	+++	++
- multiple acid hydrolase deficiency	-	-	+++	++
- abnormal radioactive sulfate kinetics	-	-	++	++
- elevated acid hydro- lases in media	-	-	+++	+++
- deficiency of -neuraminidase	+	+	+	+

S = severe form, M = mild form

field of mucolipidoses (sialidoses) before we can decide whether the term sialidosis should be carried on for all three types of diseases.

BERARD M., TOGA M., BERNARD R., DUBOIS D., MARIANI R. and HASSOUN J. (1968): Pathological findings in one case of neuronal and mesenchymal storage disease. Its relationship to lipidoses and to mucopolysaccharidoses. Path. Europ. 3, 1972-183.

BERMAN E.R., KOHN G. and STEIN H. (1974): Acid hydrolase deficiencies and abnormal glycoproteins in mucolipidosis III (Pseudohurlerpolydystrophy). Clin. Chim. Acta 52, 115-124.

CANTZ M., GEHLER J. and SPRANGER J. (1977): Mucolipidosis I: increased sialic acid content and deficiency of an α-N-acetyl-neuraminidase in cultured fibroblasts. Biochem. Biophys. Res. Commun. 74, 732-738.

DOOSE H., SPRANGER J. and WARNER M. (1975): EEG in Mucolipidosis I. Neuropädiatrie 6, 98-101.

DURAND P., GATTI R., CAVALIERI S., BORRONE C., TONDEUR M., MICHALSKI J.C. and STRECKER G. (1977): Sialidosis (Mucolipidosis I). Helv. Paediat. Acta 32, 391-400.

ENDO H., AL-SAMARRAI S., SAKAKIBARA K., NAGASHIMA K. and SHIMADA Y. (1977): A new type of mucolipidosis associated with hereditary thrombocytopathy and color blindness. Acta Path. Jap. 27, 421-434.

ERICKSON R.P., PFLENGER O.H., SANDMAN R. and HALL B.D. (1975): Placental pathology in Mucolipidosis II. Birth Defects orig. art. series 11, 365.

FARRIAUX J.P., WALBAUM R., HONGRE J.F. and DUBOIS O. (1976): La mucolipidose II on "I-cell disease". Revue critique de 5 observations personelles. Lille med. 21, 51-70.

GEHLER J., CANTZ M., STOECKENINS M. and SPRANGER J. (1976): Prenatal diagnosis of Mucolipidosis II (I-cell disease). Europ. J. Pediat. 122, 201-206.

GLASER J.H., McALISTER W.H. and SLY W.S. (1974): Genetic hetero-geneity in multiple lysosomal hydrolase deficiency: Mucolipidosis II and III. J. Ped. 85, 192-198.

GOLDBERG M.F., COTHIER E., FICHENSCHER L.G., KENYON K., ENAT R., BOROWSKY S.A. (1971): Macular cherry red spot, corneal clouding and galactosidase deficiency. Clinical, biochemical and electronmicroscopic study of a new autosomal recessive storage disease. Arch. Intern. Med. 128, 387-398.

HICKMAN S. and NEUFELD E.F. (1972): A hypothesis for I-cell disease: Defective hydrolases that do not enter lysosomes. Biochem. Biophys. Res. Commun. 49, 992-999.

KELLY T.E., THOMAS G.H., TAYLOR H.A., McKUSICK V.A., SLY W.S., GLASER J.H., ROBINSON M., LUZZATTI L., ESPIRITA C., FEINGOLD M., BULL M.J., SKENHURST E.M. and IVES E.J. (1975): Mucolipidosis III (Pseudo-Hurler Polydystrophy): clinical and laboratory studies in a series of 12 patients. J. Hopkins Med. J. 137, 156-175.

LEROY J.G., DE MARS R.I. and OPITZ J.M. (1969): I-cell disease.
  Birth Defects, Orig.art. series 4, 174-185.
LEROY J.G., SPRANGER J.W., FEINGOLD M., OPITZ J.M. and CROCKER A.C.
  (1971): I-cell disease: a clinical picture. J. Pediat. 79,
  360-365.
LEROY J.G., HO M.W., McBRINN M.C., ZIELKE K., JACOT J. and
  O'BRIEN J.S. (1972): I-cell disease: Biochemical studies.
  Ped. Res. 6, 752-757.
LEROY J.G. and MARTIN J.J. (1975): Mucolipidosis II (I-cell
  disease): present status of knowledge. Birth Defects, Orig.
  art. series 11, 283-293.
LEROY J.G. and VAN ELSEN A.F. (1975): Natural history of a Muco-
  lipidosis. Twin girl discordant for ML III.
  Birth Defects. Orig. art. series 11, 325-334.
LEROY J.G. and O'BRIEN J.S. (1976): Different residual activity
  of β-galactosidase in cultured fibroblasts. Clin. Genet. 9,
  533-539.
LIGHTBODY J., WIESMANN U., HADORN B. and HERSCHKOWITZ N.(1971):
  I-cell disease: multiple lysosomal enzyme defect. Lancet 1, 451.
MAROTEAUX P. and LAMY M. (1966): La pseudopolydystrophie de
  Hurler. Presse med. 74, 2889-2892.
MAROTEAUX P., POISSONIER M., TONDEUR M., STRECKER M. and
  LEUMONNIER M. (1978): Sialidose par déficit en α (2-6) neuramini-
  dase sans atteinte neurologique. Arch. Fr. Pédiat. 35, 290-291.
MARTIN J.J., LEROY J.G., FARRIAUX J.P., FONTAINE G., DESNICK R.J.
  and CABELLO A. (1975): I-cell disease (Mucolipidosis II).
  A report on its Pathology. Acta neuropath. 33, 285-305.
MATSUDA I., ARASHIMA S., MITSUYAMA T., OKA Y., IKENCHI T.,
  KANEKO Y., ISHIKAWA M. (1975): Prenatal diagnosis of I-cell
  disease. Humangenetik 30, 69-73.
NAGASHIMA K., SAKAKIBARA K., ENDO H., KONISHI Y., NAKAMURA N.,
  SUZUKI Y. and ABE T. (1977): I-cell disease (mucolipidosis II)
  pathological and biochemical studies of an autopsy case.
  Acta Path. Jap. 27, 255-264.
NOLTE K. and SPRANGER J. (1976): Early skeletal changes in Muco-
  lipidosis III. Ann. Radiol. 19, 151-159.
ORII T., MINAMI R., SUKEGAWA K., SATO S., TSUGAWA S., HORINO K.,
  MIURA R. and NAKAO T. (1972): A new type of mucolipidosis with
  β-galactosidase deficiency and glycopeptiduria. Tokoku J. Exp.
  Med. 107, 303-315.
PATRIQUIN H.B., KAPLAN P., KIND H.P. and GIEDION A. (1977):
  Neonatal mucolipidosis II (I-cell disease): clinical and radio-
  logical features in 3 cases. Am. J. Roentgenol. 129, 37-43.
REITHER M., ZIMMERMANN G., GEHLER J., GATHMANN H. and TULUSAN H.
  (1977): Mucolipidosis II. Paediat. Praxis 18, 601-608.
SPRANGER J., WIEDEMANN H.R., TOLKSDORF M., GRAUCOB E. and
  CAESAR R. (1968): Lipomucopolysaccharidose. Z. Kinderheilk. 103,
  285-306.

SPRANGER J.W. and WIEDEMANN H.R. (1970): The genetic mucolipidoses. Dianosis and differential diagnosis. Human. Genet. 9, 113-139.

SPRANGER J. (1975): Mucolipidosis I. Birth Defects. Orig. arts. series 11, 279-282.

SPRANGER J., GEHLER J. and CANTZ M. (1977): Mucolipidosis I. - A Sialidosis. Amer. J. Med. Genetic 1, 21-29.

STRECKER G., HONDI-ASSAH T., FOURNET B., SPIK G., MONTREUIL J., MAROTEAUX P., DURAND P. and FARRIAUX J.P. (1976): Structure of 3 major sialyloligosaccharides excreted in the urine of 5 patients with three distinct inborn diseases. I-cell disease and two new types of Mucolipidosis.Biochem. Biophys. Acta 444, 349-358.

STRECKER G., PEERS M.C., MICHALSKI J.C., HODNI ASSAH T., FOURNET B., SPIK G., MONTREUIL J., FARRIAUX J.P., MAROTEAUX P. and DURAND P. (1977): Structure of nine sialyloligosaccharides accumulated in urine of eleven patients with three different types of sialidosis. Europ. J. Biochem. 75, 391-403.

SUZUKI Y., FUKUOKA K., WEY J.J. and HANDA S. (1977): Beta-galactosidase in mucopolysaccharidoses and mucolipidoses. Deficiency of GM_1 beta-galactosidase in liver and leukocytes. Clin. Chim. Acta 75, 91-97.

THOMAS G.H., TAYLOR H.A., REYNOLDS L.W. and MILLER C.S. (1973): Mucolipidosis III (Pseudo-Hurler polydystrophy). Multiple lyso-somal enzyme abnormalities in serum and cultured fibroblast cells. Ped. Res. 7, 751-756.

THOMAS G.H., TILLER G.E., REYNOLDS L.W., MILLER C.S. and BACE J.W. (1976): Increased levels of sialic acid associated with a sialidase deficiency in I-cell disease (Mucolipidosis II) fibro-blasts. Biochem. Biophys. res. Commun. 71, 88-95.

VLADUTIU G.D. and RATTAZZI M.C. (1975): Abnormal lysosomal hydro-lases excreted by cultured fibroblasts in I-cell disease (Mucolipidosis II). Biochem. Biophys. Res. Commun. 67, 956-964.

WARREN R.J. (1973): Antenatal diagnosis of mucolipidosis II (I-cell disease). Ped. Res. 7, 343-349.

WIESMANN U.N., LIGHTBODY J., VASSELLA F. and HERSCHKOWITZ N. (1971a): Multiple lysosomal enzyme deficiency due to enzyme leakage? New Engl. J. Med. 284, 109.

WIESMANN U.N., VASSELLA F. and HERSCHKOWITZ N. (1971b). I-cell disease: leakage of lysosomal enzymes into extracellular fluids. New Engl. J. Med. 285, 1090.

WIESMANN U.N., VASSELLA F. and HERSCHKOWITZ N. (1973): Mucolipi-dosis II (I-cell disease): A clinical and biochemical study. Acta Paediat. Scand. 63, 9-18.

WIESMANN U.N. and HERSCHKOWITZ N. (1974): Studies on the patho-genetic mechanism of I-cell disease in cultured fibroblasts. Pediat. Res. 8, 865-870.

YAMAMOTO A., ADACHI S., KAWAMURA S., TAKAHASCHI M., KITANI T., OKTORI T., SHINJI Y. and NISHIKAWA M. (1974): Localized β-galactosidase deficiency. Occurrence in cerebellar ataxia with myoclonus epilepsy and macular cherry red spot. A new variant of GM_1-gangliosidosis. Arch. Intern. Med. 134, 627-631.

SIALIDASE IN BRAIN AND FIBROBLASTS IN THREE

PATIENTS WITH DIFFERENT TYPES OF SIALIDOSIS

Bernhard Pallmann and Konrad Sandhoff

Max-Planck-Institut für Psychiatrie,
Neurochemische Abteilung, 8000 München 40, FRG

Bruno Berra

Chair of Biochemistry, College of Pharmacy,
University of Milan, Italy

Tadashi Miyatake

Department of Neurology, Jichi Medical School,
Yakushiji, Tochigi, Japan

INTRODUCTION

In recent years numerous diseases have become known which are characterized by an elevated urinary excretion of sialooligosaccharides and a deficiency of sialidase in cultured fibroblasts (for literature, see SPRANGER, GEHLER and CANTZ, 1977; STRECKER and MICHALSKI, 1978; MAROTEAUX, 1978).

These diseases are collectively called sialidoses of which two main groups can be distinguished. One group, with a single enzyme deficiency, a defect of α-neuraminidase (E.C. 3.2.1.18) demonstrable in cultured fibroblasts, comprises mucolipidosis I (ML I) (CANTZ, GEHLER and SPRANGER, 1977), cherry-red spot myoclonus syndrome (GOLDSTEIN *et al.*, 1974; O'BRIEN, 1978; THOMAS *et al.*, 1978; RAPIN *et al.*, 1978) and some related but as yet unclassified cases attributable to one or more forms of this group (MAROTEAUX and HUMBEL, 1976; DURAND *et al.*, 1977; KELLY and GRAETZ, 1977; STRECKER *et al.*, 1977; WENGER, TARBY and WHARTON, 1978).

Some of the affected individuals exhibit a marked reduction of acid β-galactosidase activity (TONDEUR *et al.*, 1971; LEROY *et al.*,

1972; HOLMES *et al.*, 1975; WENGER *et al.*, 1978). Sialidase deficiency
can be readily demonstrated in patients' cultured fibroblasts by assaying with water-soluble sialooligosaccharides as substrates. Obligate heterozygotes for ML I have sialidase activities intermediate
between those of normal individuals and of patients.

The second group of sialidoses includes mucolipidosis II (ML II)
and mucolipidosis III (ML III), both of which are characterized by
multiple lysosomal enzyme deficiencies in cultured fibroblasts and
a concomitant increase of several lysosomal enzyme activities in the
culture medium (LEROY *et al.*, 1972; WIESMANN and HERSCHKOWITZ, 1974;
THOMAS *et al.*, 1973). The re-uptake of these enzymes by ML II cells,
as well as by fibroblasts of other origins, is impaired. HICKMANN
and NEUFELD (1972) have suggested that the primary defect in ML II
is a structural abnormality of the lysosomal enzymes secreted. This
abnormality may reside in a defective assembly of the enzyme carbohydrate moieties (VLADUTIU and RATTAZZI, 1975), the nature of which
is still largely unknown. It has been proposed that sialidase deficiency in fibroblasts of ML II (MICHALSKI and STRECKER 1977;
STRECKER *et al.*, 1976; THOMAS *et al.*, 1978) and ML III (MICHALSKI
and STRECKER, 1977) may result in over-sialidation of lysosomal
enzymes, and masking of their recognition site for pinocytosis by
cultured skin cells (STRECKER and MICHALSKI, 1978). However, desia-
lidation of lysosomal enzymes of ML II fibroblasts did not increase
the rate of their pinocytosis (VLADUTIU and RATTAZZI, 1978; SPRITZ,
COATES and LIEF, 1979). Recent reports suggest that mannose and/or
mannose phosphate residues of lysosomal enzymes may be involved in
their recognition and pinocytosis by cultured fibroblasts (KAPLAN,
ACHORD and SLY, 1977; ULLRICH *et al.*, 1978).

Common to both groups of sialidoses is a deficiency in cultured
skin cells of sialidase assayed with water-soluble substrates. Organ
and substrate specificity as well as severity of the enzyme defect
are still unknown. Therefore, we tried to approach experimentally
the questions as to whether the enzyme deficiency can also be demonstrated with lipophilic substrates and whether the same defect is
present in solid tissues, such as brain, thus affecting the catabolism of sialoglycoconjugates in this organ.

MATERIALS

Anion exchange resin (AG 1-X8, 140-325 mesh, formate form) of
analytical grade was obtained from Bio-Rad Laboratories (Richmond,
Calif., USA). Silica gel-60 TLC-plastic sheets (Art. 5748) were from
Merck (Darmstadt, FRG) and all 4-methylumbelliferyl derivatives from
Koch & Light Laboratories (Colnbrook, England). [^{3}H]-Sialyllactitol
(50 Ci/mol) was purchased from NEN (Dreieich, FRG) and diluted with
unlabelled sialyllactose (Boehringer, Mannheim, FRG) to yield a
specific activity of 0.5 Ci/mol. [^{3}H]-Ganglioside GD1a was labelled
in its ceramide moiety as described (SCHWARZMANN, 1978) and purified
by preparative thin layer chromatography (adsorbent: Kieselgel G-60;

solvent system: n-butanol/pyridine/water 6:3:2 v/v/v, containing
0.1 % KCl). The specific activity of the [^{3}H]-ganglioside GDla was
41 Ci/mol. All other chemicals were of the highest analytical grade
available. Autopsy brain specimens of patients (for case reports see
below) and controls without neurological involvements were stored
at -20°C for several years.

METHODS

Enzyme assays: Lysosomal hydrolases were assayed as described
earlier (CHRISTOMANOU, CAP and SANDHOFF, 1977), in the case of α-
mannosidase the procedure of MAIRE et al. (1978) was followed. Sialidase activity towards ganglioside GDla as substrate was measured
as described (SCHRAVEN et al., 1977) with slight modifications. The
incubation mixture of 100 μl contained: 10 nmol [^{3}H]-ganglioside
GDla, 80-220 μg of protein and 20 μl of 0.5 M glycolate buffer, pH
3.8, containing 25 mM CaCl$_2$ and 675 mM NaCl. The mixture was incubated for 1 h at 37°C. With sialyllactitol as substrate sialidase
activity was assayed as follows. The incubation mixture of 100 μl
contained 200 nmol [^{3}H]-sialyllactitol, 50-300 μg of protein and
20 μl of the glycolate buffer as given above. After an incubation
period of 1-5 h at 37°C the enzymic reaction was stopped by cooling
on ice. An aliquot of 80 μl was subjected to an ion-exchange chromatography with 2 ml of resin (AG 1-X8 in formate form). The [^{3}H]-
lactitol was eluted with 2 ml of 1 mM lactitol, undegraded [^{3}H]-
sialyllactitol with 2 ml of 0.5 M Na-formate. After adding 17.5 ml
of Unisolve 100 (Koch & Light Lab., Colnbrook, England) the radio-
activities of both eluates were detected in a liquid scintillation
counter (Mark II; Nuclear Chicago). Controls were run with heat-
inactivated enzyme preparations (10 min, 100°C). Homogenates of
fibroblasts, leucocytes and brain tissues (10 % w/v) were prepared
at 4°C with a glass homogeniser provided with a mechanically driven
Teflon pestle (50 strokes, 600 rev/min, radial clearance: 0.2 mm)
in 10 mM Tris/HCl buffer, pH 7.2. Aliquotes were centrifuged at
100,000 g for 1 h and the pellets resuspended in the original volume
of 10 mM Tris/HCl buffer, pH 7.2. Protein content was measured according to LOWRY et al.(1951) with crystalline bovine serum albumin
as standard.

Determination of lipid- and protein-bound sialic acid: Gangliosides were prepared from human brain. Deep-frozen brain samples
were extracted with buffered tetrahydrofuran according to TETTAMANTI
et al. (1973). After partitioning with diethyl ether and water, the
ganglioside-containing aqueous phase was concentrated, dialyzed and
purified by Silica gel H column chromatography (solvents: chloroform-
methanol-water 55:20:3 followed by chloroform-methanol-water 60:35:7).
Lipid-bound sialic acid and individual gangliosides were determined
as described (BERRA et al., 1978). Protein-bound sialic acid was
determined in the defatted residue which was hydrolyzed in four
successive steps with 0.1 N sulfuric acid at 80°C. The supernatants

Table 1. Sialidase Activity with GDla Ganglioside as Substrate in
          Postmortem Brains of Patients with ML I and ML II

	Homogenate	Pellet	Supernatant
	(nmol/h/mg protein)		
Control (n = 2, 24 & 26 y)	12.0 (9.0-14.5)	11.5 (5.6-17.5)	1.6 (0.5-2.6)
ML I	8.7 (8.2-9.0)	8.4 (7.1-9.7)	0.9 (0.4-1.4)
ML II	6.3 (2.5-9.0)	3.5 (3.3-3.7)	0.7 (0.2-1.1)

were collected and adjusted with water to a volume of 50 ml. An aliquot was subjected to chromatography on an anionic resin (Dowex 2-
X8; 0.6 x 6 cm). Sialic acid was eluted with 8 ml of 2 M acetate
buffer, pH 4.6, and quantified (WARREN, 1959). Total sialic acid
content was determined accordingly after hydrolysis of finely ground
tissue samples.

For determination of sialooligosaccharides, brain samples were extracted with 9 vol of water. Extracts were lyophilized, extracted
with a small volume of pyridine and chromatographed according to
HUMBEL and COLLART (1975).

Case Reports

1) Adult type neuronal storage disease. A detailed description of
the patient has been given by MIYATAKE *et al*. (1979b). Briefly, the
patient was a 48 year old mentally retarded male with progressively
severe action myoclonus, cerebellar ataxia, generalized convulsive
seizures and decreased visual acuity. Macular cherry-red spots were
found in both fundi. Gangliosides GM3 and GM2 were increased 49- and
23-fold, respectively, in the sympathetic ganglia. Sialooligosaccharides were excreted into the urine. Fibroblasts were cultured from a
skin biopsy of the patient and controls.

2) Mucolipidosis I. An extensive case report of the male patient
has been published by SPRANGER *et al*. (1968). In summary, the patient
exhibited gargoyle-like features, skeletal dysplasia, slow progressive
cerebral deterioration and spasms of various muscles. He died at the
age of 21 years before an enzyme analysis had been performed. Total
sialic acid content was 10-fold increased in postmortem liver. Sialidase activity in parental fibroblasts was in the range of about
50 % of the control (SPRANGER *et al*., 1977).

Table 2. Sialidase Activity with Sialyllactitol as Substrate in
          Postmortem Brains of Patients with ML I and ML II

	Homogenate	Pellet	Supernatant
	(nmol/h/mg protein)		
Control	4.3	6.9	0.55
(n = 2,	(3.4-5.3)	(5.6-7.2)	(0.35-0.75)
24 & 26 y)			
ML I	4.1	6.1	0.44
	(1.9-7.3)	(4.0-8.3)	(0.09-0.81)
ML II	1.4	2.4	0.12
	(1.0-1.9)	(1.3-3.5)	(0.03-0.2)

3) Mucolipidosis II (I-cell disease). This mentally retarded
patient had a gargoyle-like facies since the first month of life.
Sitting and upright postures were never achieved; skeletal dysplasia
was demonstrated, the patient died at 2 years of age. Cultured skin
cells had cytoplasmic inclusions which stained with PAS and Sudan-
Black. Acid hydrolases were increased in serum (COPPA *et al.*, 1979).

RESULTS

   Sialidase activities in postmortem brain samples from patients
with ML I and ML II were assayed with ganglioside GD1a and sialyllactitol as substrates and are summarized in Tables 1 and 2. Enzyme
activities observed in homogenate, pellet and supernatant of ML I
specimen were in the normal range, whereas the sialidase levels were
lower than normal in the respective fractions of the ML II sample.
For each brain only 1-7 % of the total activity was recovered in
the 100,000 g supernatant.

   Sialidase activities in leucocytes and cultured fibroblasts of
the patient with the adult type neuronal storage disease are given
in Tables 3 and 4. With sialyllactitol as substrate a pronounced
sialidase deficiency was observed in fibroblasts and a partial defect
in leucocytes.

   However, when ganglioside GD1a and GM3 were used as substrates,
sialidase activities in leucocytes and fibroblasts of the patient
fell within the normal range. Centrifugation of a 10 % fibroblast
homogenate in 10 mM Tris/HCl buffer, pH 7.2, for 1 h at 100,000 g
sedimented sialyllactitol- as well as ganglioside GD1a-sialidase
almost completely.

   In brain samples from patients with ML I and ML II, activities of

Table 3. Sialidase Activity with GM3 and GD1a Ganglioside as Substrates in Leucocytes and Fibroblasts of Patient with Adult Type Neuronal Storage Disease

	Leucocytes		Fibroblasts	
	GM3	GD1a	GM3	GD1a
	(nmol/h/mg protein)			
Patient	0.64	1.48 (1.13-1.83)	0.12	0.58 (0.42-0.74)
Control (n = 4)	0.56 (0.47-0.65)	1.69 (1.41-1.97)	0.14 (0.13-0.15)	0.49 (0.23-0.75)

Table 4. Sialidase Activity with Sialyllactitol as Substrate in Leucocytes and Fibroblasts of a Patient with Adult Type Neuronal Storage Disease

	Leucocytes	Fibroblasts
	(nmol/h/mg protein)	
Patient	0.2 (0.17-0.23)	0.32 (0.28-0.36)
Control (n = 4)	0.6 (0.5 -0.7)	6.8 (6.1 -7.5)

seven lysosomal enzymes were assayed for comparison by using the respective 4-methylumbelliferyl derivatives (Tables 5 and 6). In the ML I sample, levels of acid phosphatase, α-mannosidase, assayed at pH 6.0 and β-glucosidase were close to normal; levels of α-mannosidase assayed at pH 4.0 and activities of N-acetyl-β-D-glucosaminidase, α-fucosidase and α-galactosidase were increased, whereas the activity of β-D-galactosidase appeared to be rather low in comparison to control values. A different situation was observed in the ML II (I-cell disease) sample. In this case only the activities of α-fucosidase and α-mannosidase assayed at pH 4.0, appeared to be within the normal range, whereas α-mannosidase activity at pH 6.0 was slightly reduced. Most of the other lysosomal enzyme activities (acid phosphatase, N-acetyl-β-D-glucosaminidase, β-galactosidase and β-glucosidase) were down to about a third of the control values with the exception of β-D-galactosidase activity, which was only a fifth of the control value. Assays of α-mannosidase at pH 4.0 and 6.0 in the presence and absence of cobalt ions indicated that both lysosomal and membrane-bound mannosidases were active in the pathological samples.

Table 5. Glycosidase Activities in Postmortem Brains of Patients with ML I and ML II

	Acid Phosphatase	N-Acetyl-β-D-Glucosaminidase	β-D-Galacto-sidase	α-Fucosidase	α-Galacto-sidase	β-D-Gluco-sidase
			(nmol/h/mg protein)			
Control (24 y)	1353 (1313–1373)	867 (844–889)	53 (46–60)	15.7 (11.7–19.7)	3.5 (3.3–3.7)	3.4 (3.3–3.5)
Control (26 y)	1144 (1107–1181)	707 (688–725)	39 (35–43)	25.7 (19.7–31.7)	3.6 (3.3–3.9)	5.5 (5.3–5.7)
ML I	1076 (1048–1104)	1910 (1887–1933)	19 (12–26)	54.2 (51.2–57.2)	6.0 (5.0–7.0)	3.9 (3.6–4.2)
ML II	453 (432–474)	303 (279–327)	8 (7–9)	18.2 (14.2–22.2)	1.3 (1.0–1.6)	1.2 (1.0–1.4)

Table 6. α-Mannosidase Activity in Postmortem Brains of
Patients with ML I and ML II
(10 % Homogenates in water were centrifuged for 1 h at 100,000 g)

	At pH 4.0	At pH 4.0 10 mM Co^{2+}	At pH 6.0	At pH 6.0 10 mM Co^{2+}
		(nmol/h/mg protein)		
Control (24 y)				
Homogenate	9.6	7.4	6.6	19.1
Pellet	3.2	2.5	5.9	10.8
Supernatant	28.4	21.3	8.2	22.4
Control (26 y)				
Homogenate	14.8	10.5	10.6	25.1
Pellet	5.0	3.6	6.7	16.2
Supernatant	42.0	28.3	17.8	33.7
ML I (21 y)				
Homogenate	18.3	15.4	10.7	13.7
Pellet	9.4	7.7	8.0	15.6
Supernatant	65.0	55.3	17.5	30.7
ML II (2 y)				
Homogenate	14.3	9.8	4.5	7.7
Pellet	10.7	7.4	5.6	11.3
Supernatant	30.3	19.6	5.7	9.8

Table 7. Sialic Acid Content in Postmortem Brain Samples of
Patients with ML I and ML II

	ML I	Age matched controls (n = 5)	ML II	Age matched controls (n = 3)
		(µg/g fresh tissue ± S.D.)		
Total sialic acid	1571	1172 ±120	734	1040 ±106
Lipid-bound sialic acid	1103	862 ± 92	529	795 ± 66
Protein-bound sialic acid	474	303 ± 40	241	255 ± 32

Table 8. Ganglioside Content in Postmortem Brain Samples of Patients with ML I and ML II

	ML I	Age matched controls (n = 5)	ML II	Age matched controls (n = 5)
		(nmol/g fresh tissue [molar percentage])		
GM2	-	-	56.3 [5.8]	53.0 ± 5.4 [3.5 ±0.35]
GM1	775.6 [34.8]	580.3 ±96.8 [35.6 ±5.9]	280.3 [29.0]	469.0 ±37.8 [30.6 ±1.9]
GD3	229.8 [10.3]	145.0 ±16.3 [8.9 ±1.0]	78.5 [8.1]	108.7 ±13.9 [7.1 ±0.9]
GD1a	364.0 [16.5]	280.2 ±23.3 [17.2 ±1.4]	216.0 [22.4]	553.8 ±22.1 [36.2 ±1.3]
GD1b	459.5 [20.6]	425.1 ±78.9 [26.1 ±4.8]	165.8 [17.2]	235.8 ±10.4 [15.4 ±0.6]
GT1b	397.0 [17.8]	197.3 ±30.5 [12.2 ±1.8]	168.8 [17.5]	110.4 ±10.2 [7.2 ±0.6]

Analysis of lipid- and protein-bound sialic acid brain tissue revealed an increase of both in the ML I sample and a decrease of lipid-bound sialic acid in the ML II sample (Table 7). As expected from these data, the content of the individual gangliosides was increased in ML I with the exception of GD1b, and decreased in the ML II brain sample with the exception of ganglioside GT1b, which was apparently increased (Table 8).

Content of sialooligosaccharides in the water-extracts of the ML I and ML II brain samples was increased about 2-fold in comparison to the controls (PALLMANN and SANDHOFF, unpublished).

DISCUSSION

Fibroblasts of the patient with adult type neuronal storage disease exhibited a profound sialidase deficiency when a water-soluble substrate was used (Table 4; MIYATAKE et al., 1979a). A similar deficiency has been described for fibroblasts of patients with ML I, ML II, ML III and myoclonus cherry-red spot syndrome with and without dementia (CANTZ et al., 1977; THOMAS et al., 1978; O'BRIEN, 1978). However, with the lipophilic substrate ganglioside GD1a no deficiency of sialidase activity was found, either in the patient's fibroblasts or leucocytes (Table 3; MIYATAKE et al., 1979a). In these cells GD1a-sialidase was considerably higher (2-5 times) than the residual activity against sialyllactitol. The data suggest the existence of at least two sialidases, both of which are membrane-associated. One is predominant in fibroblasts and active on water-soluble substrates like sialyllactitol or fetuin; this enzyme is deficient in fibroblasts of all three diseases studied. The other enzyme is much more active in brain tissue than in fibroblasts. Under the assay conditions used, this enzyme apparently hydrolyzes ganglioside GD1a about 3 times faster than sialyllactitol; under such conditions, the sialyllactitol concentration in the assay mixture is 20 times higher than that of ganglioside GD1a. Therefore, most of the sialyllactitol hydrolase activity in brain tissue probably originates from the latter enzyme which remains active in all three diseases. Since its specific activity in brain is about 24 times that in fibroblasts, we were unable to show whether the fibroblast-type enzyme is active in brain as a minor enzyme, which would also be expected to be deficient in brain tissue of patients with ML I and ML II. A speculative deficiency of such a minor enzyme in brain tissue of patients with ML I and ML II might account for the moderate increase of ganglioside GT1b (Table 8) and of sialooligosaccharides. In the ML I brain sample sialidase activities measured with both substrates were in the normal range. However, in the light of the normal sialidase activity in the brain the moderate increase of lipid- and protein-bound sialic acid (Table 7) remains so far unexplained. Further measurements should show whether this increase is characteristic of the diseases in question or still within the normal biological variation. On the other hand, in the ML II brain sample a number of acid hydrolase activities were reduced to about to a half to a quarter of the control values (Tables 5 and 6), including sialidase activities assayed with ganglioside GD1a and sialyllactitol (Tables 1 and 2). It remains to be clarified by the investigation of further cases to what extent these reductions are typical for ML II. At present a loss of sialidase activity due to prolonged storage at -20°C as well as to repeated freezing and thawing cannot be ruled out. The absence of a pronounced storage of an oligosialoganglioside in ML I and ML II brain indicates that the catabolism of brain gangliosides is mainly mediated by the ganglioside GD1a-hydrolysing enzyme, which is very active in brain and not defective in the diseases investigated. In order to estimate how many isoenzymes of sialidase might exist in human tissues and to characterise their substrate specificity and their role in different forms of sialidoses, solubilization and separation of these mainly membrane-associated hydrolases seem necessary. As a possible primary defect in ML II KRESS and MILLER (1978) suggested a deficiency of Golgi-associated α-mannosidase active at pH 6.0 and stimulated by Co^{2+} ions, which might account for the abnormalities of lysosomal enzyme in this disease. This possibility is not supported by the α-mannosidase activities found in the ML II brain sample (Table 6) although the activity of the pH 6.0 enzyme was reduced like that of other enzymes. The biological defect in ML II remains unclear.

ACKNOWLEDGEMENTS

We are indebted to Prof. CANTZ for the gift of a postmortem brain sample of a patient with mucolipidosis I. We would like to thank Drs. ANZIL, CARROLL and REDDINGTON for their help with the English manuscript.

REFERENCES

BERRA B., CESTARO B., OMODEO-SALE F., VENERANDO B., BELTRAME D., and CANTONE A. (1978): Gangliosides and neuraminidases in foetal rat brain. Bull. Mol. Biol. Med. 3, 86-97.

CANTZ M., GEHLER J. and SPRANGER J. (1977): Mucolipidosis I: Increased sialic acid content and deficiency of an α-N-acetyl-neuraminidase in cultured fibroblasts. Biochem. Biophys. Res. Commun. 74, 723-738.

CHRISTOMANOU H., CAP C. and SANDHOFF K. (1977): Isoelectric focusing pattern of acid hydrolases in cultured fibroblasts, leucocytes and cell-free amniotic fluid. Neuropädiatrie 8, 238-252.

COPPA G., SANI S., PARIS D., GABRIELLI O. and MARIORANA A. (1979): Isoelectric focusing of serum lysosomal enzymes and characterization of urinary and tissutal glycosaminoglycans in I-cell disease (mucolipidosis II), in "Perspect. Inher. Metab. Dis. II", BERRA B. *et al.*, Eds., Edi Ermes Publ. (Milano), in press.

DURAND P., GATTI R., CAVALIERI S., BORRONE C., TONDEUR M., MICHALSKI
J.C. and STRECKER G. (1977): Sialidosis (mucolipidosis I). Helv.
paediat. Acta 32, 391-400.

GOLDSTEIN M.L., KOLODNY E.H., GASCON G.G. *et al.* (1974): Macular
cherry-red spot, myoclonic epilepsy, and neurovisceral storage in
a 17-year old girl. Trans. Am. Neurol. Assoc. 99, 110-112.

HICKMANN S. and NEUFELD E.F. (1972): A hypothesis for I-cell disease: defective hydrolases that do not enter lysosomes. Biochem.
Biophys. Res. Commun. 49, 992-999.

HOLMES E.W., MILLER A.L., FROST R.G. and O'BRIEN J.S. (1975): Characterization of β-D-galactosidase isolated from I-cell disease liver.
Am. J. Hum. Genet. 27, 719-727.

HUMBEL R. and COLLART M. (1975): Oligosaccharides in urine of patients with glycoprotein storage disease. Clin. Chim. Acta 60,
143-145.

KAPLAN A., ACHORD D.T. and SLY W.S. (1977): Phosphohexosyl components
of a lysosomal enzyme are recognized by pinocytosis receptors on
human fibroblasts. Proc. Natl. Acad. Sci. 74, 2026-2030.

KELLY T.E. and GRAETZ G. (1977): Isolated acid neuraminidase deficiency: A distinct lysosomal storage disease. Am. J. Med. Genet.
1, 31-46.

KRESS B.C. and MILLER A.L. (1978): Altered serum α-D-mannosidase
activity in mucolipidosis II and mucolipidosis III. Biochem.
Biophys. Res. Commun. 81, 756-763.

LEROY G.J., HO M.W., MACBRINN M.C., ZIELKE K., JAKOB J. and O'BRIEN
J.S. (1972): I-cell disease: Biochemical studies. Pediat. Res. 6,
752-757.

LOWRY O.H., ROSEBROUGH N.J., FARR A.L. and RANDALL R.J. (1951):
Protein measurement with the Folin phenol reagent. J. Biol. Chem.
193, 265-275.

MAIRE I., ZABOT M.T., MATHIEU M. and COTTE J. (1978): Mannosidosis:
tissue culture studies in relation to prenatal diagnosis. J. Inher.
Metab. Dis. 1, 19-23.

MAROTEAUX P. and HUMBEL R. (1976): Les oligosaccharidoses. Un nouveau concept. Arch. Franç. Péd. 23, 641-647.

MAROTEAUX P. (1978): Les sialidoses par déficit en α-2-6 neuraminidase: Un groupe hétérogène. Arch. Franç. Péd. 35, 815-818.

MICHALSKI J.C. and STRECKER G. (1977): in "Les Oligosaccharidoses",
FARRIAUX J.P., Ed., Crouan-Roques (Lille), pp. 127-130.

MIYATAKE T., YAMADA T., SUZUKI M., PALLMANN B., SANDHOFF K., ARIGA T.
and ATSUMI T. (1979a): Sialidase deficiency in adult-type neuronal
storage disease. FEBS Lett. 97, 257-259.

MIYATAKE T., ATSUMI T., OBAYASHI T., MIZUNO Y., ANDO S., ARIGA T., MATSUI-NAKAMURA K. and YAMADA T. (1979b): Adult type neuronal storage disease with neuraminidase deficiency. Ann. Neurol. in press.

O'BRIEN J.S. (1978): The cherry-red spot myoclonus syndrome: A newly recognized inherited lysosomal storage disease due to acid neuraminidase deficiency. Clin. Genet. 14, 55-60.

RAPIN I., GOLDFISCHER S., KATZMANN R., ENGEL J. and O'BRIEN J.S. (1978): The cherry-red spot myoclonus syndrome. Ann. Neurol. 3, 234-242.

SCHRAVEN J., CAP C., NOWOCZEK G. and SANDHOFF K. (1977): A radiometric assay for sialidase acting on ganglioside GD1a. Anal. Biochem. 78, 333-339.

SCHWARZMANN G. (1978): A simple and novel method for tritium labeling of gangliosides and other sphingolipids. Biochim. Biophys. Acta 529, 106-114.

SPRANGER J., WIEDEMANN H.R., TOLKSDORF M., GRAUCOB E. and CAESAR R. (1968): Lipomucopolysaccharidosis: a new storage disease. Z. Kinderheilk. 103, 285-306.

SPRANGER J., GEHLER J. and CANTZ M. (1977): Mucolipidosis I - a sialidosis. Am. J. Med. Genet. 1, 21-29.

SPRITZ R.A., COATES P.M. and LIEF F.S. (1979): I-cell disease: intracellular desialylation of lysosomal enzymes using an influenza virus vector. Biochim. Biophys. Acta 582, 164-171.

STRECKER G., MICHALSKI J.C., MONTREUIL J. and FARRIAUX J.P. (1976): Deficit in neuraminidase associated with mucolipidosis II (I-cell disease). Biomedicine 25, 238-240.

STRECKER G., PEERS M.C., MICHALSKI J.C., HONDI-ASSAH T., FOURNET B., SPIK G., MONTREUIL J., FARRIAUX J.P., MAROTEAUX P. and DURAND P. (1977): Structure of nine sialyloligosaccharides accumulated in urine of eleven patients with three different types of sialidosis, mucolipidosis II and two new types of mucolipidosis. Eur. J. Biochem. 75, 391-403.

STRECKER G. and MICHALSKI J.C. (1978): Biochemical basis of six different types of sialidosis. FEBS Lett. 85, 20-24.

TETTAMANTI G., BONALI F., MARCHESINI S. and ZAMBOTTI V. (1973): A new procedure for the extraction, purification and fractionation of brain gangliosides. Biochim. Biophys. Acta 296, 160-170.

THOMAS G.H., TAYLOR H.A., REYNOLDS L.W. and MILLER C.S. (1973): Mucolipidoses III (pseudo-Hurler polydystrophy): multiple lysosomal abnormalities in serum and cultured fibroblast cells. Pediat. Res. 7, 751-756.

THOMAS G.H., TIPTON R.E., CH'IEN L.T., REYNOLDS L.W. and MILLER C.S. (1978): Sialidase (α-N-acetylneuraminidase) deficiency: the enzyme defect in an adult with macular cherry-red spots and myoclonus without dementia. Clin. Genet. 13, 369-379.

TONDEUR M., VAMOS-HURWITZ E., MOCKEL-POHL S., DEREUME J.P., CREMER N. and LOEB H. (1971): Clinical, biochemical and ultrastructural studies in a case of chondrodystrophy presenting in I-cell phenotype in tissue culture. J. Pediatr. 79, 366-378.

ULLRICH K., MERSMANN G., WEBER E. and v. FIGURA K. (1978): Evidence for lysosomal enzyme recognition by human fibroblasts via a phosphorylated carbohydrate moiety. Biochem. J. 170, 643-650.

VLADUTIU G.D. and RATTAZZI M.C. (1975): Abnormal lysosomal hydrolases excreted by cultured fibroblasts in I-cell disease (mucolipidosis II). Biochem. Biophys. Res. Commun. 67, 956-963.

VLADUTIU G.D. and RATTAZZI M.C. (1978): I-cell disease: desialylation of β-hexosaminidase and its effect on uptake by fibroblasts. Biochim. Biophys. Acta 539, 31-36.

WARREN L. (1959): The thiobarbituric acid assay of sialic acids. J. Biol. Chem. 234, 1971-1975.

WENGER D.A., TARBY T.J. and WHARTON C. (1978): Macular cherry-red spots and myoclonus with dementia: coexistent neuraminidase β-galactosidase deficiencies. Biochem. Biophys. Res. Commun. 82, 589-595.

WIESMANN U.N. and HERSCHKOWITZ N.N. (1974): Studies on the pathogenetic mechanism of I-cell disease in cultured fibroblasts. Pediat. Res. 8, 865-870.

NEURAMINIDASE STUDIES IN SIALIDOSIS

Michael Cantz

Physiologisch-Chemisches Institut
Universität Münster
D-4400 Münster
Federal Republic of Germany

INTRODUCTION

In 1968, two patients were described in the literature who exhibited clinical features resembling Hurler
disease, yet had a normal urinary excretion of acid
mucopolysaccharides. Histological and ultrastructural
studies of neuronal, visceral and mesenchymal tissues
revealed the lysosomal storage of compounds thought to
be acid mucopolysaccharides and glycolipids (BÉRARD et
al., 1968; SPRANGER et al., 1968; FREITAG et al., 1971)
Together with morphologically similar disorders, these
patients were later classified as mucolipidosis I
(SPRANGER and WIEDEMANN, 1970). More recent biochemical
investigations showed an accumulation of sialic acid-
containing compounds in cultured fibroblasts and in leukocytes of such patients (CANTZ et al., 1977; SPRANGER
et al., 1977; KELLY and GRAETZ, 1977), and a massive
urinary excretion of sialyl oligosaccharides, whose
structures resembled the glycan portion of glycoproteins
MICHALSKI et al., 1977). In fibroblasts, there was a
profound deficiency of an "acid" (presumably lysosomal)
neuraminidase towards substrates such as sialyllactose,
fetuin, and methoxyphenyl neuraminic acid, whereas other
lysosomal hydrolases were within the normal range
(CANTZ et al., 1977; SPRANGER et al., 1977; KELLY and
GRAETZ, 1977). From these studies it became evident that
the metabolic defect in mucolipidosis I consists of an
impaired catabolism of glycoproteinderived sialyl oligosaccharides due to the genetic deficiency of a neuraminidase, i.e., that the disease is a sialidosis.

A neuraminidase deficiency and an excessive urinary excretion of sialyl oligosaccharides had been observed also in two "new" types of mucolipidosis (STRECKER et al., 1977; DURAND et al., 1977; STRECKER and MICHALSKI, 1978) and in the cherry-red spot - myoclonus syndrome (O'BRIEN, 1977 ; THOMAS et al., 1978), disorders which are clically different from mucolipidosis I. In addition, a similar metabolic abnormality was detected in muco-lipidosis II (I-cell disease) (THOMAS et al., 1976; STRECKER et al., 1976, 1977). However,whereas the neura-minidase deficiency was singular in the former diseases, it was associated in mucolipidosis II with a deficiency in cultured fibroblasts, and grossly increased levels in serum and other body fluids, of many lysosomal hydro-lases, thought to be due to a defect in lysosomal enzyme packaging (NEUFELD, 1974).

Lysosomal neuraminidase active towards sialoglyco-peptide and ganglioside substrates has been described in a variety of tissues (ARONSON and DE DUVE, 1968; HORVAT and TOUSTER, 1968; TULSIANI and CARUBELLI, 1970). How-ever, it is not clear if these activities are the pro-perties of one single or of several enzymes, or if there are specific neuraminidases for different types of neu-raminosyl linkages. Obviously, knowledge of the specifi-city of the neuraminidases is of great importance for a better understanding of the various forms of siali-dosis.

Therefore, the neuraminidase activities of fibro-blasts cultured from the skin of patients with mucolipi-dosis I (sialidosis), mucolipidosis II (I-cell disease), and their family members were investigated using as sub-strates sialyl oligosaccharides bearing an $\alpha 2 \rightarrow 3$ or an $\alpha 2 \rightarrow 6$ neuraminosyl linkage, as well as various gangliosides.

The results demonstrate that in mucolipidosis I there is an impaired degradation of both kinds of oligo-saccharides, but a normal hydrolysis of the gangliosides. In mucolipidosis II fibroblasts, on the other hand, oligosaccharide as well as ganglioside neuraminidase activities were found to be markedly deficient, sugges-ting a defective catabolism of both varieties of sialo-glycoconjugates. The implications of these findings with respect to the pathogenesis of the diseases and the genetics of fibroblast neuraminidases will be discussed.

METHODS

Fibroblast lines

Fibroblasts cultured from the skin of normal individuals and from patients with genetic diseases were maintained as described (CANTZ et al., 1972). Diagnoses had been established on the basis of clinical and biochemical findings including the determination of lysosomal hydrolases in fibroblasts and in the serum. Clinical data on the mucolipidosis I patients have been published in the cases of R.H. (case no. 1 of SPRANGER et al., 1968) and D.F. (SPRANGER et al., 1977).

Neuraminidase assays

Neuraminidase activity of fibroblast homogenates towards sialyllactose (mainly N-acetylneuraminosyl $\alpha 2 \rightarrow 3$lactose, Calbiochem, Los Angeles, USA) and sialyl hexasaccharides (NeuAc $\alpha 2 \rightarrow 3$Gal$\beta 1 \rightarrow 4$GlcNAc$\beta 1 \rightarrow 2$Man $\alpha 1 \rightarrow 3$Man$\beta 1 \rightarrow 4$GlcNAc and NeuAc $\alpha 2 \rightarrow 6$Gal$\beta 1 \rightarrow 4$GlcNAc$\beta 1 \rightarrow 2$Man $\alpha 1 \rightarrow 3$Man$\beta 1 \rightarrow 4$GlcNAc, kindly provided by Dr. G. Strecker, Lille) was determined as described previously (CANTZ et al. 1977), with the following modifications. After incubation at 37 $^{\circ}$C for 2 h (6 h for the NeuAc $\alpha 2 \rightarrow 6$-hexasaccharide), the samples were neutralized with 1 M NaOH and chromatographed on small columns of AG 1X10 (HORVAT and TOUSTER, 1968). After concentration by lyophilization, the liberated neuraminic acid was determined (WARREN, 1959).

Fibroblast ganglioside neuraminidase activity was determined using bovine brain gangliosides (type II, Sigma Chemical Co., St. Louis, Mo., USA) or gangliosides G_{M3}, G_{D1a} and G_{D3} (which were the gifts of Dr. (L. Svennerholm, Hisings Backa, Sweden). The gangliosides were dissolved in 0.5 % (w/v) Triton X-100 (Serva, Heidelberg) to a concentration of 5 mg/ml for the commercial preparation, or 2.5 mM for the individual gangliosides, and sonicated. Incubation mixtures consisted of 150 ul freshly prepared homogenate (600 to 850 ug protein), 10 ul 1 M sodium acetate buffer, pH 4.2 (to give a final pH of 4.5), and 40 ul ganglioside substrate. In the blanks either the homogenate was replaced by water, or Triton solution was substituted for the substrate. After 22 h incubation at 37 $^{\circ}$C, the samples were neutralized, chromatographed on AG 1X10, and the liberated neuraminic acid determined as described above.

For the determination of neuraminidase activity
towards endogenous (cell-associated) substrate, 200 ul
of fibroblast homogenate (600 to 1150 ug cell protein)
was mixed with 10 ul 1 M sodium acetate buffer, pH 4.2
(final pH 4.5) and incubated at 37 ^{O}C for 6 h. The released neuraminic acid was determined as above. Samples
with boiled homogenate served as blanks. Under these
conditions, neuraminidase activity was linear to incubation time.

Neuraminidase activity is expressed as mU/mg protein, one milliunit of enzyme activity corresponding to
one nmole of N-acetylneuraminic acid released per min
under the conditions employed. All assays were carried
out in duplicate.

RESULTS

Using sialyllactose as the substrate, the neuraminidase
activity of a normal fibroblast homogenate was linear

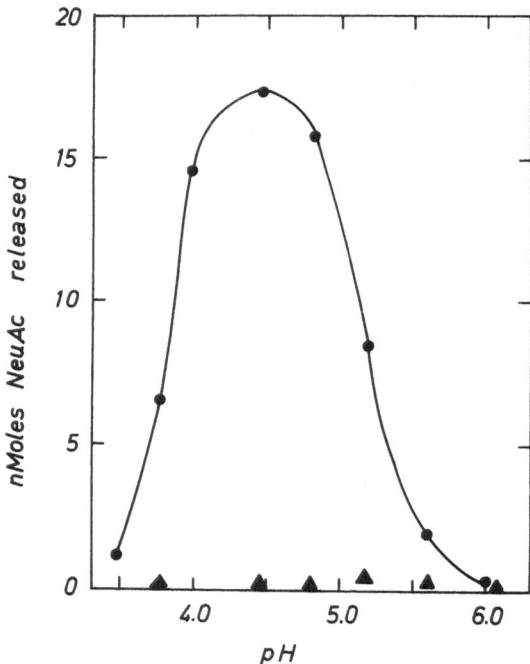

Fig. 1: pH-activity profile of fibroblast oligosacchari-
        de neuraminidase. The substrate was sialylacto-
        se; 0.05 M sodium acetate buffers were used.
        ●, Normal homogenate; ▲, mucolipidosis I homoge-
        nate.

to incubation time for up to 120 min; it was also linear with enzyme concentration up to about 0.5 mg homogenate protein per incubation. The pH-activity profile of this neuraminidase is shown in Fig. 1. In the normal homogenate, optimal activity was at pH 4.4. In a homogenate of mucolipidosis I fibroblasts, the activity was zero at any of the pH's investigated.

The oligosaccharide neuraminidase activities of fibroblasts from patients with mucolipidoses I and II and some of their family members, and from healthy controls, are presented in Table 1. In mucolipidosis I, there was a complete to nearly complete neuraminidase deficiency towards substrates containing either an $\alpha2 \rightarrow 3$ or an $\alpha2 \rightarrow 6$ neuraminosyl linkage. A mixture of normal and mucolipidosis I homogenates yielded the calculated intermediate activity, thus ruling out the presence of neuraminidase inhibitors in the mucolipidosis I samples. Fibroblasts from parents of a mucolipidosis I patient showed an intermediate activity of between 41 and 86 % of the normal mean, as expected for heterozygous individuals. In mucolipidosis II, there was likewise a severe neuraminidase deficiency towards each of the substrates. The activity of the fibroblasts of the parents of mucolipidosis II patient S.G., however, was completely normal, contrary to the findings in the mucolipidosis I heterozygotes mentioned above. In fibroblasts from patients with mucopolysaccharidosis I (Hurler disease) and mucopolysaccharidosis II (Hunter disease), this neuraminidase activity was normal.

The ganglioside activities of fibroblast homogenates were assayed using gangliosides G_{M3}, G_{D1a}, G_{D3}, and a commercial ganglioside preparation from bovine brain as substrates.

As demonstrated with the bovine brain ganglioside, a normal homogenate showed neuraminidase activity only in the presence of Triton X-100, with an optimal detergent concentration of between 1 and 2 mg/ml (Fig. 2). Fig. 3 shows the pH-optima with the bovine brain and G_{D1a} ganglioside preparations, which were at about pH 4.5 and pH 4.6, respectively; with ganglioside G_{D1a}, there was a suggestion of a shoulder around pH 5.5. Neuraminidase activity towards the bovine brain ganglioside was linear to incubation time; variation of this activity with enzyme concentration was not quite linear, however, the release of neuraminic acid being proportionately lower at low enzyme concentrations.

Table 1: Oligosaccharide neuraminidase activities of fibroblast homogenates (mU/mg protein)

Cell line		Sialyllactose	NeuAcα2 → 3-HS	NeuAcα2 → 6-HS
Mucolipidosis I	D.F.	0	0	0
	R.H.	0.007	0	0
Mother of R.H.		0.160		
Father of R.H.		0.076		
Brother of R.H.		0.210		
Mucolipidosis II	S.A.	0	0	0.0016
	A.B.	0.021	0	0
	S.G.	0		
Mother of S.G.		0.178		
Father of S.G.		0.281		
Normal controls	mean value	0.178	0.172	0.0449
	range	0.100–0.394	0.142–0.207	0.0300–0.0664
	n	12	5	7

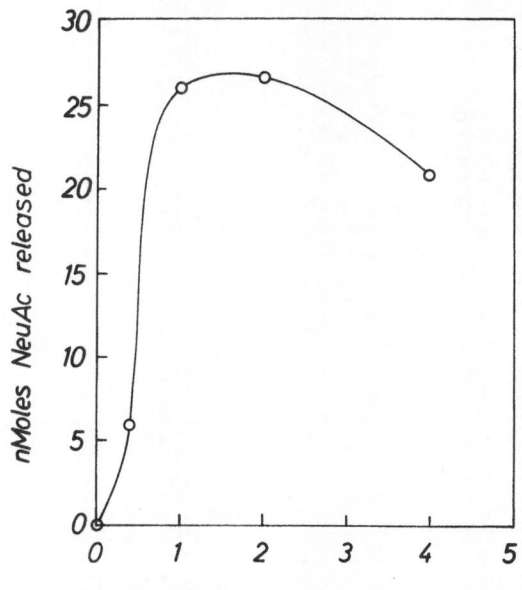

Fig. 2: Dependence of fibroblast ganglioside neuraminidase activity on the concentration of Triton
X-100. The substrate was bovine brain gangliosides.

Preincubation of the homogenates prior to the addition
of ganglioside, in order to deplete the samples of any
endogenous substrate, yielded much lower activites,
presumably due to enzyme inactivation.

Table 2 summarizes the ganglioside neuraminidase
activities found in fibroblast homogenates from mucolipidosis I and II patients, from patients with other disorders, and from healthy controls. In mucolipidosis I,
there was a slightly diminished activity with the bovine
brain and G_{D1a} gangliosides, but normal activity towards
gangliosides G_{M3} and G_{D3}. The ganglioside neuraminidase
activities of mucolipidosis II fibroblasts, on the other
hand, were markedly deficient with all of the substrates
investigated. Interestingly, there was also a drastically reduced activity in fibroblasts from patients with
mucopolysaccharidosis I (Hurler) and mucopolysaccharidosis II (Hunter), probably due to inhibition by the
stored mucopolysaccharide. Fibroblasts from patients
with G_{M1}- and G_{M2}-gangliosidoses showed normal ganglioside neuraminidase activity.

Table 2: Ganglioside neuraminidase activities of fibroblast homogenates (mU/mg protein)

Cell line	Bovine brain gangliosides	Ganglioside G_{M3}	Ganglioside G_{D1a}	Ganglioside G_{D3}
Mucolipidosis I D.F.	0.0093	0.0107	0.0065	0.0117
Mucolipidosis II S.A.		0		0.0016
A.B.	0	0.0026		0.0093
S.G.	0.0043			
Mucopolysaccharidosis I C.C.	0.0039			
Mucopolysaccharidosis II D.M.	0.0013			
G_{M1}-Gangliosidosis S.K.	0.0101			
S.S.	0.0126			
G_{M2}-Gangliosidosis variant O B.B.	0.0120			
Normal controls mean value	0.0189	0.0129	0.0107	0.0139
Range	0.0128-0.0249	0.0088-0.0181	0.0078-0.0173	0.0091-0.0203
n	10	6	6	5

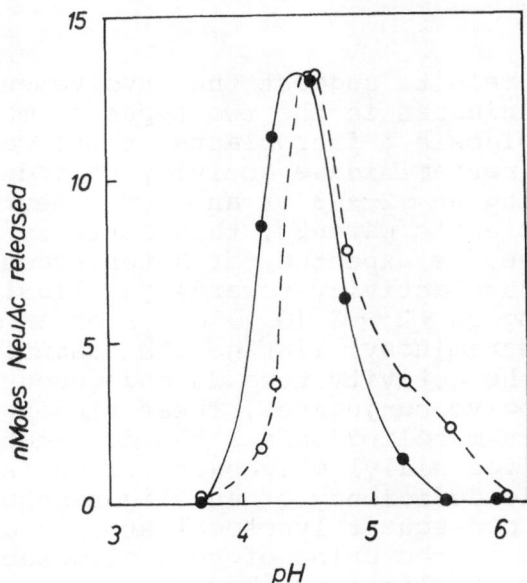

Fig. 3: pH-activity profiles of fibroblast ganglioside neuraminidase. The substrates were bovine brain ganglioside (●) and ganglioside G_{Dla} (o). 0.05 M sodium acetate buffers were used.

Fibroblast homogenates were also assayed for the release of neuraminic acid from endogenous substrate. As shown in Table 3, this activity was in the normal range in the mucolipidosis I cell line tested, but was markedly reduced in mucolipidosis II fibroblasts.

Table 3

Neuraminidase activity towards endogenous substrate

Cell line		mU/mg protein
Mucolipidosis I	D.F.	0.0277
Mucolipidosis II	S.G.	0.0085
	A.B.	0.0029
Normal controls	mean	0.0201
(n = 5)	range	(0.0131 - 0.0359)

DISCUSSION

The present results suggest the involvement of different neuraminidases in the two types of mucolipidosis. In mucolipidosis I fibroblasts, there was a complete absence of neuraminidase activity towards oligosaccharides bearing an $\alpha 2 \rightarrow 3$ or an $\alpha 2 \rightarrow 6$ neuraminosyl linkage. In a patient's parents, this activity was in an intermediate range, as expected for heterozygous individuals. Neuraminidase activity towards ganglioside substrates possessing an $\alpha 2 \rightarrow 3$ (G_{M3}, G_{D1a}) or an $\alpha 2 \rightarrow 8$ (G_{D3}) terminal neuraminosyl linkage was essentially normal, as well as the activity towards endogenous (cell-associated) sialoglycoconjugates. These findings strongly suggest that in mucolipidosis I the degradation of glycoprotein-derived sialyl oligosaccharides is impaired due to the genetic deficiency of an oligosaccharide neuraminidase, with consequent lysosomal storage and excessive excretion in the urine of such oligosaccharides. The catabolism of gangliosides, however, seems unaffected.

In mucolipidosis II fibroblasts, a deficiency of a neuraminidase specific for both $\alpha 2 \rightarrow 3$ and $\alpha 2 \rightarrow 6$ types of neuraminosyl linkages was likewise observed, confirming the recent findings of STRECKER and MICHALSKI (1978). In addition, however, neuraminidase activity towards gangliosides and endogenous substrate was markedly reduced. Thus, in mucolipidosis II, the defective catabolism of sialyl oligosaccharides seems to be associated with an inability to degrade gangliosides. Indeed, an abnormal accumulation of gangliosides G_{M3} and G_{D3} has been noted earlier in mucolipidosis II fibroblasts (DAWSON et al., 1972).

The simultaneous deficiency of both oligosaccharide and ganglioside neuraminidase activities in mucolipidosis II fibroblasts is not surprising in view of the fact that many lysosomal hydrolases are deficient in these cells (e.g., LEROY et al., 1972; WENGER et al., 1976). The multiple deficiency in fibroblasts, and the excessive elevation of these enzymes in the extracellular fluids, are assumed to result from a defect in a recognition marker on the enzymes required for their receptor-mediated transport into lysosomes (NEUFELD, 1974). While it was shown that phosphomannosyl residues on lysosomal enzymes function as a signal in their specific endocytosis by cultured fibroblasts (KAPLAN et al., 1977; SANDO and NEUFELD, 1977), it is not clear whether these groups are involved in the pathogenesis of the

disease. Recently, it was postulated that a deficiency
of a neuraminidase specific for the $\alpha 2 \rightarrow 3$ neuraminosyl
linkage of glycoproteins would be the primary lesion in
mucolipidosis II, thus leaving masked the structure of
the recognition marker by not removing neuraminic acid
residues (STRECKER and MICHALSKI, 1978). Although in
line with earlier findings in mucolipidosis II of an
oversialylation of extracellular hydrolases (VLADUTIU
and RATTAZZI, 1975), this view appeared less likely when
it was shown that neuraminidase treatment of such enzymes did not enhance their abnormally low rate of pinocytosis (VLADUTIU and RATTAZZI, 1978). Further evidence
against this hypothesis comes from the present findings
of an $\alpha 2 \rightarrow 3$ oligosaccharide neuraminidase deficiency
not only in mucolipidosis II but also in mucolipidosis I,
as well as a normal, rather than an intermediate, activity of this neuraminidase in obligate heterozygotes
(parents) of a mucolipidosis II patient. It is therefore likely that the oligosaccharide neuraminidase deficiency in mucolipidosis II fibroblasts results from the
same pathogenetic mechanism which is responsible for a
deficiency of the other lysosomal hydrolases. Although
such mechanism may also cause the low levels of ganglioside neuraminidase activity in mucolipidosis II cells,
it is possible that this activity is inhibited by the
excess sulfated glycosaminoglycans present in these
cells (TONDEUR et al., 1971), as very low ganglioside
neuraminidase activities have now been found in fibroblasts from patients with mucopolysaccharidoses. Further
work is in progress to clarify this point.

The present findings also provide insight into the
genetics of human neuraminidases. Thus, the deficiency
of both an $\alpha 2 \rightarrow 3$ and an $\alpha 2 \rightarrow 6$ oligosaccharide neuraminidase in mucolipidosis I suggests that the two acti-
vities are genetically related and are either the properties of a single enzyme, or of isoenzymes sharing a
common subunit. The presence in mucolipidosis I of essentially normal neuraminidase activities towards gangliosides G_{M3}, G_{D1a} and G_{D3}, on the other hand, indicates
that ganglioside and oligosaccharide neuraminidases are
genetically distinct. However, the remote possibility
of a single mutant neuraminidase with deficient activity
towards oligosaccharides, but normal activity towards
gangliosides, cannot at present be excluded.

The genetic relationship of mucolipidosis I to
other types of sialidosis remains as yet unclear. While
a similar deficiency of an oligosaccharide neuraminidase

specific for both $\alpha 2 \rightarrow 3$ and $\alpha 2 \rightarrow 6$ neuraminosyl linkages was observed in the cherry-red spot - myoclonus syn-
drome, a clinically much milder condition (O'BRIEN,
1977), an "isolated" $\alpha 2 \rightarrow 6$ oligosaccharide neuraminidase deficiency was found in several patients with sialidosis, whose clinical pictures also differed from mucolipidosis I (STRECKER and MICHALSKI, 1978 ; DURAND et
al., 1977; MAROTEAUX et al., 1978). Possible explanations include mutations in neuraminidase isoenzymes
sharing a common polypeptide subunit, analogous to the
situation in Tay-Sachs and Sandhoff diseases, as well as
allelic mutations of a single neuraminidase with resul-
tant different effects on substrate specificity and
kinetics.

CONCLUSIONS

Mucolipidosis I is a sialidosis due to the genetic
deficiency of a neuraminidase specific for both $\alpha 2 \rightarrow 3$
and $\alpha 2 \rightarrow 6$ neuraminosyl linkages of oligosaccharides
(glycoproteins). The ganglioside catabolism seems unaffected.

In mucolipidosis II, the deficiencies in cultured
fibroblasts of oligosaccharide and ganglioside neuraminidase activities are considered to represent another
example of the multiple lysosomal hydrolase deficiency
observed in these cells. As obligate heterozygotes
(parents) exhibited a normal, rather than an intermediate, oligosaccharide neuraminidase activity, a primary
genetic defect of this enzyme appears unlikely.

In fibroblasts, the $\alpha 2 \rightarrow 3$ and $\alpha 2 \rightarrow 6$ oligosaccharide neuraminidase activities appear to be genetically
related, yet seem to be distinct from the neuraminidase
activities towards gangliosides possessing an $\alpha 2 \rightarrow 3$
(G_{M3}, G_{D1a}) or an $\alpha 2 \rightarrow 8$ (G_{D3}) terminal neuraminosyl
linkage.

Acknowledgement

The expert technical assistance of Miss Helgard
Messer and Mrs. Renate Hellmann is gratefully acknow-
ledged. This work was supported by grants from the
Deutsche Forschungsgemeinschaft and the Stiftung
Volkswagenwerk.

References

ARONSON N.N. and DE DUVE C. (1968): Digestive activity of lysosomes. II. The digestion of macromolecular carbohydrates by extracts of rat liver lysosomes. J. biol. Chem. 243, 4564-4573

BÉRARD M., TOGA M., BERNARD R., DUBOIS D., MARIANI R. and HASSOUN J. (1968): Pathological findings in one case of neuronal and mesenchymal storage disease. Its relationship to lipidoses and to mucopoly-saccharidoses. Path. europ. 3, 172-183

CANTZ M., KRESSE H., BARTON R.W. and NEUFELD E.F. (1972): Corrective factors for inborn errors of mucopolysaccharide metabolism, in "Methods in Enzymology", GINSBURG V., Ed., Vol. XXVIII, Part B, Academic Press (New York and London), pp. 884-897

CANTZ M., GEHLER J. and SPRANGER J. (1977): Mucolipidosis I: Increased sialic acid content and deficiency of an α -N-acetylneuraminidase in cultured fibroblasts. Biochem. Biophys. Res. Commun. 74, 732-738

DAWSON G., MATALON R. and DORFMAN A. (1972): Glycosphingolipids in cultured human skin fibroblasts. II. Characterization and metabolism in fibroblasts from patients with inborn errors of glycosphingolipid and mucopolysaccharide metabolism. J. biol. Chem. 247, 5951-5958

DURAND P., GATTI R., CAVALIERI S., BORRONE C., TONDEUR M., MICHALSKI J.C. and STRECKER G (1977): Sialidosis (mucolipidosis I). Helv. paediat. Acta 32, 391-400

FREITAG F., BLÜMCKE S. and SPRANGER J. (1971): Hepatic ultrastructure in mucolipidosis I (lipomucopoly-saccharidosis). Virchows Arch. Path. B7, 189-204

HORVAT A. and TOUSTER O. (1968): On the lysosomal occurence and the properties of the neuraminidase of rat liver and of Ehrlich ascites tumor cells. J. biol. Chem 243, 4380-4390

KAPLAN A., ACHORD D.T. and SLY W.S. (1977): Phosphohexosyl components of a lysosomal enzyme are recognized by pinocytosis receptors on human fibroblasts. Proc. Nat. Acad. Sci. USA 74, 2026-2030

KELLY T.E. and GRAETZ G. (1977): Isolated acid neurami-
    nidase deficiency: a distinct lysosomal storage
    disease. Amer. J. Med. Genet. 1, 31-46

LEROY J.G., HO M.W., MACBRINN M.C., ZIELKE K., JACOB J.
    and O'BRIEN J.S. (1972): I-cell disease: Biochemi-
    cal studies. Pediat. Res. 6, 752-757

MAROTEAUX P., POISSONNIER M., TONDEUR M., STRECKER G.
    and LEMONNIER M. (1978): Sialidose par deficit en
    alpha (2-6) neuraminidase sans atteinte neurolo-
    gique. Mucolipidose de type I ? Arch. Franç. Péd.
    35, 280-291

MICHALSKI J.C., STRECKER G., FOURNET B., CANTZ M. and
    SPRANGER J. (1977): Structures of sialyl-oligosac-
    charides excreted in the urine of a patient with
    mucolipidosis I. FEBS Lett. 79, 101-104

NEUFELD E.F. (1974): The biochemical basis for mucopoly-
    saccharidoses and mucolipidoses, in "Progress in
    Medical Genetics", STEINBERG A.G. and BEARN A.G.,
    Eds., Vol. X, Grune and Stratton Inc. (New York
    and London), 81-101

O'BRIEN J.S. (1977): Neuraminidase deficiency in the
    cherry red spot - myoclonus syndrome. Biochem.
    Biophys. Res. Commun. 79, 1136-1141

SANDO G.N. and NEUFELD E.F. (1977): Recognition and re-
    ceptormediated uptake of a lysosomal enzyme, α-L-
    iduronidase, by cultured human fibroblasts.
    Cell 12, 619-627

SPRANGER J., WIEDEMANN H.R., TOLKSDORF M., GRAUCOB E.
    and CAESAR R. (1968): Lipomucopolysaccharidose.
    Eine neue Speicherkrankheit. Zschr. Kinderheilk.
    103, 285-306

SPRANGER J. and WIEDEMANN H.R. (1970): The genetic
    mucolipidoses. Humangenetik 9, 113-139

SPRANGER J., GEHLER J. and CANTZ M. (1977): Mucolipi-
    dosis I - a sialidosis. Amer. J. Med. Genet. 1,
    21-29

STRECKER G., MICHALSKI J.C., MONTREUIL J. and FARRIAUX
    J.P. (1976): Deficit in neuraminidase associated
    with mucolipidosis II (I-cell disease). Biomedi-
    cine 25, 238-239

STRECKER G., PEERS M.C., MICHALSKI J.C., HONDI-ASSAH T.,
    FOURNET B., SPIK G., MONTREUIL J., FARRIAUX J.P.,
    MAROTEAUX P. and DURAND P. (1977): Structure of
    nine sialyl-oligosaccharides accumulated in urine
    of eleven patients with three different types of
    sialidosis, mucolipidosis II and two new types of
    mucolipidosis. Eur. J. Biochem. 75, 391-403

STRECKER G. and MICHALSKI J.C. (1978): Biochemical basis
    of six different types of sialidosis. FEBS Lett.
    85, 20-24

THOMAS G.H., TILLER G.E., REYNOLDS L.W., MILLER C.S. and
    BACE J.W. (1976): Increased levels of sialic acid
    associated with a sialidase deficiency in I-cell
    disease (mucolipidosis II) fibroblasts. Biochem.
    Biophys. Res. Commun. 71, 188-195

THOMAS G.H., TIPTON R.E., CH'IEN T., REYNOLDS L.W. and
    MILLER C.S. (1978): Sialidase (α-N-acetyl neura-
    minidase) deficiency: the enzyme defect in an adult
    with macular cherry-red spots and myoclonus without
    dementia. Clin. Genet. 13, 369-379

TONDEUR M., VAMOS-HURWITZ E., MOCKEL-POHL S., DEREUME
    J.P., CREMER N. and LOEB H. (1971): Clinical, bio-
    chemical, and ultrastructural studies in a case of
    chondrodystrophy presenting the I-cell phenotype in
    tissue culture. J. Pediat. 79, 366-378

TULSIANI D.R.P. and CARUBELLI R. (1970): Studies on the
    soluble and lysosomal neuraminidases of rat liver.
    J. biol. Chem. 245, 1821-1827

VLADUTIU G.D. and RATTAZZI M.C. (1975): Abnormal lyso-
    somal hydrolases excreted by cultured fibroblasts
    in I-cell disease (mucolipidosis II). Biochem. Bio-
    phys. Res. Commun. 67, 956-964

VLADUTIU G.D. and RATTAZZI M.C. (1978): I-cell disease.
    Desialylation of ß-hexosaminidase and its effect
    on uptake by fibroblasts. Biochim. Biophys. Acta
    539, 31-36

WARREN L. (1959): The thiobarbituric acid assay of
    sialic acids. J. biol. Chem. 234, 1971-1975

WENGER D.A., SATTLER M., CLARK C. and WHARTON C. (1976) :
    I-cell disease: Activities of lysosomal enzymes
    toward natural and synthetic substrates.
    Life Sci. 19, 413-420

FUNCTION OF GANGLIOSIDES (RECEPTOR PROPERTIES, INVOLVEMENT IN NEUROTRANSMISSION)

SACCHARIDE TRAFFIC SIGNALS IN RECEPTOR-MEDIATED

ENDOCYTOSIS AND TRANSPORT OF ACID HYDROLASES

William S. Sly

The Edward Mallinckrodt Department of Pediatrics
Washington University School of Medicine, and
Division of Medical Genetics, St. Louis Children's Hospital
St. Louis, Missouri 63110

INTRODUCTION TO ENDOCYTOSIS AS A TRANSPORT PROCESS

Endocytosis is a transport process which allows cells to interiorize extracellular material (1). Endocytic vesicles form when segments of the plasma membrane invaginate, pinch off, and enclose a volume of extracellular fluid. Fusion of plasma membrane to plasma membrane seals the neck of the vesicles (2) and the sites from which they invaginate. Fusion of the endocytic vesicle with another membrane permits the transport of the contents of the vesicle to another cellular compartment, or to the cell exterior.

Two general types of vesicles have been distinguished on morphological grounds, smooth vesicles, and vesicles which have a filamentous coat on their cytoplasmic surface. Smooth vesicles have been associated mainly with nonselective fluid and solute transport. Coated endocytic vesicles appear to arise from specialized regions of the plasma membrane called coated pits, and have been associated with specific or receptor-mediated endocytosis. However, coated vesicles have also been associated with exocytosis in the secretory pathway for milk protein in mammary gland (3), with delivery of lysosomes to secondary lysosomes in the vas deferens (4), with plasma membrane biogenesis, and with retrieval of synaptic membrane from plasma membrane following nerve discharge (5). Thus coated vesicles appear to play an important part in vesicular transport of a number of specific macromolecules in more than one direction.

In this paper we will review the general types of endocytosis, discuss several well defined examples of receptor-mediated adsorptive pinocytosis, and then focus on those examples of adsorptive pinocytosis in which carbohydrate moieties on glycoproteins are recognized in the ligand-receptor interaction leading to pinocytosis. This class of pinocytosis receptors plays a role in receptor-mediated uptake and transport of acid hydrolases in several mammalian cell types.

TYPES OF ENDOCYTOSIS

Two general types of endocytosis have been distinguished. Phagocytosis refers to the uptake or ingestion of large particles with relatively little fluid. It occurs mainly in specialized phagocytic cells. Pinocytosis refers to the uptake of fluid and fluid contents, and is a property of all cells. In both processes, the endocytosed material enters the cell in a membrane-enclosed vesicle whose ultimate cellular destination and fate varies depending on the cell type and, to some extent, on the material ingested. In both cases, the endocytic process may involve interiorization of large amounts of plasma membrane components which can enter intracellular components and recycle back to plasma membrane. Pinocytosis is further divided into adsorptive or receptor-mediated pinocytosis, which is selective, and fluid phase pinocytosis, which is nonselective.

Phagocytosis

Phagocytosis refers to ingestion of large particles such as erythrocytes or latex spheres (6). It differs from pinocytosis in that it occurs mainly in specialized phagocytic cells, it is clearly triggered by the particle ingested, it has different energy requirements than pinocytosis, and it is much more dependent on the cytoskeleton (is cytochalsin B sensitive).

Pinocytosis

Pinocytosis is a ubiquitous process found in all cells that leads to the interiorization of fluid, solutes, and small particles. Pinocytosis can be selective (adsorptive) in which specific macromolecules are interiorized preferentially, or it may be nonselective (7). Both types of pinocytosis can transport macromolecules from the exterior to intracellular compartments, or can be coupled with exocytosis to produce transcellular transport of vesicle contents.

Adsorptive Pinocytosis. Adsorptive pinocytosis refers to the uptake of macromolecules for which there are binding sites on the plasma membrane. The uptake system is saturable because the binding sites are finite in number. At low ligand concentrations, high affinity receptors permit pinocytosis to concentrate bound ligands from the medium with little intake of fluid or of soluble unbound macromolecules relative to the receptor-bound ligand. It is important to remember that not all saturable binding sites are physiological receptors. To mention a few examples, low density lipoprotein binds to glass beads,

antibodies bind to plastic surfaces, lectins bind to cell surface carbohydrates, and insulin binds to talcum powder. Saturable binding
proves only a limited number of binding sites for a particular ligand,
not that the binding sites represent physiological receptors.

Fluid Phase Pinocytosis. Fluid phase pinocytosis refers to uptake in endocytic vesicles of fluid and of soluble unbound markers in
proportion to their concentration in the extracellular environment.
This process has been studied quantitatively by Steinman et al. (8), and
delivers fluid (0.1 μl/mg/hr), electrolytes, soluble macromolecules,
and membrane components to the cell interior. Horseradish peroxidase
(HRP) is a commonly used marker for studies of fluid phase pinocytosis,
since there are no binding sites for this enzyme on many types of cells.
However, HRP is subject to adsorptive pinocytosis by certain cells of
the mononuclear phagocytic cell system and should be used with cau-
tion as a fluid phase marker for these cell types. The quantitative
studies mentioned above led to the conclusion that vesicular transport
of HRP containing vesicles from the medium to the cell interior is a very
active process which utilizes around 50% of the entire plasma membrane
of the cell surface of mouse L cells per hour. The rapid interiorization
of plasma membrane components in endocytic vesicles, and the slow
turnover calculated for membrane proteins, led to the conclusion that
plasma membrane components may be internalized and returned to the
plasma membrane surface (recycled) many times in their life cycle.

Vesicular Transport Across Endothelial Cells. A specialized
form of fluid phase endocytosis is seen in endothelial cells where endocytosis is coupled with exocytosis to transfer fluid from the luminal
face to the tissue face of endothelial cells and back (9). This vesicular
transport pathway can be viewed as an extension of the circulatory
system. It produces high volume, bi-directional, nonselective move-
ment of fluid and solutes in endocytic vesicles. Because endocytosis
is followed directly by exocytosis, the lysosomal system is largely by-
passed.

NONCARBOHYDRATE SIGNALS FOR ADSORPTIVE PINOCYTOSIS

As indicated earlier, adsorptive pinocytosis refers to uptake of
macromolecules for which there are binding sites on the plasma membrane. A number of high affinity receptors that mediate adsorptive
pinocytosis of physiologically important molecules have been identified.
Some of the well known examples which do not appear to involve carbohydrate recognition by the pinocytosis receptor are listed in Table 1.
The best studied of these systems is the low-density lipoprotein (LDL)
uptake system (10). LDL is internalized as coated pits containing
receptor-bound ligand invaginate to form coated vesicles (2). The
endocytic vesicles deliver LDL to lysosomes where LDL protein is degraded, LDL cholesterol ester is hydrolyzed, and the released cholesterol influences the regulation of cholesterol biosynthesis (by enzyme

Table 1. Transport of Macromolecules by Adsorptive Pinocytosis
          Systems

Ligand	Cell Types	Role
LDL	Fibroblasts, etc.	Cholesterol transport through lysosomes
Transcobalamin II	Fibroblasts, etc.	Cobalamin transport through lysosomes
Immunoglobulins	Epithelial cells	1) Secretion, IgA, IgM 2) Transepithelial uptake of humoral immunity
α_2-Macroglobulin-protease	Fibroblasts	Protease clearance?
Polypeptide hormones		
Insulin	Lymphocytes, etc.	Receptor down-regulation
EGF	Fibroblasts, etc.	"
hCG	Leydig cells	"

repression), cholesterol esterification (by enzyme induction), and
LDL receptor synthesis. Kinetic analysis of internalization suggests
that the receptor is long lived (t$\frac{1}{2}$ many hours), and that it is capable
of being reutilized many times (recycled).

Another physiologically important pinocytosis system is that which
mediates the uptake of transcobalamin II, a cobalamin transport protein
which is delivered by adsorptive pinocytosis to the lysosomes of most,
perhaps all, cells (11). Cobalamin is released from lysosomes by a
mechanism still unclear to become available to enzymes that require
cobalamin as a cofactor. This pinocytosis system does not appear subject to regulation, i.e. the receptor is expressed constitutively. This
is reasonable for a transport system designed to trap trace quantities
of an essential vitamin which is present in insufficient quantities normally to saturate the carrier protein.

Adsorptive pinocytosis is involved in immunoglobulin transport
(12), both in IgG and IgA secretion, and in the uptake of humeral

immunity in the newborn intestine. These systems largely bypass the lysosomal apparatus to provide transepithelial transport, a traffic pattern analagous to that described above for nonselective vesicular transport across endothelial cells. A system for adsorptive pinocytosis of α_2 macroglobulin-protease complex has recently been defined (13). This system may be important for removal of proteases from the extracellular environment. Considerable interest has developed also in the adsorptive pinocytosis of polypeptide hormones (14-16). The internalization of receptor-bound hormones leads to degradation of the hormone and to disappearance of the high affinity receptors from the cell surface, a phenomenon which has been called "down regulation." While this phenomenon may be important to controlling the hormone receptor level of the cell surface, it does not appear essential for the major metabolic consequences of hormone binding, i.e. for hormone action.

From analyses of these examples of adsorptive pinocytosis systems, it is clear that adsorptive pinocytosis receptors differ in more ways than their binding specificity and their cellular distribution. Some are long lived and reused, i.e. they recycle in the absence of new protein synthesis (the LDL and the TcII receptors). Some disappear following internalization and are presumed to be metabolized (EGF and hCG receptors). At least one receptor, the TcII receptor is synthesized constitutively. In fact, the TcII receptor is neither subject to repression nor "down regulation." At least one receptor, the LDL receptor, is subject to feedback regulation even though long lived. The LDL receptor has a long half-life (20 hours) and recycles, but its synthesis can be repressed by cholesterol after which receptor number falls gradually as expected from its half-life. Note the difference between "down regulation" of polypeptide hormone receptors, which is rapid and implies receptor catabolism following endocytosis, and the repression of receptor synthesis by LDL. Receptor synthesis is not regulated directly by LDL pinocytosis. It is regulated by the sterol released by enzymatic hydrolysis following delivery of LDL containing cholesterol ester to lysosomes. In other words, LDL receptor regulation is not dependent on receptor utilization per se, but on a secondary or tertiary consequence of pinocytosis of LDL.

LEVELS OF STUDY OF ADSORPTIVE PINOCYTOSIS SYSTEMS

Whole animal studies have provided the initial evidence for several pinocytosis systems which were recognized by rapid clearance of an infused ligand. In these studies, one infuses the ligand and measures the rate of clearance of the ligand from plasma; the cell type(s) which mediate its clearance; the fate of the cleared ligand-urine, bile, etc? Intact or degraded?; and the effects of agents which modify its clearance. Such studies led to the discovery and characterization of the clearance system for asialo-ceruloplasmin, which led to the elegant studies defining the galactose receptor on hepatocytes (17).

The next level of analysis is the isolated cell preparation which permits direct analysis of binding and endocytosis. Here one studies a) The kinetics of binding to cells at low temperature to estimate the receptor number and affinity and to define the requirements for binding; b) The kinetics of endocytosis at higher temperatures to study the rate of internalization, receptor turnover, and receptor regulation; c) The effects of agents which modify or fail to modify uptake such as competitive inhibitors, drugs which inhibit protein synthesis such as cycloheximide, or drugs which affect the cytoskeleton such as colchicine; and d) receptor-ligand visualization to determine the receptor geography. Are they diffuse, clustered, in coated pits?

The final level of study is the isolated receptor which permits direct analysis of ligand receptor interactions. Here one determines the requirements for and the kinetics of binding and compares them to those determined for whole cell preparations. A receptor binding assay may also permit analysis of the subcellular distribution of receptors which may in turn shed light on the biological function of the adsorptive pinocytosis system.

CARBOHYDRATE RECOGNITION MARKERS IN ADSORPTIVE PINOCYTOSIS OF GLYCOPROTEINS

Table 2 lists the carbohydrate recognition signals that have been identified in adsorptive pinocytosis of glycoproteins. Carbohydrate involvement in recognition by these pinocytosis receptors has been established by three general methods. One method is to modify the ligands chemically (eg, by periodate treatment) or enzymatically (eg, by glycosidase treatment). Loss of uptake following these treatments implicates carbohydrate in the recognition marker on the ligand, and may define the sugar recognized, assuming the glycosidase is pure and specific. A second method is to study the effects of sugars or sugar containing compounds as competitive inhibitors of binding, uptake, or clearance of the ligand. The third method is to use synthetic ligands (neoglycoproteins) in which sugars are attached to proteins and the characterized glycoconjugates are studied either as ligands themselves or as competitive inhibitors of ligand binding or uptake.

Of those examples listed in Table 2, the galactose receptor on hepatocytes is clearly the best defined (17). It functions in clearance of asialo-glycoproteins from the circulation and mediates clearance of cobalamin bound to a cobalamin carrier protein that is found in granulocytes (18). Recent evidence from Brady's laboratory (19) suggests that it may also mediate uptake by hepatocytes of at least one lysosomal enzyme (β-glucocerebrosidase) which has oligosaccharide side chains of the complex type (i.e. GlcNAc, Gal, sialic acid containing). For an extensive review of the galactose receptor on hepatocytes, see the review by Ashwell and Morell (17).

Table 2. Carbohydrate Recognition Signals for Adsorptive
            Pinocytosis of Glycoproteins

Signal on ligand	Cell expressing the receptor	References
Gal (GalNAc, Glc)	Hepatocyte, mammalian	17
GlcNAc	Hepatocyte, avian	21
L-Fuc (α1→3) GlcNAc	Hepatocyte, mammalian	51
Man (GlcNAc, Glc)	Reticuloendothelial	24,25 23,27,29 22 28,52
Man-6-P	Fibroblasts	34-37

We will discuss in detail the mannose/N-acetylglucosamine uptake system and the phosphomannosyl-glycoprotein uptake systems,
which have been defined more recently, and which appear to play
important roles in the transport of acid hydrolases in mammalian cellular systems.

THE MANNOSYL/N-ACETYLGLUCOSAMINE CLEARANCE SYSTEM ON RETICULOENDOTHELIAL CELLS (CELLS OF THE MONONUCLEAR PHAGOCYTIC SYSTEM)

A mammalian system for clearance of infused N-acetylglucosamine (GlcNAc) terminal glycoproteins was proposed by Stockert et al.
(20). Such a clearance system was shown in pigeons and a GlcNAc
receptor was isolated from avian liver but little binding activity was
found in rabbit liver (21). A mannose glycoprotein clearance system
was also suggested by the rapid clearance of several mannose terminal
glycoproteins (22). Stahl et al. (23) presented evidence for a carbohydrate mediated recognition system for clearance of infused rat lysosomal enzymes in the rat and, based on the inhibition of clearance by
agalacto-orosomucoid (terminal N-acetylglucosamine), concluded that
the enzymes were cleared by the GlcNAc recognition system mentioned
above. The studies summarized below demonstrated that lysosomal
hydrolases are cleared by a reticuloendothelial uptake system that
recognizes both mannose-terminal and GlcNAc-terminal glycoproteins.

We used the rat animal model system for study of the fate of infused human β-glucuronidase from different organ sources (24,25).

Infused human placental enzyme which was predominantly a "low uptake" form for fibroblasts, i.e. not subject to adsorptive endocytosis by fibroblasts, was rapidly cleared from rat plasma and localized predominantly in rat liver. Prior treatment of the enzyme with sodium periodate converted the enzyme into a slow clearance form (Fig. 1). Clearance of the enzyme was inhibited by both GlcNAc-terminal (Fig. 1) and mannose-terminal glycoproteins (Fig. 2). These studies suggested either that two different systems contributed to clearance of the enzyme, or that one system with relaxed specificity led to clearance of both mannose-terminal and GlcNAc-terminal glycoproteins. Evidence for the latter came from experiments in which yeast mannans (mannose-terminal) and agalacto-orosomucoid (GlcNAc-terminal) were found to cross-compete with each other for clearance (26), suggesting a single clearance system for glycoproteins with either mannose or N-acetylglucosamine as the terminal, nonreducing sugar.

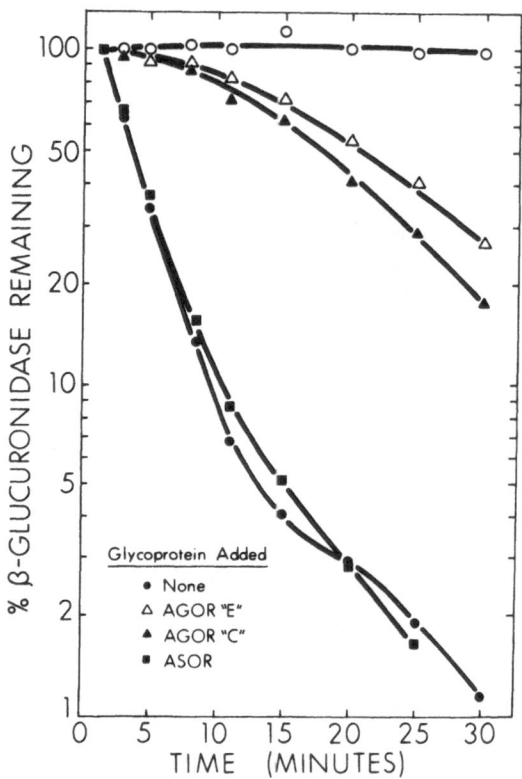

Figure 1. Inhibition of Clearance of Infused Placental β-Glucuronidase by Orosomucoid Derivatives and by Periodate Treatment. All rats received 0.5 ml of 0.15 M NaCl containing 200,000 units of human placental β-glucuronidase, with or without the indicated glycoproteins. Tail vein infusions required 30 sec. Arterial blood samples were taken at the times indicated following the start of the infusion. Results

are expressed as a percentage of the first post-infusion sample taken. Clearance of β-glucuronidase only (●), +5 mg asialo-orosomucoid (■), +2 mg chemically prepared agalacto-orosomucoid (▲), +2.2 mg enzymatically prepared agalacto-orosomucoid (△) and periodate--treated β-glucuronidase (○). From Achord et al., reference 25.

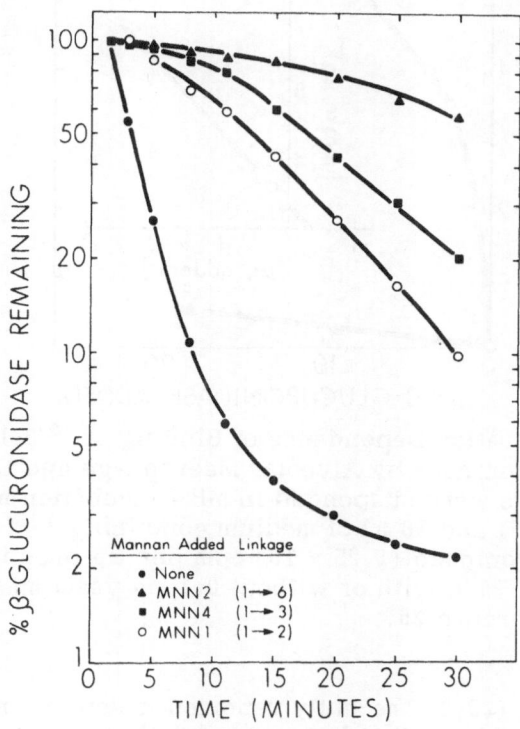

Figure 2. Clearance of Infused Placental β-Glucuronidase in the Presence of Various S. Cerevisiae Mutant Mannans. Rats were infused and blood samples taken as described in Figure 1. Animals received 200,000 units of human placental β-glucuronidase only, or in combination with 1 mg of either S. cerevisiae mannan mnn1, mnn2 or mnn4. From Achord et al., reference 25.

An uptake system with these properties was described by Stahl et al. (27) in isolated rat alveolar macrophages. This preparation was found to mediate uptake of human placental β-glucuronidase (Fig. 3) and the adsorptive pinocytosis system in isolated cells was inhibited by the same competitive inhibitors that inhibit clearance of infused enzyme in the whole rat (Fig. 4). Alveolar macrophages from the rat, sheep, human, and mouse all have a similar uptake system for mannose glycoconjugates.

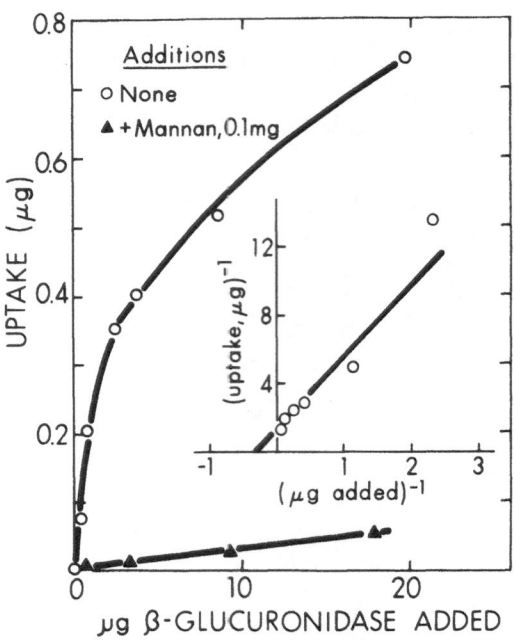

Figure 3. Concentration Dependence of Binding of ^{125}I-Labeled Human Placental β-Glucuronidase by Alveolar Macrophage and Effect of Mannan. Alveolar cells were suspended in MEM incubation medium. 50 μl of cells (5×10^5) and 50 μl of medium containing ^{125}I-labeled human placental β-glucuronidase (1.25×10^4 cpm per μg protein) were incubated for 30 min at $37°$C with or without 0.1 mg yeast mannan. From Achord et al., reference 25.

Two groups (28,29) recently reported isolating a mannan-binding activity from rat liver that is present in other organs containing macrophages. The isolated receptor has the carbohydrate specificity predicted from the whole animal and isolated cell studies. The aggregate of these in vivo and in vitro studies indicate that there is a pinocytosis receptor on fixed tissue mononuclear phagocytic cells (including alveolar macrophages and Kupffer cells of the liver) that leads to adsorptive pinocytosis of glycoproteins with terminal mannose or N-acetylglucosamine. This system leads to capture of many lysosomal enzymes and explains the rapid clearance of enzymes isolated from liver and placenta following infusion into experimental animals, or into enzyme deficient human patients.

The main physiological role of this pinocytosis system is presently unclear. It probably does function to clear the circulation of lysosomal enzymes released from dying cells or from circulating granulocytes and platelets. It may have a role in host defense as it may mediate phagocytosis of microorganisms with exposed mannose

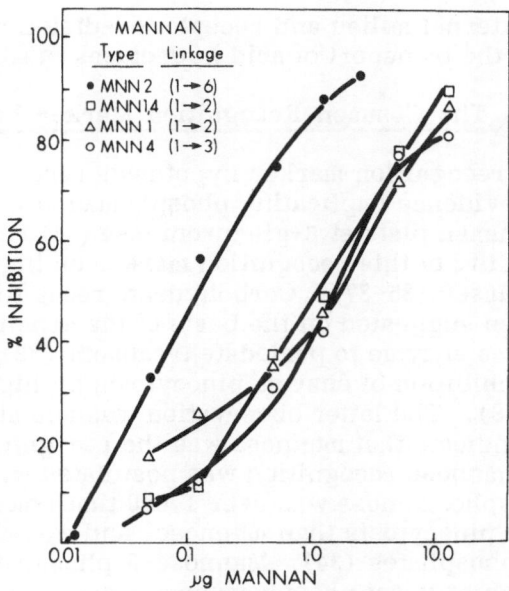

Figure 4. Effect of Various Mutant Mannans on the Binding of ^{125}I-labeled Human Placental β-Glucuronidase by Alveolar Macrophages. Assay conditions were used as described in Figure 4 with 4.0 μg of ^{125}I-labeled human placental β-glucuronidase (1.25 x 10[5] cpm). The mutant mannan concentration was varied in the incubation mixture, and inhibition of uptake of ^{125}I-β-glucuronidase was calculated by comparison to a control incubation without mannan. From Achord et al., reference 25.

or N-acetylglucosamin residues. It may provide a means of directing enzymes or other agents to reticuloendothelial target organs where that is desirable, as in enzyme therapy for Gaucher's disease (30). On the other hand, directing infused lysosomal enzymes or other glycoproteins with exposed mannose or N-acetylglucosamine residues to non-reticuloendothelial target sites may require a specific strategy to prevent capture by this highly efficient pinocytosis system (25).

THE PHOSPHOMANNOSYL RECOGNITION SYSTEM ON FIBROBLASTS

Adsorptive pinocytosis of acid hydrolases by fibroblasts was first recognized as uptake of corrective factors by enzyme deficient fibroblasts (31). With a number of enzymes, the process displayed the selectivity and saturability of a receptor-mediated process (32). The finding that I-cell disease fibroblasts were deficient for many enzymes but secreted large amounts of recognition defective enzymes led to two hypotheses: 1) that many acid hydrolases share a common recognition marker for uptake that is missing or masked on hydrolases from I-cell disease fibroblasts (33), and 2) that secretion of acid hydrolases into the external milieu and receptor-mediated recapture are obligatory steps in the transport of acid hydrolases in fibroblasts (32).

Phosphomannose Is The Common Recognition Marker For Uptake

The common recognition marker hypothesis received strong support from indirect evidence implicating phosphomannose in the recognition marker for human platelet β-glucuronidase (34), and subsequent evidence for generality of this recognition marker on high-uptake forms of many acid hydrolases (35-37). Carbohydrate recognition in enzyme pinocytosis had been suggested on the basis of the sensitivity of the uptake activity of one enzyme to periodate treatment (38), and later on the basis of weak inhibition of enzyme pinocytosis by high concentrations of mannose (39). The latter observation was mistakenly interpreted initially to indicate that mannose was the recognition marker (39,40). Phosphomannose recognition was postulated when we discovered that 6-phosphomannose was over 1,000 times more potent an inhibitor of enzyme pinocytosis than mannose, and was also more potent than other hexose phosphates (34). Mannose-6-phosphate in certain macromolecules such as yeast phosphomannans was even more inhibitory (100 times the potency of free 6-phosphomannose) (41). We recently argued from this and other data that the greater efficiency of the polyphosphomonoester macromolecules resulted from their ability to interact as multivalent ligands (42). The presence of the phosphate as a phosphomonoester on the enzyme was deduced from the observation that alkaline phosphatase converted the enzymes to low uptake forms, no longer susceptible to adsorptive pinocytosis by fibroblasts (34). More recent data shows that treatment of high uptake enzymes with endo-β-N-acetylglucosaminidase H also converts them to low uptake forms, indicating that the 6-phosphomannose is present on high-mannose oligosaccharide chains (43). The generality of these observations led us to suggest that 6-phosphomannose, or close structural analogue of 6-phosphomannose, was an essential component of the common recognition marker on acid hydrolases that was missing or masked in enzymes secreted by I-cell fibroblasts (35). We subsequently reported ^{32}P incorporation into high-uptake glycoproteins that were secreted by fibroblasts maintained in ^{32}P containing serum-free medium (44). Recently Hasilik et al. (42) reported ^{32}P incorporation into immunoprecipitable acid hydrolases secreted by normal fibroblasts, and the failure to observe ^{32}P incorporation into I-cell secretion enzymes.

The indirect evidence leading to predictions of 6-phosphomannose on high-uptake enzymes was finally supported by direct evidence when Marvin Natowicz recently demonstrated 6-phosphomannose release on acid hydrolysis of high-uptake human β-glucuronidase from spleen. He further showed in a comparison of several purified enzymes which differed in susceptibility to pinocytosis, a direct correlation between mannose-6-phosphate content and the uptake activity of the enzyme (M. Natowicz and W. Sly, submitted).

Phosphohexose As The Signal For Intracellular Segregation Of Acid Hydrolases From Secretory Glycoproteins Destined For Export

Before 6-phosphomannose was identified as the recognition marker on enzymes taken up by fibroblasts, Neufeld et al. (32) proposed the provocative hypothesis that secretion and recapture were obligatory steps in enzyme transport to lysosomes. The finding that 6-phosphomannose could inhibit enzyme uptake made it possible to test this hypothesis in its simplest form (44,45). When fibroblasts were grown for 10 days in 6-phosphomannose sufficient to completely block adsorptive pinocytosis of added enzyme (10 mM Man-6-P), no significant reduction in internal enzymes was seen (46). One would have predicted a substantial reduction if newly synthesized enzyme is normally secreted, equilibrates with the medium, and is then recaptured. Its reuptake should be blocked by mannose-6-phosphate. Failure to see this effect makes it clear that, although some high uptake enzyme is secreted into the medium of growing fibroblasts and can be taken up, secretion of enzyme, equilibration with the media, and recapture from the media, are not obligatory steps for the bulk of newly synthesized enzymes.

An alternate hypothesis was proposed by Lloyd (48) who suggested that acid hydrolases do not normally leave cells enroute to lysosomes, but are packaged as lysosomes form by budding off the endoplasmic reticulum. However, he proposed that the recognition marker was required for enzymes to remain in lysosomes. Endocytic vesicles continually enter the cell, fuse with lysosomes, pinch off, and return plasma membrane components to the plasma membrane as exocytic vesicles. In Lloyd's view, the recognition marker is essential to anchor the enzymes to lysosomal membranes and prevent their loss from lysosomes to the medium via exocytic vesicles. Loss of enzyme to the medium in I-cell disease results from absence of the anchor in this view. Two observations are incompatible with Lloyd's hypothesis. First, it was shown by Glaser et al. (49) that when high uptake enzyme is taken up by fibroblasts, the recognition marker essential for uptake is rapidly removed (within hours). However, the half-life of β-glucuronidase taken up by fibroblasts is greater than 2 weeks. Thus the recognition marker required for entry to lysosomes by pinocytosis is removed rapidly but the enzyme remains a very long time, indicating that retention of the recognition marker is not involved in retention of lysosomal enzymes in lysosomes. The recognition marker appears required only for delivery of lysosomes, not for retention in lysosomes.

A third hypothesis was presented by von Figura and Weber (47). They proposed that acid hydrolases normally are delivered to plasma membrane in vesicles that arise from Golgi or GERL, but the enzyme is tightly bound to receptors on the plasma membrane and does not equilibrate with the cell culture medium. This is a modification of the

original secretion-recapture hypothesis. It postulates that the true
intermediate in transport to lysosomes is not enzyme secreted into the
medium, but enzyme bound to specific receptors on the plasma membrane.
Loss of enzymes to the medium in I-cell disease in this model results
from absence of the recognition marker which allows release rather
than retention of enzymes as vesicles bearing newly synthesized enzyme
fuse with the plasma membrane.

We proposed a somewhat different hypothesis to explain the biogenesis of acid hydrolases and the findings in I-cell disease (46). We
might call it the phosphomannosyl-enzyme segregation hypothesis.
We suggested that the 6-phosphomannose recognition marker on acid
hydrolases is an intracellular traffic signal that directs high-uptake
forms of lysosomal enzymes to lysosomes and prevents their secretion.
It does so by binding to a specific receptor in the rough endoplasmic
reticulum or GERL (at least to a pre Golgi membrane). Evidence cited
earlier suggests that this phosphomannosyl recognition marker is present on high-mannose oligosaccharide chains. Its presence appears
to prevent processing of high-mannose oligosaccharide chains in the
Golgi apparatus to complex oligosaccharide chains. Absence of the
recognition marker, as in I-cell disease, does not lead simply to secretion of enzyme lacking 6-phosphomannose. Rather, it leads to secretion
of nearly all newly synthesized enzymes as qualitatively different species of enzymes. The I-cell secretion enzymes bind quantitatively to
Ricin-sepharose columns (43) (i.e. contains galactose) and have been
reported to contain sialic acid (50). Thus, it appears that absence of
the recognition marker is associated not only with diversion of nearly
all newly synthesized enzymes to the outside, but also with secretion
of enzymes which have gone through a different oligosaccharide processing pathway in the Golgi. The oligosaccharides on I-cell enzymes
have been converted into complex type (i.e. GlcNAc, Gal, sialic acid-
containing oligosaccharide chains).

This hypothesis which is consistent with most of the data says
that 6-phosphomannose is the intracellular traffic signal which allows
the segregation of one class of glycoproteins, the acid hydrolases,
from those to be processed for secretion. Presumably all glycoproteins
are secreted into the cisternal space of the rough endoplasmic reticulum
by the co-translational transfer via a ribosome-membrane junction (the
signal hypothesis). The 6-phosphomannose is a "second signal" used
to segregate one special class of glycoproteins, the acid hydrolases,
from true secretory glycoproteins destined for export. The 6-phos-
phomannose "second signal" which is present on the high-mannose
oligosaccharide chains of acid hydrolases, acts by binding to receptors
on membranes of the rough ER, GERL, or Golgi, thus segregating the
acid hydrolases from the soluble glycoproteins. Receptor-bound enzyme could segregate physically intracellularly by budding off as a
specialized vesicle of receptor-bound enzymes (a primary lysosome)
or segregate from soluble glycoproteins only after both are delivered

to the plasma membrane with release (secretion) of the soluble glyco-proteins, and retrieval of receptor-bound enzyme by endocytosis. Enzyme lacking the recognition marker would fail to be segregated and would proceed through the Golgi where the oligosaccharide chains would be processed to complex type, and the acid hydrolases would be secreted with other soluble secretory glycoproteins and would resemble them in their carbohydrate structure.

Judging from the Ricin-binding properties of lysosomal enzymes secreted by normal fibroblast enzymes, this processing pathway for acid hydrolases (bearing complex type oligosaccharide chains) is nor-mally not prominent in fibroblasts. However, it may be a prominent pathway of acid hydrolase secretion in certain specialized cells (e.g. those contributing to semen). This pathway probably produces the low level of sialated forms of hydrolases that are found in normal serum.

This hypothesis differs from the hypothesis proposed by Vladutiu and Ratazzi (50) to explain the enzyme findings in I-cell disease. They explain the excess sialation of I-cell enzymes as a consequence of "pref-erential exocytosis" which leads to failure of removal of sialic acid residues from lysosomal enzymes that normally occurs following uptake into lysosomes. We suggest that the sialated acid hydrolases do not reflect a failure to arrive at lysosomes and be desialated, but rather a change in oligosaccharide processing that results from absence of the recognition marker that normally segregates them from secretory gly-coproteins.

REFERENCES

1. S. C. Silverstein, R. M. Steinman, and Z. A. Cohn, Endocyto-sis, Ann. Rev. Biochem. 46: 669-722 (1977).
2. R. G. W. Anderson, M. S. Brown, and J. L. Goldstein, Role of the Coated Endocytic Vesicle in the Uptake of Receptor-bound Low Density Lipoprotein in Human Fibroblasts, Cell 10: 351-364 (1977).
3. W. W. Franke, M. R. Luder, J. Kortenbeck, H. Zerban, and T. W. Keenen, Involvement of vesicle coat material in casein secretion and surface regeneration, J. Cell Biol. 69: 173-195 (1976).
4. D. S. Friend, and M. G. Farquahr, Functions of coated vesicles during protein absorption in the rat vas deferens, J. Cell Biol. 35: 357-376 (1967).
5. J. E. Heuser, and T. S. Reese, Evidence for recycling of syn-aptic vesicle membrane during transmitter release at the frog neuromuscular junction, J. Cell Biol. 57: 315-344 (1973).
6. S. C. Silverstein, J. Michl, and S. S. J. Sung, Phagocytosis, in: Transport of Macromolecules in Cellular Systems, S. C. Silverstein, ed., Dahlem Konferenzen, Berlin, pp. 245-264 (1978).

7.   P. J. Jacques, Endocytosis, in: Lysosomes in Biology and Path-
      way, J. T. Dingle and H. B. Fel, eds., North-Holland Pub-
      lishing Co., Amsterdam, pp. 395-420 (1969).

8.   R. M. Steinman, J. M. Silver, and Z. A. Cohn, Pinocytosis in
      fibroblasts: Quantitative studies in vitro, J. Cell Biol.
      63: 665-687 (1974).

9.   G. E. Palade, M. Simionescu, and N. Simionescu, Transport of
      solutes across vascular endothelium, in: Transport of Macro-
      molecules in Cellular Systems, S. C. Silverstein, ed., Dahlem
      Konferenzen, Berlin, pp. 145-166 (1978).

10.   J. L. Goldstein, and M. S. Brown, The LDL pathway in human
      fibroblasts: a receptor mediated mechanism for the regulation
      of cholesterol metabolism, in: Current Topics in Cellular
      Regulation, 11, B. L. Horecher and E. R. Stadtman, eds.,
      Academic Press, New York, pp. 147-181 (1976).

11.   P. Youngdahl-Turner, L. E. Rosenberg, and R. H. Allen, Bind-
      ing and uptake of transcobalmin II by human fibroblasts, J.
      Clin. Invest. 61: 133-141 (1978).

12.   J. P. Kraehenbuhl, and L. Kuhn, Transport of Immunoglobulins
      Across Epothelia, in: Transport of Macromolecules in Cellu-
      lar Systems, S. C. Silverstein, ed., Dahlem Konferenzen,
      Berlin, pp. 213-228 (1978).

13.   F. Van Leuven, J. J. Cassimon, and H. Van Den Berghe, Uptake
      and degradation of α_2-Macroglobulin-protease complexes in
      human cells in culture, Expt. Cell Res. 117: 273-282 (1978).

14.   G. Carpenter, and S. Cohen, ^{125}I-labeled human epidermal
      growth factor: Binding, internalization and degradation in
      human fibroblasts, J. Cell Biol. 71-159-171 (1976).

15.   M. Ascoli, and D. Puett, Degradation of receptor-bound human
      choriogonadotropin by murine Leydig tumor cells, J. Biol.
      Chem. 253: 4892-4899 (1978).

16.   F. R. Maxfield, J. Schlessinger, Y. Schecter, I. Pastan, and
      M. C. Willingham, Collection of insulin, EGF, and α_2-Macro-
      globulin in the same patches on the surface of cultured fibro-
      blasts and common internalization, Cell 14: 805-810 (1978).

17.   G. Ashwell, and A. G. Morell, The role of surface carbohydrates
      in the hepatic recognition and transport of circulating gly-
      coproteins, Adv. Enzym. 41: 99-128 (1974).

18.   R. L. Burger, R. J. Schneider, C. S. Mehlman, and R. H. Allen,
      Human R-type vitamin B_{12} binding proteins. II. The role of
      transcobalamin I, transcobalamin III, and the normal granu-
      locyte vitamin B_{12} binding protein in the plasma transport of
      vitamin B_{12}, J. Biol. Chem. 250: 7703-7713 (1975).

19.   F. S. Furbish, C. J. Steer, J. A. Barringer, E. A. Jones, and
      R. O. Brady, The uptake of native and desialylated gluco-
      cerebrosidase by rat hepatocyte and Kupffer cells, Biochem.
      Biophys. Res. Commun. 81: 1047-1053 (1978).

20.   R. O. Stockert, A. G. Morell, and I. H. Scheinberg, The exis-
      tence of a second route for transfer of certain glycoproteins

from the circulation into the liver, Biochem. Biophys. Res. Commun. 68: 988-993 (1976).

21.   J. Lunney, and G. Ashwell, A hepatic receptor of avian origin capable of binding specifically modified glycoproteins, Proc. Nat. Acad. Sci. USA 73: 341-343 (1976).

22.   T. L. Brown, L. A. Henderson, S. R. Thorpe, and J. W. Baynes, The effect of alpha-mannose-terminal oligosaccharides on the survival of glycoproteins in the circulation, Arch. Biochem. Biophys. 188: 418-428 (1978).

23.   P. Stahl, H. Six, J. S. Rodman, P. Schlesinger, D. R. P. Tulsani, and O. Touster, Evidence for specific recognition sites mediating clearance of lysosomal enzymes in vivo, Proc. Nat. Acad. Sci. USA 73: 4045-4049 (1976).

24.   D. T. Achord, F. E. Brot, A. Gonzalez-Noriega, W. S. Sly, and P. Stahl, Human β-glucuronidase II. Fate of infused human placental β-glucuronidase in the rat, Pediat. Res. 11: 816-822 (1977).

25.   D. T. Achord, F. E. Brot, C. E. Bell, and W. S. Sly, Human β-glucuronidase: In vivo clearance and in vitro uptake by a glycoprotein recognition system on reticuloendothelial cells, Cell 15: 269-278 (1978).

26.   D. T. Achord, F. E. Brot, and W. S. Sly, Inhibition of the rat clearance systems for agalacto-orosomucoid by yeast mannans and by mannose, Biochem. Biophys. Res. Commun. 77: 409-415 (1977).

27.   P. D. Stahl, J. S. Rodman, M. J. Miller, and P. H. Schlesinger, Evidence for receptor-mediated binding of glycoproteins, glycoconjugates, and lysosomal glycosidases by alveolar macrophages, Proc. Nat. Acad. Sci. USA 75: 1399-1403 (1978).

28.   T. Kawasaki, Y. Mizuno, and I. Yamashima, Mannan binding proteins of rat tissues, Fed. Proc. 38: 468 Abst. (1979).

29.   P. Stahl, and P. Schlesinger, Mannose/N-Acetylglucosamine receptor: Plasma clearance and macrophage uptake of glycoconjugates and lysosomal glycosidases. Fed. Proc. 38: 467 Abst. (1979).

30.   C. J. Steer, F. S. Furbish, J. A. Barringer, R. O. Brady, and E. A. Jones, The uptake of agalacto-glucocerebrosidase by rat hepatocytes and Kupffer cells, FEBS Lett. 91: 202-205 (1978).

31.   E. F. Neufeld, T. W. Lim, and L. J. Shapiro, Inherited disorders of lysosomal metabolism, Ann. Rev. Biochem. 44: 357-376 (1975).

32.   E. F. Neufeld, G. N. Sando, A. J. Garvin, and L. H. Rome, The transport of glycosomal enzymes, J. Supramol. Struct. 6: 95-101 (1977).

33.   S. Hickman, and E. F. Neufeld, A hypothesis for I-cell disease: defective hydrolases that do not enter lysosomes, Biochem. Biophys. Res. Commun. 49: 992-999 (1972).

34.   A. Kaplan, D. T. Achord, and W. S. Sly, Phosphohexosyl com-
          ponents of a lysosomal enzyme are recognized by pinocytosis
          receptors on human fibroblasts, Proc. Nat. Acad. Sci. 74:
          2026-2030 (1977).

35.   A. Kaplan, D. Fischer, D. Achord, and W. S. Sly, Phosphohexo-
          syl recognition is a general characteristic of pinocytosis of
          lysosomal glycosidases by human fibroblasts, J. Clin. Inv.
          60: 1088-1093 (1977).

36.   G. N. Sando, and E. F. Neufeld, Recognition and receptor-medi-
          ated uptake of a lysosomal enzyme α-L-iduronidase, by cul-
          tured fibroblasts, Cell 12: 619-627 (1977).

37.   K. Ullrich, G. Mersmann, E. Weber, and K. von Figura, Evi-
          dence for lysosomal enzyme recognition by human fibroblasts
          via a phosphorylated carbohydrate moiety, Biochem. J.
          170: 643-650 (1978).

38.   S. Hickman, L. J. Shapiro, and E. F. Neufeld, A recognition
          marker required for uptake of a lysosomal enzyme by cul-
          tured fibroblasts, Biochem. Biophys. Res. Commun. 57:
          55-61 (1974).

39.   V. Hieber, J. Distler, R. Myerowitz, R. D. Schmickel, and C.
          W. Jourdian, The role of glycosidically bound mannose in the
          assimilation of β-galactosidase by β-galactosidase deficient
          fibroblasts, Biochem. Biophys. Res. Commun. 73: 710-717
          (1976).

40.   G. Sahagian, J. Distler, V. Hieber, R. Schmickel, and G. W.
          Jourdian, Role of mannose-6-phosphate in β-galactosidase
          assimilation, Fed. Proc. 38: 467 Abst. (1979).

41.   A. Kaplan, D. Fischer, and W. S. Sly, Correlation of structural
          features of phosphomannans with their ability to inhibit pino-
          cytosis of human β-glucuronidase by human fibroblasts, J.
          Biol. Chem. 253: 647-650 (1978).

42.   H. D. Fischer, M. Natowicz, W. S. Sly, and R. K. Bretthauer,
          Fibroblast receptor for lysosomal enzyme mediates uptake of
          phosphomannans, Fed. Proc. 38: 467 Abst. (1979).

43.   W. S. Sly, A. Gonzalez-Noriega, M. Natowicz, H. D. Fischer,
          and J. P. Chambers, Role of the phosphomannosyl recogni-
          tion marker in the uptake and transport of lysosomal enzymes.
          Fed. Proc. 38: 467 Abst. (1979).

44.   W. S. Sly, D. T. Achord, and A. Kaplan, Phosphohexose on
          lysosomal enzymes is the common recognition marker for
          pinocytosis receptor on fibroblasts, in: Protein Turnover
          and Lysosome Function, H. L. Segal and D. J. Doyle, eds.,
          Academic Press, New York (1978).

45.   A. Hasilik, L. H. Rome, and E. F. Neufeld, Processing of lyso-
          somal enzymes in human skin fibroblasts, Fed. Proc. 38:
          467 Abst. (1979).

46.   W. S. Sly, and P. Stahl, Receptor mediated uptake of lysosomal
          enzymes, in: Transport of Macromolecules in Cellular Sys-
          tems, S. C. Silverstein, ed., Dahlem Konferenzen, Berlin,

pp. 229-244 (1978).

47.   K. von Figura, and E. Weber, An alternative hypothesis of cellular transport of lysosomal enzymes in fibroblasts, Biochem. J. 176:943-956 (1978).

48.   J. B. Lloyd, Cellular transport of lysosomal enzymes-an alternate hypothesis, Biochem. J. 164:281-282 (1977).

49.   J. H. Glaser, K. J. Roozen, F. E. Brot, and W. S. Sly, Multiple isoelectric and recognition forms of human β-glucuronidase activity, Arch. Biochem. Biophys. 166:536-542 (1975).

50.   G. D. Vladutiu, and M. C. Rattazzi, Abnormal lysosomal hydrolases excreted by cultured fibroblasts in I-cell disease (mucolipidosis II), Biochem. Biophys. Res. Commun. 67:956-964 (1975).

51.   J. P. Prieels, S. V. Pizzo, L. R. Glasgow, J. C. Paulson, and R. L. Hill, Hepatic receptor that specifically binds oligosaccharides containing fucosyl $\alpha 1 \rightarrow 3$ N-acetylglucosamine linkages, Proc. Nat. Acad. Sci. USA 75:2215-2219 (1978).

52.   J. Kawasaki, R. Etoh, and I. Yamashima, Isolation and characterization of a mannan-binding protein from rabbit liver, Biochem. Biophys. Res. Commun. 81:1018-1024 (1978).

ACKNOWLEDGEMENTS:

The author is supported by National Institute of General Medical Sciences grant GM 21096 and by the Ranken Jordan Trust Fund for Crippling Diseases of Children. The author gratefully acknowledges the collaboration of Drs. Philip Stahl and Paul Schlessinger over the past several years. Work from the author's laboratory which was summarized in this paper was that of Drs. Frederick Brot, Kenneth Roozen, Janet Glaser, Daniel Achord, and Arnold Kaplan, as well as that of David Fischer, Alfonso Gonzalez-Noriega, Marvin Natowicz, and Jeffrey Grubb. Their contributions to the work discussed are gratefully acknowledged.

GANGLIOSIDES AS RECEPTORS FOR BACTERIAL TOXINS AND SENDAI VIRUS

J. Holmgren[1], H. Elwing[1], P. Fredman[2], Ö. Strannegård[1] and L. Svennerholm[2]

Institute of Medical Microbiology[1] and Department of Neurochemistry[2], University of Göteborg, S-413 46 Göteborg, Sweden

INTRODUCTION

Gangliosides have recently attracted considerable interest as candidate receptor structures for various bacterial toxins, viruses, interferon and glycoprotein hormones. Speculations about a receptor role for gangliosides have occurred for more than two decades but it was first with the more recent work on cholera toxin that such a function could be shown.

In this article we will review in some detail the current understanding of GM1 ganglioside as receptor for cholera toxin, since this system can be regarded as a prototype for ganglioside receptor studies. We will then describe the results of recent studies of the interactions of tetanus toxin and Sendai virus, respectively, with gangliosides. The use of highly purified gangliosides and newly developed quantitative binding assays (HOLMGREN et al., previous article in this volume) has provided information which allows rather detailed predictions about the recognition specific structures of the biological receptor for these agents.

CHOLERA TOXIN

It was in 1973 that GM1, from independant studies by three groups, was first recognized as the probable receptor for cholera toxin.

A few years earlier VAN HEYNINGEN and co-workers (1971) had

described that a crude ganglioside mixture could inactivate cholera toxin. We demonstrated that the inactivation was due to a specific reaction with a single ganglioside, i.e. GM1 (HOLMGREN et al., 1973). GM1 inhibited the biological effects of cholera toxin in concentrations down to equimolar with toxin. Of all other gangliosides and glycolipids only GD1a and GA1 reacted with cholera toxin; their activity was, however, 500 times less than that of GM1. We found also that GM1, in contrast to all the other substances, gave rise to a specific precipitation band with cholera toxin in Ouchterlony type double diffusion-in-gel tests, thus providing direct evidence for specific binding between toxin and GM1 in a system independent of biological assays. Independently, CUATRECASAS (1973 a), and KING and VAN HEYNINGEN (1973) described GM1 to be the most active ganglioside in reacting with cholera toxin although at an apparent less specificity than in our studies probably due to the fact that they worked with impure ganglioside preparations.

Subsequent studies in several laboratories have provided much further, practically indisputable, evidence that GM1 is the natural biological receptor for cholera toxin:

1) Our studies with various cell types, including small intestinal mucosal cells of different species, demonstrated a direct relationship between the cell content of GM1 and the number of toxin molecules that the cell can bind (HOLMGREN et al., 1975; HANSSON et al., 1977).

2) Chemical modifications of cholera toxin by means of various reagents were consistently found to affect binding to cells and to isolated GM1 to the same extent (HOLMGREN and LÖNNROTH, 1976).

3) MULLIN et al. (1976) observed that pretreatment of cell membranes with cholera toxin specifically blocked the membrane GM1 ganglioside from reacting with galactose oxidase. This, and the surface protein radiolabelling experiments by HART (1974) seem to rule out the possibility that cholera toxin would bind to a structurally related cell membrane glycoprotein receptor rather than to GM1 itself.

4) A very strong support for the receptor role of GM1 comes from experiments showing that exogenous GM1 can be incorporated into the cell membrane and then act as functional receptors. This was first indicated by CUATRECASAS (1973 b) who observed increased binding capacity and lipolytic responsiveness of fat cells which had been soaked in GM1. Using ^{3}H-labelled GM1 we could unambiguously demonstrate incorporation of GM1 into small intestinal epithelial membrane and show that the increase in GM1 was associated with a corresponding increase in the capacity of the intestine to bind cholera toxin as well as in an increased susceptibility of the gut to the diarrheagenic action of the toxin (HOLMGREN et al., 1975). Consistent with these findings MOSS and co-workers (1976) recently

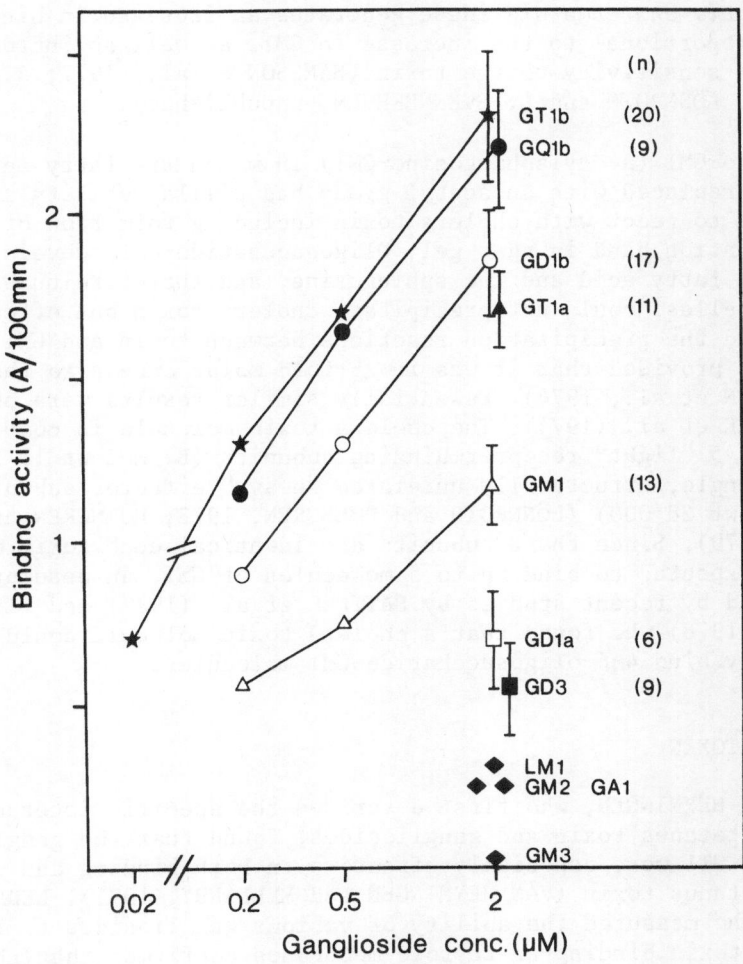

Fig. 1 Binding of tetanus toxin to various gangliosides as
        studied with the ganglioside-ELISA method. The mean and
        SEM (bars) values of n experiments, each performed in
        dupli- or triplicates, are shown.

showed that incorporation of GM1 into ganglioside-deficient transformed cells specifically restored cell responsiveness to cholera
toxin.

5) Finally, we have found that in enzyme-susceptible tissues (not
intestinal mucosa) hydrolysis of the higher membrane gangliosides
with *Vibrio cholerae* sialidase generates an increase in binding
sites proportional to the increase in GM1, as well as increased
cellular sensitivity to the toxin (HANSSON et al., 1977; J. HOLM-
GREN, I. LÖNNROTH and L. SVENNERHOLM, unpublished).

Lyso-GM1 (acetylsphingosine-GM1) in which the fatty acid of
GM1 was replaced with an acetyl group had similar ability as intact GM1 to react with cholera toxin including formation of a
precipitation band in agar gel. Oligosaccharide-GM1, devoid of
both the fatty acid and the sphingosine, and therefore unable to
form micelles, could not precipitate cholera toxin but effectively
inhibited the precipitation reactions between toxin and GM1 or
lyso-GM1 provided that it was in ⩾5-fold molar excess to the toxin
(HOLMGREN et al., 1974). Essentially similar results were obtained
by STAERK et al. (1974). The cholera toxin molecule is composed of
probably 5 "light" receptor-binding subunits (B: mol wt 11 500 each)
and a single, structurally unrelated "heavy" effector subunit
(A: mol wt 28 000) (LÖNNROTH and HOLMGREN, 1973; HOLMGREN and LÖNN-
ROTH, 1979). Since the B subunits are identical each toxin molecule
can be expected to bind up to 5 molecules of GM1, an assumption
supported by recent studies by SATTLER et al. (1977) and FISHMAN
et al. (1978) who found that a cholera toxin molecule could
maximally bind 4-6 oligosaccharide-GM1 molecules.

TETANUS TOXIN

VAN HEYNINGEN, who first described the specific interaction
between tetanus toxin and gangliosides, found that the gangliosides
GD1b and GT1 were especially effective in both binding and neutralizing tetanus toxin (VAN HEYNINGEN and MELLANBY, 1971). LEDLEY et al.
(1977) who measured the ability of various gangliosides to inhibit
tetanus toxin binding to thyroid membranes confirmed that the order
of reactivity of the various tested gangliosides was GT1≃GD1b>GM1≃
GD1a>GM2. In contrast to this HELTING et al. (1977) reported GM1 to
be equally effective as GD1b in binding to tetanus toxin.

The aim of our studies (HOLMGREN, J., ELWING, H., FREDMAN, P.,
and SVENNERHOLM, L., in press) was to further define the degree
of structure specificity in the binding of tetanus toxin to gangliosides and thus identify the presumed oligosaccharide recognition structure of the receptor. To avoid possible unspecific interactions by the ceramide moiety of the gangliosides, binding was
determined to plastic-attached gangliosides using either immuno-

enzyme (ganglioside-ELISA) or vapour condensation-on-surface (VCS)
techniques for quantitation of the bound toxin (see HOLMGREN et al.,
this volume).

    Fig. 1 shows the results obtained with the ganglioside-ELISA
method. In contrast to cholera toxin, which in the same system
bound only to GM1, tetanus toxin bound significantly to several
gangliosides. The strongest reactions were obtained with GT1b, GQ1b
and GD1b closely followed by GT1a. Distinctly lower but still strong
activity was seen with GM1 (difference to e.g. GD1b: p<0.01), and
also GD1a and GD3 were capable of binding tetanus toxin to further
lesser extent (Fig. 1). A decrease in ganglioside concentration
used for coating, from e.g. 2 to 0.2 µM, resulted in markedly reduced
binding to tetanus toxin (Fig. 1); this differed from cholera toxin
the binding of which to GM1 as studied with this method was
practically unchanged within the concentration range 0.02 - 2 µM.
Also a plot for absorbance vs toxin concentration differed markedly
between tetanus and cholera toxin in that considerably more material
was required for detectable binding and the slope was less steep for
tetanus toxin.

    Essentially the same reaction patterns were demonstrated with
the VCS method where the bound toxin was directly visualized, thus
avoiding the possible influence of any subsequent immunoreaction
steps. Tetanus toxin gave positive, "wet" reactions with the same
gangliosides that were positive in the ELISA method, i.e. GT1b,
GQ1b, GD1b and less strongly with, in order GT1a, GM1, GD1a and
GD3 (Fig. 2). The critical dependence on a relatively high ganglioside coating concentration for tetanus but not cholera toxin
was evident also with the VCS technique (Fig. 2 insert).

    Our results by either method that gangliosides of the G1b
series were more effective than GM1 in binding tetanus toxin are
in contrast to those of HELTING et al. (1977) recently. We therefore undertook experiments with enzymatic hydrolysis of the more
complex gangliosides to clarify this matter further. Gangliosides
GT1b, GD1b or GM1 were attached to plastic tubes, and half of the
tubes were subsequently incubated with *V. cholerae* sialidase to
hydrolyse the higher gangliosides to GM1. The binding of tetanus
and cholera toxins to the tubes before and after hydrolysis was
then measured with the ganglioside-ELISA method. Fig. 3 shows
that sialidase treatment of GT1b as well as GD1b decreased the
binding of tetanus toxin to the same level as that obtained in the
GM1-coated tubes; the binding of cholera toxin was instead increased, also to the GM1 tube level, supporting that the presumed
hydrolysis of GT1b and GD1b to GM1 did really take place. From
these studies the following structural conclusions may be drawn:

1) Tetanus toxin has special affinity for the G1b series of
gangliosides.

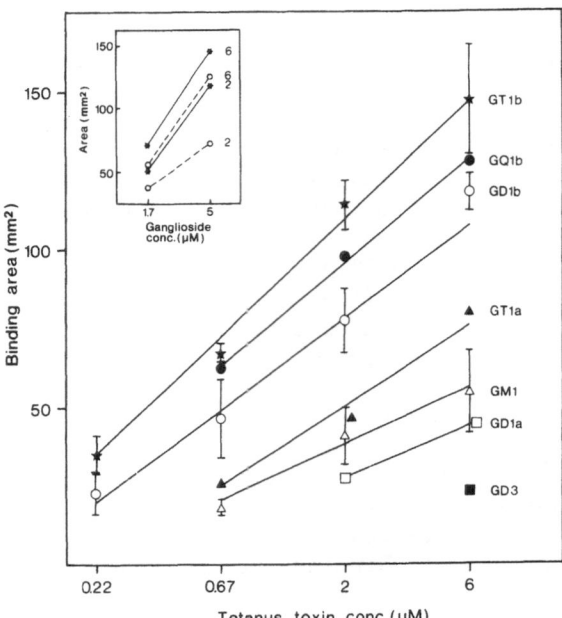

Fig. 2   Binding of tetanus toxin to various gangliosides as deter-
         mined with the VCS method; the ganglioside concentration
         used for coating was 5 μM. The <u>insert</u> shows the effect of
         decreasing the coating concentration for GTlb and GDlb as
         studied with two different concentrations of tetanus toxin
         (6 and 2 μM). Mean values of two or three (here SEM are also
         shown) experiments performed in duplicates are shown.

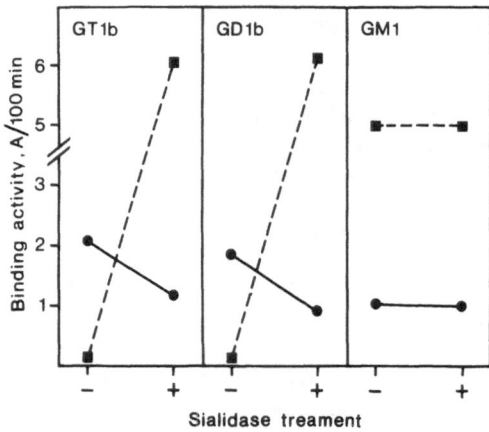

Fig. 3   Effect of hydrolysis of plastic-attached gangliosides
         with <u>V. cholerae</u> sialidase on their binding capacity for
         tetanus (●) and cholera (■) toxins as determined by
         the ganglioside-ELISA method.

2) The number and position of the sialic acid residues are critical for the binding affinity. Thus there is a minimal requirement for one sialosyl residue linked to the inner galactose for detectable binding (compare GM1 and GA1), but for optimal affinity a disialosyl group, preferentially linked to the inner galactose as in the G1b series, is needed (compare e.g. GD1b with GM1 and GD1a). Additional sialic acid residues do not seem to contribute much further to the recognition structure (GQ1b≃GT1b≃GD1b).

3) The oligosaccharide backbone is also of critical importance. GD3 which lacks the terminal galactose and N-acetylgalactosamine residues has little binding activity despite the proper disialosyl linkage to a galactose residue, and GM2 in contrast to GM1 has no binding activity supporting the important role of the terminal galactose.

Are the G1b gangliosides the biological receptors for tetanus toxin? This question remains to be answered by experiments of the type described for cholera toxin. Tetanus toxin seems to bind specifically to presynaptic nerve endings (PRICE et al., 1977). It is very suggestive therefore that in the central nervous system the gangliosides are concentrated in the pre- and postsynaptic membranes of the synaptic junctions, and that it is the high affinity gangliosides for tetanus toxin - GD1b, GT1b and GQ1b - that are particularly enriched in these membranes (HANSSON et al., 1977; SVENNERHOLM, this volume).

SENDAI VIRUS

The initial event in the entry of viruses into cells is the attachment of the virus to specific receptors on the cell membrane. The chemical structures of the receptors for animal viruses are in most instances poorly defined. Cell surface glycoprotein, glyco-lipids, and phospholipids have been implicated. Very recently HELENIUS et al. (1978) could identify human HLA and murine H-2 histocompatibility antigens as receptors for Semliki Forest virus; these antigens are well-defined membrane glycoproteins.

Sendai virus, like other myxo- and paramyxovirus have surface glycoprotein spikes which adsorb to specific receptors on erythrocytes of most mammalian and fowl species and cause hemagglutination. The receptors on erythrocyte membranes contain neuraminic acid as indicated by the fact that they are destroyed by neuraminidase. HAYWOOD (1974) demonstrated that liposomes containing gangliosides could inhibit the agglutination of erythrocytes by Sendai virus. Liposomes without ganglioside had no inhibitory effect suggesting that the Sendai virus receptor might contain or resemble gang-liosides. She further noticed that the effect seemed to require

a   **b**

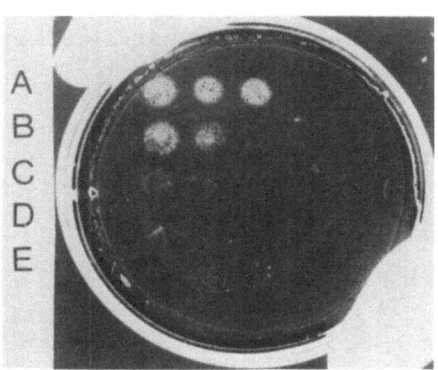   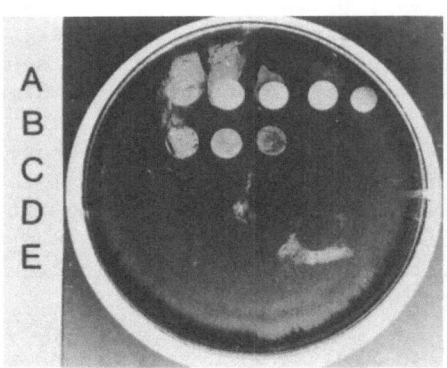

Fig. 4.   Specific binding of Sendai virus to GQlb ganglioside and
          its inhibition by antiserum as demonstrated with (a) the
          VCS method and (b) hemadsorption (see text). A: virus in
          4-fold serial dilutions in buffer; B: in low-titer anti-
          serum, 1%; C: in the same serum, 10% D: in higher titer
          antiserum, 1%; E: this serum, 10%.

ganglioside with more than one sialic acid since commercial di-
and trisialoganglioside preparations could inhibit Sendai virus
hemagglutination in contrast to a monosialoganglioside preparation.

     We have developed simple methods allowing direct visualization
of the specific binding of Sendai virus to gangliosides attached
to a plastic surface. Using these methods and highly purified gangliosides of defined structures we have tried to identify the re-
cognition-specific structure in Sendai virus receptor (HOLMGREN, J.,
ELWING, H., FREDMAN, P., STRANNEGÅRD, Ö. and SVENNERHOLM, L., in
press).

     Sendai virus, purified after growth in the allantoic cells of
fertilized eggs, was applied spotwise in serial dilutions onto the
surface of polystyrene Petri dishes coated with the various gangliosides in different concentrations (virus spot assay). After
incubation for three hours the dishes were washed and exposed to
water vapour (vapour condensation-on-surface (VCS) method, see
HOLMGREN et al., this volume). This resulted in characteristic
"wet" spots with a few of the gangliosides (Fig. 4 a). The

TABLE I. HIGH-AFFINITY BINDING OF SENDAI VIRUS TO G1 GANGLIOSIDES
          WITH A TERMINAL DISIALOSYL GROUP

| | Minimal effective concentration (μM) | | |
| | Virus spot VCS assay | Ganglioside spot assay | |
		VCS	Hemadsorption
GA1	>5	>25	>25
GM1	>5	>25	>25
GM2	>5	>25	25
GM3	>5	>25	n.t.[1]
GD1a	5	2.8	0.9
GD1b	>5	>25	25
GD3	>5	>25	25
GT1a	0.1	n.t.	n.t.
GT1b	5	2.8	2.8
GQ1b	0.1	0.05	0.01
GP1c[2]	0.1	0.05	0.02

[1]n.t. = not tested

[2]from dogfish. Approx 10% of NeuAc was also O-acetylated.

strongest reactions were obtained with ganglioside GTa, GQ1b and
GP1b which gave positive results in concentrations as low as
0.1 μM (Table I). Two other gangliosides, GD1a and GT1b, also had
slight binding capacity but only when they were used in about
100-fold higher concentration. Other gangliosides tested were
completely negative (Table I).

Sendai virus attaches to erythrocytes, and as a means to
support that the wet spots truly represented specifically bound
virus it was investigated whether the spot forming material would
bind erythrocytes secondarily (hemadsorption). This was shown to
be the case. In each instance in which the water condensation
method gave positive results a parallelly titrated plate displayed
specific hemadsorption for the same positions (Fig. 4 b). The
hemadsorption method was slightly more sensitive than the water
condensation technique allowing detection of a 4- to 16-fold
higher virus dilution. Immune serum was shown to specifically
inhibit the binding of Sendai virus to e.g. GQ1b as examined by
either of the two methods (Fig. 4).

A quantitatively more precise method to compare the binding affinity of Sendai virus for the various gangliosides was to attach the gangliosides spotwise in serial dilutions to the plastic and then incubate the whole plate with virus (ganglioside spot assay). The minimal effective coating concentration of each ganglioside for binding Sendai virus could then be determined with either the water condensation or the hemadsorption methods (Table I). The results confirmed those obtained with the virus spot assay showing that the virus affinity for GQ1b and GP1 was about 50- to 100-fold higher than for GD1a and GT1b and >500-fold higher than for any of the other tested substances (Table I).

Thus it is clear that Sendai virus has a very strong binding tendency to GQ1b which seems to exceed that of tetanus toxin to its "receptor" gangliosides and actually approach the binding strength of cholera toxin to GM1. Conversely to the situation with tetanus toxin the sialic acid residues extending from the *terminal* galactose are the critical ones for binding. One such residue is an absolute requirement (compare GT1b and GD1a with GD1b) and a disialosyl group in this position apparently confers maximal binding capacity (GQ1b≈GT1a≈GP1). However, also the N-acetylgalactosamine residue (or the chain length as such) in the backbone seems to contribute markedly to the "receptor" structure, as indicated by the fact that GD3 had only minimal binding capacity in spite of possessing a disialosyl group linked to a terminal galactose. The "receptor" structure for Sendai virus thus seems to be "NeuAc - NeuAc - gal - galNAc" as it is present in GQ1b ganglioside.

MODE OF ACTION OF GANGLIOSIDE RECEPTORS?

In the last few years it has become increasingly apparent in many systems (perhaps clearest here too for cholera toxin) that receptor-induced triggering of biological events is generally much more complex than the mere stereospecific binding of the agonist. It is likely that such events, when elucidated in greater detail in the future, will be shown to involve conformational changes in either or all of the receptor, other membrane components and the agonist itself.

In the case of ganglioside receptors it is easy to envisage how these amphipathic molecules through their complex oligosaccharide regions can bind e.g. toxins, viruses or hormones with sufficient strength and selectivity. Could they also have a further role in promoting the biological effect of the agonist on the cell? From our own and others´ studies of the cholera toxin system we are tempted to speculate that gangliosides, in their receptor function, not only mediate specific binding but also induce a conformational change in the agonist facilitating its interaction with other membrane structures.

Cholera toxin binds with high affinity ($K_A \approx 1 \times 10^9$ M^{-1}) to the cell membrane by means of the B subunits, while the A subunit (or rather the A_1 fragment) is the effector structure mediating the toxic-biologic action (reviewed by HOLMGREN and LÖNNROTH, 1979). The A_1 fragment activates adenylate cyclase by catalyzing an ADP-ribose transfer reaction from intracellular NAD onto the GTP-binding regulatory component of adenylate cyclase complex. This modifies the GTP-binding protein so that the normal adenylate cyclase "turn-off" mechanism that is mediated through GTPase is inhibited (reviewed by GILL and ENOMOTO, 1979).

The binding of cholera toxin to cells is almost instantaneous, yet there is a lag of 10-120 minutes before adenylate cyclase exhibits increased activity. The lag period is the time it takes for subunit A to penetrate through the membrane and release A_1, because in broken cells adenylate cyclase is immediately activated by A_1 in presence of NAD and in less than one minute by unreduced A. Knowledge is still very incomplete about this cell penetration process. Immunohistochemical studies have shown that on most cells the GM1 ganglioside receptors appear to be evenly distributed on the membrane surface giving an initial "smooth" binding pattern for cholera toxin. The cell-bound toxin soon, however, moves laterally to form patches and also capping. These findings in themselves suggest that toxin must bind at least bivalently to cells, and since each B subunit is capable of binding one GM1 molecule, up to 5 attachment points are possible. In fact, it seems that even when toxin is added in excess concentration to cells, which should favour uni- or few-valence binding, almost all toxin molecules that bind do so with all their B subunits. This is indicated by experiments showing that mouse thymocytes naturally contain 5-6 times as many GM1 molecules as the number of toxin molecules they can bind, and, most strikingly, on incorporation of exogenous GM1 in various quantities the ratio between the number of GM1 and toxin molecules that can maximally bind remains constant at 5-7 (Fig. 5). This fully saturated orientation, which occurs at both 4 and 37°C, may be explained by positive cooperation between the B subunits in binding to GM1. No doubt, it must also involve lateral diffusion of toxin and GM1 in the membrane. Thus, we may deduce that when the true entry steps are about to take place the toxin molecules are firmly anchored on to the membrane by all the B subunits. Nevertheless the binding is still reversible meaning that the bound

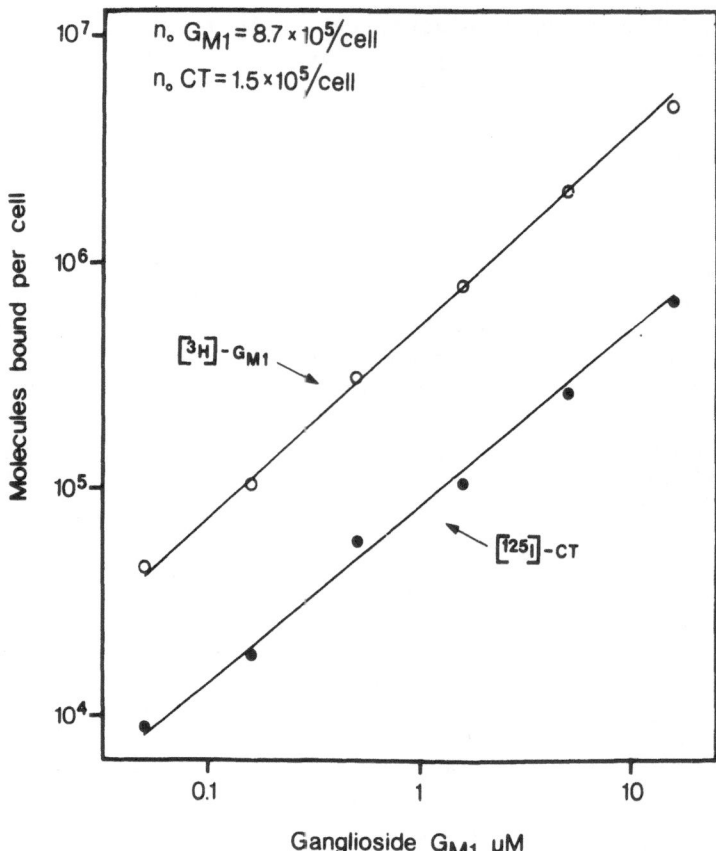

Fig. 5 Incorporation of GMl ganglioside into cells and its effect
on cholera toxin binding. Mouse thymocytes (10^7 cells in 1
ml) were incubated with various concentrations of ^{3}H-label-
led GMl at 37°C for 1 h, and then washed carefully to remove unincorporated ganglioside. Half of the cells were then
subjected to scintillation determination of the amount of
incorporated (^{3}H)-GMl and the other half examined for binding capacity of ^{125}I-labelled cholera toxin (CT) (by incubation with an excess concentration of labelled toxin at
room temperature for 20 min, followed by washing on a fil-
ter and gammacounting). The numbers of ^{3}H-GMl molecules
incorporated per cell and of extra toxin molecules that can
bind per cell (as compared to cells not treated with ganglioside) can be seen to change almost in parallel, the
ratio being 5-7 over a wide concentration range. The n_o values refer to the mean number of GMl molecules per cell
prior to incorporation (chemically determined) and to the
number of cholera toxin molecules that each such cell could
maximally bind. All data are the means of three experiments
performed in duplicates (J. HOLMGREN and L. SVENNERHOLM,
unpublished).

toxin can be displaced by addition of excessive concentrations
of either toxin or B subunits. This situation persists for several
hours if the cell temperature is kept below the melting point of
the membrane lipids. However, if the temperature is increased to
above 20°C we find that with time a progressively larger fraction
of the toxin becomes irreversibly associated with the cell membrane.
Since this irreversibilization is obtained also with purified B
subunit and the temperature-dependence is similar to that noticed
for toxin stimulation of adenylate cyclase we believe that the
B subunits must have a role in the entry process for the A subunit
in addition to that of merely anchoring the toxin to the membrane
GM1 receptors. We have suggested that the change from reversible
to irreversible binding is due to a more intimate association of
the B subunits with the membrane leading to their partial embedding
in the membrane thickness, and that this in its turn might be due
to a conformational change in the B subunits induced by the binding
to GM1 (HOLMGREN and LÖNNROTH, 1976). This conformation change
could result in exposure of otherwise hidden hydrophobic subunit
regions capable of fusion with hydrophobic protein or lipid components of the plasma membrane as outlined in steps 1-3 in Fig. 6.
Recent fluorescence and circular dichroism analyses by FISHMAN
et al. (1978) and nuclear magnetic resonance studies (J. STAERK,
personal communication) are consistent with such exposition of
hydrophobic regions in the B subunit on binding to GM1.

Fig. 6 outlines three possible ways in which the proposed
partial burying of the B subunit ring into the membrane may lead
to entry of the A subunit:

In the first model (4 a) the B subunits extend themselves
through the full thickness of the membrane thus creating a hydrophilic channel through which A could pass. The channel must
essentially be unidirectional because our studies with radio-
labelled toxin indicate that very little A ever leaves the cell
which would be expected if free random diffusion of A subunit took
place. The ability of B subunits to form a channel all the way
through at least an artificial membrane was recently indicated in
experiments by MOSS et al. (1977) who found that both cholera-
genoid and cholera toxin caused release of trapped glucose from
liposomes containing GM1, and by TOSTESON and TOSTESON (1978) who
demonstrated channel formation in black lipid membranes.

A second possibility (4 b) is that the channel is built only
in part of the B subunits, the inner portion being created of an
integral membrane protein. That a membrane protein may become
attached to the GM1-B subunit complex in viable cells has some
indirect experimental support in the "capping" of both toxin and
aggregated B subunits on the cell surface. In contrast to the
spontaneous cross-linking of the initial toxin-GM1 complex into
small patches, the long-distance capping is a process which

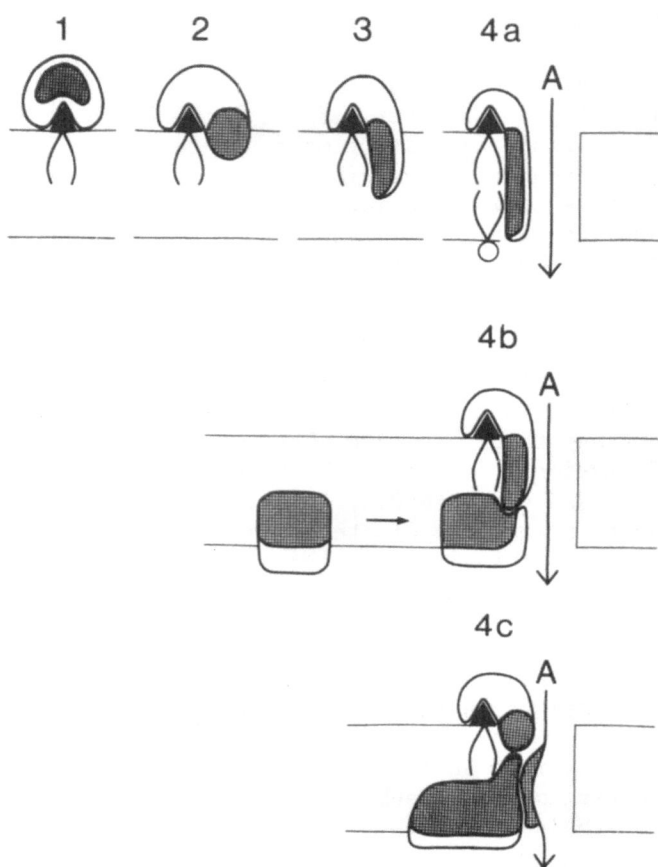

Fig. 6 The speculative scheme of events that may be involved in
the cholera toxin A subunit entry process. 1: binding of a
B subunit to a GM1 ganglioside receptor in the cell membrane (in a short time all the five B subunits in the toxin
establish such binding with GM1 receptors). 2: the binding
to GM1 induces conformational changes in the B subunits
leading to exposure of normally hidden hydrophobic subunit
structures (shaded region) which associate with hydrophobic
membrane constituents. 3: this internalization has progres-
sed so the B subunits are partially embedded in the membrane. 4: three possible ways throgh which the A subunit
might get intracellularly are outlined. In (a) and (b) the
B subunits, with or without the assistance of an integral
membrane protein, form a hydrophilic channel through which
A (released through the strain put on the toxin molecule)
can pass; in (c), in contrast, hydrophobic A (A_2) subunit
regions are brough in contact with an integral membrane
protein, which through a conformational change translocates
the A subunit across the membrane. For discussion, see text.

requires metabolic energy and is mediated by filamentous cyto-
mechanical structures underneath the membrane. BOURGUIGNON and
SINGER (1977) have proposed that capping is mediated by actin
(and myosin) and can occur because actin is attached to one or
more specific integral membrane proteins to which cross-linked
receptor aggregates, but not free receptors, become linked. Since
the cholera toxin receptor is a glycolipid which does not span
the membrane (which glycoprotein receptors may do) it seems almost
inevitable that capping of this toxin must involve some bridging
membrane component, probably a protein. The proposed channel is
still hydrophilic.

Finally, although perhaps thermodynamically less likely
than the hydrophilic channel models, one might also consider a
translocation based on a hydrophobic interaction between the A
subunit and an integral membrane component (4 c). It is conceivable that through steps 1-3 the hydrophobic A_2 region is inserted
into the membrane matrix where it could bind to an integral
protein. This might induce a sufficient conformational change in
this protein to induce a "revolving" translocation of A to the
membrane inside, where A_1 could then be released by intracellular
glutathione reducing the A_1-A_2 disulphide linkage.

A variety of experiments can be designed to support or
refute any or all of these hypothetical penetration models. It
is likely that continued work on this aspect of the cholera toxin
action will yield important information applicable to other cell
penetration systems, e.g. viruses and other toxins with an
intracellular action.

In comparison with cholera toxin very little is known about
the interactions of tetanus toxin and Sendai virus with the cell
membrane. Tetanus toxin, similar to cholera toxin, has separate
binding and effector regions but an important difference is that
there is only one binding site per molecule (HELTING et al., 1977).
Sendai virus has separate hemagglutinin-neuraminidase (HN) and
fusion (F) proteins (GETHING et al., 1978). A comparison of the
possible effects on the conformation of the isolated cholera
toxin B subunit, tetanus toxin β (heavy) chain and Sendai virus
HN protein by, respectively, gangliosides GM1, GT1b and GQ1b
should increase our understanding of the receptor function of
gangliosides.

REFERENCES

BOURGUIGNON L.Y.W. and SINGER S.J. (1977): Transmembrane inter-
    actions and the mechanism of capping of surface receptors by
    their specific ligands. Proc. Natl. Acad. Sci. (USA), 74,
    5031-5035.

CUATRECASAS P. (1973a): Interaction of *Vibrio cholerae* entero-
    toxin with cell membranes. Biochemistry 12, 3547-3558.

CUATRECASAS P. (1973b): Gangliosides and membrane receptors for
    cholera toxin. Biochemistry 12, 3558-3566.

FISHMAN P.H., MOSS J. and OSBORNE J.C. (1978): Interaction of
    choleragen with the oligosaccharide of ganglioside GM1:
    Evidence for multiple oligosaccharide binding sites. Bio-
    chemistry 17, 711-716.

GETHING M.J., WHITE J.M. and WATERFIELD M.D. (1978): Purification
    of the fusion protein of Sendai virus: Analysis of the NH_2-
    terminal sequence generated during precurson activation.
    Proc. Natl. Acad. Sci. (USA), 75, 2737-2740.

GILL D.M. and ENOMOTO K. (1979): Intracellular, enzymic action
    of enterotoxins: The biochemical basis of cholera. In "43rd
    Nobel Symposium: Cholera and related diarrheas. Molecular
    aspects on a global health problem", Ouchterlony Ö. and Holm-
    gren J., eds., Karger, Basel.

HANSSON H.-A., HOLMGREN J. and SVENNERHOLM L. (1977): Ultra-
    structural localization of cell membrane G_{M1} ganglioside by
    cholera toxin. Proc. Natl. Acad. Sci. 74, 3782-3786.

HART D.A. (1975): Evidence for the non-protein nature of the
    receptor for the enterotoxin of *Vibrio cholerae* on murine
    lymphoid cells. Infect. Immun. 11, 742-747.

HAYWOOD A.M. (1974): Characteristics of Sendai virus receptors in
    a model membrane. J. Mol. Biol. 83, 427-436.

HELENIUS A., MOREIN B., FRIES E., SIMONS K., ROBINSON P.,
    SCHIRRMACHER V., TERHORST C. and STROMINGER J.L. (1978):
    Human (HLA-A and HLA-B) and murine (H-2K and H2-D) histo-
    compatibility antigens are cell surface receptors for Semliki
    Forest virus. Proc. Natl. Acad. Sci. (USA), 75, 3846-3850.

HELTING T.B., ZWISLER O. and WIEGANDT H. (1977): Structure of
    tetanus toxin. II. Toxin binding to ganglioside. J. Biol.
    Chem. 252, 194-198.

HOLMGREN J. and LÖNNROTH I. (1976): Cholera toxin and the
adenylate cyclase-activating signal. J. Infect. Dis. 133,
64-74.

HOLMGREN J. and LÖNNROTH I. (1979): Structure and function of
enterotoxins and their receptors. In "43rd Nobel Symposium:
Cholera and Related Diarrheas. Molecular aspects on a global
health problem", Ouchterlony Ö. and Holmgren J., eds.,
Karger, Basel.

HOLMGREN J., LÖNNROTH I. and SVENNERHOLM L. (1973): Tissue
receptor for cholera exotoxin: Postulated structure from
studies with G_{M1} ganglioside and related glycolipids. Infect.
Immun. 8, 208-214.

HOLMGREN J., MÅNSSON J.-E. and SVENNERHOLM L. (1974): Tissue
receptor for cholera exotoxin: Structural requirements of
G_{M1} ganglioside in toxin binding and inactivation. Medical
Biology 52, 229-233.

HOLMGREN J., LÖNNROTH I., MÅNSSON J.E. and SVENNERHOLM L. (1975):
Interaction of cholera toxin and membrane G_{M1} ganglioside of
small intestine. Proc. Natl. Acad. Sci. (USA), 72, 2520-2524.

KING C.A. and VAN HEYNINGEN W.E. (1973): Deactivation of cholera
toxin by a sialidase-resistant monosialosyl-ganglioside. J. Infect. Dis. 127, 639-647.

LEDLEY F.D., LEE G., KOHN L.D., HABIG W.H. and HARDEGREE M.C.
(1977): Tetanus toxin interactions with thyroid plasma membranes. Implications for structure and function of tetanus
toxin receptors and potential pathophysiological significance.
J. Biol. Chem. 252, 4049-4055.

LÖNNROTH I. and HOLMGREN J. (1973): Subunit structure of cholera
toxin. J. Gen. Microbiol. 76, 417-427.

MOSS J., FISHMAN P.H., MANGANIELLO V.C., VAUGHAN M. and BRADY
R.O. (1976): Functional incorporation of ganglioside into in-
tact cells: Induction of choleragen responsiveness. Proc. Natl.
Acad. Sci. (USA), 73, 1034-1037.

MOSS J., RICHARDS R.L., ALVING C.R. and FISHMAN P.H. (1977):
Effect of the A and B protomers of choleragen on release of
trapped glucose from liposomes containing or lacking ganglioside G_{M1}. J. Biol. Chem. 252, 797-798.

MULLIN B.R., FISHMAN P.H., LEE G., ALOJ S.M., LEDLEY F.D.,
   WINAND R.J., KOHN L.D. and BRADY R.O. (1976): Thyrotropin-
   ganglioside interactions and their relationship to the
   structure and function of thyrotropin receptors. Proc. Natl.
   Acad. Sci. (USA), 73, 842-846.

MULLIN B.R., PACUSZKA T., LEE G., KOHN L.D., BRADY R.O. and
   FISHMAN P.H. (1978): Thyroid gangliosides with high affinity
   for thyrotropin: Potential role in thyroid regulation. Science
   199, 77-79.

PRICE D.L., GRIFFIN J.W. and PECK K. (1977): Tetanus toxin:
   evidence for binding at presynaptic nerve endings. Brain Res.
   121, 379-384.

SATTLER J., SCHWARZMANN G., STAERK J., ZIEGLER W. and WIEGANDT H.
   (1977): Studies of the ligand binding to cholera toxin. Hoppe-
   Seyler´s Z. Physiol. Chem. 358, 159-163.

STAERK J., RONNEBERGER H.J., WIEGANDT H. and ZIEGLER W. (1974):
   Interaction of ganglioside G_{Gtet1} and its derivatives with
   choleragen. Eur. J. Biochem. 48, 103-110.

TOSTESON M.T. and TOSTESON D.C. (1978): Bilayers containing gangliosides develop channels when exposed to cholera toxin. Nature
   (London), 275, 142-144.

VAN HEYNINGEN W.E., CARPENTER C.C.J., PIERCE N.F. and GREENOUGH
   W.B. (1971): Deactivation of cholera toxin by ganglioside.
   J. Infect. Dis. 124, 415-418.

VAN HEYNINGEN W.E. and MELLANBY J. (1971): Tetanus toxin. In
   "Microbiol. toxins, vol 2A, Bacterial protein toxins". Eds.
   S. Kadis, T.C. Montie and S.J. Ajl, 69-108, Academix Press
   (New York).

ISOLATION OF CHOLERA TOXIN BY AFFINITY CHROMATOGRAPHY

ON POROUS SILICA BEADS WITH COVALENTLY COUPLED GANGLIOSIDE G_{M1}

Jean-Louis Tayot - Michel Tardy

Institut Mérieux
Marcy l'Etoile
69260 Charbonnières les Bains, France

The great interest of ion exchange and affinity chromatography
for protein purification has prompted us to design new particles
able to meet the technical requirements for industrial scale fractionation. Porous silica beads "Spherosil ® " were impregnated with
DEAE Dextran. Like other aminated macromolecules with a positive
electric charge, it has a very strong affinity for the silica surface. In its presence, the strong and irreversible adsorption
properties of silica completely disappear and DEAE Dextran gives
the support the physicochemical properties of anion exchangers.
After cross-linking with bisepoxy reagents, the DEAE Dextran coating
is completely stable even in acid or alkaline solutions. Furthermore
this new chromatographic support has outstanding mechanical properties which makes large columns very easy to prepare. The bed
height is completely incompressible and allows very high flow-
rates from 100 to 400 ml/cm^2/h under reasonable pressures. For all
these reasons, Spherosil-DEAE Dextran is well adapted to the large
scale fractionation of proteins by ion-exchange (Tayot et al.,
1978-a) and immunoaffinity chromatography (Tardy et al., 1978).
We now describe the technique used for the preparation and attachment of the ganglioside G_{M1} to this spherosil derivative and its
application to cholera toxin purification by affinity chromatography. A part of this work is described in a preliminary report
(Tayot et al., 1978-b).

1. SPHEROSIL-DEAE DEXTRAN MICROBEADS

Spherosil is manufactured by Rhône Poulenc, 21, rue Jean
Goujon, 75008, Paris. Two types of derivatives have been prepared.

471

Table 1. Characteristics of some Spherosil particles 100-200 μ

Type	Specific surface	Mean porous diameter	Internal volume
XOB 015	25 m^2/g	125 nm	1 ml/g
XOC 005	10 m^2/g	300 nm	1 ml/g

For ion exchange applications, we use the XOB 015 type (table 1), it is coated with the maximum amount of DEAE Dextran (15 per cent w/w) to get optimal fixation of negatively charged macromolecules (250 mg albumin per g) and small anions (0.25 meq Cl^- per g). This is the support reported by Dr Fredman in this symposium and which we also used for preparation of large amounts of gangliosides.

For affinity chromatography, we use the XOC 005 type. It is then coated with the minimum amount of DEAE Dextran just needed to suppress the adsorption properties of the silica surface (between 1 and 2 per cent w/w). The fixation capacity of negatively charged albumin is reduced to 25 mg per g and the anion exchange capacity is 0.03 meq Cl^- per g).

Both types of composite particles have a widely opened structure which makes the internal surface freely accessible for macromolecules. If restricted to research purposes and small quantities, these Spherosil-DEAE Dextran particles can be obtained from our laboratory.

2. SEROLOGICAL METHODS OF ANALYSIS

An oral presentation of these methods has been added in this symposium. Because of the editor's technical difficulties, the details are not printed in this book but will be published later on. These new serological methods are based on hemagglutination tests of red cells artificially coated with ganglioside GM_1. The coupling conditions of gangliosides on sheep red blood cells (SRBC) are the following. To a 10 per cent suspension of washed SRBC, is added 0.5 per cent of glutaraldehyde in phospnate buffered saline (PBS). After 30 mn incubation, the suspension is washed in PBS. Glycine 0.05M is added and the pH adjusted to 11. This overnight treatment neutralizes the residual aldehyde groups to prevent further covalent associations. SRBC are then washed in 0.1N NaOH. To a 10 per cent suspension of these alkaline SRBC, crude or pure gangliosides are then added to get a final concentration of 250 n.mole/ml. After 6 H incubation at 45°C, the sensitized SRBC are washed in PBS and adjusted to a 1 per cent concentration. Storage is possible several months at + 4°C with added bovine albumin (3 mg/ml) and sodium azide (0.1 per cent). Sialic acid titrations by the resorcinol method (Svennerholm, 1957) demonstrate a 50 per cent integration of the gangliosides into or on the red cell membrane. This integration does not occur significantly at neutral pH.

The presence of 10^7 molecules of G_{M1} per cell is so obtained and gives agglutination figures with cholera toxin (hemagglutination test). This interaction between cholera toxin and G_{M1} coated red cells is specific because we have demonstrated that :

* After hemagglutination, choleragen or choleragenoïd are present on the red cell membrane giving new particles : SRBC-G_{M1}-Tox.
* It is possible to completely eliminate the toxic activity of a cholera crude culture filtrate by adsorption on G_{M1}-coated red cells.
* The immunization of animals with SRBC-G_{M1}-Tox directly give mono-specific choleragen antibodies at high concentration.

- This hemagglutination is specifically inhibited by ganglioside G_{M1} and not at all by G_{D1a}, G_{D1b} or G_{T1b}.
- This hemagglutination is not inhibited by choleragen antibodies. On the contrary, choleragen antibodies specifically enhance the agglutination titer. This fifty fold potentiation is probably explained by the fact that soluble toxin-antitoxin aggregates are larger than cholera toxin alone and better agglutinogens.

The hemagglutination test (HA) allows us to measure choleragen or choleragenoïd condentration in 1 hour (minimum concentration : 8 ng/ml i.e. : 0.4 ng in the sample). The inhibition test (IHA) allows us to measure ganglioside G_{M1} concentration in 2 hours (minimum concentration : 6 ng/ml i.e. : 200 femtomole in the sample).

3. PREPARATION OF THE G_{M1} DERIVATIVE : LYSO G_{M1}

Pure ganglioside G_{M1} was given by Pr L. Svennerholm. Crude gangliosides preparations can also be used. As previously described by Holmgren et al.(1974), an alkaline treatment in butanol-KOH 10 M (90-10) under reflux heating for 2 H, hydrolyses the N-acyl and N-acetyl functions of ganglioside G_{M1} giving 3 amino groups. An equal volume of water is then added. After overnight decantation, the lower aqueous phase is adjusted to pH 9 and clarified by membrane filtration before coupling. The bound sialic acid is completely protected in this step. Pure lyso G_{M1} given by Pr L. Svennerholm was used for coupling experiments in parallel.

4. COUPLING CONDITIONS OF THE LYSOGANGLIOSIDE ON SPHEROSIL XOC 005 DEAE DEXTRAN

The DEAE Dextran coating is oxydised in sodium metaperiodate 0.02M. The resulting aldehyde groups are then reacted with amino groups of lyso G_{M1} in 0.01M Na carbonate pH 9 containing NaCl 10g/l. Concentrations of lyso G_{M1} used for coupling vary from 2 to 10 μmole per g of particles. After 15 H incubation at room temperature, sodium borohydride 0.2M is added. The formed imino-derivatives are so reduced to stable secondary amines. The particles are then poured into a column. The supernatant is recovered by filtration. Resorcinol reactions demonstrate less than 5 per cent of initial

sialic acid present in the filtrate and indicate that the coupling
yield is about 100 per cent. No traces of sialic acid are detected
in the following washings.

- 100 column volumes of 0.1 N NaOH overnight,
- 10 column volumes of Chloroform-Methanol-HCl 0.1 N (30-
  60-25) for 1 H.

This coupling method has already been proposed by Parikh et
al. (1974) for agarose and Wilson et al. (1976) for sephadex gels.
Agarose, however, has few α-glycol functions and consequently is
not very efficiently derivatized by this method. On the other hand,
Sephadex beads are severely degraded or have not the adequate porosity to interact with macromolecules.

In our case, the DEAE Dextran coating is not degraded by the
NaIO$_4$ oxydation because of its high degree of cross-linkage. These
numerous cross-links do not decrease the porosity of the particles
because DEAE Dextran represents only a monolayer coating the inside
cavities.

5. AFFINITY CHROMATOGRAPHY ON SPHEROSIL-G$_{M1}$ PARTICLES

The column is equilibrated in 0.01 M phosphate buffer containing 10 g/1 NaCl at pH 6.8. NaCl is used to entirely suppress ionic
adsorption of proteins on the DEAE Dextran monolayer. This is demonstrated in the following experiment : *Fig. 1*, where an artificial
sample containing three different proteins is used. The electro-
positive gamma-globulin is not adsorbed on the modified DEAE Dextran
layer while the negatively charged albumin is fixed and then quantitatively eluted by NaCl addition. The low ionic strength of the
buffer and the numerous positive charges of the support do not
hinder the affinity between lyso G$_{M1}$ and cholera toxin which is
completely fixed. Cholera toxin is only eluted by decreasing the
pH at 2.8.

Crude culture filtrates of vibrio cholerae INABA 569 B are
produced by the Bacteriology department of Institut Merieux. We
use a solution (C.C.F.) which was previously concentrated by ultrafiltration and adjusted to 10 g/1 NaCl. Cholera toxin concentration
is 200 µg/ml as estimated by radial immunodiffusion in comparison
with the Schwarz-Mann reference lot GZ 2790. The specific choleragen antibodies are prepared in our laboratory (TAYOT, this symposium). C.C.F. is continuously pumped through the column until
saturation. The effluent is registered by U.V. analysis and hemagglutination reactions (TAYOT, this symposium). After rinsing with
the chromatographic buffer, cholera toxin is eluted by 0.05 M Na
citrate buffer pH 2.8. The eluted solution is neutralized by NaOH
addition. Filtration and ultrafiltration with an Amicon PM 10 membrane give a clear, concentrated solution of purified cholera toxin.

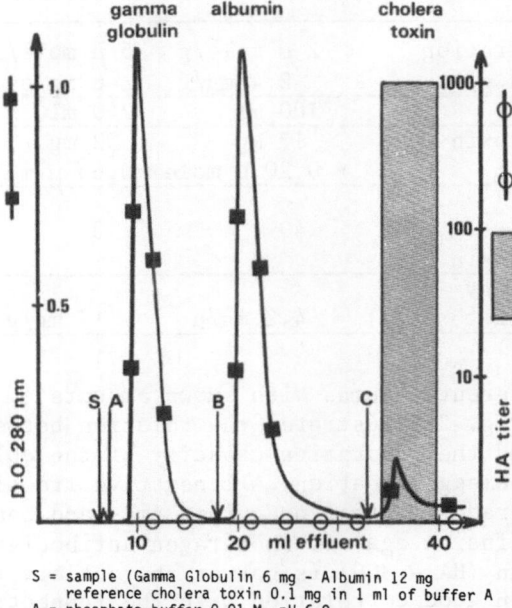

Fig. 1. Association of ion exchange and affinity chromatography
on 1 g of Spherosil-DEAE-Dextran-Lyso G_{M1}.

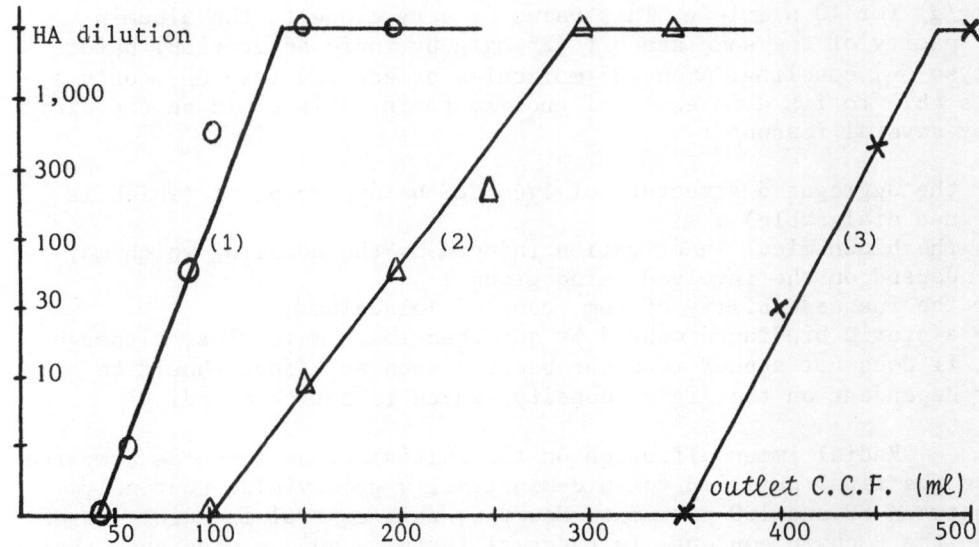

Fig. 2. Specific adsorption of cholera toxin (200 μg/ml) on 3 different lyso-G_{M1} columns (4 g each).
(1)=2 μ mole/g ; (2)= 5 μ mole/g ; (3)= 10 μ mole/g

Table 2. Comparison of 3 different affinity columns for cholera toxin

Lyso G_{M1} concentration on the column (4 g)	2 μ mole/g 2.4 mg/g	5 μ mole/g 6 mg/g	10 μ mole/g 12 mg/g
"V 10" (see text)	100 ml	240 ml	440 ml
Eluted cholera toxin (M.W. = 84,000)	17 mg 0.20 μ mole	52 mg 0.63 μ mole	91 mg 1.08 μ mole
$\dfrac{\mu \text{ mole lyso } G_{M1}}{\mu \text{ mole cholera toxin}}$	40	32	37
Adsorption capacity of the column (mg toxin/g)	4.2 mg/g	13 mg/g	22.7 mg/g

Three different columns with known amounts of coupled lyso-G_{M1} were prepared. Fig. 2 illustrates the relation between the ligand concentration and the adsorption capacity of the column. The outlet is analysed by hemagglutination. All negative fractions are non toxic in the intradermal test on guinea pigs and non reactive by radial immunodiffusion against choleragen antibodies. When 10 per cent of the toxin (HA = 300) is got at the outlet, the reached volumes "V 10" are in good correlation with the respective amounts of lyso G_{M1} : Table 2. After saturation, each column was rinsed and eluted. The amounts of purified toxin thus recovered were estimated by U.V. analysis : $E_{280 nm}^{1 \%}$ = 11.4 (Holmgren et al., 1977). These results were confirmed by radial immunodiffusion and are summarized in Table 2. The specific adsorption capacity of the column is directly proportional to the lyso-G_{M1} content and reaches a maximum (22.7 mg/g) for 10 μ mole/g. This value is very close to the albumin capacity of the same support (25 mg/g by ionic adsorption) before lyso G_{M1} coupling. Among 35 molecules of coupled lyso G_{M1}, only 1 is able to fix a molecule of cholera toxin. This could be explained by several reasons :

- the aggregated structure of lyso G_{M1} before coupling (which is non dialysable) ;
- the biochemical inactivation induced by the coupling which may depend on the involved amino group ;
- the inaccessibility of some coupled molecules ;
- a steric hindrance caused by adsorbed toxin molecules, although it does not appear to occur because such an effect should be dependent on the ligand density, which is not observed.

Radial immunodiffusion on the initial crude toxin as compared to the final purified toxin demonstrate a good yield, routinely between 80 and 100 per cent. However, this control is mainly based on the B chain content. Intradermal tests on guinea pigs show that the purified material keeps more than 50 per cent of its toxicity in the best conditions. That is very dependent on the concentration of the purified toxin after column elution. For lowest concentra-

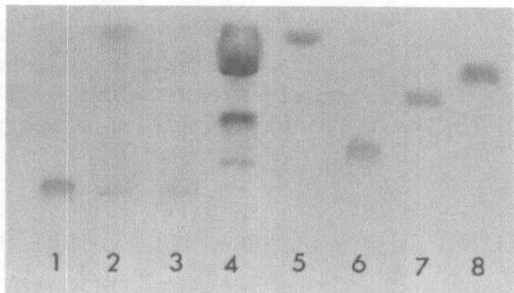

Fig. 3. SDS-polyacrylamide gel electrophoresis of purified toxin
in comparison with different markers of known M.W. (in
parenthesis).
Buffer : Tris-Borate-EDTA pH 8.35 SDS 2 g/1 ; polyacrylamide gel : acrylamide : 7.5 % ; bisacrylamide 0.25 %
1 : Albumin (65,000) ; 2 : Initial toxin ; 3 : Column
filtrate (the B chain is absent) ; 4 : Purified toxin at
high concentration (1 mg in 20 µl) with 2 major bands :
B chain (11,000) and A chain (27,000). A 3rd band (about
50,000) and other minor impurities are present. 5 : Purified
B chain (11,000) ; 6 : Ovalbumin (43,000) ; 7 : Chrymotryp-
sinogen (25,000) ; 8 : Ribonuclease (13,700).

tions the reaggregation of A and B subunits after pH neutralization
is not good. For highest concentrations the A subunit precipitates
at acid pH and is eliminated by filtration. The relative purity
of the preparation obtained in one single step is illustrated in
Fig. 3.

Our results contrast with the almost irreversible binding
and the low yields obtained by Cuatrecasas et al. (1973), using a
G_{M1}-sepharose column. The beneficial influence of our support and
lyso-G_{M1} as a ligand as well as the above observations could explain
these differences.

Repetitive cycles on the same column demonstrate the stability
of the coupling which is not influenced by repetitive acid washings
even with HCl 0.1 N. Further experiments in collaboration with Pr
L. Svennerholm should indicate if the coupled lyso-G_{M1} is (or not)
progressively transformed in lyso G_{A1} by sialic acid elimination
in the acid environment.

Scaling up of this technique is possible : a 1 kg column
allows us to treat 1,000 1 of crude toxin (20 g choleragen) per
cycle. An acid gel filtration of the purified toxin according to
Holmgren et al.(1977) opens the way to prepare large quantities of
B subunits in view of future vaccines.

ACKNOWLEDGEMENTS

    This work is part of a progress report in collaboration with
Dr J. Holmgren and Pr L. Svennerholm from the University of Göteborg,
Sweden. We thank Mrs G. Favre for her excellent cooperation.

REFERENCES

CUATRECASAS P., PARIKH I. and HOLLENBERG M.D. (1973) : Affinity
    chromatography and structural analysis of vibrio cholerae entero-
    toxin - Ganglioside agarose and the biological effects of gan-
    glioside containing soluble polymers. Biochemistry 12, 4253-4264.
FREDMAN P. (1979) : Quantitative isolation of gangliosides on a new
    form of glass bead ïon-exchanger. This symposium.
HOLMGREN J., MÅNSSON J.E. and SVENNERHOLM L. (1974) : Tissue recep-
    tor for cholera exotoxin : structural requirements of G_{M1} gan-
    glioside in toxin binding and inactivation. Medical Biology 52,
    229-233.
HOLMGREN J., SVENNERHOLM A.M., LÖNNROTH I., FALL-PERSSON M.,
MARKMAN B. and LUNDBECK H. (1977) : Development of improved cholera
    vaccine based on subunit toxoïd. Nature 269, 602-604.
PARIKH I., MARCH S. and CUATRECASAS P. (1974) : Topics in the
    methodology of substitution reactions with agarose. In "Methods
    in Enzymology" JAKOBY-WILCHEK Eds, Vol. XXXIV, Academic Press
    (N.Y. and London), pp. 77-102.
SVENNERHOLM L. (1957) : Quantitative estimation of sialic acids.
    II. A colorimetric resorcinol-hydrochloric acid method. Biochim.
    Biophys. Acta 24, 604-611.
TARDY M., TAYOT J.L., ROUMIANTZEFF M. and PLAN R. (1978) : Immuno-
    affinity chromatography on derivatives of porous silica beads.
    Industrial extraction of antitetanus antibodies from placental
    blood and plasma. In "Chromatography of Synthetic and Biological
    Polymers", EPTON R. Ed, Vol. II, Ellis Horwood, Chichester U.K.,
    pp. 298-313.
TAYOT J.L. (1979) : Specific agglutination between gangliosides-
    coated red cells and cholera toxin. This symposium (not printed).
TAYOT J.L., TARDY M., GATTEL P., PLAN R. and ROUMIANTZEFF M. (1978-a)
    Industrial ïon exchange chromatography of proteins on DEAE-Dextran
    derivatives of porous silica beads. In "Chromatography of Syn-
    thetic and Biological Polymers", EPTON R. Ed, Vol. II, Ellis
    Horwood, Chichester U.K., pp. 95-110.
TAYOT J.L., TARDY M. and MYNARD M.C. (1978-b) : Biospecific chroma-
    tography on new derivatives of porous silica beads. Coupling of
    ganglioside G_{M1} or anticholeragen antibody for purification of
    cholera toxin. In "Affinity Chromatography", HOFFMANN-OSTENHOF O.
    et al. Eds, Pergamon Press (Oxford and N.Y.), pp. 265-269.
WILSON M.B. and NAKANE P.K. (1976) : The covalent coupling of pro-
    teins to periodate oxydized sephadex : a new approach to immuno-
    adsorbent preparation. J. Immunol. Methods 12, 171-181.

THE INTERACTION BETWEEN GANGLIOSIDES AND INTERFERON

V.E. Vengris*, B.F. Fernie and P.M. Pitha

Food and Drug Administration, Bureau of Veterinary
Medicine, Rockville, Maryland 20857, and The Johns
Hopkins University School of Medicine, Baltimore,
Maryland 21205

INTRODUCTION

The mechanism by which interferon induces an antiviral state
in cells has not been completely understood. All the experimental
evidence suggests that for the manifestation of the effect contact
with the sensitive cells is sufficient. The existence of interferon
specific membrane receptor site has been postulated[1,2]. Treatment
of cells with phytohemagglutinin (PHA) blocked the development of
antiviral activity following interferon treatment[3]. Also, sepharose-
bound interferon lost its antiviral activity after preincubation
with gangliosides, especially GM_2 and GT_{1},[4] and soluble interferon
was bound to sepharose-ganglioside beads.[4] Recently, it has been
reported that cholera toxin (choleragen) added together with interferon inhibited the establishment of antiviral activity.[5] Since
ganglioside GM_1 functions as a specific membrane receptor and mediates the action of cholera toxin[6,7], it is not unreasonable to
suggest, in view of the interferon-ganglioside interaction, that
membrane ganglioside may also play a role in the binding and action
of interferon. This work further explores the relationship between
gangliosides and interferon action in normal human fibroblasts and
in ganglioside-deficient, transformed mouse fibroblasts.[7,8,9]

MATERIALS AND METHODS

Cells, virus and induction of interferon

Human foreskin fibroblast (HFC), mouse cells (L cells ,

---

*This work was done while the senior author was a postdoctoral fel-
low at the Johns Hopkins University School of Medicine.

Balb/c-3T3, K Balb/c-3T3[8], SVS AL/N[7], TAL/N[7] and NCTC 2071[9]) were grown as described before[9,10]. Vesicular stomatitis virus (VSV), New Jersey serotype and Encephalomyocarditis virus (EMC) were propagated in L cells at low multiplicity.

Interferon was induced in the cell monolayers by treatment with poly rI poly rC complex (P.L. Biochemicals) as described previously[10].

Interaction of Gangliosides with Interferon and Cells

Interferon and gangliosides (equal volumes) were incubated in MEM or PBS at 37° for 1 hr before application to cells. In order to determine whether the binding of gangliosides to cells affects VSV yield or the induction of interferon by poly rI poly rC, HFC were incubated overnight with a solution of gangliosides before virus challenge or poly rI poly rC treatment. To measure the effect of ganglioside uptake on the antiviral activity of interferon, cell monolayers were incubated with varying concentrations of ganglioside in PBS for 1 hr at 37°. The cell monolayers were washed free of ganglioside solution, and were then incubated for another 30 min in maintenance medium before treatment with interferon.

Interaction of choleragen with mouse interferon

Choleragen (a gift of Dr. M. Hollenberg) and mouse interferon were mixed in equal volumes before being applied to the cells. The cells were incubated at 37° for 24 hr, washed, and then infected, at a moi of 10, with either EMC or VSV, and the virus yield was determined after 15 hrs.

RESULTS AND DISCUSSION

When human fibroblast interferon was incubated with a ganglioside mixture before being applied to cells, the antiviral effect was neutralized (Fig 1). Neutralization was dependent on the concentration of ganglioside, with half-maximal neutralization of 10,000 units of interferon at 50 µg/ml, and maximal neutralization (99.9%) at 300 µg of crude ganglioside/ml. The rate of neutralization of human fibroblast interferon by ganglioside was rapid with maximal inactivation achieved after a 30 min. preincubation period (Fig 2) Human leukocyte interferon (10,000 units) could also be completely neutralized by preincubation with gangliosides (500 µg/ml). The crude ganglioside mixture was as effective as the individual purified gangliosides in neutralizing human fibroblast interferon. The antiviral effect of human fibroblast interferon was not affected by pretreatment of HFC with exogenous gangliosides. The incubation of HFC with gangliosides before or during treatment with poly rI poly rC did not affect the induction of interferon. Treatment of HFC with gangliosides was without any detectable effect on VSV replication.

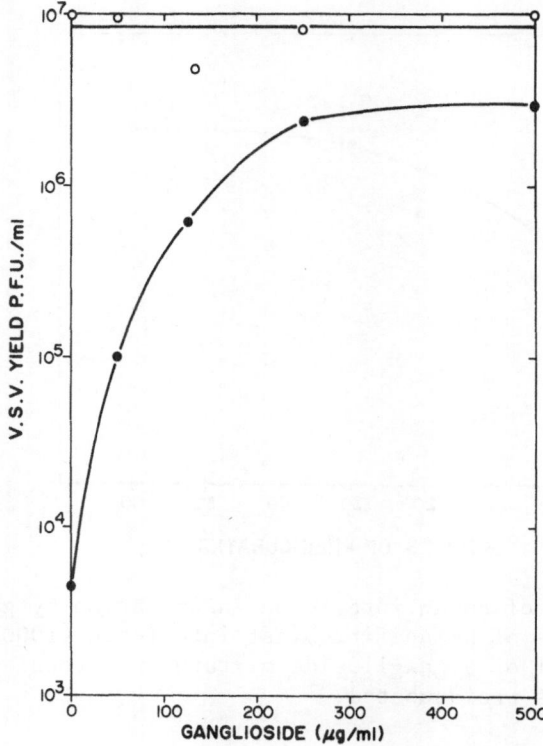

Fig. 1.   The inactivation of human fibroblast interferon by gang-
          lioside. Human fibroblast interferon (10,000 units/ml)
          and different amounts of gangliosides were preincubated.
          o—o : VSV yield with ganglioside alone; •—• : interferon
          and ganglioside mixture.

     Transformed mouse cells, i.e. K Balb/c-3T3, SVS AL/N and TAL/N,
which are deficient in some of the more complex gangliosides, were
less sensitive to the antiviral effect of exogenouse mouse interferon than were the nontransformed Balb/c-3T3 cells which contain
the normal complemetn of gangliosides (Table 1). Both the virus
transformed (SVS AS/N) and spontaneously transformed cells
(TAL/N, p>200) are deficient in ganglioside GM_1[11]. SVS AL/N cells
have showed the lowest sensitivity to mouse interferon. Thus 100
times more interferon was required to reduce VSV replication in
these cells than in nontransformed Balb/c-3T3 cells. The K Balb/c-3T3
cells transformed by the Kirsten mouse sarcoma virus (lacking GM_1[8])
were 25 times less sensitive to interfreron than the parent cell line.

     It has been shown that the incubation of ganglioside-deficient
cells with exogenous ganglioside leads to the membrane uptake of
gangliosides[10], however, no increase in the amount of ganglioside.

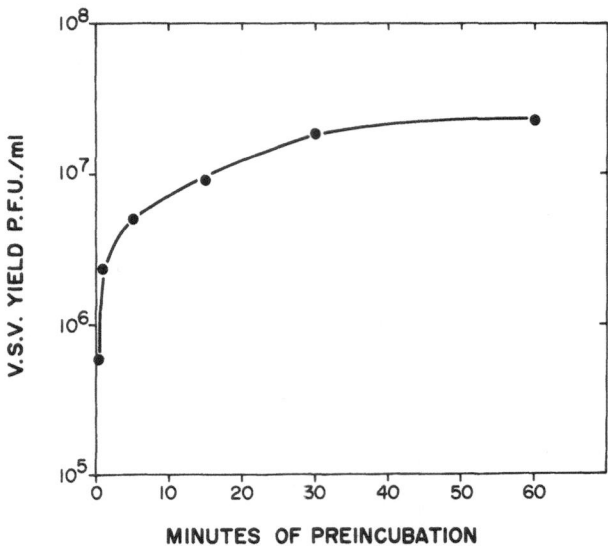

Fig. 2.   Kinetics of human interferon inactivation by ganglioside.
           A mixture of human fibroblast interferon (1000 units) and
           100 µg/ml of a ganglioside mixture in a total volume of
           1 ml were preincubated.

uptake was detected with the untransformed cells[10]. Table 1 indicates that ganglioside treatment (10 µg/ml) increased the sensitivity
of the SVS AL/N and TAL/N cells to interferon two-to-fourfold. No
further increase in the sensitivity of the cells was obtained when
the concentration of gangliosides was increased to 100 µg/ml. The
mixture of gangliosides gave the largest increase in sensitivity.
In SVS AL/N and TAL/N cells, treatment with gangliosides GT_1 and
GM_2 increased sensitivity to interferon, while GD_{1a} and GM_1 were
without effect. No increase in sensitivity was observed in
K Balb/c-3T3 cells, HFC and Balb/c-3T3 cells. The relationship between ganglioside content and interferon sensitivity was further examined using the chemically transformed NCTC 2071 cell line. These
cells contain undetectable amounts of the gangliosides GM_1, GD_{1a}
and only 20 picomoles of GM_2/mg cell protein[9]. They are not sensitive
to choleragen and are able to grow in serum free medium. However,
these cells proved to be sensitive to mouse interferon. Furthermore, incubation of NCTC 2071 cells with exogenous GM_1, GM_2 or GD_{1a}
gangliosides failed to increase the sensitivity of these cells to
interferon (Fig 3), while the treatment of these cells with GM_1 res-
tores their sensitivity to cholera toxin[9]. It was reported that
choleragen inhibited the establishment of antiviral activity in
mouse cells and it was suggested that this inhibition resulted from
a specific interaction between choleragen and interferon. Our data
indicate that the observed inhibition depends on the challenging

Table 1. Increase of Sensitivity of Ganglioside-Deficient Mouse Cells to Mouse Interferon by Incubation With Various Gangliosides[a]

Gangliosides	Interferon[b] (units/ml)			
	Balb/ c-3T3	SVS AL/ N	TAL/ N	K Balb/ c-3T3
Control	1	100	50	25
Mixture GT_1, GM_2, GD_{1a}, GM_1	1	12	25	25
GT_1, GM_2	–	25	25	–
GT_1	–	25	50	–
GD_{1a}	–	100	50	–
GM_1	–	100	50	–
GM_2	–	25	25	–

[a] Cell cultures were preincubated with gangliosides (10 µg/ml), then washed and incubated with interferon overnight. Cells were washed and infected with VSV for virus yield assay. The assay of interferon sensitivity was performed by an investigator unaware of the previous treatment of cells with ganglioside.

[b] The lowest concentration of interferon necessary to reduce VSV yield by 50%.

virus used (Fig 4). When EMC was used to assay the antiviral state the action of interferon was inhibited by cholera toxin. However, when VSV was used, the inhibition of cholera toxin was marginal. This indicates that the interaction between cell, virus, cholera toxin, and interferon is more complex than previously postulated.

The antiviral state was not inhibited in NCTC 2071 cells when interferon was preincubated with choleragen and assayed using VSV virus (data not shown). Unfortunately we were not able to make a direct comparison with EMC since this virus did not propogate well in NCTC 2071 cells.

The present study indicates strong interaction between gang-liosides and interferon. Similarly as with mouse interferon[4], both

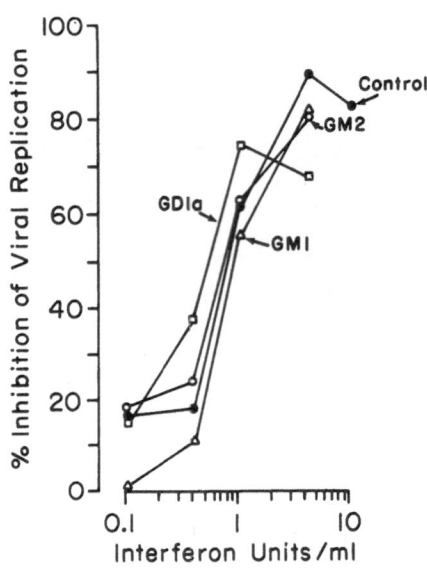

Fig. 3.   Effect of gangliosides on the interferon sensitivity of
NCTC 2071 cells. Cells were grown in NCTC 135 medium.[9]
Monolayers were incubated with (10µg/ml) or without gangliosides. Cultures were washed and incubated overnight
with interferon. VSV was used as the challenging virus.[10]
●—● no gangliosides; o—o GM_2; Δ—Δ GM_1; □ — □ GD_{1a}.

human fibroblast and leukocyte interferons can be neutralized by
preincubation with gangliosides.

The biological significance of the interferon-ganglioside interaction was examined in ganglioside deficient cell lines. Unfortunately, no ganglioside deficient human cell line was available and
therefore these studies were done using mouse system. Our data indicate that transformed cell lines, deficient in GM_2 ganglioside,
were less sensitive to interferon than the parental cells. (Table 1).
The incubation of these cells with gangliosides partially restored
their sensitivity to interferon. The fact that NCTC 2071 cells are
sensitive to interferon and the sensitivity is not increased by the
additional incorporation of GM_2 indicates that if the presence of
ganglioside is needed, it may be satisfied by picomoles of ganglioside in the membrane. The data also indicate that ganglioside-in-
terferon interaction is not analogic to the ganglioside-choleragen

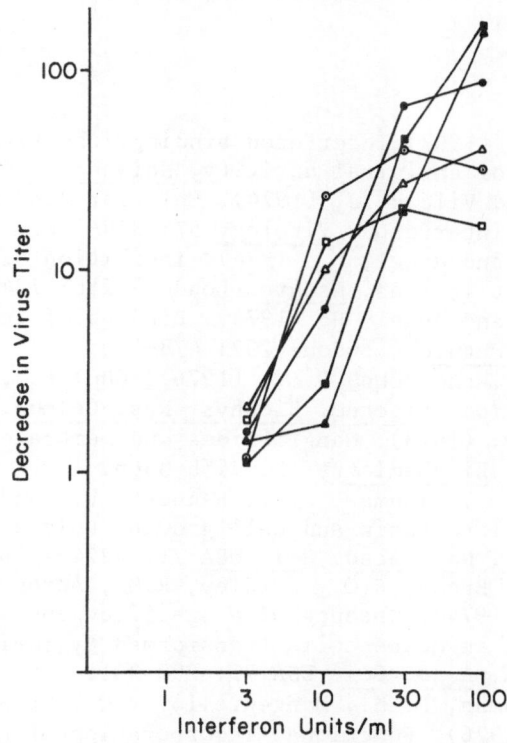

Fig. 4.   Effect of virus on choleragen mediated inhibition of interferon action. L cells were treated with choleragen
and mouse interferon. After 24 hr, the cells were washed
and infected with VSV or EMC virus. Virus was assayed 16
hr later. o—o, VSV, no choleragen; ▫ _ ▫ , VSV, 10^{-9} M
choleragen; Δ–Δ, VSV, 10^{-7} M choleragen; ●–● EMC, no choleragen; ◼ _ ◼ , EMC, 10^{-9} M choleragen; ▲–▲ EMC, 10^{-7} M
choleragen.

interaction — preincubation of NCTC 2071 cells with GM_1 makes them
sensitive to choleragen[9], while sensitivity to interferon remains
unchanged after incubation with exogenous GM_1.

The blocking of the interferon action by choleragen seems to be
more complex than a simple competition for the receptor. While the
choleragen is biologically active at 10^{-12} M concentration, the
inhibition of the interferon effect was demonstrated with 10^3-10^5
fold higher concentration (unpublished). Our data also shows that
the inhibition is virus specific, and can be demonstrated only
when EMC but not VSV is used as a challenging virus.

Thus, this data and apparent lack of species specificity in

the interferon-ganglioside interaction indicates that membrane gangliosides are not "the" specific and only receptors for interferon.

REFERENCES

1. Friedman, R.M. (1967) Interferon binding: the first step in establishment of antiviral activity, Science 156: 1760-1761.
2. Berman, B., and Vilček, J. (1974), Cellular binding characteristics of human interferon, Virology 57: 378-386.
3. Besancon, F., and Ankel, H. (1974), Inhibition of interferon action by plant lectins, Nature (London) 250: 784-786.
4. Besancon, F., and Ankel, H. (1974), Binding of interferon to gangliosides, Nature (London) 252: 478-480.
5. Friedman, R.M., and Kohn, D.L., (1976), Cholera toxin inhibits interferon action, Biochem. Biophys. Res. Commun., 70: 1078-1084.
6. Cuatrecasas, P. (1973). Gangliosides and membrane receptors for cholera toxin, Biochemistry 12: 3558-3566.
7. Hollenberg, M.D., Fishman, P.H., Bennett, V., and Cuatrecasas, P. (1974), Cholera toxin and cell growth: Role of membrane gangliosides. Proc. Nat. Acad. Sci. USA 71: 4224-4228.
8. Fishman, P.H., Brady, R.O., Bradley, R.M., Aaronson, S.A., and Todaro, G.J. (1974). Absence of a specific ganglioside galactosyltransferase in mouse cells transformed by murine sarcoma virus. Proc. Nat. Acad. Sci. USA 71: 298-301.
9. Moss, J., Fishman, P.H., Mangeniello, V.C., Vaughan, M., and Brady, R.O. (1976). Functional incorporation of gangliosides into intact cells: Induction of choleragen responsiveness. Proc. Nat. Acad. Sci. USA 73: 1034-1037.
10. Vengris, V.E., Reynolds, F.H.Jr., Hollenberg, M.D., and Pitha, P. M.(1976). Interferon action: Role of membrane gangliosides. Virology, 72: 486-493.
11. Brady, R.O., and Fishman, P.H. (1974). Biosynthesis of glycolipids in virus-transformed cells. Biochim. Biophys. Acta 355: 121-148.

THYROTROPIN RECEPTORS AND GANGLIOSIDES

Leonard D. Kohn[1], Eduardo Consiglio[1,2], Marc J. S. DeWolf[1,3], Evelyn F. Grollman[1], Fred D. Ledley[1], George Lee[1], and Nicholas P. Morris[1]

[1]Section on Biochemistry of Cell Regulation, Laboratory of Biochemical Pharmacology, National Institute of Arthritis, Metabolism, and Digestive Diseases, National Institutes of Health, Bethesda, Maryland 20014, U.S.A., [2]Centro di Endocrinologia ed Oncologia Sperimentale del C.N.R., Naples, Italy, and [3]RUCA-Laboratory for Human Biochemistry, University of Antwerp, B2020, Antwerp, Belgium

INTRODUCTION

Recent studies (1, 2, 28-30, 33-35) have implicated gangliosides in the receptor recognition process for thyrotropin (TSH) and have led to the hypothesis (15-17, 19) that the TSH receptor is structurally and functionally related to receptors for other ligands which interact with gangliosides *in vitro*, i.e., cholera toxin, tetanus toxin, and interferon. This report summarizes the accumulated evidence which implicates both a glycoprotein (4, 38-40, 43) and a ganglioside in the structure and function of the TSH receptor. In the context of this symposium, particular attention will be addressed to the role of the ganglioside.

THE GLYCOPROTEIN COMPONENT OF THE TSH RECEPTOR

Studies using thyroid cell cultures (43) showed that limited tryptic digestion of functioning thyroid cells resulted in a coincident loss and coincident return of TSH binding and cell function. Tryptic digestion did not, however, destroy the receptor but rather released the TSH binding component into the media (39, 43). The binding component could be radiolabeled by pulsing cultures with [^{14}C]glucosamine (43); it was purified by applying affinity chromatographic procedures to tryptic digests of bovine thyroid

membranes solubilized with lithium diiodosalicylate (38-40). The
purified receptor fragment contained 30% carbohydrate by weight
and had a molecular weight of 24,000. It had the same pH optima,
hormone specificity, and salt inhibition phenomena as the crude
TSH receptor preparations. Antisera prepared against the purified receptor component were able to precipitate the TSH binding
activity and reacted with the receptor fragment which was lost
from the plasma membrane after trypsinization of the thyroid
cells. In sum, then, early evidence pointed to a membrane glycoprotein as the TSH receptor.

GANGLIOSIDES AND THE TSH RECEPTOR

Studies of the properties of the TSH receptor fragment indicated that sialic acid was an important receptor determinant (38),
i.e., the binding activity of the solubilized fragment was found
to be sensitive to neuraminidase. In addition, when the antibody
to the bovine thyroid TSH receptor fragment was reacted with
solubilized TSH receptor preparations from other sources than the
thyroid, i.e., guinea pig retro-orbital tissue or human fat cell
membranes, an immunoprecipitin line was visualized (40). This
suggested that there was a component, perhaps an oligosaccharide
in the bovine thyroid TSH receptor, which was the same as that
in the TSH receptor in guinea pig retro-orbital tissue or the TSH
receptor in human fat cells.

With these results in mind, the mechanism by which cholera
toxin was believed at that time to act with cells was recalled,
i.e., the evidence that the ganglioside G_{M1} was the receptor for
cholera toxin (see reference 5 for a recent review) led to the
idea that a ganglioside or ganglioside-like structure could be
involved in or be an analog of the receptor for TSH. Accordingly,
a variety of these gangliosides were evaluated for their effects
on TSH binding to bovine thyroid receptors (33).

These studies showed that the gangliosides G_{D1b} and G_{T1} were
potent inhibitors of TSH binding to its specific membrane receptor; that this inhibition was caused by an interaction of these
gangliosides with the hormone rather than the membrane; and that
the inhibition was hormone-specific. Most important, it appeared
that gangliosides which were good inhibitors of TSH binding imposed on the TSH molecule a conformation distinct from that resulting from the interaction of TSH with gangliosides which were poor
inhibitors of TSH binding, i.e., G_{D1a}, the structural isomer of
G_{D1b}. These results led us to consider the following possibilities (15-17, 19, 33): (i) Gangliosides might be a structural and
functional component of the TSH receptor despite our evidence
implicating a membrane glycoprotein. (ii) TSH had a receptor
structure analogous to those for other ligands which interacted
with gangliosides *in vitro* and that each of these ligands, cholera

toxin, tetanus toxin, and interferon, as examples, might have features in common in their receptor interaction and message transmission process (15-17, 19). It is to be emphasized, however, that we considered it intuitively obvious that cholera toxin was not TSH; hence, differences should emerge in comparative studies. These differences should distinguish a toxin and hormone and could be in the ligand rather than the receptor.

Among the experimental evidence which accumulated to support these predictions were the following: Thyroid plasma membranes were not only rich in gangliosides with the chromatographic characteristics of G_{D1b}, G_{T1}, and G_{M1}, but also in a higher order ganglioside which (i) is an even more potent inhibitor of TSH binding and (ii) does not appear to be present in the brain (33, 35). A TSH receptor defect in a rat thyroid tumor was correlated with an alteration in the ganglioside content of the membranes of this tumor and with a defect in the biosynthetic enzymes concerned with the synthesis of the higher order gangliosides implicated as components of the TSH receptor structure (29).

Cholera toxin interacted with thyroid plasma membranes; these interactions affected the ability of TSH both to bind to the membranes and to stimulate the adenylate cyclase activity of the membranes (34). TSH did not, however, have the ability of cholera toxin to hydrolyze NAD (32).

Sequence homologies have been demonstrated to exist between cholera toxin, TSH, and the other glycoprotein hormones (20, 23). In regard to the sequence homology noted between the A protein of cholera toxin and α subunit of the glycoprotein hormones (10, 17, 23), this region was startlingly similar in sequence to the nonapeptide neurohypophyseal hormones, oxytocin and vasopressin.

Tetanus toxin was shown to interact with thyroid plasma membranes (11, 12, 22, 26, 27). The properties of this interaction were shown to closely resemble the properties of the TSH interaction with TSH receptors on these membranes; the interaction of tetanus toxin with thyroid membranes exhibited several characteristics of its interaction with neural tissue; tetanus toxin would not bind to membranes of a thyroid tumor with a TSH receptor defect; and tetanus toxin could stimulate thyroid hyperfunction in mice subjected to conditions known to precipitate thyroid storm.

Cholera toxin and TSH inhibited the establishment of the antiviral state by interferon (8, 18). *In vitro* binding experiments indicated that this inhibition could be correlated with the effect of interferon on cholera toxin and TSH binding to plasma membranes.

THE TWO-COMPONENT MODEL FOR THE TSH RECEPTOR

In order to reconcile data implicating a glycoprotein and other data implicating a ganglioside, we evolved (10, 14, 15) the hypothesis that the receptor contained both components, that each component contributed to the *in vivo* properties of the receptor, and that both components were necessary for the hormonal message to be transmitted to the cell machinery. Support for this view has come from several sources.

Trypsinized thyroid cells which have lost their TSH receptor and functional TSH response were shown to have normal amounts of membrane gangliosides and the usual distribution of gangliosides seen in membranes from nontrypsinized cells (manuscript in preparation).

The existence in membranes of two distinct components which can interact with TSH was clarified when lithium diiodosalicylate solubilization studies of the membrane were repeated (30) in order to insure that the glycoprotein component had not been simply contaminated with a ganglioside component.

Most important, however, support for this view has come from studies of the interferon and TSH receptor in interferon-sensitive and interferon-insensitive cell lines (6, 10, 17). As noted above, TSH binding could block interferon bioactivity in mouse Ly cells, but TSH binding was not accompanied by increases in cell cAMP levels. The question which evolved, therefore, was what was the structure of a specific TSH receptor on these membranes which could block the biological response of interferon yet have no target tissue-specific response. A corollary question concerned interferon and human KB-3 cells. These cells were insensitive to mouse or human interferon yet had receptors for both.

As in the 1-8R thyroid tumor, the absence of a biological response in the mouse Ly or human KB-3 cells was correlated with absolute or relative ganglioside deficiencies (6, 10, 17). Preliminary experiments with gangliosides have been able to partially reconstitute the biological response in human KB-3 cells for cholera toxin but not for TSH or interferon. In addition, although ganglioside resynthesis experiments with 1-8R thyroid membranes (the tumor with the abnormal TSH receptor and a ganglioside-deficiency) have partially reconstituted the TSH response, it is not clear that the ganglioside alone is the reconstituting factor. These experiments, in fact, have only emphasized the relationship between the ganglioside and glycoprotein receptor components.

Concerning the equally pertinent question of specific binding, the possibility was raised that interferon, TSH, and cholera toxin binding to the human KB-3 membranes could be accounted for by a glycoprotein receptor component in these membranes. This hypothesis was confirmed when membranes of human KB-3 cells and mouse Ly cells were extracted with lithium diiodosalicylate, and the lithium diiodosalicylate supernatant of both membrane preparations was incorporated into artificial membranes (liposomes) (6, 10, 17). Liposomes containing the lithium diiodosalicylate supernatant from either membrane preparation were able to bind ^{125}I-labeled TSH or cholera toxin. Binding was specific, was inhibited by interferon at concentrations similar to those effective using membrane preparations, and had appropriate temperature optima for each ligand. The existence of a glycoprotein component capable of interacting with cholera toxin was confirmed when toxin binding to KB-3 cell membranes was shown to be as trypsin-sensitive as TSH binding (10, 17).

The best direct evidence for the importance of the glycoprotein component to bioactivity also comes from experiments with interferon (21). Tryptic digestion of the lithium diiodosalicylate-solubilized glycoprotein preparation from mouse Ly cells resulted in the formation of several glycoprotein component fragments which could bind TSH directly or inhibit TSH binding to membrane or liposome preparations (21). Two of these fragments enhance interferon bioactivity *100-fold* when tested *in vivo*. The enhancement is specific for fragments from the mouse Ly lithium diiodosalicylate extract; requires preincubation with the membranes; and is associated with specific binding of the mouse Ly fragment to the cells (manuscript in preparation).

FUNCTIONAL ROLES OF THE TWO COMPONENTS

In regard to the TSH receptor, the data indicate that the glycoprotein component serves as the high affinity recognition site (38-40). Gangliosides, in contrast, are low affinity components (1, 2).

The glycoprotein may not, however, act independent of the ganglioside in affinity and capacity measurements or the ganglioside independent of the glycoprotein. Thus at low levels of gangliosides in liposomes, i.e., at concentrations where the ganglioside alone has no measurable TSH binding, the presence of the ganglioside markedly enhances TSH binding to liposomes containing the glycoprotein receptor component. At high ganglioside concentrations, the presence of both components can restrict TSH binding (manuscript in preparation).

The glycoprotein component alone or in a liposome can account for hormonal specificity in regard to nonglycoprotein hormone analogs, such as insulin, glucagon, ACTH, and prolactin (10, 17, 38-40). It cannot, however, distinguish between glycoprotein hormone analogs as well as the intact membrane (10, 17, 38-40). In contrast, the ganglioside interactions with glycoprotein hormone analogs appear unique in each case. For example, fluorescence studies show that gangliosides induce different perturbations in each ligand; a case in point is the comparative effects on TSH, luteinizing hormone (LH), and human chorionic gonadotropin (hCG) (15-17, 19, 24, 25, 33). Cholera toxin interactions with G_{M1} are irreversible, and tetanus toxin gives positive cooperativity (27). The tremendous difference between TSH and cholera toxin binding to liposomes as a function of ganglioside concentration (manuscript in preparation) is a case in point, as is the observation that TSH binds extremely well to liposomes with mixed brain gangliosides, whereas hCG will not (manuscript in preparation).

In addition, however, to helping discriminate in the binding recognition among glycoprotein hormones, toxins, interferon, and related ligands, we suggest that the ganglioside induces in the ligand a particular conformation necessary for it to continue the message transmission process. We also suggest that the ganglioside also brings the ligand or a portion of it into the hydrophobic portion of the lipid bilayer, where in its appropriate conformation it can interact with other membrane components important in message transmission.

The initial indication that different gangliosides induced different specific conformation changes in TSH came from fluorescence studies of native TSH (33).

Studies with dimethylaminonaphtalenesulphonate (DNS)-TSH have amplified this point (2). First, the interaction of gangliosides with DNS-TSH results in a large increase in the fluorescence intensity of the hormone-bound dye and in a blue shift in the fluorescence peak of DNS-TSH. G_{M1}, G_{D1a}, and G_{D1b} do not induce the same increases in the fluorescence intensity of DNS-TSH, nor do they cause identical blue shifts (2).

Second, when DNS-TSH and native TSH fluorescence changes were analyzed as a function of ganglioside concentration, the data indicated that the affinity of TSH for G_{M1} and G_{D1a} was essentially the same, despite their different effects on the fluorescence of the tyrosine residues or the DNS chromophore and despite their different effects on inhibition of TSH binding to thyroid plasma membranes (2). Clearly, differences in the affinity of G_{M1} or G_{D1a} for TSH could not account for differences in their inhibitory abilities.

Third, the polarization of fluoresence of DNS-TSH increased upon the addition of G_{M1}, G_{D1a}, or G_{D1b}, and in each case the increase in polarization quantitatively paralleled the increase in

quantum yield each ganglioside effected on the DNS-TSH molecule
(2). Since the effect of gangliosides on DNS-TSH results in an
increase of the quantum yield, the polarization should have decreased in the absence of a change of molecular configuration of
the hormone. The results which show a coincident increase in
quantum yield and polarization thus support the interpretation
that the ganglioside--DNS-TSH interaction results not only in
a higher molecular weight complex but also in a change in the
molecular conformation of the TSH molecule.

Fourth, an examination of the effect of ganglioside on the
circular dichroism of TSH shows that in the near UV all the gangliosides effected a significant change in the circular dichroism
spectra of TSH (2). This was not simply a "lipid" effect since
phosphatidylcholine, as an example, had no effect. It is recognized that the perturbation specific for each ganglioside may only
be a small portion of the overall perturbation. This conclusion
is evident in the far UV circular dichroism data which indicate
that the ganglioside-induced perturbation in the TSH molecule may
not be restricted to the microenvironment of one or more tyrosine
residues but also involves elements of the secondary structure of
the hormone.

Last, acid titration data show that the effect of G_{D1b} on
the conformation of the TSH molecule is significantly different
from that of G_{D1a} in that G_{D1b} can prevent subunit dissociation
in acid, whereas G_{D1a} cannot (2).

In sum, the gangliosides clearly induce a conformation change;
this change can be profound in terms of TSH molecular structure
and specific for different gangliosides (2).

The question of the role of the ganglioside in affecting an
approximation of the hormone with the bilayer is a result of
studies concerning the effect of salts on TSH interactions with
the gangliosides (1, 2). The binding of TSH to thyroid plasma
membranes is strongly inhibited by neutral salts at relatively low
concentrations (4, 39). If gangliosides, as normal constituents
of thyroid plasma membranes, can either participate in the structure of the receptor for TSH or serve as a model of the TSH receptor, their interaction with the hormone should be similarly sensitive to neutral salts *in vitro*.

We previously reported that the enhancement of TSH tyrosine
fluorescence caused by the gangliosides G_{D1a} and G_{D1b} was markedly
reduced in the presence of low concentrations of NaCl (1) and that
this effect of NaCl paralleled its effect on TSH binding to thyroid plasma membranes (1). The enhanced fluorescence of DNS-TSH
(2) induced by G_{D1a} and G_{D1b} was also strongly reduced by the
presence of increasing NaCl concentrations. This effect was seen

with other mono- and divalent salts (2) and was also evident when
the DNS-TSH fluorescence changes induced by G_{M1} and G_{T1} were
measured (2). The salt perturbations were additionally notable
in several respects.

First, the ability of salts to prevent the ganglioside-
induced fluorescence perturbation of the DNS-TSH molecule was
associated with a failure to form the DNS-TSH : ganglioside adduct.
Evidence for this was obtained on gel filtration columns, in
ultracentrifugation experiments, and in fluorescence polarization
studies (2).

Second, whereas salts prevented the ganglioside-induced
change in DNS-TSH fluorescence, salts would not reverse the ganglioside perturbation of the DNS-TSH molecule after the ganglio-
side--DNS-TSH complex had been formed (2). This effect was independent of the ganglioside tested or of the salt used. Thus, 40
mM NaCl or calcium acetate was able to significantly block the G_{M1}
or G_{T1} perturbation of the DNS-TSH molecule, and 80 mM NaCl or
calcium acetate completely blocked the effect of G_{M1} or G_{T1} on
DNS-TSH. In contrast, the addition of 80 mM NaCl or calcium
acetate after the complex had formed had no effect on the ganglioside-induced change in DNS-TSH fluorescence nor on the stability of the preformed adduct in gel filtration experiments or during
polarization studies.

Third, EDTA reversed the inhibitory effect of Ca^{2+} on the
ganglioside perturbation of the DNS-TSH molecule (2). In this
experiment the ganglioside perturbation of the DNS-TSH molecule
was first performed in the presence of Ca^{2+}, and then EDTA at the
noted concentrations was added. The ability of EDTA to remove
Ca^{2+} from a ganglioside-Ca^{2+} adduct was, however, in striking
contrast to the inability of Ca^{2+} to perturb a preformed ganglioside : DNS-TSH adduct. This result would predict that coordina-
tion sites important for the Ca^{2+} interaction with sialic acid
are no longer available to Ca^{2+} in the ganglioside : DNS-TSH
adduct (3).

The studies showing that salts can prevent but not reverse
the formation of the ganglioside-TSH adduct are of direct mechanistic interest to the forces approximating the two molecules.
In regard to the formation of the adduct, the ability of salts to
prevent complex formation suggests that the initial association
reaction involves electrostatic interactions. The inability of
salts to reverse the ganglioside-TSH interaction indicates that
short-range interactions, most likely hydrophobic, dominate the
kinetic dissociation constant. The hydrophobic nature of the
short-range interactions are also indicated by the observation
that dansylation of the TSH molecule increased its affinity for
gangliosides in all cases and by the observed fluorescence changes

of dansyl-TSH in each instance. That *specific* short-range interactions, as opposed to a *generalized "hydrophobic effect"*, are
involved is adduced from the observation that while the addition
of Ca^{2+} could not disrupt a preformed ganglioside-TSH adduct, it
could disrupt a preformed phospholipid-TSH adduct (2).

Studies of Ca^{2+}-sialic acid adducts (14) offer a potential
explanation for the specificity of the short-range interactions
insofar as gangliosides are concerned. Ca^{2+} interacts not only
with the negatively-charged carboxylate ion in the C1 position but
also with the glycerol-like moiety at the C7, C8, and C9 positions
of the molecule. For example, modification of the C8 position of
sialic acid totally alters the Ca^{2+}-sialic acid interaction (14).
In an apolar environment resulting from the approximation of the
TSH-ganglioside by short-range hydrophobic forces, the glycerol
moiety of the sialic acid might well be involved in hydrogen
bonding to amino acid residues of the TSH molecule. The ability
of Ca^{2+} to interact with sialic acid would then be significantly
reduced; this phenomenon would not be evident with phospholipids.

Gangliosides are amphiphilic molecules which form micelles
of 250,000 to 350,000 molecular weight (44) in aqueous solutions.
Previous reports suggested that the critical micelle concentration
for mixed brain gangliosides was approximately 10^{-4} M; individual
brain gangliosides were reported to aggregate at 10^{-5} M concentrations. According to the available data, at the concentrations
used by ourselves in TSH studies (10^{-6} M to 10^{-5} M), the gangliosides should have been predominantly in the monomeric form; *this
was, however, not true* (2). Thus the micellar state of the gangliosides was evident at concentrations at least as low as 3×10^{-8}
M in ultracentrifuge and polarization data (3). More detailed
studies of the micellar nature of the gangliosides are the subject
of a separate report by Formisano et al. (7).

In regard to inhibition of TSH binding, it is clear, then,
that differences between gangliosides cannot be accounted for by
differences in monomer-micelle relationships. More important, in
regard to the short-range hydrophobic interactions indicated to
be important in the formation of the TSH-ganglioside adduct, the
micellar state of the gangliosides would appear to preclude a
simple nonspecific association of a portion of the TSH molecule
with a lipid moiety of the ganglioside. Since this lipid moiety
is buried within the micellar matrix, this association, if it
occurs, must result subsequent to the initial electrostatic
interaction involving the negatively-charged polar head groups of
the carbohydrate moiety of the ganglioside. The specific nature
of this interaction is further emphasized in a report by Poss et
al. (36). These workers showed that TSH interactions with G_{T1},
but not G_{D1a} and G_{M1}, resulted in transmembrane ion fluxes across
lipid bilayers and that various sugars could inhibit this response.

Analogous specificity has also been recently described for cholera
toxin-G_{M1} interactions in lipid bilayers (41).

In the context of data indicating that the TSH receptor might
involve both a glycoprotein and a ganglioside component (10, 15-17,
19), the present study supports the possibility that the short-
range hydrophobic interactions which dominate the ganglioside-TSH
dissociation reaction are highly specific by comparison to other
membrane lipids; can be different from ganglioside to ganglioside; and result in an exclusion of water and salts from approximating surfaces of the TSH and ganglioside. The possibility is
thus raised that a portion of the TSH molecule may be roughly
approximated with the hydrophobic portion of the lipid bilayer
as a result of the ganglioside interaction.

GANGLIOSIDES AND MESSAGE TRANSMISSION

In previous reports (9, 10, 12, 15-17, 19, 37), we have
emphasized the possibility that a common mechanistic link among
all the agents interacting with gangliosides might be their ability to perturb ion fluxes across the membrane; that these actions
would be a cell signal independent of cAMP; and that these signals might even be important in eliciting the cAMP response.

This view is beginning to take on experimental flesh. In our
initial studies (9) we showed that cultured thyroid cells accumulated the lipophilic cation triphenylmethylphosphonium, indicating
that there was an electrical potential (interior negative) across
the plasma membrane. The stimulatory effect of TSH on $TPMP^+$ accumulation was not mimicked by human chorionic gonadotropin, a glycoprotein hormone with a similar structure (the target organ of
which is not the thyroid), and the effect was abolished when the
TSH receptor activity of the cells was destroyed by treatment with
trypsin. Analogous effects were observed with thyroid plasma membrane vesicles which are essentially devoid of mitochondrial and
soluble enzyme activities. TSH-Stimulated $TPMP^+$ uptake in the
vesicle preparations reached a quasi-steady state within 3 minutes;
in contrast, TSH-stimulated adenylate cyclase activity was negli-
gible during this period of time, became measurable after about 4
minutes, and was optimal after 12 to 15 minutes. Most important,
we could see this effect of TSH when cyclase was nonfunctional,
i.e., in vesicles where none of the cofactors necessary for cyclase
stimulation were present.

Studies with interferon (10) have also defined ligand-induced
early changes in membrane potential, changes in sodium-dependent
amino acid transport, and changes in the proton flux across the
membrane.

Our studies with tetanus toxin have recently defined *in vitro* effects on the electrical potential, as well as specific effects on Ca^{2+} and K^+ flux (12, 37).

These data raised the possibility, in conjunction with the data of Poss et al. (36), that the ganglioside offers to the membrane an appropriately oriented molecule capable of causing changes in transmembrane ion fluxes which are either capable of initiating or are part of the message transmission process. Recent liposome reconstitution experiments using tetanus as the effector suggest that the specific tetanus-induced ion fluxes cannot be induced in ganglioside-liposomes, whereas they can in ganglioside-liposomes containing other membrane components (manuscript in preparation).

GANGLIOSIDES AND THE PHOSPHOLIPID BILAYER

Vibrational Raman spectroscopy provides a sensitive technique for specifically probing lipid conformations in bilayer assemblies. Raman spectroscopy has been recently used to investigate, first, the spectral behavior of dipalmitoyl phosphatidylcholine (DPPC)-water liposomes perturbed by the intercalation of the G_{M1} ganglioside within the lipid matrix and then the spectral changes which arise on exposure of the model receptor system, thus created, to cholera toxin (13). The data indicated that G_{M1} is inserted within the bilayer and does not exist as separate micellar aggregates. Further, G_{M1} increases the intermolecular disorder of the DPPC bilayer while concomitantly increasing its intramolecular order. Finally for the G_{M1} containing liposomes to which cholera toxin has been added, the data demonstrate that while the intermolecular disorder parameters are not significantly altered, changes arise in the intramolecular order characteristics which are best described as a generally enhanced bilayer order containing localized patches of disordered lipids.

The interaction of cholera toxin with G_{M1} imbedded in black lipid films (41) and the TSH interaction with G_{T1} in similar systems (36) induce ion fluxes across the membrane. In addition, the cholera toxin-G_{M1} interaction in bilayer systems has resulted in glucose leaks from within the liposome (31). These data raised the possibility that the effector-ganglioside interaction caused a major change in the phospholipid bilayer which results in a generalized change in membrane permeability. From the Raman spectroscopic data we conclude this is not true. The important bilayer changes involve only a discrete or localized intramolecular and intermolecular chain disordering within the phospholipid matrix around the ganglioside. Further, this relatively small disordering effect is superimposed upon a general increase in intrachain (trans-gauche) order experienced by the lipid chains upon adding either G_{M1} or G_{M1} plus cholera toxin.

Experiments using Raman spectroscopy to evaluate thyrotropin-
G_{D1a}, thyrotropin-G_{M1}, thyrotropin-G_{T1}, or thyrotropin-G_{D1b}
interactions are in progress in order to see which of the changes
observed in the ordering of the phospholipid bilayer and induced
by the effector-ganglioside interaction can be directly correlated
with the previously reported ion fluxes from black lipid films.

GANGLIOSIDE INTERACTIONS WITH TSH COMPARED WITH OTHER ACIDIC
PHOSPHOLIPIDS IN THE MEMBRANE

Phosphatidylinositol, phosphatidylserine, and phosphatidylethanolamine interact with ^{125}I-TSH and inhibit its binding to
thyroid plasma membranes; phosphatidylcholine is not similarly
effective (3). The interaction has been monitored by column
chromatography on Sephadex G-100 which shows, for example, that
^{125}I-labeled TSH forms an adduct with phosphatidylinositol but
not with phosphatidylcholine. Formation of the ^{125}I-labeled TSH-
phosphatidylinositol adduct is dependent on the phosphatidylinositol concentration but can be reversed by both unlabeled TSH and
excess membranes (3).

The efficacy of the phospholipid interaction and the phospholipid inhibition of TSH binding to thyroid membranes is paralleled
by changes in fluorescence and fluorescence polarization imposed
on the dansyl derivative of TSH (3). These changes are reversed
by unlabeled TSH but not by prolactin, placental lactogen, or
growth hormone; similar changes are not observed when phospholipids are incubated with dansylated growth hormone, prolactin,
and placental lactogen (3). Monovalent potassium, sodium, and
lithium salts neither prevent nor reverse the formation of the
phospholipid : dansyl-TSH adduct; these results contrast with
the effects of the same salts on the formation of ganglioside
adducts with dansyl-TSH (3).

Despite their ability to interact with ^{125}I-TSH in solution,
neither phosphatidylinositol, phosphatidylserine, nor phosphatidylethanolamine, when incorporated in a liposome, bind the ^{125}I-
labeled ligand (3). These same phospholipids have no effect on
ganglioside binding of ^{125}I-labeled TSH when gangliosides are
incorporated in a liposome (3). These phospholipids do, however,
modulate the expression of the glycoprotein component of the TSH
receptor when it is imbedded in a liposome (3). The phosphatidylinositol in this case serves as a negative modulator, both by decreasing the incorporation of the glycoprotein component of the
receptor into the liposome, and by inhibiting the binding activity
of the glycoprotein component which is incorporated.

In regard to comparative ganglioside-TSH as opposed to phospholipid-TSH interactions, the following comments can be made:

First, we suggested earlier that the TSH interaction with
gangliosides involved an initial charge-charge interaction,
followed by a subsequent hydrophobic interaction. The key data in
this hypothesis were the salt sensitivity of the TSH interactions
with gangliosides but the lack of sensitivity of the TSH-ganglioside adduct to reversal by salts. In this report we show that
formation of the phospholipid-TSH adduct is not prevented by the
same monovalent salts which *can* prevent formation of the ganglioside-TSH adduct. These observations confirm the importance of the
sialic acid-containing carbohydrate moiety in the ganglioside-TSH
interaction and reaffirm the importance of an initial charge-
charge interaction between the ganglioside and the TSH effector.

Second, these data indicate that ganglioside interactions
with glycoprotein hormones do not simply reflect nonspecific
hydrophobic interactions between the hormone and the lipid moiety
of the ganglioside. Thus, phospholipids, membrane lipids with
hydrophobic properties similar to gangliosides, exhibit different
interactive properties from gangliosides, i.e., different salt
sensitivities. Further, there is no interaction between phosphatidylcholine and TSH. Phosphatidylcholine is the major "structural"
lipid of the bilayer; is similar in its hydrophobic properties to
a ganglioside; and has similarly complicated processes of self-
association in aqueous solvents.

Third, it is clear that although alone the "head group" of
the lipid has no effect, the head group plays a major specificity
determinant which contributes to the nature and stability of the
complex. In short, although the interaction is clearly a function
of both the hydrophobic lipid moiety and the charged "head group"
of the phospholipid, the assumption that the properties of the
whole molecule will be the simple sum of its parts or that hydrophobic interactions alone account for the present results or results with gangliosides, is invalid.

CONCLUSION

At this time we envision a TSH receptor with glycoprotein and
ganglioside components. We see the glycoprotein as a specific
sticky "flypaper" which concentrates ligands at the cell surface.
We speculate that the ganglioside is helping in the discrimination
of similarly-structured ligands, is inducing specific conformational changes in the ligand which are important to message transmission, and is bringing a portion of the ligand into the lipid
bilayer where it can contribute to alterations in ion fluxes and
can interact with other intramembrane components. At this time
our data do distinguish between gangliosides and other membrane
lipids. Unfortunately, the data do not yet unequivocally prove
any aspect of the above hypothesis in real physiological terms.
Experiments to further define the ganglioside role are thus in

progress and should at the very least enhance our understanding of the structure and function of the TSH receptor.

BIBLIOGRAPHY

1. ALOJ S.M., KOHN L.D., LEE G. and MELDOLESI M.G. (1977): The binding of thyrotropin to liposomes containing ganglio-sides. Biochem. Biophys. Res. Commun. 74, 1053-1059.
2. ALOJ S.M., LEE G., CONSIGLIO E., FORMISANO S., MINTON A.P. and KOHN L.D. (1979): Dansylated thyrotropin as a probe of hormone receptor interactions. J. Biol. Chem. 254, in press.
3. ALOJ S.M., LEE G., GROLLMAN E.F., BEGUINOT F., CONSIGLIO E. and KOHN L.D. (1979): Role of phospholipids in the struc-ture and function of the thyrotropin receptor. J. Biol. Chem. 254, in press.
4. AMIR S.M., CARRAWAY T.F. Jr., KOHN L.D. and WINAND R.J. (1972): Thyrotropin binding to thyroid plasma membranes. J. Biol. Chem. 248, 4092-4100.
5. BENNETT V. and CUATRECASAS P. (1977): Membrane gangliosides and activation of adenylate cyclase, in "The Specificity and Action of Animal, Bacterial, and Plant Toxins", CUATRECASAS P. and GREAVES M.F., Eds., Chapman and Hall (London), pp. 3-66.
6. CHANG E.H., GROLLMAN E.F., JAY F.T., LEE G., KOHN L.D. and FRIEDMAN R.M. (1978): Membrane alterations following inter-feron treatment, in "Human Interferon", STINEBRING W.R. and CHAPPLE P.J., Eds., Plenum (New York), pp. 85-99.
7. FORMISANO S., LEE G., JOHNSON M.L. and EDELHOCH H. (1979): Critical micelle concentrations of gangliosides. Biochem-istry 18, in press.
8. FRIEDMAN R.M. and KOHN L.D. (1976): Cholera toxin inhibits interferon action. Biochem. Biophys. Res. Commun. 70, 1078-1084.
9. GROLLMAN E.F., LEE G., AMBESI-IMPIOMBATO F.D., MELDOLESI M.F., ALOJ S.M., COON H.G., KABACK H.R. and KOHN L.D. (1977): Effects of thyrotropin on the thyroid cell membrane: hyper-polarization induced by hormone-receptor interaction. Proc. Natl. Acad. Sci. U.S.A. 74, 2352-2356.
10. GROLLMAN E.F., LEE G., RAMOS S., LAZO P.S., KABACK H.R., FRIEDMAN R.M. and KOHN L.D. (1978): Relationships of the structure and function of the interferon receptor to hormone receptors and establishment of the antiviral state. Cancer Res. 38, 4172-4185.
11. HABIG W.H., GROLLMAN E.F., LEDLEY F.D., MELDOLESI M.F., ALOJ S.M., HARDEGREE M.C. and KOHN L.D. (1978): Tetanus toxin interactions with the thyroid: decreased toxin binding to membranes from a thyroid tumor with a thyrotropin receptor defect and *in vivo* stimulation of thyroid function. Endocrinology 102, 844-851.

12. HABIG W.H., HARDEGREE M.C., GROLLMAN E.F., LEE G., RAMOS S. and KOHN L.D. (1979): Interaction of tetanus toxin with thyroid membranes, in "Proceedings of the 5th International Tetanus Conference, Stockholm, Sweden, June 18-23, 1978", in press.

13. HILL I.R., LEVIN I.W., LAZO P.S. and KOHN L.D. (1979): Effect of ganglioside and ganglioside-cholera toxin interactions on the vibrational Raman spectra of the bilayer lipid in dipalmitoyl phosphatidylcholine liposomes. Biochemistry 18, in press.

14. JAQUES L.W., BROWN E.B., BARRETT J.M., BREY W.S. and WELLNER W. (1977): Sialic acid: a calcium binding carbohydrate. J. Biol. Chem. 252, 4533-4538.

15. KOHN L.D. (1977): Characterization of the thyrotropin receptor and its involvement in Graves' disease, in "Horizons in Biochemistry and Biophysics", QUAGLIARIELLO E., Ed., Vol. 3, Addison-Wesley (Reading, Massachusetts), pp. 123-163.

16. KOHN L.D. (1977): Relationships in the structure and function of cell surface receptors for glycoprotein hormones, bacterial toxins, and interferon, in "Annual Reports in Medicinal Chemistry", CLARKE F.H., Ed., Vol. 12, Academic Press (New York), pp. 211-222.

17. KOHN L.D. (1978): Relationships in the structure and function of receptors for glycoprotein hormones, bacterial toxins, and interferon, in "Receptors and Recognition", CUATRECASAS P. and GREAVES M.F., Eds., Series A, Vol. 5, Chapman and Hall (London), pp. 134-212.

18. KOHN L.D., FRIEDMAN R.M., HOLMES J.M. and LEE G. (1976): Use of thytrotopin and cholera toxin to probe the mechanism by which interferon initiates its antiviral activity. Proc. Natl. Acad. Sci. U.S.A. 73, 3695-3699.

19. KOHN L.D., LEE G., GROLLMAN E.F., LEDLEY F.D., MULLIN B.R., FRIEDMAN R.M., MELDOLESI M.F. and ALOJ S.M. (1978): Membrane glycolipids and their relationship to the structure and function of cell surface receptors for glycoprotein hormones, bacterial toxins, and interferon, in "Cell Surface Carbohydrate Chemistry", HARMAN R.E. (Ed.), Academic Press (New York), pp. 103-133.

20. KUROSKY A., MARKEL D.E., PETERSON J.W. and FITCH W.M. (1977): Primary structure of cholera toxin β-chain: a glycoprotein hormone analog? Science 195, 299-301.

21. LAZO P.S., GROLLMAN E.F., LEE G. and FRIEDMAN R.M. (1978): Structure:function studies of the interferon receptor. Fed. Proc. 37, 1822.

22. LEDLEY F.D., LEE G., KOHN L.D., HABIG W.H. and HARDEGREE M.C. (1977): Tetanus toxin interactions with thyroid plasma membranes: implications for structure and function of tetanus toxin receptors and potential pathophysiological significance. J. Biol. Chem. 252, 4049-4055.

23. LEDLEY F.D., MULLIN B.R., LEE G., ALOJ S.M., FISHMAN P.H.,
     HUNT L.T., DAYHOFF M.O. and KOHN L.D. (1976): Sequence
     similarity between cholera toxin and glycoprotein hormones:
     implications for structure activity relationship and mecha-
     nism of action. Biochem. Biophys. Res. Commun. 69, 852-859.

24. LEE G., ALOJ S.M., BRADY R.O. and KOHN L.D. (1976): The struc-
     ture and function of glycoprotein hormone receptors: gang-
     lioside interactions with human chorionic gonadotropin.
     Biochem. Biophys. Res. Commun. 73, 370-377.

25. LEE G., ALOJ S.M. and KOHN L.D. (1977): The structure and
     function of glycoprotein hormone receptors: ganglioside
     interactions with luteinizing hormone. Biochem. Biophys.
     Res. Commun. 77, 434-441.

26. LEE G., CONSIGLIO E., HABIG W.H., DYER S., HARDEGREE M.C. and
     KOHN L.D. (1978): Structure:function studies of receptors
     for thyrotropin and tetanus toxin: lipid modulation of
     effector binding to the glycoprotein receptor component.
     Biochem. Biophys. Res. Commun. 83, 313-320.

27. LEE G., GROLLMAN E.F., DYER S., BEGUINOT F., KOHN L.D., HABIG
     W.H. and HARDEGREE M.C. (1979): Tetanus toxin and thyrotro-
     pin interactions with rat membrane preparations. J. Biol.
     Chem. 254, in press.

28. LEE G., GROLLMAN E.F., ALOJ S.M., KOHN L.D. and WINAND R.J.
     (1977): Abnormal adenylate cyclase activity and altered
     gangliosides in thyroid cells from patients with Graves'
     disease. Biochem. Biophys. Res. Commun. 77, 139-146.

29. MELDOLESI M.F., FISHMAN P.H., ALOJ S.M., KOHN L.D. and BRADY
     R.O. (1976): Relationship of gangliosides to the structure
     and function of thyrotropin receptors: their absence on
     plasma membranes of a thyroid tumor defective in thyrotro-
     pin activity. Proc. Natl. Acad. Sci. U.S.A. 73, 4060-4064.

30. MELDOLESI M.F., FISHMAN P.H., ALOJ S.M., LEDLEY F.D., LEE G.,
     BRADLEY R.M., BRADY R.O. and KOHN L.D. (1977): Separation
     of the glycoprotein and ganglioside components of thyro-
     tropin receptor activity in plasma membranes. Biochem.
     Biophys. Res. Commun. 75, 581-587.

31. MOSS J., FISHMAN P.H., RICHARDS R.L., ALVING C.R., VAUGHAN M.
     and BRADY R.O. (1976): Choleragen-mediated release of
     trapped glucose from liposomes containing ganglioside G_{M1}.
     Proc. Natl. Acad. Sci. U.S.A. 73, 3480-3483.

32. MOSS J., MANGANIELLO V.C. and VAUGHAN M. (1977): Hydrolysis
     of nicotinamide adenine dinucleotide by choleragen and its
     A protomer: possible role in the activation of adenylate
     cyclase. Proc. Natl. Acad. Sci. U.S.A. 73, 4424-4427.

33. MULLIN B.R., ALOJ S.M., FISHMAN P.H., LEE G., KOHN L.D. and
     BRADY R.O. (1976): Cholera toxin interactions with thyro-
     tropin receptors on thyroid plasma membranes. Proc. Natl.
     Acad. Sci. U.S.A. 73, 1679-1683.

34. MULLIN B.R., FISHMAN P.H., LEE G., ALOJ S.M., LEDLEY F.D., WINAND R.J., KOHN L.D. and BRADY R.O. (1976): Thyrotropin-ganglioside interactions and their relationship to the structure and function of thyrotropin receptors. Proc. Natl. Acad. Sci. U.S.A. 73, 842-846.

35. MULLIN B.R., PACUSZKA T., LEE G., KOHN L.D., BRADY R.O. and FISHMAN P.H. (1978): Thyroid gangliosides with high affinity for thyrotropin: potential role in thyroid regulation. Science 199, 77-79.

36. POSS A., DELEERS M. and RUJSSCHERT J.M. (1978): Evidence for a specific interaction between G_{T1} ganglioside incorporated into bilayer membranes and thyrotropin binding. FEBS Letts. 86, 160-162.

37. RAMOS S., GROLLMAN E.F., LAZO P.S., HABIG W.H., HARDEGREE M.C., KABACK H.R. and KOHN L.D. (1979): Effect of tetanus toxin on the accumulation of a permanent lipophilic cation ($TPMP^+$) by guinea pig synaptosomes. Proc. Natl. Acad. Sci. U.S.A. 76, in press.

38. TATE R.L., HOLMES J.M., KOHN L.D. and WINAND R.J. (1975): Characteristics of a solubilized thyrotropin receptor from bovine thyroid plasma membranes. J. Biol. Chem. 250, 6527-6533.

39. TATE R.L., SCHWARTZ H.I., HOLMES J.M., KOHN L.D. and WINAND R.J. (1975): Thyrotropin receptors in thyroid plasma membranes. J. Biol. Chem. 250, 6509-6515.

40. TATE R.L., WINAND R.J. and KOHN L.D. (1976): Solubilization and partial purification of the thyrotropin receptor, in "Thyroid Research", ROBBINS J. and BRAVERMAN L., Eds., International Congress Series No. 378, Excerpta Medica (Amsterdam), pp. 57-60.

41. TOSTESON M.T. and TOSTESON D.C. (1978): Bilayers containing gangliosides develop channels when exposed to cholera toxin. Nature 275, 142-144.

42. WADELEUX P.A., ETIENNE-DECERF J., WINAND R.J. and KOHN L.D. (1978): Effects of thyrotropin on iodine metabolism of dog thyroid cells in tissue culture. Endocrinology 102, 889-902.

43. WINAND R.J. and KOHN L.D. (1975): Thyrotropin effects on thyroid cells in culture: effects of trypsin on the thyrotropin-medicated cyclic 3':5'-AMP changes. J. Biol. Chem. 250, 6534-6540.

44. YOHE H.C., ROARK D.E. and ROSENBERG A. (1976): C_{20}-Sphingosine as a determining factor in aggregation of gangliosides. J. Biol. Chem. 251, 7083-7087.

GANGLIOSIDES AND THERMAL ADAPTATION

Hinrich Rahmann

Zoological Institute of the University

Stuttgart-Hohenheim, West Germany

SUMMARY

The concentration and pattern of brain gangliosides of lower and higher vertebrates were compared. Only when considering the eco-factor temperature clear correlations between systematical position of species, thermal adaptation status and brain ganglioside composition can be shown: lowering of the environmental temperature induces a long-term formation of more polar ganglioside fractions. In fishes additional long-term compensatory effects to changes in temperature were demonstrated with regard to motor activity, bioelectrical activity of the CNS (post-synaptic potential amplitude) and avoidance conditioning-learning. All phenomena taken together reflect a mechanism of modulating the thermo-sensitivity of the membrane-mediated processes of transmission.

INTRODUCTION

In contrast to the fast growing knowledge on the chemical structure, metabolism and distribution of gangliosides in neuronal as well as non-neuronal tissues only little is known about the possible role of these compounds especially on neuronal functions. The high accumulation of gangliosides at synaptic terminals, the binding capacity for neurotropic substances, the accretion during synaptogenesis, the participation in mental disorders, all these factors can be taken as evidence for the functional significance of gangliosides

in neuronal processes (Rahmann 1978). These factors and
first experiments using electrophysiological methods
(Römer and Rahmann 1979) indicate the gangliosides being involved in the process of synaptic transmission.

In order to get more precise insight in the biological function of gangliosides with respect to the
above referred phenomena, it was of outstanding interest to investigate whether there are any correlations
between the brain ganglioside metabolism and physiological parameters which are well known to be directly
dependent on the process of synaptic transmission. The
mechanism of temperature adaptation in vertebrates (acclimatization = thermal adaptation of an organism to
seasonal fluctuations in environmental temperature
within its natural eco-system, and acclimation = thermal
adaptation to well defined temperature changes under
laboratory conditions) is assumed to be based on adaptive changes in the synapse as the primary site of all
kind of adaptive processes in the nervous system (Hazel
and Prosser 1974; Katz and Miledy 197o; Lagerspetz 1974).
Up to now, the molecular mechanism of these changes is
still unknown. But according to the high developed a-
bility of gangliosides to form more or less stable complexes with Ca^{2+}-ions (Probst et al. 1979), for which
sialic acid obviously is the functional group, we regard gangliosides as the extra-cellular Ca^{2+}-storage-
system of the synaptic complex. And since Ca^{2+} is known
to be essential for the release of transmitter (Katz and
Miledi 197o) and since this transmitter release is known
to be extremely sensitive to changes in the environmental temperature, we therefore carried out various
experiments concerning the participation of gangliosides in process of thermal adaptation (Rahmann 1976,
1978).

MATERIALS AND METHODS

The brain gangliosides of in total 15 new species
belonging to all 5 vertebrate classes, especially to
fishes, were analyzed and compared with those of 3o
previously described species (fig. 1). The gangliosides
were extracted according to Tettamanti et al. (1973).
The concentration of ganglioside-bound sialic acid
(NeuAc) was determined according to Jourdian et al.
(1971). Aliquots of purified gangliosides containing 5
µg NeuAc were separated by thin-layer-chromatography
(Merck-silica gel 6 plates: chloroform-methanol-water-
NH_3; 6o:35:7:1), visualized with resorcinol according to

Svennerholm (1957), identified and quantified by their
relative migration rates in comparison to standards by
means of densitometric scanning. The ganglioside fractions were numbered according to their increasing migration rates and compared with standards. By this procedure the concentration (nmol NeuAc/gr.fr.wt.) or the mol
%-levels of the main ganglioside fractions within the
ganglioside pattern and their variation within phylo-
genetical position or thermal adaptation phenomena was
determined. In order to make the great variety of
ganglioside patterns more intuitive in some of the given
figures the relative proportion of the different
ganglioside fractions with a higher or lower migration
rate than the trisialo-ganglioside G_{Gtet3a} NeuAc (=
G_{T1a}) was drawn.

RESULTS

When comparing the brain ganglioside content of the
varies vertebrates it has to be stated that the concentration obviously correlates to the level of nervous
organization: in the brain of the higher vertebrates
(mammals and birds) it is in the range of about 5oo to
1ooo μg ganglioside bound NeuAc/gr.fr.wt. in comparison
to about 13o up to 5oo μg in the case of the lower
vertebrates (reptiles, amphibes and fishes). Already on
this level of discussion it has to be emphasized that
with regard to thermal adaptation only mammals and birds
belong to the homoiothermic animals while the lower
vertebrates, as poikilotherms, do not have the ability
of temperature-regulation. From this it can be concluded
that the increasing enrichment of the nervous system
with gangliosides must have been obviously an evolutionary advantage during the phylogeny of vertebrates.

Besides these differences in ganglioside concentration there exists a remarkable variability in the
brain ganglioside pattern between higher and lower
vertebrates. With respect to the %-distribution of the
ganglioside-bound NeuAc in the three major ganglioside
groups it has to be stated that in total the brain
ganglioside pattern of the homeothermic vertebrates is
characterized by the proponderance of oligo-sialo-
gangliosides and that of the poikilothermic vertebrates
by that of multi-sialo gangliosides (fig. 1).

Since no clear correlations between the ganglioside pattern and the taxonomical position of the investigated species could be shown, the question was raised,
whether the differences in the ganglioside concentra-

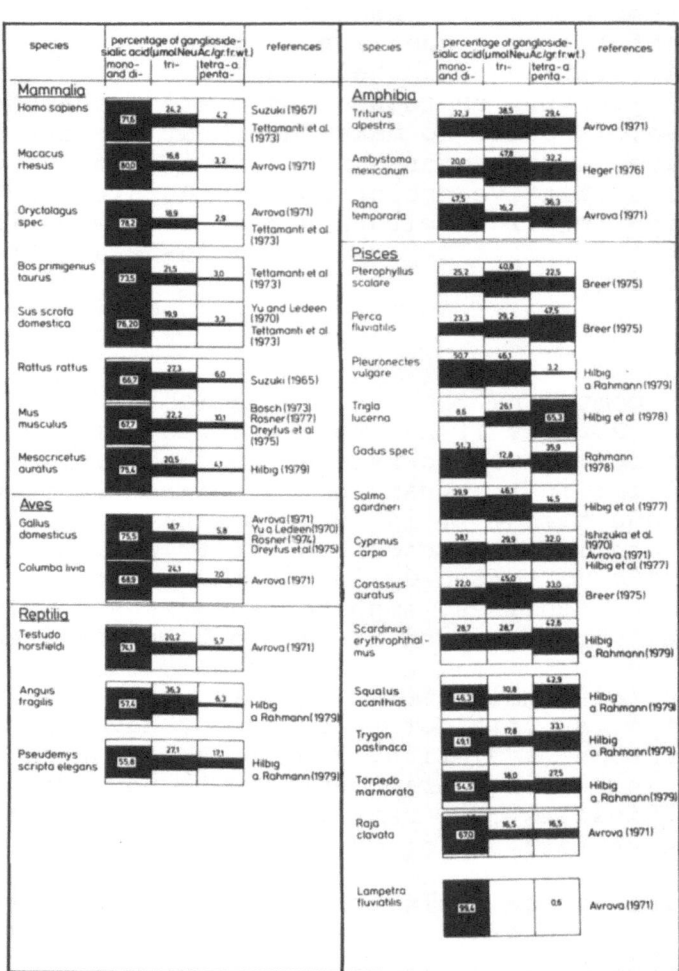

Fig. 1. Determination of the %-distribution of the ganglioside-bound NeuAc in the three major brain ganglioside fraction groups of vertebrates.

tion and pattern might be correlated with the eco-factor
temperature.

Therefore in poikilotherms we investigated whether
the brain ganglioside pattern of taxonomically close and
distant related species remains constantly or not when
the environmental temperature was changed. In the case
of the homeothermic animals, so-called heterothermic
phases were analyzed, during which these animals are not
able to regulate their own body temperature on a constant level: hibernation, neonatal phase.
a) Acclimatization and acclimation in poikilotherms:
The first approach was done with single poikilothermic
fish-species during the individual long-term process of
seasonal acclimatization. For instance, the ganglioside
patterns of rainbow trout, caught in winter and summer
were compared: during adaptation to winter conditions
the slow migrating polar multi-sialoganglioside fractions dominate, while during summer the less polar
bands, especially the hematoside fraction become domi-
nant. This phenomenon of poly-sialisation effect of brain
gangliosides was not only observed in rainbow trout but
also in carps, gold-fishes and plaice (Hilbig et al.
1979). Similar effects were obtained during experimentally induced thermal adaptation (acclimation) (Rahmann
1976, 1978; Rösner et al. 1979); the lowering of the
environmental temperature induced an accumulation of
multi-sialogangliosides in the CNS.
b) Heterothermic phases in vertebrates with thermo-
regulation: although birds and mammals as homeotherms
have developed the thermo-regulation ability, which, of
course, is the most effective procedure in order to a-
dapt to changing temperatures, they show heterothermic
phases during which they behave like poikilotherms. This
occurs during perinatal development and in some species
during hibernation. During the perinatal development of
the rat, e.g. (fig. 2), the concentration of gangliosides in the cortex increases significantly. Parallel
the percentage, especially of the disialoganglioside
G_{D1a} steaply increases too. Concomittantly the values
for the more polar trisialoganglioside G_{T1a} decreases,
while those for the tetra-sialoganglioside G_Q did not
change. Similar data were obtained for the neonatal development of mice, chicken and guinea pigs. Quite comparable results to those of the early development were
obtained when analyzing the brain ganglioside patterns
of hibernating with normothermic (non-hibernating)
golden hamsters. Here again in adaptation to lowered
environmental temperatures a temporary polysialisation

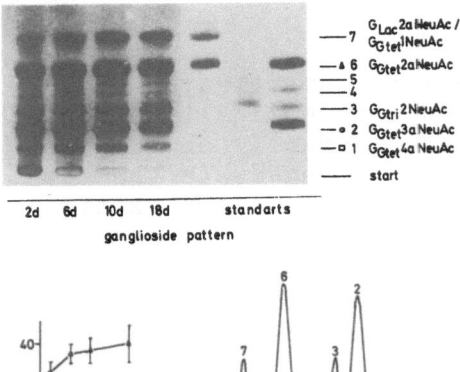

	G_{Lac} 2a NeuAc /
7	G_{Gtet} 1 NeuAc
▲ 6	G_{Gtet} 2a NeuAc
5	
4	
3	G_{Gtri} 2 NeuAc
○ 2	G_{Gtet} 3a NeuAc
□ 1	G_{Gtet} 4a NeuAc
—	start

2d 6d 10d 18d   standarts

ganglioside pattern

Fig. 2. Changes in ganglioside pattern and concentration of rat cortex during neo-natal development.

mol % ganglioside

2 6 10 18d

216 241 280 351

µg NeuAc /g fresh wt.

chromatogram
C:M:NH₄:H₂O:Ca⁺⁺ 60:36:0.4:8:0.01

RELATIVE DISTRIBUTION OF BRAIN GANGLIOSIDES AND THERMAL ADAPTATION

thermal classification		species	percentage of ganglioside-sialic acid (µmol NeuAc/gr. fr. wt.)		
			mono-a.di-	tri-	tetra-a. penta-
			5 - 14	4	1 - 3
A. homeothermic		Mus, Rattus, Sus, Gallus			
B. heterothermic	early ontogeny	Gallus or Mus			
1. ontogenetic	postnatal				
	juvenile				
2. hibernation	active sleeping	Mesocricetus aur. (gold. hamster)			
C. poikilothermic					
1. stenothermic	summer(19±4°C)	Salmo (trout)			
	winter (4±3°C)				
2. eurythermic	summer(28±5°C)	Carpio (carp), or Carassius (goldfish)			
	winter (4±3°C)				

4 ♀ G_{Gtet} 3a NeuAc
1 - 3 ⇒ fractions migrating faster than 4
5-14 ⇒ fractions migrating slower than 4

Fig. 3. %-distribution of ganglioside-bound NeuAc in the three major brain ganglioside fraction groups of vertebrates according to their thermal classification and thermal adaptation.

of neuronal gangliosides was shown (Hilbig and Rahmann 1979).

DISCUSSION

     The results of long-term poly-sialisation effects of brain gangliosides as well during early postnatal development, during hibernation and also during ac- climatization and acclimation show that the molecular mechanism of thermal adaptation in vertebrates is quite similar as well in poikilotherms and homeotherms (fig. 3). We suppose that these changes in the molecular com- position of neuronal membranes may reflect an important and more general mechanism of modulating the thermo- sensitivity of membrane-mediated processes, as e.g. transmission. This assumption is supported by some recent physiologically relevant findings in teleost fishes (fig. 4), which are in full agreement with the long-term compensatory events on the ganglioside meta- bolism:
1. The transfer of goldfishes to warmth (fig. 4a) or coldness (fig. 4b) caused initial changes of photic evoked field potential amplitudes, especially in the postsynaptic amplitude, which reflected the high thermal sensitivity of this system. This remained for about 8 days. Then slow compensation effects oc- curred and finally after about 4 to 6 weeks the postsynaptic amplitude was again on its original level.
2. These results correspond with the regain of motor activity after a rapid transfer of goldfishes from warm to cold water. Here again after a sedation phase of about 8 days a regain of mobility was observed.
3. Furthermore these results correspond with the regain of learning ability (electrical avoidance condi- tioning): after transfer from cold water, where these fishes cannot perform any learning, to warm water, after about 8 days a compensation began so that the full learning-capacity was regained 5 to 6 weeks after the transfer (fig. 4d).

     Finally some concluding remarks: the results of long-term changes in the molecular composition of neu- ronal gangliosides in adaptation to changes in the environmental temperature and the compensatory phenome- na of motor activity, bioelectrical activity of the CNS and of higher associative functions (learning ability) reflect a mechanism of modulating the thermo-sensitiv- ity of the membrane mediated process of transmission,

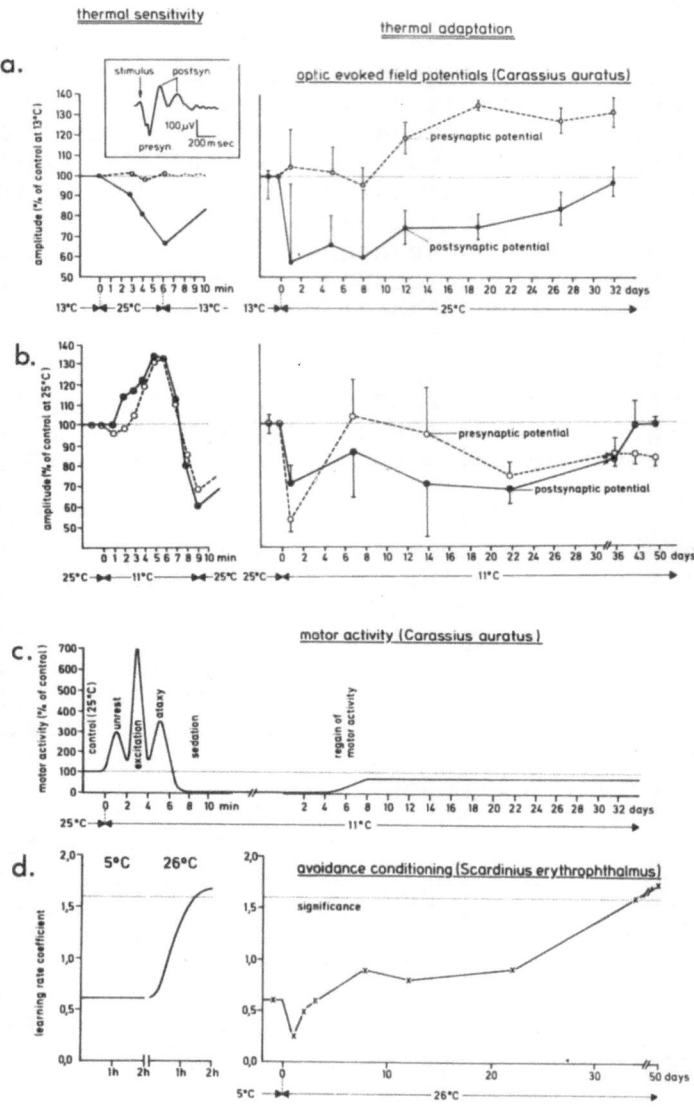

Fig. 4. Correlations between thermal sensitivity/adaptation and optic evoked potentials (a, b)/motor activity (c)/avoidance conditioning (d) of fishes to changes in the environmental temperature.

which ought to be discussed with regard to the high complexation-ability of gangliosides together with Ca^+ ions (Rahmann et al. 1976).

REFERENCES

HAZEL, J.R. and PROSSER, C.L. (1974): Molecular mechanisms of temperature compensation in poikilotherms. Physiol. Rev. 54, 62o-677.
HILBIG, R. and RAHMANN, H. (1979): Changes in brain ganglioside composition of hormothermic and hibernating golden hamsters (Mesocricetus auratus). Comp. Biochem. Physiol. (in press).
HILBIG, R., RAHMANN, H. and RÖSNER, H. (1979): Brain gangliosides and temperature adaptation in eury- and stenothermic teleost fish (carp and rainbow trout). J. Therm. Biol. 4, 29-34
JOURDIAN, G.W., DEN, L. and ROSEMAN, S. (1971): The sialic acids: XI. A periodate-resorcinol method for the quantitative estimation of free sialic acids and their glycosides. J. biol. Chem. 246, 43o-435.
KATZ, B. and MILEDI, R. (197o): Further study of the role of calcium in synaptic transmission. J. Physiol.(Lond.) 2o7, 789-8o1.
LAGERSPETZ, K.Y.H. (1974): Temperature acclimation and the nervous system. Biol. Rev. 49, 477-514.
PROBST, W., RÖSNER, H., WIEGANDT, H. and RAHMANN, H. (1979): Das Komplexationsvermögen von Gangliosiden für Ca^{2+}. I. Einfluß mono- und divalenter Kationen sowie von Acetylcholin. Hoppe-Seyler's Arch. (in press).
RAHMANN, H. (1976): Possible functional role of gangliosides, in "Ganglioside Function", PORCELLATI, G., CECCARELLI, B. and TETTAMANTI, G., Eds., Advances in Experimental Biology, Vol. 71, Plenum Press (New York and London), pp. 151-161.
RAHMANN, H. (1978): Gangliosides and thermal adaptation in vertebrates.Japan.J.Exp.Med. 48, 85-96.
RAHMANN, H., RÖSNER, H. and BREER, H. (1976): A functional model of sialo-glyco-macromolecules in synaptic transmission and memory formation. J. Theor.Biol. 57, 231-237.
RÖMER, H. and RAHMANN, U. (1979): Effects of exogenous neuraminidase on unit activity in frog spinal cord and fish optic tectum. Exp.Brain Res. 34, 49-58.
RÖSNER, H., BREER, H., HILBIG, R. and RAHMANN, H. (1979): Temperature effects on the incorporation of sialic acid into gangliosides and glycoproteins of fish brain. J.Therm.Biol. 4, 69-73.

SVENNERHOLM, L. (1957): Quantitative estimation of sialic aciels. Biochim. Biophys. Acta <u>24</u>, 604-611.

TETTAMANTI, G., Bonalli, F., Marchesini, S. and Zambotti, V. (1973): A new procedure for the extraction, purification and fractionation of brain gangliosides. Biochim. Biophys. Acta <u>296</u>, 160-170.

NEURONAL AND GLIAL CELL CULTURES, A TOOL FOR INVESTIGATION OF GANGLIOSIDE FUNCTION

P. Mandel, H. Dreyfus[1], A.N.K. Yusufi, L. Sarlième[1],
J. Robert, N. Neskovic[1], S. Harth[2] and G. Rebel[2]

Centre de Neurochimie du CNRS, and Unité 44 de l'INSERM,
11, Rue Humann, 67085 Strasbourg Cedex, France

Gangliosides are membrane constituents that are partly embedded in the bilayer structure of the membrane and partly exposed to the external environment by negative-charged polysaccharide chains. One may expect that they are involved in structural plasticity and in the functional activity of the plasma membrane. Great interest has been devoted to ganglioside structure and function during cell differentiation, maturation and ageing (see for review HAKOMORI, 1973). Changes associated with cell transformation, reduction of the levels of the most complex sphingolipids, a decrease in the activity of glycosyl- and/or sialyltransferases have raised the question of involvement of gangliosides in contact inhibition (HAKOMORI, 1973). Binding properties of gangliosides to exotoxins of *Vibrio cholerae* and *Clostridium tetani* drew attention to a function of sialoglycolipids as cell surface receptors and to their interactions with proteins and glycoproteins localized on cell surfaces or close to cell surfaces (SHAROM and GRANT, 1978; YAKAMAWA and NAGAI, 1978). Finally, an increasing interest in gangliosides of the nervous system arose when alterations of gangliosides in some genetic diseases was discovered (for review see SUZUKI, 1976) and when the abundance of gangliosides in plasma and synaptic membrane was established (MORGAN *et al.*, 1971; LEDEEN, 1978). Moreover, attention became focused on the role of gangliosides in cation binding, transport and release (ABRAMSON, YU and ZABY, 1972; BEHR and LEHN, 1973; HAYASHI and KATAGIRI, 1974) and in neurotransmission (SVENNERHOLM, this book).

However, elucidation of molecular mechanisms involving gangliosides are difficult to achieve in any organ and *à fortiori* in the

---

[1] Chargés de Recherche à l'INSERM; [2] Chargés de Recherche au CNRS.

central nervous system (CNS). In view of the morphological and functional complexity of the CNS and the difficulty in elucidating the
sequence of molecular events in which gangliosides participate,
workers in several laboratories have attempted to study ganglioside
structure and metabolism in nerve cell cultures (see for review
REBEL, ROBERT and MANDEL, this book). Actually, cell cultures afford
model systems for studying molecular mechanisms of potential ganglioside functions on rather homogeneous cell populations and in a
well-defined medium. It should be stressed that cell cultures offer
an unique advantage for investigations of surface components as has
been shown, for instance, in our former studies on ectoenzymes
(MANDEL, CIESIELSKI-TRESKA and STEFANOVIC, 1977).

Until recently, three types of nerve cell cultures were available:

(1) Homogeneous cell cultures of tumor and transformed cells
which maintained their neuronal and glial characteristics;

(2) Primary cultures of a mixed population of neuronal and glial
cells;

(3) Homogeneous populations of astrocytes produced by keeping
cultures of dissociated brain cells from new-born animals for longer
periods during which neurons disappear.

Recently, YAVIN and YAVIN (1974) made the observation that the
use of polylysine favors the production of pure neuronal cell cultures. This method was used for investigation of gangliosides in
chick and mouse brain neurons.

GANGLIOSIDES IN CLONAL CELL LINES

Effect of Neuronal-Glial Interaction in Cell Culture

As reported by several authors (see for review REBEL, ROBERT and
MANDEL, this book) the ganglioside pattern of clonal cell lines from
either tumor or transformed nerve cells differs strikingly from that
of neurons, synaptic membranes, or glial cells isolated from brain
by cellular or subcellular fractionation. Even after acquisition of
the morphological aspect of adult neurons, usually induced by removal
of serum or by addition of bromodeoxyuridine or dibutyryl cyclic AMP,
tri- or tetra-sialogangliosides that are abundant in synaptic membranes could not be detected in clonal cultures. Although high
amounts of GD1a could be found in some cultures, only very small
amounts of GD1b could be detected (Tables 1 to 3; REBEL, ROBERT and
MANDEL, this book). However, when neuronal and glial type cells from
tumor or transformed cells were cocultured, low quantities of trisialogangliosides appeared. Coculturing of neuronal and glial type
cells also produces striking changes in the whole ganglioside pattern.

Table 1. Distribution of Gangliosides in Glial and Neuronal
          Cell Lines and in their Coculture

	NN[a]	MT17[a]	NN + MT17[b]
	All cells in proliferation		4 days in coculture
GM3	66.8 ±3.3	–	30.4 ±3.5
GM2	–	7.4 ±0.3	5.3 ±1.0
GM1	1.3 ±0.2	33.9 ±0.6	6.5 ±0.5
GD3	30.6 ±2.1	13.0 ±0.5	14.9 ±0.8
GD1a	2.3 ±0.2	44.2 ±0.9	40.9 ±4.0
GD1b	–	–	2.0 ±0.3
GT1b	–	–	–
GQ1	–	–	–

Results are expressed as the percentage of ganglioside-sialic acid.
(a) NN and MT17: means ± S.D. (10 experiments) are indicated. Ganglioside analysis was as previously described (DREYFUS *et al.*, 1975).
(b) NN + MT17: means ± deviations (2 experiments) are indicated.
Total lipids were purified through a Sephadex G-25 superfine column
and HPTLC (Fertig-platten Kieselgel 60, Merck, art. 5641) of gangliosides was performed as described elsewhere by TLC (HARTH *et al.*,
1978). Using this technique 2 or 3 spots could be seen in the region
of GM3, GM2, GD3 and GD1a. Distribution of gangliosides was determined by the densitometric technique.

Table 2. Distribution of Gangliosides in Neuronal and Glial
          Cell Lines and in their Coculture

	M1[a]	MT16[a]	M1 + MT16[b]
			4 days in coculture
GM3	5.7	11.1	11.7[c]
GM2	23.9	7.4	9.1[c]
GM1	12.6	24.8	13.5
GD3	–	20.0	13.8[c]
GD1a	53.1	36.7	46.1[c]
GD1b	4.7	–	3.3
GT1L[d]	–	–	2.0
GT1b	–	–	0.5
GQ1	–	–	–

Results are expressed as in Table 1. (a) M1 and MT16: means of 10
experiments. For ganglioside analysis see legend of Table 1 (a).
(b) M1 + MT16: means of 2 determinations. For ganglioside analysis
see legend of Table 1 (b). (c) Two spots in the region of this
ganglioside. (d) GT1L: this ganglioside is located between GD1b and
GT1b; it could represent an isomeric form of GT1 or a lactonic form
of tri- and tetra-sialoganglioside.

Table 3. Distribution of Gangliosides in Glial and Neuronal Cell
Lines and in the Re-isolated Clones after Coculture

Gangliosides	NN[a]	M1[a]	NN[b]	M1[b]
			Re-isolated	
GM3	66.8 ±3.3	5.7 ±0.9	81.1 ±0.4[c]	13.8 ±2.6[c]
GM2	–	23.9 ±3.1	4.5 ±0.7[c]	13.4 ±1.5[c]
GM1	1.3 ±0.2	12.6 ±4.1	1.5 ±0.2[c]	4.9 ±0.6
GD3	30.6 ±2.1	–	9.8 ±0.2[c]	26.8 ±1.2[c]
GD1a	2.3 ±0.2	53.1 ±3.9	3.0 ±0.5[c]	47.7 ±5.0[c]
GD1b	–	4.7 ±0.3	0.1	3.1 ±0.6
GT1b	–	–	–	0.3 ±0.2

All cells were in the period of proliferation. Results are expressed
as in Table 1. (a) NN and M1: means ± S.D. of 10 experiments. For
ganglioside analysis see legend of Table 1 (a). (b) NN and M1 re-
isolated after coculture, means ± deviations of 2 experiments. For
ganglioside analysis, see legend of Table 1 (b). (c) See legend of
Table 2 (c).

In a coculture of spontaneously transformed astroblasts of clone NN
and transformed mouse brain cells of neuronal type MT17 (Table 1)
the percentage of GM1 decreased from 34 % in MT17 to 6.5 % in the
coculture whereas GD1a represented 44 % in MT17 and 40 % in the coculture. A simple "dilution" of gangliosides of the neuronal clone
MT17 by NN glial cells cannot explain these changes considering the
percentage of these gangliosides in the NN cells: 1.3 for GM1 and
2.3 for GD1a. Moreover, GD1b did not exist in either of the two
clonal cell lines and appeared in the cocultures. Similarly, GT1L
(migrating between GD1b and GT1b; it is perhaps a lactonic form containing GT1) and GT1b did not exist in the M1 and MT16 cultures
(Table 2), but appeared when both clonal lines were cultured together.

Finally, in the M1 clone, after coculture with the glial cell
line NN and re-isolation (Table 3), significant changes occurred in
re-isolated M1 culture and, in particular, GT1b was detected.

GANGLIOSIDES IN PURE NEURONAL AND PURE GLIAL PRIMARY CULTURES

Ganglioside Patterns of Chick and Mouse Brain Neurons or Glia

The ganglioside pattern of these cells resembles that usually
found for normal cultures. Tri- and tetra-sialogangliosides are
present in rather high amounts and monosialogangliosides at a low
percentage in a 5-day culture of neurons derived from 8-day-old
chick embryo. In primary glial cells of chick embryo, more GM3 and
less GD1b and trisialogangliosides were found. Only traces of tetra-
sialogangliosides could be detected (Table 4).

Table 4. Distribution of Gangliosides in Cultured Neuronal
and Glial Cells from Chick Embryo Hemispheres

Gangliosides	Neuronal cells[b] (5 days in culture)		Glial cells[b] (25 days in culture)	
GM3[c]	2.4		29.9	
GM2[c]	0.8		4.8	
GM1	1.3		1.6	
GD3	21.5[c]	(4.5) (17.0)	27.5[a]	(9.8) (9.2) (8.5)
GD1a[c]	24.8		17.4	
GD1b[c]	12.4		8.1	
GT1L[d]	7.4		5.8	
GT1b	12.0		4.3	
GQ1	10.9		0.6	
GQ'[e]	6.5		–	

Results are expressed as in Table 1. (a) Three spots in the regions
of this ganglioside. (b) Values and ganglioside analyses as (b) in
legend of Table 1. (c) and (d) See (c) and (d) in legend of Table 2.
For GD3, percentage of each band is given in parenthesis. (e) GQ':
this ganglioside is located below GQ1.

No neurons exist in our glial cell culture, and therefore the
presence of tri- and traces of tetra-sialogangliosides in glial
cells should be accepted. Neurons obtained from 14-15-day-old mouse
embryo brain (Table 5) and maintained in culture by the same method
as chick embryo neurons also have high amounts of polysialogangliosides as seen in neurons isolated from adult rat and rabbit brain
by cellular fractionation (NORTON *et al.*, 1975; URBAN *et al.*, this
book).

Similar amounts of GD1b and of total tri- and tetra-sialogangliosides were found in cultured isolated neurons from chick and mouse
brains (Tables 4, 5). In mouse astroblasts, as in chick astroblasts,
polysialogangliosides (including tri- and tetra-sialogangliosides)
were present although in lower amounts than in neurons.

Moreover, when glia and neurons were cultured together, striking
changes occurred which could not be explained by a simple dilution
of one cell type by the other, as appears from the striking increase
of GM3 from 5.9 and 9.4 in isolated cultures to 25.1 in the coculture and the increase of GM1 from 7.1 in neurons and 6.3 in glial
cells to 11.3 in the coculture.

Table 5.   Distribution of Gangliosides in Cultured Neuronal and
Glial Cells and in a Coculture of both of Mouse Hemispheres

Gangliosides	Neuronal cells[b]	Glial cells[b]	Glia + neurons in coculture[b]
GM3	5.9 ±1.2	9.4 ±0.2	25.1[c]
GM2	3.3 ±1.1	2.6 ±0.6	1.6
GM1	7.1 ±3.4[c]	6.3 ±0.2	11.3[a]
GD3	16.9 ±2.9[c]	27.4 ±1.5[c]	8.1[c]
GD1a	21.6 ±2.7[c]	35.0 ±0.8[c]	41.0[c]
GD1b	10.8 ±1.1	4.7 ±1.6[c]	2.4[c]
GT1L[d]	3.6 ±1.4	6.3 ±0.1	–
GT1b	25.0 ±5.3	7.9 ±2.5[c]	8.1
GQ1	5.9 ±2.3	0.4 ±0.3	2.4

Expression of the results, ganglioside analysis and significance of
(a), (b), (c), (d), see (a), (b), (c), (d) in legend of Table 4.

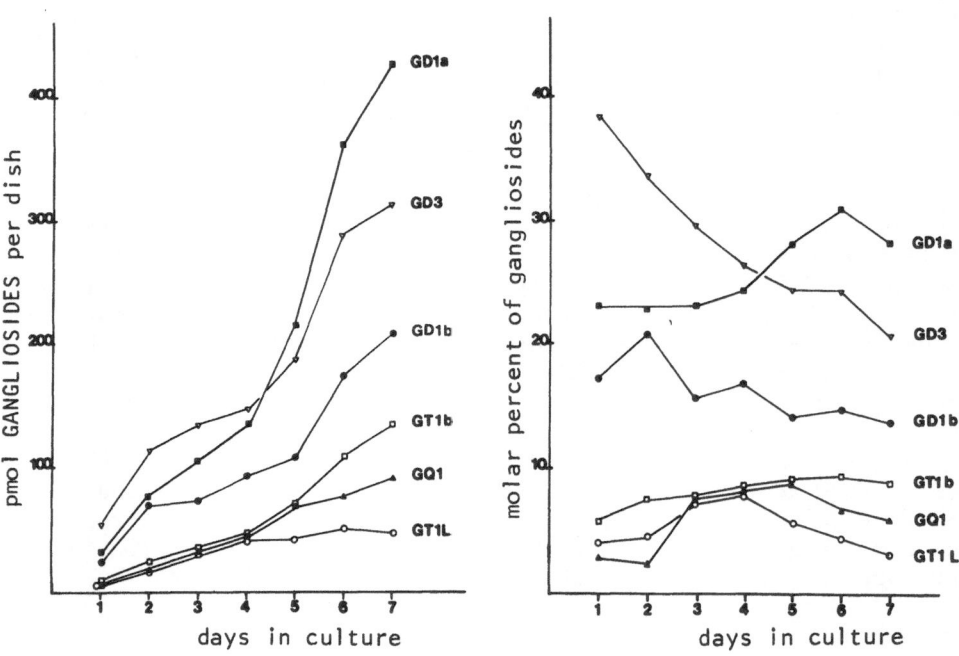

Fig. 1.   Left: Ganglioside amounts during the maturation of cultured
neuronal cells from chick embryo hemispheres. Right: Ganglioside
patterns during the maturation of cultured neuronal cells from chick
embryo hemispheres. Gangliosides are named according to SVENNERHOLM.
For GT1L and GQ' see legend of Table 4. Amounts of GQ' are included
in GQ1.

Table 6. Sialyltransferase Activities in Cultured Neurons and Astroblasts from Mice Hemispheres

Substrate		Compound labeled by NeuAc	A		A + N		N	
			Act.[a]	% Inc.[b]	Act.	% Inc.	Act.	% Inc.
Endo-genous	(1)	GM3	132	(87)	N.D.		165	(90)
Lac-Cer	(2)	GM3	458	(96)	534	(96)	636	(97)
GM3	(3)	GD3	165	(30)	177	(26)	91	(24)
GM1	(4)	GD1a	190	(51)	236	(48)	260	(54)
		GD1b	7	(2)	6	(1)	15	(3)
GD1a	(5)	GD1a	40	(37)	25	(25)	33	(25)
		GT1b	-	-	-	-	2	(2)
GD1b	(6)	GT1b	187	(80)	101	(49)	129	(54)

A: Astroblasts (15 days of culture). N: neurons (7-8 days of culture). A + N: mixture of both cells. (a) Act.: specific activity expressed as pmol NeuAc incorp./h/mg protein. (b) % Inc.: percent of total radioactivity recovered in each ganglioside. N.D.: Not determined. Incubation conditions: 300 µg Triton CF-54; (1) without exogenous substrate; (2) + 25 nmol Lac-Cer; (3) + 50 nmol GM3; (4) + 50 nmol GM1; (5) + 50 nmol GD1a; (6) + 50 nmol GD1b; 500 mM sodium cacodylate buffer pH 6.35 (1,2,4,5,6) and 7.0 (3); 5 mM $MnCl_2$; 60 nmol (1.68 mCi/mmol) CMP-NeuAc; 400-500 µg protein cell homogenate (A) and (N); half of (A) and (N) for (A + N). Final volume 0.1 ml, 37°C, 90 min, incubation (1,2,4,5,6) and 60 min (3). Total lipids were purified through a Sephadex G-25 superfine column; TLC of gangliosides was performed as described elsewhere (HARTH et al., 1978). Radioactivity content in each spot was determined using a Packard-Prias liquid scintillation spectrometer with the Kieselgel dispersed in 50 µl H_2O, 1 ml ethanol and 5 ml of 0.4 % scintimix 3 in toluol.

Evolution of Ganglioside Patterns in Chick Brain Neurons

The evolution of the absolute quantity as well as the pattern of gangliosides in neurons during the period of culture (Fig. 1) resembles that which occurs *in vivo* (DREYFUS et al., 1975). Between 1 and 7 days of culture a 13-fold increase occurred in GD1a, a 6-fold increase in GD3, a 9-fold increase in GD1b and a 14-fold increase in GT1b + GT1L (Fig. 1, left). The relative amount of tri- and tetra-sialogangliosides in neurons increased until the 5th day of culture and decreased afterwards. The relative amount of GD3 decreased while that of GD1a increased (Fig. 1, right).

Table 7. Effect of Anti-GM3[a] on NN Cell Growth

	μg Protein/Petri dish			
	Control	Normal γ-globulin	Anti-GM3	
0 day	90 ±4	-	-	
1 day	125 ±8	-	140 ±12	
2 days	385 ±32	-	330 ±18	p < 0.01
3 days	887 ±23	900 ±15	720 ±46	p < 0.01
Change of medium after 3 days				
5 days	1860 ±120	1800 ±40	1140 ±28	p < 0.001
7 days	1330 ±60	1460 ±45	800 ±14	p < 0.001

(a) Antibodies against GM3 and GM1 were produced in rabbits by two intradermal injections of 5 mg of each ganglioside in total at a 10-day interval using methylated bovine serum albumin (BSA) and complete Freund's adjuvant as carrier. An immunoglobulin fraction (40 % ammonium sulphate saturation) passed through a BSA/methylated-BSA affinity column (CUATRECASAS and ANFINSEN, 1971) was prepared. A normal immunoglobulin fraction prepared in the same way from a non-treated rabbit was used for controls. Stock hamster astroblasts, NN cell lines (SHEIN et al., 1970) and rat glioblastoma C6 cell lines (BENDA et al., 1968) were removed by trypsinization and seeded in 6 cm Petri dishes with 5 ml Dulbecco's modified Eagle medium containing 10 % fetal calf serum. The cells were allowed to attach to Petri dish for 2 h and then 0.5 ml preparations of antibodies or normal γ-globulins (12-14 mg immunoglobulin protein/ml) were added per Petri for their respective groups.

SIALYLTRANSFERASE ACTIVITIES IN CULTURED NEURONS AND ASTROBLASTS FROM MOUSE HEMISPHERES

Sialyltransferases that produce GM3 from endogenous substrates and GM3, GD3, GD1a, GD1b and GT1b from exogenous substrates were found in cultured astroblasts and neurons. The enzymatic activity for lactosylceramide→GM3, GM1→GD1a seemed to be higher in neurons, whereas that catalyzing sialylation of GD1b to form GT1b was slightly higher in glial cells (Table 6).

When astroblasts and neurons were used together in equal amounts, GD1b→GT1b sialylation seemed to decrease, whereas that producing GD3 from GM3 was higher than expected in a coculture consisting of equal population of neurons and glial cells (Table 6).

Table 8. Effect of Anti-GM3 on C6 Cell Growth

	μg Protein/Petri dish				
	Control	Normal γ-globulin	Medium without γ-glob.	Anti-GM3	Medium without anti-GM3
0 day	105 ±5	-		-	
2 days	925 ±65	995 ±48		740 ±10	
3 days	1195 ±47	1120 ±14		900 ±36	
Change of medium after 3 days					
4 days	1550 ±36	1435 ±10	1540 ±124	1130 ±72	1125 ±26
5 days	1730 ±66	1690 ±74	1580 ±77	1535 ±19	1485 ±19

The results are expressed as mean ± S.D. of 4 separate determinations. Stock cells were trypsinized before antibodies treatment. For methods see Table 7.

EFFECT OF ANTI-GANGLIOSIDE ANTIBODIES ON CELL GROWTH

Cell cultures of a clone of spontaneously transformed hamster astrocytes (clone NN) and a clone from rat astrocytoma (C6) were treated with anti-GM1 or anti-GM3 antibodies during their exponential growth phases (see legend of Table 7, for methodology). The results revealed that both antibodies were able to reduce the growth rate on both protein (Tables 7 to 10) and cell count bases (Table 11). An immunoglobulin fraction prepared in the same way from a non-treated rabbit did not inhibit cell growth. Whether the stock cells were or not treated with trypsin before the culture starts, antibodies inhibited cell growth. The antibodies seemed to be more effective when the stock culture was removed without trypsin before initiating subcultures (Tables 11, 12). Moreover, even 3 days after antibodies had been withdrawn from the medium of C6 cells, the cells continued to grow similarly as in anti-GM3 containing medium (Table 8).

When anti-GM3 antibodies were added 24 h after initiating the culture there was no change in C6 cell growth (data not shown). This led us to conclude that the consequences of exposure of the cells to anti-GM3 antibodies might occur during the first few hours and that the effect may involve the plasma membranes.

EFFECT OF ANTI-GM3 ON ADENYLATE AND GUANYLATE CYCLASES OF C6 CELLS

Since an increase of cAMP produces a block of cell proliferation while cGMP is usually high in proliferating cells (see for review RASMUSSEW and GOODMAN, 1977), we determined the effect of anti-GM3,

Table 9. Effect of Anti-GM1 on NN Cell Growth

	μg Protein/Petri dish		
	Control	Anti-GM1	
1 day	512 ±8	493 ±11	
2 days	815 ±55	707 ±32	p < 0.01
3 days	1025 ±52	858 ±66	p < 0.01

Stock cells were trypsinized before antibodies treatment. The results are expressed as mean ± S.D. of 4 separate determinations. For methods see Table 7.

————*————

Table 10. Effect of Anti-GM1 on C6 Cell Growth

μg Protein/Petri dish	
Control	Anti-GM1
340 ±15 (6)	225 ±15 (6)

Stock cells were trypsinized before antibodies treatment. The results after 3 days of treatment are expressed as mean ± S.D. for 6 separate determinations. For methods see Table 7.

————*————

Table 11. Effect of Anti-GM1 and Anti-GM3 on C6 Cell Growth

	Number of cells/ml x 10^4			
	With trypsin		Without trypsin	
Control	315 ±25		150 ±30	
Anti-GM1	220 ±20	p < 0.01	42 ±25	p < 0.001
Anti-GM3	265 ±15	p < 0.05	87 ±12	p < 0.001

The results after 3 days of treatment are expressed as mean ± S.D. for 5 separate determinations.

————*————

Table 12. Effect of Anti-GM3 and Anti-GM1 on C6 Cell Growth

	μg Protein/Petri dish			
	Control	Anti-GM3	Anti-GM1	
Experiment 1	823 ±29	563 ±65		p < 0.001
Experiment 2	240 ±15		160 ±10	p < 0.01

The results after 3 days of treatment are expressed as mean ± S.D. for 5 separate determinations. Stock cells were not trypsinized.

Table 13. Effect of Anti-GM3 on Adenylate and Guanylate Cyclases[a] of C6 Cells

	pmol/min/mg protein		
	1 day (12)	2 days (6)	3 days (6)
Adenylate cyclase			
Control	3.39 ±0.46	11.68 ±0.59	14.27 ±0.90
Anti-GM3	8.63 ±1.41	14.70 ±1.00	23.24 ±0.47
	$p < 0.001$	$p < 0.01$	$p < 0.001$
Guanylate cyclase			
Control	11.38 ±0.46	10.89 ±0.32	9.07 ±0.67
Anti-GM3	5.45 ±1.31	8.48 ±0.85	7.13 ±0.81
	$p < 0.001$	$p < 0.01$	$p < 0.01$

(a) Adenylate and guanylate cyclases were determined according to RAMCHANDRAN and LEE (1970), and HELWIG *et al.* (1975) and GORIDIS and REUTTER (1975), respectively. Cells were treated in the same way as described earlier (see legend of Table 7). The results are expressed as mean ± S.D. for the number of experiments given in parentheses.

which inhibits cell growth, on adenylate and guanylate cyclases in C6 cells. The activities of the enzymes were determined after 1, 2 and 3 days of growth both in treated and control cells (Table 13).

The activity of adenylate cyclase was increased during the culture time with respect to cell growth in controls; the addition of anti-GM3 further increased the activity of the enzyme at the different culture periods. The effect was maximum after one day of treatment. On the second day, there was quite a parallel increase in enzyme activity in treated and non-treated cells. After 3 days of culture, more increase in adenylate cyclase was observed in anti-GM3 treated cells as compared to respective control.

The activity of guanylate cyclase in control culture was reduced during successive days of growth. In the anti-GM3 treated cells, it was further decreased although the differences between control and treated were lower during the later days. As in case of adenylate cyclase the effect on guanylate cyclase was also more marked after one day of treatment.

DISCUSSION

Until recently, knowledge of the function of gangliosides in the nervous system was extremely limited. In spite of great progress in

methodology, the variety of cell types and differences in the level
of their functional activity even in a limited small brain area make
it difficult to obtain clear cut conclusions. On the other hand,
molecular events can be easily investigated in model systems and
lead to unequivocal conclusions, but extrapolation to phenomena *in
vivo* is not easy. Thus it seems that between the "minimal" level of
complexity and the integrated brain loci the investigation of either
isolated neurons and glials cells or of coculture may offer an interesting intermediate step.

Among techniques available for separating neuronal and glial
cells, brain microdissection gives only minimal amount of material;
centrifugation techniques provide neuronal fractions deprived of
nerve endings and of a great part of plasma membranes or glial cell
fractions contaminated by synaptosomes.

Theoretically cell lines provide optimal experimental conditions
in view of the homogeneity of cell types and the possibility of obtaining enough material for qualitative and quantitative biochemical
investigation.

However, transformed cells or neuroblastoma cells grown in culture fail to synthetize tri- and tetra-sialogangliosides even in
presence of various differentiation inducers and after acquisition
of morphological aspects of mature neurons (CIESIELSKI-TRESKA *et
al.*, 1977). Recently, DAWSON and STOOLMILLER (1976) have shown that
neuroblastoma cells grown *in vivo* acquire the capacity to produce
GT1a. In this paper we have shown that in a simple coculture of glial
and neuronal type transformed cells the capacity to produce low
amounts of GT1L and GT1b may appear, and that in re-isolated neuronal type but not in glial type cells, a compound running in thin
layer chromatography as a tri-sialoganglioside is detected.

This finding suggests that the lack of polysialoganglioside synthesis is due to the conditions of tissue culture or the absence of
communications with glial cells rather than to a loss of genotypic
expression.

One should keep in mind that in spite of the lack of tri- and
tetra-sialogangliosides in neuroblastoma cells, these cells exhibit
a number of electrophysiological and enzymic properties, including
the synthesis of putative neurotransmitters, associated with neurons.
This raises the question whether GD1a and GD1b abundant in synaptic
membranes (LAPETINA, SOTO and DE ROBERTIS, 1968; BRECKENRIDGE,
GOMBOS and MORGAN, 1972; AVROVA, CHENYKAEVA and OBUKHOVA, 1973) and
present in neuroblastoma cells may contribute to elementary electrophysiological properties and tri- and tetra-sialogangliosides are
not absolutely necessary. We have reported that the presence of
fundamental cellular constituents like some isoenzymes of lactate
dehydrogenase can be suppressed in neuroblastoma cells but exist in

a ratio close to normal in primary neuronal and glial cell cultures.
Some neuronal markers like choline acetyltransferase (CAT) and
acetylcholinesterase (AChE) are present in a mixed culture of chick
embryo neurons and glia in a ratio close to that *in situ*. High affinity uptake, and enzymes involved in catecholamine metabolism have
also been observed in primary cultures in appreciable amounts. Several glial-specific proteins coexist in primary glial cell cultures
and only very rarely in cultures of transformed or tumoral glial
cells (see for review MANDEL, CIESIELSKI-TRESKA and SENSENBRENNER,
1976).

The present report has shown that primary cultures of isolated
neurons possess a ganglioside pattern very close to that of neuronal cell fractions containing tri- and tetra-sialogangliosides. The
evolution of the pattern of gangliosides is also similar to that of
ontogenesis *in vivo*: increase of GD1a, GT1L, GT1b, GQ1 and decrease
of GD3.

The distribution of gangliosides in primary cultures of glial
cells offers interesting information concerning GD1b and tri-sialo-
gangliosides in glial cells. The presence of these gangliosides in
glial cells obtained by cellular fractionation could be considered
doubtful, since the glial fraction contains contaminants including
plasma membrane and synaptic membranes, as appears evident from the
presence of catecholamines, acetylcholinesterase, and other neuronal
markers (HEMMINKI, HEMMINKI and GIACOBINI, 1973; FREYSZ, personal
communication). In our glial cell culture there are no neurons, and
therefore the presence of tri- and tetra-sialogangliosides in glial
cells should be accepted.

There are appreciable differences in ganglioside patterns in neuronal and glial cells from different sources (e.g. chick brain, mouse
brain). The differences seem to be greater for glial cell patterns
than for those of neurons. Analysis of cultured cells where an effect
of functional circuits is excluded show pronounced differences between neurons of the two species examined concerning GT1b and GQ1;
the difference concerning GD1a in glial cells is even stronger
(17.4 % in chick glial cells, 35.0 % in mouse glial cells).

Considering specific neuronal function of gangliosides it is interesting to stress that gangliosides GD1b, GT1b and GQ1 are in a
higher amount in neuronal cells than in glial cells. A role of GD1b
in the synaptic cleft was already suggested. It seems also possible
to postulate a similar role for GT1b and GQ1 unless these two gangliosides are more specific for the whole neuronal plasma membrane. Let
us stress that the amount of these gangliosides is slightly higher
in mouse brain cells compared to chicken brain cells in 5-day-old
cultures able to produce synapses. GD1a and GD3 exist in similar
amounts in chick neurons and glia and were at a higher amount in
mouse glial cells. The time course of increase of these gangliosides

in brain and cultured neurons is also similar. This would suggest
for GD1a and GD3 another function than for GD1b, GT1L and GQ1 (involvement in membrane function such as transport and cell interactions). The significance of the high amount of GM3 in chick glial
cells and in mouse neuron-glia cocultures remains to be explored.

As in transformed or tumoral cells, coculture of astroglia and
neurons from mouse brain induces changes in the ganglioside pattern
which cannot be explained by the existence of a simple mixture of
neurons and astroglia. Thus, GM3 in the mixed culture is present in
much higher amount than in neurons and astroglia, whereas GD3 or
GD1b are present in much lower amounts; GD1a is present in much
higher amount than in neurons or astroglia. This suggests again an
effect of cell interactions on ganglioside pattern.

Appreciable differences in sialyltransferase activities could be
found when neurons and glial cells of primary cultures were compared.
However, the most interesting finding seems to us to be the presence
of sialyltransferase activities on the external surface of neurons
(DREYFUS *et al.*, this book). The presence of sialyltransferases on
the cell surface acting either on endogenous or on exogenous substrates suggests the possibility that these enzymes are able to
modify the charge of their own cell or of interacting cells.

Antibodies against GM1 or GM3 gangliosides inhibited nerve cell
growth whether these cells contained the respective gangliosides or
not. Thus for C6 cells in which 98 % of the total ganglioside is accounted for by GM3, growth is inhibited either by GM1 or GM3 antibodies. Similarly, the growth of NN cells which contain about 66 %
GM3 and 1.3 % GM1 is inhibited by the same antibodies. γ-Globulin
from untreated rabbits is without effect. This lack of specificity
may be explained by the cross reactivity of anti-GM1 with GM3.
Growth inhibition occurs only when the antibodies are added at the
time of replication. Addition of antibodies after 1 day of culture
is without effect. LINGWOOD and HAKOMORI (1977) have shown that
monovalent antibodies directed against N-acetylhematoside inhibit
the growth of Balb/3T3 and NIL hamster fibroblasts but not that of
their transformed counterparts. In our experiments transformed and
tumoral cell growth was inhibited, suggesting that cell transformation does not prevent the effect of ganglioside antibodies. Ganglioside antibodies seem to be less effective on cells pretreated with
trypsin. This could be the consequence of the well known growth activation effect of proteases which counteract the inhibition by the
antibody (SEFTON and RUBIN, 1970; HOVI and VAHERI, 1975).

Several hypotheses could be formulated in order to explain the
mechanism of cell growth inhibition by ganglioside antibodies. In
view of the fact that arrest of cell growth is usually accompanied
by an increase of cAMP and a decrease of cGMP, the effect of

ganglioside antibodies on the respective synthetizing enzymes was
explored. It appears that antibodies actually produce an increase
of adenylate cyclase and a decrease of guanylate cyclase activity.
Thus, increase and decrease of cyclases could be at least one of
the causes of cell growth inhibition by ganglioside antibodies.

In summary, it appears that while tumor and transformed nerve
cells in culture are lacking tri- and tetra-sialogangliosides, primary cultures even of isolated neurons and glia do contain these
gangliosides and thus offer an interesting model for studies on
regulation of ganglioside metabolism. Moreover, the study of surface
enzymes, toxins and antibodies has provided an interesting approach
to the role of gangliosides in cell-cell interactions and further
progress regarding the function of neuronal circuits is expected.

ACKNOWLEDGEMENTS — The authors wih to thank Mrs. P. GUERIN, Mrs. A.
GOMEZ DE GRACIA, Miss G. ULRICH and M. F. HOG for their excellent
technical assistance. We are grateful to Dr. J.C. LOUIS who provided
primary cultures of chick neurons. This work was supported by the
Centre National de la Recherche Scientifique, and grants from the
Institut National de la Santé et de la Recherche Médicale (Contracts
n° 78-4-017-6 and 78-5-260-6).

REFERENCES

ABRAMSON M.B., YU R.K. and ZABY V. (1972): Ionic properties of beef
    brain gangliosides. Biochim. Biophys. Acta 280, 365-372.

AVROVA N.F., CHENYKAEVA E.Y. and OBUKHOVA E.L. (1973): Ganglioside
    composition and content of rat brain subcellular fractions. J.
    Neurochem. 20, 997-1004.

BEHR J.P. and LEHN J.M. (1973): The binding of divalent cations by
    purified gangliosides. FEBS Lett. 31, 297-300.

BENDA P., LIGHTBODY J., SATO G., LEVINE L. and SWEET W. (1968): Dif-
    ferentiated rat glial cell strain in tissue culture. Science 161,
    370-371.

BRECKENRIDGE W.C., GOMBOS G. and MORGAN I.G. (1972): The lipid com-
    position of adult brain synaptosomal plasma membranes. Biochim.
    Biophys. Acta 266, 695-707.

CIESIELSKI-TRESKA J., ROBERT J., REBEL G. and MANDEL P. (1977):
    Gangliosides of active and inactive neuroblastoma clones. Differ-
    entiation 8, 31-37.

CUATRECASAS P. and ANFINSEN C.B. (1971): Affinity chromatography,
    in "Methods in Enzymology", JACOBY W.B., Ed., Vol. 22, Academic
    Press (New York), pp. 345-378.

DAWSON G. and STOOLMILLER A.C. (1976): Comparison of the ganglioside composition of established mouse neuroblastoma cell strains grown *in vivo* and in tissue culture. J. Neurochem. 26, 225-226.

DREYFUS H., HARTH S., URBAN P.F. and MANDEL P. (1975): Developmental patterns of gangliosides and phospholipids in chick retina and brain. J. Neurochem. 25, 245-250.

GORIDIS C. and REUTTER W. (1975): Plasma membrane-associated increase in guanylate cyclase activity in regenerating rat liver. Nature 257, 698-700.

HAKOMORI S.I. (1973): Glycolipids. Their chemical pattern, synthesis and degradation in normal and tumor cells, in "Tumor Lipids. Biochemistry and Metabolism", WOOD R., Ed., American Oil Chemist's Society Press (Champaign, Ill.), pp. 269-284.

HARTH S., DREYFUS H., URBAN P.F. and MANDEL P. (1978): Direct thin layer chromatography of gangliosides of a total lipid extract. Anal. Biochem. 86, 543-551.

HAYASHI K. and KATAGIRI A. (1974): Studies on the interaction between gangliosides, protein and divalent cations. Biochim. Biophys. Acta 337, 107-117.

HELWIG J.J., BOLLACK C., MANDEL P. and GORIDIS C. (1975): Renal cortex guanylate cyclase: preferential enrichment in glomerular membranes. Biochim. Biophys. Acta 377, 463-472.

HEMMINKI K., HEMMINKI E. and GIACOBINI E. (1973): Activity of enzymes related to neurotransmission in neuronal and glial fraction. Int. J. Neurosci. 5, 87-90.

HOVI T. and VAHERI A. (1975): Reversible release of chick embryo fibroblast cultures from density dependent inhibition of growth. J. Cell Physiol. 87, 245-252.

LAPETINA E.G., SOTO E.F. and DE ROBERTIS E. (1968): Lipids and proteolipids in isolated subcellular membranes of rat brain cortex. J. Neurochem. 15, 437-445.

LEDEEN R.W. (1978): Ganglioside structures and distribution: are they localized at the nerve ending? J. Supramol. Structure 8, 1-17.

LINGWOOD C.A. and HAKOMORI S. (1977): Selective inhibition of cell growth and associated changes in glycolipid metabolism induced by monovalent antibodies to glycolipids. Exp. Cell Res. 108, 385-391.

MANDEL P., CIESIELSKI-TRESKA J. and SENSENBRENNER M. (1976). Neurons *in vitro*, in "Molecular and Functional Neurobiology", GISPEN W.H., Ed., Elsevier (Amsterdam), pp. 111-157.

MANDEL P., CIESIELSKI-TRESKA J. and STEFANOVIC V. (1977): Neuroblastglioblast interactions: ectoenzymes, in "Cell, Tissue and Organ Cultures in Neurobiology", FEDOROFF S. and HERTZ L., Eds., Academic Press (New York), pp. 593-615.

MORGAN I.G., WOLFE L.S., MANDEL P. and GOMBOS G. (1971): Isolation of plasma membranes from rat brain. Biochim. Biophys. Acta 241, 737-751.

NORTON W.T., ABE T., PODUSLO S.E. and DE VRIES G.H. (1975): The lipid composition of isolated brain cells and axons. J. Neurosci. Res. 1, 57-75.

RAMCHANDRAN J. and LEE W. (1970): Divergent effects of O-nitro-phenyl-sulfenyl ACTH on rat and rabbit fat cell adenyl cyclases. Biochem. Biophys. Res. Commun. 41, 358-366.

RASMUSSEW H. and GOODMAN D.B.P. (1977): Relationship between calcium and cyclic nucleotides in cell activation. Physiol. Rev. 57, 421-509.

SEFTON B.M. and RUBIN H. (1970): Release from density dependent growth inhibition by proteolytic enzymes. Nature 227, 843-845.

SHAROM F.J. and GRANT C.W.M. (1978): A model for ganglioside behaviour in cell membranes. Biochim. Biophys. Acta 507, 280-293.

SHEIN H.M., BRITAVA A., HESS H.H. and SELKOE D.J. (1970): Isolation of hamster brain astroglia by *in vitro* cultivation and subcutaneous growth, and content of cerebroside, ganglioside, RNA and DNA. Brain Res., 19, 497-501.

SUZUKI K. (1976): Neuronal storage disease: a review, in "Progress in Neuropathology", ZIMMERMAN H.M., Ed., Vol. III, Grune and Stratton (New York), pp. 173-202.

TETTAMANTI G., PRETI A., LOMBARDO A., BONALI F. and ZAMBOTTI V. (1973): Parallelism of subcellular location of major particulate neuraminidase and gangliosides in rabbit brain cortex. Biochim. Biophys. Acta 306, 466-477.

YAMAKAWA T. and NAGAI Y. (1978): Glycolipids at the cell surface and their biological functions. Trends Biochem. Sci. 3, 128-131.

YAVIN E. and YAVIN Z. (1974): Attachment and culture of dissociated cells from rat embryo cerebral hemispheres on polylysine-coated surface. J. Cell Biol. 62, 540-546.

GANGLIOSIDES AND SYNAPTIC TRANSMISSION

Lars Svennerholm

Psychiatric Research Centre, Department of Neurochemistry,
University of Göteborg, St. Jörgen Hospital,
S-422 03 HISINGS BACKA, Sweden

Gangliosides usually make up only a small proportion of the
lipids constituting the plasma cell membrane. It is also characteristic that the concentration of gangliosides and their pattern
in a given organ vary widely with species and between different
organs in a species. Wide variation in ganglioside pattern of the
red cell membranes of various species has been known for some time
(YAMAKAWA and NAGAI, 1978) and we could recently demonstrate a
large difference in ganglioside pattern of intestinal mucosa
between three species, viz man, pig and cow (HOLMGREN et al., 1975).
On the other hand, the ganglioside patterns of the intestinal
muscular layer, which reflects mainly the gangliosides of the
myenteric plexus, showed only small differences. The ganglioside
concentrations and patterns of mammalian brains from primates to
small animals, such as rodents are similar in adults and also show
similar changes during development (SUZUKI, 1965). The gangliosides
also belong to the same series, almost 99% to .the ganglio series,
and 80-90% of them have the basic gangliotetraose structure. In
our experience only one species does not conform to this general
brain ganglioside picture, namely the pig; in a certain form of
mini-pigs fucosylgangliotetraosylceramide was the basic structure
of 15-20% of the gangliosides (KLINGHARDT G., FREDMAN P. and
SVENNERHOLM L., unpublished results).

The high concentration and the uniformity in composition of
the gangliosides in the nervous tissue compared with that in other
organs (Table I) suggests that the gangliosides serve a special
purpose in neural tissue. The early speculations regarding the
role of brain gangliosides were influenced by KLENK´s (1939)
assumption that the gangliosides were an integral component of
neuronal cytoplasm, since he first isolated gangliosides from a

TABLE I. Concentration of gangliosides in some human organs

Source	Gangliosides nmole NeuAc/g fresh weight
Brain: Cerebral cortex	3,000-3,500
Cerebral white matter	1,000-1,250
Retina	100-150
Peripheral nerve	35-80
Skeletal muscle	50-80
Liver	50-100
Spleen	200-300
Placenta	100-200
Thyroid	60-100
Kidney	30-60
Skin	30-35
Fat tissue	10-15
Small intestine, mucosa	5-8
muscular layer	50-80

From L.SVENNERHOLM, P.FREDMAN and O.NILSSON, unpublished results.

Tay-Sachs brain, where the gangliosides could be shown to distend the neuronal cell perikaryon. Later McIlwain and associates (MCILWAIN, 1963) demonstrated that the gangliosides restored the excitability of cerebral tissue sections *in vitro*. Apart from these studies very little evidence has come forth in support of the assumption that the gangliosides play a role in the nerve impulse propagation and/or synaptic transmission.

CHEMICAL STRUCTURES OF GANGLIOSIDES

The ganglio- and neolacto-series contain most of the gangliosides hitherto characterized. Their structures and designations are summarized in Table II. The vast majority of the brain gangliosides belong to the ganglio series, and in synaptosomal membranes five gangliosides constitute more than 90% (Fig. 1). When NeuAc is attached by the α-ketosidic linkage, the sialic acid of GM1 is pro-

TABLE II. Ganglioside designation

LIPID DOCUMENT (1977)	SVENNERHOLM (1963)
II^3NeuAc-LacCer	GM3
II^3NeuGc-LacCer	
II^3(NeuAc)$_2$-LacCer	GD3
II^3NeuAc,NeuGc-LacCer	
II^3(NeuGc)$_2$-LacCer	
II^3NeuAc-GgOse$_3$Cer	GM2
II^3(NeuAc)$_2$-GgOse$_3$Cer	GD2
II^3NeuAc-GgOse$_4$Cer	GM1
IV^3NeuAc-nLcOse$_4$Cer	LM1
IV^3NeuAcII^{3}NeuAc-GgOse$_4$Cer	GD1a
II^3(NeuAc)$_2$-GgOse$_4$Cer	GD1b
IV^3(NeuAc)$_2$-nLcOse$_4$Cer	LD1
IV^3(NeuAc)$_2$II^{3}NeuAc-GgOse$_4$Cer	GT1a
IV^3NeuAc,II^3(NeuAc)$_2$-GgOse$_4$Cer	GT1b
II^3(NeuAc)$_3$-GgOse$_4$Cer	GT1c
IV^3(NeuAc)$_2$,II^3(NeuAc)$_2$-GgOse$_4$Cer	GQ1b
IV^3NeuAcII^3(NeuAc)$_3$-GgOse$_4$Cer	GQ1c
IV^3(NeuAc)$_3$II^3(NeuAc)$_2$-GgOse$_4$Cer	GP1b
IV^3(NeuAc)$_2$II^3(NeuAc)$_3$-GgOse$_4$Cer	GP1c

tected from the adjacent sugars and is not degraded by sialidase (KUHN and WIEGANDT, 1963; SVENNERHOLM, 1963). The other four gangliosides GD1a, GD1b, GT1b and GQ1b are all hydrolysed to GM1, but at different rates. The disialosyl group NeuAc(α2-8)NeuAc linked to the terminal galactose is hydrolysed at a slightly slower rate than NeuAcα(2-3)Gal present in GD1a and GM3. When the same disialosyl group is linked to the internal galactose the hydrolysis is much

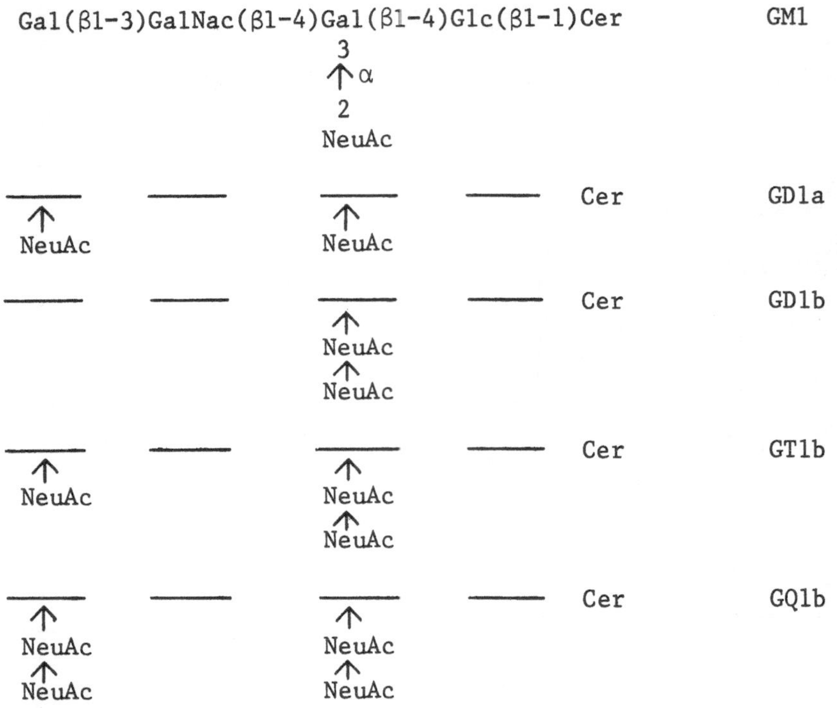
Gal(β1-3)GalNac(β1-4)Gal(β1-4)Glc(β1-1)Cer
GM1
3
α
2
NeuAc
Cer
GD1a
NeuAc
NeuAc
Cer
GD1b
NeuAc
NeuAc
Cer
GT1b
NeuAc
NeuAc
NeuAc
Cer
GQ1b
NeuAc
NeuAc
NeuAc
NeuAc

Fig. 1. Schematic structure of five major gangliosides of human brain synaptosomes.

slower, an observation of considerable importance in the elucidation of the ganglioside structure. When GQ1 and GP1 gangliosides are partially hydrolysed by *Vibrio cholerae* sialidase, GT1c can be isolated from both, although GQ1c and GP1c are only minor components of human brain gangliosides compared with GQ1b' and GP1b.

DISTRIBUTION

An almost exclusive neuronal localization of gangliosides has been inferred from the concentration of gangliosides within different regions of the nervous system (KLENK and LANGERBEINS, 1941; LOWDEN and WOLFE, 1964). The association of gangliosides with the axonic layers in the unmyelinated brain suggested that not only neuronal perikarya, but also axons and dendrites contained gangliosides (SVENNERHOLM, 1957). Since the advent of thin-layer chromatographic methods for separating lipids, the occurrence of gangliosides has been demonstrated in virtually every vertebrate cell type of tissue. The concentration in brain is, however, so many times higher than in other organs that a special function of the gangliosides of the nervous system, particularly in their neurons, was suspected. The highest concentration of gangliosides in the subcellular fractions was found in synaptosomes and microsomes (WOLFE, 1961; LAPETINA

et al, 1967; WIEGANDT, 1967; HAMBERGER and SVENNERHOLM, 1971;
and KORNGUTH et al., 1974) indicating association of gangliosides
with cell membranes. Various laboratories reported similar figures
for the ganglioside content of the synaptosomes in different
species, but widely varying figures for ganglioside content of
synaptic plasma membranes. AVROVA et al. (1973) and BRECKENRIDGE
et al. (1972) reported a ganglioside content corresponding to
150 mmol NeuAc/kg of protein, which is 3-5 times higher than
that generally accepted (WHITTAKER, 1969). In their very careful
study AVROVA et al. (1973) emphasize that the high enrichment of
gangliosides has always been found in low-density membrane fractions
with a density similar to that of myelin. They also pointed out that
the relative GM1 content was lower than that in the homogenate.
DERRY and WOLFE (1968), who used isolated cells, found the ganglioside content of neurons to be several times higher than that in
clumps of glial cells. The neuropil teased from the immediate
proximity of the neurons, which are thought to be rich in nerve
terminals, had an even higher ganglioside content than the neurons.
These findings might seem to be in contrast with the results of bulk
prepared neuronal cell bodies or glial cell enriched fractions
reported by NORTON and PODUSLO (1971) and HAMBERGER and SVENNERHOLM
(1971). The ganglioside concentrations in isolated rat and rabbit
brain neuronal perikary were respectively 4.0 and 8.0 mmol NeuAc/kg
protein. Corresponding values for the ganglioside content of the
glial cell enriched fractions were 11.2 and 16.0 mmol NeuAc/kg
protein, respectively. These values can be compared to the ganglioside content of the whole homogenates 19.5 and 26.0 mmol NeuAc/kg

TABLE III. Concentration of gangliosides in human and rat
synaptosomes

Fraction	Human cerebral cortex	Rat brain AVROVA et al. 1973
	mmol ganglioside/kg of protein	
Homogenate	32	28
Synaptosomes	32	30
Synaptosomal membrane		
0.8 M-sucrose	124	146
1.0 M-sucrose	48	96

protein, respectively. The ganglioside content of rabbit synaptosomes
was 44.5 mmol NeuAc/kg protein. HAMBERGER and SVENNERHOLM (1971)
interpreted their values as indicating that the concentration of
gangliosides is much higher in the plasma membrane of the neurons
than of glial cells since the ratio of surface membrane to organelle
content (mass) is severalfold lower in the neuronal cell bodies
shorn of dendrites than in the glial cells.

GANGLIOSIDE CONCENTRATION AND PATTERN OF SYNAPTOSOMAL MEMBRANES

Preparation of synaptic membranes (WHITTAKER, 1969) and
synaptic junctions (COTMAN and TAYLOR, 1972) were made from the
cerebral cortex of the frontal lobe from newborns to old aged
subjects (L. SVENNERHOLM, unpublished results). In Tables III and
IV the values from a one-year-old child was chosen since his values
were most similar to those found in the rat brain. AVROVA et al.
(1973) drew attention to the proportional reduction of ganglioside
GM1 in the synaptosomal membranes, but, in my opinion, the most
important finding is the increased proportion of ganglioside GD1b
and GT1b most evident in the synaptic junction.

LEDEEN (1978) has calculated the distribution of the gangliosides in the rat brain and has arrived at the conclusion that
nerve endings contribute less than 12% of the total cerebral
cortical ganglioside. He based his calculation on the reported
values for axon terminal density, but it seems that he did not take
into account that the largest number of synaptosomes are derived

TABLE IV.   Ganglioside patterns in human and rat synaptosomes
            (molar proportion of NeuAc)

| Ganglioside | Human frontal cerebral cortex | | | Rat forebrain | |
	Homogenate	Synaptosome membrane	Synaptic junction	Homogenate	Synaptosome membrane
GM1	17	15	11	18	12
GD1a	39	40	31	32	34
GD1b	11	14	20	16	19
GT1b	16	21	26	21	23
GQ1b	3	3	4	3	3

The ganglioside values of rat fractions have been taken from AVROVA,
CHENYKAEVA and OBUKHOVA, J. Neurochem. 20, 997-1004, 1973.

from interdendritical connections.

ULTRASTRUCTURAL LOCALIZATION OF CELL MEMBRANE GM1-GANGLIOSIDE BY CHOLERA TOXIN

The determinations of the gangliosides in the subcellular fractions suggested that the concentration of gangliosides were particularly high in the synaptic junctions. To check the validity of this assumption we tried to visualize the enrichment of the ganglioside in this structure with an immunoelectron microscopic method. But all efforts to raise a potent ganglioside antiserum failed. We therefore decided to utilize the discovery that cholera toxin is specifically bound to membrane GM1-ganglioside. The idea was to incubate tissue sections or subcellular fractions with cholera toxin and then to localize the bound toxin with specific peroxidase-conjugated antibody and enzyme substrate (HANSSON et al., 1977). Thin sections were examined for electron-opaque precipitates in a transmission microscope. Semiquantitative data were obtained by titrating the limiting concentration of cholera toxin producing specific precipitates. With this method was shown that membrane GM1-ganglioside was positioned exclusively on the external side. In the central nervous system GM1-ganglioside was concentrated in the pre- and postsynaptic membranes of the synaptic terminals, less ganglioside was distributed over the other sections of the neuronal surface and only minor amounts of gangliosides on the astroglial cell surface. A further increase in reactivity in the pre- and postsynaptic junctions was noted after incubation of the sections with *V. cholerae* sialidase. The combined results of the electron microscopical study of the distribution of gangliosides and the assay of the gangliosides in subcellular fractions suggested that the density of gangliosides is largest in the synaptic junction and that the proportion of the three gangliosides GD1b, GT1b and GQ1b is higher in synaptic junction than in the total homogenates. In the synaptic junction isolated from the human adult forebrain 70-80% of the total ganglioside NeuAc is derived from these three gangliosides. This suggests that they might serve some special purpose.

LACTON FORMATION OF GANGLIOSIDES

The terminal sialic acid of the internal galactose of the three gangliosides GD1b, GT1b and GQ1b spontaneously undergoes lacton formation at neutral or weakly acidic pH. At stronger acidity lacton formation might be induced also in sialic acids other than the terminal sialic acid in a disialosyl linkage. It has been found that Ca^{++} stabilizes the terminal sialic acid and prevents lacton formation (P. FREDMAN, O. NILSSON and L. SVENNERHOLM, unpublished results). When the three gangliosides were dialysed against distilled water,

weak solutions of sodium or potassium chloride (0.01 mol/1) partial
lacton formation of the sialic acid occurred. It was further increased when the gangliosides were dialysed against K_3-EDTA but was
completely restored in free acid form when dialysed against Ca^{++} at
pH 7.0.

DYNAMIC NATURE OF NERVE CELLS

     Neurons are by no means rigid, stationary and durable fixtures
but are highly unstable and reproductively active cellular units
(WEISS, 1976). It was also PAUL WEISS who introduced the term axonal
or more generally "neuroplasmic" flow. With this term he did not
mean a convection of fluid within a stable stationary axon, but a
continuous outgrowth of the whole axonal body from the nerve cell
perikaryon. This "axonal flow proper" proceeds at a rate of about
1-2 mm/day, in contrast with the fast axonal transport of several
100 mm/day. The neuronal gangliosides are most likely synthesized in
the Golgi apparatus and then transported to the outer site of the
plasma membrane of the neuronal cell perikaryon but also to the
dendrites and axonal nerve endings, which they reach by fast
transport (LEDEEN et al., 1978b). Several findings support this
assumption.

     It has not been possible to demonstrate *in vivo* any precursor-
product relationship between the major gangliosides which suggests
that the gangliosides are formed in a small pool and then transported to another part of the cell, where they cannot act as substrate
for any glycosyltransferase e.g. a sialyltransferase (HOLM and
SVENNERHOLM, 1972). Our electron-microscopical study has shown an
even distribution of the gangliosides over the neuronal surface
(HANSSON, HOLMGREN and SVENNERHOLM, 1977) except for a local increase in the synaptic junction. The concentration of gangliosides
and the ganglioside pattern are similar in grey matter homogenates
and the synaptosomes of the same brain region, and the ganglioside
patterns of the plasma membranes of neuronal cell perikaryon and
synaptosomes are very similar (HAMBERGER and SVENNERHOLM, 1971).
LEDEEN (1978) has also arrived at the same conclusion that the
gangliosides are evenly distributed over a large part of the neuronal
surface.

     The degradation of the gangliosides seems to occur at two sites.
The oligosialogangliosides of the gangliotetraose series are degraded to ganglioside GM1 in the nerve endings with the highest
local concentration in brain of sialidase. Ganglioside GM1 is then
transported by retrograde axonal flow to the lysosomes of the
neuronal cell perikaryon, where the final degradation will occur.

A MODEL FOR THE ROLE OF GANGLIOSIDES AT THE SYNAPTIC TRANSMISSION

The mechanism by which the impulse at a presynaptic nerve terminal causes transmitter release has been studied in great detail at neuromuscular junctions, and the synaptic transmission in the central nervous system has been assumed to be similar. The stimulus for secretion of the transmitter is depolarization of the presynaptic ending produced either by an impulse or by artificially applied current. Invariably, a delay of approximately 0.5 msec at room temperature intervenes between the voltage change and transmittor secretion. For release to occur, Ca^{++} must be present in the bathing fluid at the time of the depolarization and there is now generally accepted that Ca^{++} enters the terminal to trigger the release. Ca^{++} has long been known as an essential link in the process of transmission and when its concentration is decreased, the release of transmittor is reduced and eventually abolished. The smallest physiological unit of transmitter release is a quantum which in muscle usually produces a potential change of less than 1 mV and a quantum is made up of more than 1000 transmittor molecules perhaps 10,000 molecules. At the neuromuscular junction the normal number of quanta released at each impulse has been estimated to 300. At a neuromuscular junction several thousands of impulses will reach the presynaps per minute. In a neuronal synapse the distance between pre- and postsynapse is not more than 20 nm compared to 60 nm at the neuromuscular junction, and it is probable that the number of transmittor molecules per quantum is less than in a neuromuscular junction but it is still a challenge to understand how the multimolecular packets of transmittor can be released in the synaptic cleft and the majority of the transmittors reabsorbed again to the presynapse within less than 10 msec.

Gangliosides are enriched at the synaptic junction and they bind Ca^{++} with high affinity. They would therefore be suitable candidates as carriers for the Ca^{++} needed for each transmittor release. The action potential releases Ca^{++} which requires only a weak electrical signal (ADEY and REYS, 1975).

When Ca^{++} is replaced by Na^+ as counter ion the stability of the synaptic lipid bilayer membrane will diminish and the bi-molecular sheet membrane can easily switch over in the micell form and thereby facilitate the coalescence of the transmittor vesicle membrane with the synaptic membrane which occur at the release of transmittor. When the Ca^{++}-concentration of the synapse reaches a certain level an interaction between the proteins of the micro-filaments occurs and results in a contraction of the microfila-ments which keep pre- and postsynapse in close contact. By this contraction of the microfilaments, the transmittor substances will be thrown like by a catapult to the postsynapse and the subsynaptical receptor. The contraction of the microfilaments

will persist until Ca^{++} is removed from the interior of the presynapse. These Ca^{++} will then be taken up by the mitochondria and
secreted in the intercellular space or to the astroglial cells.
The transmittor substances, released to the synaptic cleft, which
are positively charged, acetylcholine and the biogenic amines, and
which are not attached directly to their postsynaptic receptor
will be immediately taken up again by the emtied vesicles – this
retransport is facilitated by the negatively charged gangliosides
which have a stronger charge when Na^+ is counter ion instead for
Ca^{++}. When Ca^{++} enters the synaptic cleft again they will be
attached to the strongly acidic sialic acids of the gangliosides,
which will result in a reconstitution of the presynapse plasma
membrane and it will be ready to receive the next impulse.

In my model the role of gangliosides at the synaptic junctions is to bind Ca^{++}, which is released at each impulse and
transported in specific Ca^{++} channels into the presynapse. The
removal of Ca^{++} from the gangliosides at each impulse will lead
to an unstability of the presynaptical membrane and facilitate a
fusion with the membrane of the synaptic vesicles. The increased
concentration of Ca^{++} *within* the presynapse will result in the
contraction of the microfilaments and an ejection of transmittor
substance to the postsynapse and the subsynaptical receptor. The
gangliosides will also participate in the reabsorption of the
positively charged transmittors since the gangliosides on the
presynapse with Na^+ as counter ion have a stronger negative charge
than those of the postsynapse, which are in form of Ca^{++} salts.

REFERENCES

ADEY W.R. and REES D.A. (1975): Evidence for cooperative mechanisms
     in the susceptibility of cerebral tissue to environmental and
     intrinsic electric fields, in "Functional Linkage in Biomolecular
     Systems", SCHMITT F.O., SCHNEIDER D.M. and CROTHERS D.M., Eds.,
     Raven Press (New York).

AVROVA N.F., CHENYKAEVA E.YU. and OBUKHOVA E.L. (1973): Ganglioside
     composition and content of rat-brain subcellular fractions.
     J. Neurochem. 20, 997-1004.

BRECKENRIDGE W.L., GOMBOS G. and MORGAN I.G. (1972): The lipid
     composition of adult rat brain synaptosomal plasma membranes.
     Biochim. Biophys. Acta 266, 695-707.

COTMAN C.W. and TAYLOR D. (1972): Isolation and structural studies
     on synaptic complexes from rat brain. J. Cell Biol. 55, 696-711.

DERRY D.M. and WOLFE L.S. (1968): Ganglioside analyses of serial cryostat sections through ammon´s horn and cerebellar folia. Exper. Brain Res. 5, 32-44.

HAMBERGER A. and SVENNERHOLM L. (1971): Composition of gangliosides and phospholipids of neuronal and glial cell enriched fractions. J. Neurochem. 18, 1821-1829.

HANSSON H.-A., HOLMGREN J. and SVENNERHOLM L. (1977): Ultra-structural localization of cell membrane GM1 ganglioside by cholera toxin. Proc. Natl. Acad. Sci. USA 74, 3782-3786.

HOLM M. (1972): Gangliosides of the optic pathway: biosynthesis and biodegradation studies *in vivo*. J. Neurochem. 19, 623-629.

HOLM M. and SVENNERHOLM L. (1972): Biosynthesis and biodegradation of rat brain gangliosides studied *in vivo*. J. Neurochem. 19, 609-622.

HOLMGREN J., LÖNNROTH I., MÅNSSON J.-E. and SVENNERHOLM L. (1975): Interaction of cholera toxin and membrane GM1-ganglioside of small intestine. Proc. Natl. Acad. Sci. 72, 2520-2524.

KLENK E. (1939): Beiträge zur Chemie der Lipidosen. 3. Mitteilung Niemann-Picksche Krankheit und amaurotische Idiotie. Hoppe-Seyler´s Z. physiol. Chem. 262, 128-143.

KLENK E. and LANGERBEINS H. (1941): Über die Verteilung der Neuraminsäure im Gehirn (mit einer Mikromethode zur quantitativen Bestimmung der Substanz im Nervengewebe). Hoppe-Seyler´s Z. physiol. Chem. 270, 185-193.

KORNGUTH S.S. (1974): The synapse: a perspective from *in situ* and *in vitro* studies. Rev. Neurosci. 1, 63-114.

KORNGUTH S., WANNAMAKER B., KOLODNY E., GEISON R., SCOTT G. and O´BRIEN J.F. (1974): Subcellular fractions from Tay-Sachs brains: ganglioside, lipid, and protein composition and hexosaminidase activities. J. Neurol. Sci. 22, 383-406.

KUHN R. and WIEGANDT H. (1963): Die Konstitution der Ganglio-N-tetraose und des Gangliosides G_I. Chem. Ber. 96, 866-880.

LAPETINA E.G., SOTO E.F., DE ROBERTIS E. (1968): Lipids and proteolipids in isolated subcellular membranes of rat brain cortex. J. Neurochem. 15, 437-445.

LEDEEN R.W. (1978): Ganglioside structures and distribution: Are they localized at the nerve ending? J. Supramol. Struct. 8, 1-17.

LEDEEN R.W., SKRIVANEK J.A., TIRRI L.J., MARGOLIS R.K. and
    MARGOLIS R.U. (1976b): Gangliosides of the neuron: Localization
    and origin, in "Ganglioside Function: Biochemical and Pharma-
    cological implications", PORCELATTI G., CECCARELLI B. and
    TETTAMANTI G., Eds., Advances Experimental Medicine and Biology,
    Vol. 71, Plenum Press (New York), pp. 83-103.

LEDEEN R.W. and YU R.K. (1976): Gangliosides of the nervous system,
    in "Glycolipid Methodology", WITTING L.A., Ed., American Oil
    Chemists´ Society (Champaign, Illinois), pp. 187-214.

LOWDEN J.A. and WOLFE L.S. (1964): Evidence for the location of
    gangliosides specifically in neurones. Canad. J. Biochem. 42,
    1587-1594.

MCILWAIN H. (1963): Chemical exploration of the brain. Elsevier
    Publ. Co (Amsterdam), 207 pp.

NORTON W.T. and PODUSLO S.E. (1971): Neuronal perikarya and
    astroglia of rat brain: chemical composition during myelination.
    J. Lipid Res. 12, 84-90.

SUZUKI K. (1965): The pattern of mammalian brain gangliosides
    III: Regional and developmental differences. J. Neurochem. 12,
    969-979.

SVENNERHOLM L. (1957): Quantitative estimation of gangliosides in
    senile human brains. Acta Soc. Med. Upsaliensis 62, 1-16.

SVENNERHOLM L. (1963): Chromatographic separation of human brain
    gangliosides. J. Neurochem. 10, 613-623.

WEISS P. (1976): Neurobiology in statu nascendi, in "Perspectives
    in Brain Research", CORNER M.A. and SWAAB D.F., Eds., Progress
    in Brain Research, Vol. 45, Elsevier Scientific Publ. Co
    (Amsterdam), pp. 7-38.

WHITTAKER V.P. (1969): The Synaptosome, in "Handbook of Neuro-
    chemistry", LAJHTA A., Ed., Plenum Press (New York), pp. 327-361.

WIEGANDT H. (1967): The subcellular localization of gangliosides
    in the brain. J. Neurochem. 14, 671-674.

WOLFE L.S. (1961): The distribution of gangliosides in subcellular
    fractions of guinea-pig cerebral cortex. Biochem. J. 79, 348-355.

YAMAKAWA T. and NAGAI Y. (1978): Glycolipids at the cell surface and
    their biological functions. Trends Biochem. Sci. 3, 128-131.

PARTICIPANTS

J. ARMAND
   Institut Mérieux, Marcy-L'Etoile, 69260 CHARBONNIERES-LES BAINS,
   FRANCE

N.F. AVROVA
   Academy of Sciences of the U.S.S.R., Sechenov Institute of
   Evolutionary Physiology and Biochemistry, Thorez pr., 52
   LENINGRAD, K-223, U.S.S.R.

Y. BARENHOLZ
   Department of Biochemistry, The Hebrew University, Hadassah
   Medical School, P.O.B. 1172, JERUSALEM, ISRAEL

S. BASU
   Department of Chemistry, University of Notre Dame, College of
   Science, NOTRE DAME, Indiana 46556, U.S.A.

N. BAUMANN
   Laboratoire de Neurochimie, INSERM U134-CNRS ERA 421, Hôpital de
   la Salpêtrière, 47, Boulevard de l'Hôpital, 75634 PARIS CEDEX 13,
   FRANCE

B. BERRA
   Istituto di Chimica Biologica, Facoltà di Medicina e Chirurgia
   dell'Università di Milano, Via C. Saldini 50, 20133 MILANO, ITALY

M. CANTZ
   Westfälische Wilhelms-Universität, Physiologisch-Chemisches
   Institut, Waldeyerstrasse 15, 4400 MÜNSTER (Westf.), GERMANY

B. CESTARO
   Istituto di Chimica Biologica, Facoltà di Medicina e Chirurgia
   dell'Università di Milano, Via C. Saldini 50, 20133 MILANO, ITALY

E. CONZELMANN
   Max-Planck-Institut für Psychiatrie, Deutsche Forschungsanstalt
   für Psychiatrie, Kraepelinstrasse 2, 8000 MUNCHEN 40, GERMANY

J.A. DAIN
  Department of Biochemistry and Biophysics, University of Rhode
  Island, KINGSTON, Rhode Island 02881, U.S.A.

L. DOUSTE-BLAZY
  Unité de Recherches de l'INSERM sur la Biochimie des Lipides
  (n° 101), Centre Hospitalier Purpan, 31052 TOULOUSE CEDEX, France

H. DREYFUS
  Centre de Neurochimie du CNRS, 11 rue Humann, 67085 STRASBOURG
  CEDEX, FRANCE

J. FINNE
  Department of Medical Chemistry, University of Helsinki,
  Siltavuorenpenger 10A, 00170 HELSINKI 17, FINLAND

P. FREDMAN
  Psychiatric Research Centre, Department of Neurochemistry,
  University of Göteborg, St. Jörgen Hospital, 42203 HISINGS BACKA 3,
  SWEDEN

L. FREYSZ
  Centre de Neurochimie du CNRS, 11 rue Humann, 67085 STRASBOURG
  CEDEX, FRANCE

S. GATT
  Laboratory of Neurochemistry, Department of Biochemistry, The
  Hebrew University, Hadassah Medical School, JERUSALEM, ISRAEL

S.I. HAKOMORI
  Biochemical Oncology Division, Department of Pathobiology School
  of Public Health, Department of Microbiology School of Medicine,
  Fred Hutchinson Cancer Research Center, and University of
  Washington, 1124 Columbia Street, SEATTLE, Washington 98104, U.S.A.

S. HARTH
  Centre de Neurochimie du CNRS, 11 rue Humann, 67085 STRASBOURG
  CEDEX, FRANCE

J. HOLMGREN
  Department of Bacteriology, Institute of Medical Microbiology,
  University of Göteborg, Guldhedsgatan 10, 41346 GÖTEBORG, SWEDEN

J.N. KANFER
  Department of Biochemistry, Faculty of Medicine, The University
  of Manitoba, 770 Bannatyne Avenue, WINNIPEG, Manitoba R3E 0W3,
  CANADA

K.A. KARLSSON
  Department of Medical Biochemistry, Faculty of Medicine, University
  of Göteborg, Medicinaregatan 9, 40033 GÖTEBORG 33, SWEDEN

L.D. KOHN
  Section on Biochemistry of Cell Regulation, National Institute of
  Arthritis, Metabolism and Digestive Diseases, National Institutes
  of Health, Bdg 4, Room B1-32, BETHESDA, Maryland 20014, U.S.A.

R. LEDEEN
  The Saul R. Korey Department of Neurology, Albert Einstein College
  of Medicine of Yeshiva University, 1300 Morris Park Avenue,
  BRONX, N.Y. 10461, U.S.A.

Y.T. LI
  Department of Biochemistry, School of Medicine, Tulane University
  NEW ORLEANS, Louisiana 70112, U.S.A.

R. MAGET-DANA
  Centre de Biophysique Moléculaire du CNRS, Avenue de la Recherche
  Scientifique, 45045 ORLEANS CEDEX, FRANCE

P. MANDEL
  Centre de Neurochimie du CNRS, 11 Rue Humann, 67085 STRASBOURG
  CEDEX, FRANCE

J.E. MANSSON
  Department of Neurochemistry, Psychiatric Research Centre,
  University of Göteborg, St. Jörgen Hospital, 42203 HISINGS BACKA 3,
  SWEDEN

D.M. MARCUS
  Department of Medicine, Division of Rheumatic Disease and Immunol-
  ology, Albert Einstein College of Medicine of Yeshiva University,
  1300 Morris Park Avenue, BRONX, N.Y. 10461, U.S.A.

T. MOMOI
  Institut für Physiologische Chemie I, Institutsgruppe Lahnberge
  der Medizinischen Fakultät der Philipps-Universität, Lahnberge,
  355 MARBURG (Lahn), GERMANY

M. MONSIGNY
  Centre de Biophysique Moléculaire du CNRS, Avenue de la Recherche
  Scientifique, 45045 ORLEANS CEDEX, FRANCE

J. MONTREUIL
  Laboratoire de Chimie Biologique, Université des Sciences et
  Techniques de Lille I, B.P. 36, 59650 VILLENEUVE D'ASCQ, FRANCE

Y. NAGAI
Department of Biochemistry, Tokyo Metropolitan Institute of
Gerontology, 35-2 Sakaecho, Itabashiku, TOKYO 173, JAPAN

M. NAIKI
Hokkaido University, Faculty of Veterinary Medicine, Hokkaido 060
SAPPORO, JAPAN

N. NESKOVIC
Centre de Neurochimie du CNRS, 11 rue Humann, 67085 STRASBOURG
CEDEX, FRANCE

I. OBERLE
Centre de Neurochimie du CNRS, 11 rue Humann, 67085 STRASBOURG
CEDEX, France

J. POLONOVSKI
Laboratoire de Biochimie, Hôpital St-Antoine, 184 Rue du Faubourg
Saint-Antoine, 75012 PARIS, FRANCE

A. PRETI
Istituto di Chimica Biologica, Facoltà di Medicina e Chirurgia
dell'Università di Milano, Via C. Saldini 50, 20133 MILANO, ITALY

H. RAHMANN
Universität Hohenheim, Institut für Zoologie, Lehrstuhl für
Allgemeine und Systematische Zoologie, 7000 STUTTGART 70 (Hohenheim),
GERMANY

M.M. RAPPORT
Division of Neuroscience, Office of Mental Health, New York State
Psychiatric Institute, 722 West 168th Street, NEW YORK, N.Y. 10032,
U.S.A.

H. RAUVALA
Department of Medical Chemistry, University of Helsinki,
Siltavuorenpenger 10, 00170 HELSINKI 17, FINLAND

G. REBEL
Centre de Neurochimie du CNRS, 11 rue Humann, 67085 STRASBOURG
CEDEX, FRANCE

J. ROBERT
Laboratoire de Biochimie Médicale A, Université de Bordeaux II,
146, Rue Léo-Saignat, 33076 BORDEAUX CEDEX, FRANCE

A. ROSENBERG
Department of Biological Chemistry, The Milton S. Hershey Medical
Center, The Pennsylvania State University, HERSHEY, Pennsylvania
17033, U.S.A.

R. SALVAYRE
  Unité de Recherches de l'INSERM sur la Biochimie des Lipides
  (n° 101), Centre Hospitalier Purpan, 31052 TOULOUSE CEDEX, FRANCE

B.E. SAMUELSSON
  Department of Medical Biochemistry, Faculty of Medicine, University of Göteborg, Medicinaregatan 9, 40033 GÖTEBORG 33, SWEDEN

K. SANDHOFF
  Institut für Organische Chemie und Biochemie der Universität Bonn,
  Gerhard-Domagk-Strasse 1, 5300 BONN 1, GERMANY

L. SARLIEVE
  Centre de Neurochimie du CNRS, 11 rue Humann, 67085 STRASBOURG
  CEDEX, FRANCE

R. SCHAUER
  Biochemisches Institut in Fachbereich Medizin, Christian-Albrechts-
  Universität Kiel, Neue Universität, Otto-Meyerhof-Haus, Gebäude
  N 11, Olshausenstrasse 40-60, 2300 KIEL 1, GERMANY

G. SCHWARZMANN
  Institut für Physiologische Chemie I, Institutsgruppe Lahnberge
  der Medizinischen Fakultät der Philipps-Universität, Lahnberge,
  355 MARBURG (Lahn), GERMANY

W.S. SLY
  Division of Medical Genetics, The Edward Mallinckrodt Dept. of
  Pediatrics, Washington University School of Medicine, St. Louis
  Children's Hospital, 500 S. Kingshighway Blvd., P.O. Box 14871,
  ST. LOUIS, Missouri 63178, U.S.A.

G. SPIK
  Laboratoire de Chimie Biologique, Université des Sciences et
  Techniques de Lille I, B.P. 36, 59650 VILLENEUVE D'ASCQ, FRANCE

G. STRECKER
  Laboratoire de Chimie Biologique, Université des Sciences et
  Techniques de Lille I, B.P. 36, 59650 VILLENEUVE D'ASCQ, FRANCE

K. SUZUKI
  The Saul R. Korey Department of Neurology, Albert Einstein College
  of Medicine of Yeshiva University, 1300 Morris Park Avenue,
  BRONX, N.Y. 10461, U.S.A.

L. SVENNERHOLM
  Department of Neurochemistry, Psychiatric Research Centre,
  University of Göteborg, St. Jörgen Hospital, 42203 HISINGS BACKA 3,
  SWEDEN

S.B. SVENSSON
  University of Lund, Department of Clinical Chemistry, University
  Hospital, 22185 LUND, SWEDEN

T. TAKETOMI
  Department of Biochemistry, Institute of Adaptation Medicine,
  Shinshu University, MATSUMOTO 390, JAPAN

M. TARDY
  Institut Mérieux, Marcy-L'Etoile, 69260 CHARBONNIERES-LES BAINS,
  FRANCE

J.L. TAYOT
  Institut Mérieux, Marcy-L'Etoile, 69260 CHARBONNIERES-LES BAINS,
  FRANCE

G. TETTAMANTI
  Istituto di Chimica Biologica, Facoltà di Medicina e Chirurgia
  dell'Università di Milano, Via C. Saldini 50, 20133 MILANO, ITALY

P.F. URBAN
  Centre de Neurochimie du CNRS, 11 Rue Humann, 67085 STRASBOURG
  CEDEX, FRANCE

M.T. VANIER
  Laboratoire de la Fondation Gillet, Hôpital Sainte-Eugénie,
  69230 SAINT-GENIS-LAVAL, FRANCE

R. VEH
  Institut für Anatomie, Ruhr-Universität Bochum, Universitätsstrasse
  150, Postfach 2148, 463 BOCHUM, GERMANY

V.E. VENGRIS
  Department of Health, Education and Welfare, Public Health Service,
  Food and Drug Administration, Bureau of Veterinary Medicine,
  ROCKVILLE, Maryland 20857, U.S.A.

G.VINCENDON
  Centre de Neurochimie du CNRS, 11 rue Humann, 67085 STRASBOURG
  CEDEX, FRANCE

J.F.G. VLIEGENTHART
  Rijksuniversiteit Utrecht, Organisch Chemisch Laboratorium,
  Croesestraat 79, 3522 AD UTRECHT, THE NETHERLANDS

H. WIEGANDT
  Institut für Physiologische Chemie I, Institutsgruppe Lahnberge
  der Medizinischen Fakultät der Philipps-Universität, Lahnberge,
  355 MARBURG (Lahn), GERMANY

U.N. WIESMANN
  Inselspital Bern, Medizinische Universitäts-Kinderklinik und
  Poliklinik, 3010 BERN, SWITZERLAND

R.K. YU
  Department of Neurology, Yale University, The School of Medicine,
  333 Cedar Street, NEW HAVEN, Connecticut 06510, U.S.A.

A.N.K. YUSUFI
  Centre de Neurochimie du CNRS, 11 rue Humann, 67085 STRASBOURG
  CEDEX, FRANCE

B. ZALC
  U.134 de l'INSERM-Neurobiologie Expérimentale et Clinique,
  Hôpital de la Salpêtrière, 47, Boulevard de l'Hôpital,
  75634 PARIS CEDEX 13, FRANCE

CONTRIBUTOR INDEX